Springer Series in Materials Science

Volume 289

The Springer Series in Materials Science covers the complete spectrum of materials research and technology, including fundamental principles, physical properties, materials theory and design. Recognizing the increasing importance of materials science in future device technologies, the book titles in this series reflect the state-of-the-art in understanding and controlling the structure and properties of all important classes of materials.

More information about this series at http://www.springer.com/series/856

Zundu Luo · Yidong Huang

Physics of Solid-State Laser Materials

Science Press
Beijing

Springer

Zundu Luo
Fujian Institute of Research
on the Structure of Matter
Chinese Academy of Sciences
Fuzhou, Fujian, China

Yidong Huang
Fujian Institute of Research
on the Structure of Matter
Chinese Academy of Sciences
Fuzhou, Fujian, China

ISSN 0933-033X ISSN 2196-2812 (electronic)
Springer Series in Materials Science
ISBN 978-981-32-9667-1 ISBN 978-981-32-9668-8 (eBook)
https://doi.org/10.1007/978-981-32-9668-8

Jointly published with Science Press
The print edition is not for sale in China (Mainland). Customers from China (Mainland) please order the print book from: Science Press.
ISBN of the Co-Publisher's edition: 978-7-03-064678-1

This Springer imprint is published by the registered company Springer Nature Singapore Pte Ltd.
The registered company address is: 152 Beach Road, #21-01/04 Gateway East, Singapore 189721, Singapore

Preface

Solid-state laser systems are compact, easy to use, easy to tune, and have good beam quality; they have unique advantages comparing with lasers of gas, liquid, and other media so that have a very wide and important application in many fields. Therefore, it is necessary to search for suitable materials according to different performance requirements. This requires the material researchers deeply grasp the physical knowledge about the main performance of solid-state laser materials.

Many excellent monographs have been published in this field; however, we feel necessary to update them by including new published literatures as well as our own research work. The subject of this book is spectroscopic principle of rare earth ions and transition metal ions doped in solid. In addition to many general principles of spectroscopy and calculation methods, to determine the symmetry of the energy level, a unique group chain method is introduced, while a method from traditional crystal field theory based on a comprehensive crystal field quantum number table is also described. The stimulated non-radiative transition theory and the apparent crystal field model introduced by us are discussed.

Comparing with many published books, this book has the following advantages. That is, it derives all the calculation formulas from the first principles of physics instead of direct citation in some books, so that readers can understand the physics insight clearly and exploit them correctly in their own work. Furthermore, readers could develop related formulas by themselves, which have not been given in this book, following the method provided in this book.

We hope this book will provide useful references for researchers in solid-state laser materials, luminescent materials, solid-state spectroscopy, and laser science and technology, as well as senior undergraduates and postgraduates having basis knowledge of optics, quantum mechanics, spectroscopy, and condensed matter physics.

The focus of this book is on the physical fundamentals. The spectral, laser, and physical properties of the materials mentioned in the book are just scales and claws, especially without reference to the rapidly developing and widely used ceramic laser materials and new activated ions such as bismuth in recent years. On the other hand, many excellent research works of peer experts have not been quoted. It is unavoidable that there are mistakes or omissions in the book, and readers are welcome to criticize and correct them.

We express gratitude to the Fujian Institute of Research on the Structure of Matter, Chinese Academy of Sciences, the Key Laboratory of Research on Chemistry and Physics of Optoelectronic Materials, Chinese Academy of Sciences, and the National Natural Science Foundation of China for their support and funding, also appreciate the help from editor Zhou Han of China Science Publishing House and editors of Springer Nature Press.

Fuzhou, China

Zundu Luo
Yidong Huang

About This Book

This book discusses in detail the spectral properties of solid-state laser materials, including emission and absorption of light, the law of radiative and non-radiative transitions, the selection rule of optical transition, and the calculation methods for the spectral parameters. To determine the symmetry of the energy level, a unique group chain method is introduced, while a method from traditional crystal field theory based on a comprehensive crystal field quantum number table is also described. The research works of the authors on this field, especially the stimulated non-radiative transition theory and the apparent crystal field model, are introduced. As such, the book provides essential tools for researchers and graduate students in the fields of solid spectroscopy and solid laser material physics, while also offering a valuable reference guide for instructors and advanced students of physics.

Contents

About the Authors

Zundu Luo is a professor at Fujian Institute of Research on the Structure of Matter (FIRSM), Chinese Academy of Sciences. He has been working in the field of spectroscopic physics and laser crystal research since 1965, serving as the leading scientist of the research group of new laser crystal materials from 1982 to 2003. He has published more than 100 articles in SCI indexed journals and has been awarded the Second Class Prize of China's National Science and Technology Progress and the First Class Prize of Scientific and Technological Progress of Chinese Academy of Sciences.

Yidong Huang got his Ph.D. from Strathclyde University in 1997. He became a professor at Fujian Institute of Research on the Structure of Matter (FIRSM), Chinese Academy of Sciences in 1999. He is currently a member of the council of Chinese Optical Society. His research interests include laser and luminescence properties of rare earth and transition metal ions in solids; new laser and non-linear crystals, laser glasses, solid luminescence materials, and related devices. He has been working in the field of laser crystal and solid state laser since 1987, as the leading scientist of the research group of laser crystal and solid state laser from 2004. He has published more than 200 articles in SCI indexed journals, and owns more than 20 patents. As a partner, he got one Second Class Prize of China's national science and technology progress and one First Class Prize of scientific and technological progress of Chinese Academy of Sciences.

About the Authors

Chapter 1
Energy Level of Free Ions

Most of the solid-state laser materials consist of insulators doped with a small amount of rare earth ion or transition-metal ion (optically active ions, which will be simply called active ions in all the discussions of this book). The spectra of all these materials have very close relationship with the energy level structure of free ion (atom) and the transition rules between these energy levels. Therefore, the basic knowledge of the atomic spectroscopy is very important for the investigation of laser materials. There are many good reference books in this field, for example, Herzberg [1], Condon and Shortley [2], Slater [3], and Condon and Odabasi [4]. In this chapter, only some basic knowledge necessary to the discussion in this book will be covered. The books mentioned can be referred for further understanding.

1.1 Energy Levels of the Single Electron in Atoms (Free Ions)

In the hydrogen atom as well as hydrogen-like atoms, for example, He^{+}, Li^{2+}, Be^{3+}, and so on, there is only one electron around a nucleus with positive charge. The spectroscopy of single electron atoms is the only problem which can be dealt with by quantum mechanics strictly. However, it is the most important background for the analysis of the other problems of atomic spectroscopy. The quantum mechanics is used to analyze hydrogen atom and hydrogen-like atoms and solve a Schrödinger equation for a charged particle in a central electric field. Its Hamiltonian can be written as

$$H = H_0 + H_{\text{so}} + H_Z + \text{other small terms} \tag{1.1}$$

where H_0 is the Hamiltonian describing the interaction of the electron with the nucleus. H_{so} is the spin–orbital interaction Hamiltonian

Z. Luo and Y. Huang, *Physics of Solid-State Laser Materials*, Springer Series in Materials Science 289, https://doi.org/10.1007/978-981-32-9668-8_1

$$H_{\text{so}} = \xi(r)\boldsymbol{l} \cdot \boldsymbol{s} \tag{1.2}$$

where $\xi(r)$ is the coefficient of the spin–orbital coupling, $\boldsymbol{l} = \boldsymbol{r} \times \boldsymbol{p}$ denotes the orbital vector operator of the electron, and $\boldsymbol{s}$ is the spin vector operator of the electron. The spin–orbital interaction is actually a relativistic effect. Besides, the other relativistic effects will not be considered here. H_Z is the interaction Hamiltonian of the atoms in external magnetic field. If the magnetic induction of external magnetic field is represented by $\boldsymbol{B}$, then H_Z can be written as

$$H_Z = \frac{e}{m_e}\boldsymbol{s} \cdot \boldsymbol{B} \tag{1.3}$$

where e denotes the electric charge of electron and m_e is the mass of electron.

The general method to deal with the above problem in quantum mechanics is to consider at first the main term in the Hamiltonian and then introduce the small terms as a perturbation. Without magnetic field and spin–orbital interaction, we have

$$H = H_0 = -\frac{\hbar^2}{2m_e}\nabla^2 - \frac{Ze^2}{4\pi\varepsilon_0 r} = -\left(\frac{\hbar^2}{2m_e}\right)\left[\frac{\partial^2}{\partial r^2} + \left(\frac{2}{r}\right)\frac{\partial}{\partial r}\right] + \frac{\hbar^2 l^2}{2m_e r^2} - \frac{Ze^2}{4\pi\varepsilon_0 r} \tag{1.4}$$

In spherical coordinate system, the components of the orbital angular momentum operator can be written as

$$l_x = i\hbar\left(\sin\phi\frac{\partial}{\partial\theta} + \cot\theta\cos\phi\frac{\partial}{\partial\phi}\right)$$
$$l_y = -i\hbar\left(\cos\phi\frac{\partial}{\partial\theta} - \cot\theta\sin\phi\frac{\partial}{\partial\phi}\right)$$
$$l_z = -i\hbar\frac{\partial}{\partial\phi}$$

Therefore

$$l^2 = -\hbar^2\left[\frac{1}{\sin\theta}\frac{\partial}{\partial\theta}\left(\sin\theta\frac{\partial}{\partial\theta}\right) + \frac{1}{\sin^2\theta}\frac{\partial^2}{\partial\phi^2}\right] \tag{1.5}$$

The properties of the angular momentum operator will be discussed later in more detail. The solution of the Schrödinger equation should be presented at first. The steady-state Schrödinger equation is expressed as

$$-\frac{\hbar^2}{2m_e}\left[\frac{1}{r^2}\frac{\partial}{\partial r}\left(r^2\frac{\partial}{\partial r}\right) + \frac{1}{r^2\sin\theta}\frac{\partial}{\partial\theta}\left(\sin\theta\frac{\partial}{\partial\theta}\right) + \frac{1}{r^2\sin\theta}\frac{\partial^2}{\partial\phi^2}\right]\psi - \frac{Ze^2}{4\pi\varepsilon_0 r}\psi = E\psi \tag{1.6}$$

The wave function Ψ can be solved by means of variable separation method. One should first separate the radial part and angular part as

$$\psi(r,\theta,\phi) = R(r)Y(\theta,\phi)$$

Then substituting into (1.6), we obtain

$$\frac{1}{R}\frac{\mathrm{d}}{\mathrm{d}r}\left(r^2\frac{\mathrm{d}R}{\mathrm{d}r}\right) + \frac{2m_e r^2}{\hbar^2}\left(E + \frac{Ze^2}{4\pi\varepsilon_0 r}\right) = -\frac{1}{Y}\left[\frac{1}{\sin\theta}\frac{\partial}{\partial\theta}\left(\sin\theta\frac{\partial Y}{\partial\theta}\right) + \frac{1}{\sin^2\theta}\frac{\partial^2 Y}{\partial\phi^2}\right] \tag{1.7}$$

One can see in the above equation that the terms in the left-hand side depend only on the radial coordinate r, while those in the right-hand side depend only on the angular variables θ and ϕ. The validity of the equation means that both sides should be equal to a constant λ. In this way, one equation of the variable r is obtained as

$$\frac{1}{r^2}\frac{\mathrm{d}}{\mathrm{d}r}\left(r^2\frac{\mathrm{d}R}{\mathrm{d}r}\right) + \left[\frac{2m_e}{\hbar^2}\left(E + \frac{Ze^2}{4\pi\varepsilon_0 r}\right) - \frac{\lambda}{r^2}\right]R = 0 \tag{1.8}$$

The other equation of angular variables θ and ϕ will be

$$\frac{1}{\sin\theta}\frac{\partial}{\partial\theta}\left(\sin\theta\frac{\partial Y}{\partial\theta}\right) + \frac{1}{\sin^2\theta}\frac{\partial^2 Y}{\partial\phi^2} + \lambda Y = 0 \tag{1.9}$$

Suppose $Y(\theta,\phi) = \Theta(\theta)\Phi(\phi)$, then the equation for angular variables can be further separated as

$$\frac{\mathrm{d}^2\Phi(\phi)}{\mathrm{d}\phi^2} + m^2\Phi(\phi) = 0 \tag{1.10}$$

with solution of $\Phi(\phi) = (2\pi)^{-1/2}\exp(im\phi)$ and $m = 0, \pm1, \pm2, \ldots$. The other equation is

$$\frac{1}{\sin\theta}\frac{\mathrm{d}}{\mathrm{d}\theta}\left(\sin\theta\frac{\mathrm{d}\Theta(\theta)}{\mathrm{d}\theta}\right) + \left(\lambda - \frac{m^2}{\sin^2\theta}\right)\Theta(\theta) = 0 \tag{1.11}$$

Let $\xi = \cos\theta$, then

$$\frac{\mathrm{d}}{\mathrm{d}\xi}\left[(1-\xi^2)\frac{d\Theta}{d\xi}\right] + \left(\lambda - \frac{m^2}{1-\xi^2}\right)\Theta = 0 \tag{1.12}$$

This is the associated Legendre equation. Solution of this equation is valid in the range of $-1 \le \xi \le 1$, and the constant λ can only be $\lambda = l(l+1)$, l = 0, 1, 2, 3, …, and $|m| \le l$. Its solution is

$$\Theta(\theta) = (-1)^{-(m+|m|)/2}\left[\frac{(2l+1)(l-|m|)!}{2(l+|m|)!}\right]^{1/2} P_{|m|}^{(l)}(\cos\theta) \tag{1.13}$$

Therefore

$$Y_m^{(l)}(\theta,\phi) = (2\pi)^{-1/2}(-1)^{(m+|m|)/2}\left[\frac{(2l+1)(l-|m|)!}{2(l+|m|)!}\right]^{1/2} P_{|m|}^{(l)}(\cos\theta)\exp(im\phi) \tag{1.14}$$

The spherical harmonic function $Y_{lm}(\theta,\phi)$, designated by quantum numbers l and m, is orthogonal and normalized, that is

$$\int_0^{2\pi}\int_0^{\pi} Y_{m'}^{(l')}(\theta,\phi)Y_m^{(l)}(\theta,\phi)\sin\theta \mathrm{d}\theta \mathrm{d}\phi = \delta_{l'l}\delta_{m'm} \tag{1.15}$$

For the problem of single electron in hydrogen atom, the equation for radial variable can be solved and expressed by Laguerre polynomials as

$$R_{nl}(r) = N_{nl}\exp(-Zr/na_1)\left(\frac{2Zr}{na_1}\right)^l \mathrm{L}_{n+1}^{2l+1}(2Zr/na_1) \tag{1.16}$$

where the constant of normalization N_{nl} is

$$N_{nl} = -\left\{\left(\frac{2Z}{na_1}\right)^3 \frac{(n-l-1)!}{2n[(n+l)!]^3}\right\}^{1/2} \tag{1.17}$$

where a_1 is the first Bohr radius of the hydrogen atoms

$$a_1 = \frac{4\pi\varepsilon_0\hbar^2}{m_e e^2}$$

Integer n called principal quantum number should satisfy $n \geq l+1$ and the energy eigenvalues will be given by

$$E_n = -\frac{Z^2e^2}{4\pi\varepsilon_0 \times 2a_1n^2} = -\frac{m_eZ^2e^4}{2(4\pi\varepsilon_0)^2n^2\hbar^2} \tag{1.18}$$

Here SI unit system is used. Changing to CGS unit system, the electron charge e should be multiplied by a factor of $(4\pi\varepsilon_0)^{1/2}$.

It can be seen from (1.18) that the energy of single electron in the hydrogen and hydrogen-like atoms only depends on the principal quantum number n, if the spin–orbit interaction and other relativistic effects are neglected. Thus the states with different angular momentum and spin momentum quantum number but the same principal quantum number have the same energy, which correspond to the degeneracy of the electronic states. The degree of the degeneracy f is found to be

$$f = 2\sum_{l=0}^{n-1}(2l+1) = 2n^2 \tag{1.19}$$

Note that (1.4) shows the Hamiltonian without perturbation and possesses coordinates inversion symmetry, that is, the Hamiltonian is invariant with respect to the transformation of $\boldsymbol{r} \to -\boldsymbol{r}$. In the spherical coordinate system, it corresponds to the transformations of $r \to r$, $\theta \to \pi - \theta, \phi \to \pi + \phi$. If the inversion operator P_i is introduced, then

$$P_i f(\boldsymbol{r}) = f(-\boldsymbol{r})$$

It is easy to show that $[H_0, P_i] = 0$. Then, by a basic principle of quantum mechanics, the eigenstates of H_0 are also the eigenstates of P_i. In this way, the eigenstates can be classified as even parity states and odd parity states. It can be demonstrated that

$$Y_l^m(\pi - \theta, \pi + \phi) = (-1)^l Y_l^m(\theta, \phi) \tag{1.20}$$

namely

$$P_i|nlm\rangle = (-1)^l|nlm\rangle$$

where l if even or odd corresponds to even or odd parity state, respectively.

The energy correction by the spin–orbit interaction effect is

$$\Delta E_{ls} = -\frac{Z^4 m_e e^4 \alpha^2}{(4\pi\varepsilon_0)^2 \times 2\hbar^2 n^3 l\left(l+\frac{1}{2}\right)(l+1)} \frac{j(j+1) - l(l+1) - s(s+1)}{2} \tag{1.21}$$

where $\alpha = 1/137$ is the fine structure constant and j is the total angular momentum quantum number involving the spin–orbital interaction. By calculating the energy correction due to relativistic effect of mass but not the other relativistic effects, Summerfield obtained the following result

$$\Delta E_{\text{rel}} = -\frac{Z^4 m_e e^4 \alpha^2}{(4\pi\varepsilon_0)^2 \times 2\hbar^2 n^4}\left(\frac{n}{l+\frac{1}{2}} - \frac{3}{4}\right) \tag{1.22}$$

Dirac established a relativistic equation of the electron in 1928, taking into account simultaneously the spin–orbital interaction and the other relativistic effects, and he obtained

$$E_{nj} = -\frac{Z^2e^2}{(4\pi\varepsilon_0)\times 2a_1n^2} - \frac{Z^4e^2\alpha^2}{(4\pi\varepsilon_0)\times 2a_1n^3}\left(\frac{1}{j+\frac{1}{2}} - \frac{3}{4n}\right) \quad (1.23)$$

Owing to the fact that $\alpha^2 = 5 \times 10^{-5}$, the corrections introduced by the spin–orbital interaction and the relativistic effect of mass are very small and have a magnitude of 10^{-1} – 10^{-2} cm^{-1}.

There is one kind of atoms, in which the electrons are like that in the hydrogen atom. For example, the atoms of alkali metal, such as Li, Na, and K, have one valence electron with very weak coupling to its atomic core. The atomic cores have the electronic structure like those of the inert gases He, Ne, and Ar. The energy eigenvalue of the valence electron without the effect of spin–orbital interaction has a similar expression as that of the hydrogen atom

$$E_{nl}^0 = -\frac{Z^2e^2}{(4\pi\varepsilon_0)\times 2a_1n^{*2}} \quad (1.24)$$

where the principal quantum number n is replaced by effective quantum number n^* and $n^* = n - \Delta$. The quantum defect Δ has only a slight variation with the change of principal quantum number n. However, it decreases obviously even if the angular momentum quantum number has only a slight increase. It can be seen that the energy eigenvalue E_{nl} depends not only on n but also on l. If the spin–orbital effect is also taken into account, one has

$$E_{nljs} = E_{nl}^0 - \frac{Z^{*4}m_ee^4\alpha^2}{(4\pi\varepsilon_0)^2\times 2\hbar^2n^3l(l+\frac{1}{2})(l+1)} \times \frac{j(j+1) - l(l+1) - s(s+1)}{2} \quad (1.25)$$

where the effective charge $Z^* = Z - \sigma$ and σ is the shielding constant. It means that the effect of interaction of the valence electron studied with the electrons of filled shell in the atom is to reduce the effect of nucleus charge. The spin–orbital interaction is proportional to Z^4 and so the splitting of yellow D line of sodium is only 17.2 cm^{-1} while the corresponding value for the main line system of cesium atom reaches 554 cm^{-1}.

The basic but most useful result of the theory of single electron atom is that an electronic state is labeled by a principal quantum number n, an orbital angular momentum quantum number, a magnetic quantum number m, and a spin quantum number s. This is the basis for the discussion of multi-electron problems. Characters *s, p, d, f, h, i, k, l, m, n,* o, *q* … are used to denote l = 0, 1, 2, 3, 4, 5, 6, 7, 8, 9, 10, 11, 12 … in spectroscopy.

1.2 General Properties of Energy Level in Multi-electron of Free Ions

Multi-electron systems are complicate systems in quantum mechanics. The atoms can be treated as N electrons with mass m_e and charge $-e$ moving around a nucleus with infinite mass and charge of $+Ze$ and other electrons in the atom. The non-relativistic Hamiltonian without spin–orbital interaction is

$$H = \sum_{i=1}^{N} \left(\frac{p_i^2}{2m_e} - \frac{Ze^2}{4\pi\varepsilon_0 r_i} \right) + \sum_{i>j=1}^{N} \frac{e^2}{4\pi\varepsilon_0 r_{i,j}} \tag{1.26}$$

The second term in the above expression involves variable $r_{i,j}$, the distance between electron i and electron j; therefore, it is impossible to use the method of variable separation. Furthermore, this term is large enough and so perturbation method cannot be used. It is impossible to solve its energy level structure strictly and analytically. Fortunately, just like many other problems in quantum mechanics, this problem can be solved by approximate method neglecting less important interactions. Among these, the central field approximation is the most important one; it assumes that each electron moves independently in the field of nucleus and the central field is made up of spherically averaged potential fields of the other electrons described by a function $U(r_i)/e$. The non-spherical part of the electronic interactions is treated as a perturbation, so the approximated Hamiltonian can be expressed as

$$H' = \sum_{i=1}^{N} \left(\frac{p_i^2}{2m_e} + U(r_i) \right) \tag{1.27}$$

Then Schrödinger equation will be

$$\sum_{i=1}^{N} \left(-\frac{\hbar^2 \nabla_i^2}{2m_e} + U(r_i) \right) \psi = E' \psi \tag{1.28}$$

Obviously, the N variables r_i in the above equation can be separated, that is the wave function and eigenvalue can be expressed as

$$\psi = \prod_{i=1}^{N} \psi_i(k_i), E' = \sum_{i=1}^{N} E'(k_i) \tag{1.29}$$

where n_i, l_i, m_i are simply represented by k_i. For different variables r_i, the equations have the same form. Thus the subscript i can be eliminated

$$\left(\frac{-\hbar^2\nabla^2}{2m_e} + U(r)\right)\psi(k) = E'\psi(k) \tag{1.30}$$

The difference between the above equation and that of the hydrogen atom is the potential function and so there is a difference in the radial wave function but the wave functions of the angular part are the same. The wave function can still be written as

$$\psi(k) = R_{nl}(r)\mathrm{Y}_m^{(l)}(\theta,\phi) \tag{1.31}$$

The energy $E'(k)$ is degenerated for different quantum number m.

Taking into account the electron spin, the wave function $\psi(k)$ should be multiplied by a spin wave function α or β, which corresponds respectively to $m_s = +1/2$ or $m_s = -1/2$. According to Pauli exclusion principle, no two electrons can have the same set of four quantum numbers n, l, m, and m_s simultaneously. As a result, the mathematic expression for the wave functions must be a determinant. For example, the wave function for three electrons will be

$$\psi' = (3!)^{-1/2}\begin{vmatrix} \psi_1(k_1)\alpha_1 & \psi_1(k_2)\alpha_1 & \psi_1(k_2)\beta_1 \\ \psi_2(k_1)\alpha_2 & \psi_2(k_2)\alpha_2 & \psi_2(k_2)\beta_2 \\ \psi_3(k_1)\alpha_3 & \psi_3(k_2)\alpha_3 & \psi_3(k_2)\beta_3 \end{vmatrix} \tag{1.32}$$

and can be simplified as

$$\psi' = \begin{Bmatrix} + & + & + \\ k_1 & k_2 & k_3 \end{Bmatrix} \tag{1.33}$$

where + is used to denote spin function α, and $-$ denotes the spin function β.

In an atom, there are at most $2(2l + 1)$ electrons that can simultaneously have the same n and l, and a completely filled electron shell is in which all the states with given n, l, m, and ms are occupied.

From expressions of (1.20) and (1.29), it can be shown that the parity of multi-electron system is $\sum_{i=1}^{N} l_i$, odd of this summation corresponds to a odd parity state and even of this summation corresponds to a even parity state, because each term in the determinant has the following relation

$$\prod_i\left(R_{nl}(r_i)Y_m^{(l)}(\theta-\pi,\pi+\phi_i)\right) = (-1)^{\sum_i l_i}\prod_i\left(R_{nl}(r_i)Y_m^{(l)}(\theta,\varphi_i)\right)$$

therefore

$$P\Psi = (-1)^{\sum_i l_i}\Psi$$

There is also an important simplification in the studies of the multi-electron problem of atoms (ions), that is, neglecting the effect of completely filled electron shells. In these shells, the total angular momentum and spin momentum of the electrons are equal to zero and so their effect is the same on the electrons of different energy states in the partially filled shells. Therefore, in the discussion of the energy level splitting of the multi-electron system, the effect of the completely filled shells is not necessary to be taken into account. For example, the electronic configuration of Nd^{3+} ions is $1s^22s^22p^63s^23p^63d^{10}4s^24p^64d^{10}5s^25p^64f^3$. If the completely filled electron shells are not considered, only three $4f$ electrons should be taken into account. In this way, the work of the energy level calculation will be greatly reduced. However, even in this situation, the number of the energy levels is still very large.

In the $4f^3$ configuration, $l = 3$, $-3 \leq m \leq 3$, and $m_s = \pm\frac{1}{2}$. There are 14 states with the same n and l but different m and m_s. Then, the three f electrons can occupy any of the 14 states. Therefore, taking into account of the spin, the total number of states will be $\frac{14!}{(14-3)!3!} = 364$. The quantum numbers n and l are all equal to 4 and 3, respectively; thus it is enough to use only quantum number m and m_s to designate the energy state as $\{3^+2^+1^+\}$, $\{3^+-3^+1^+\}$, $\{0^--1^--3^+\}$, and so on. Even in this case, there are $364 \times 364 = 132,496$ matrix elements to be calculated, if one wants to study their energy levels. Fortunately, the symmetry of the system can be used to simplify the calculation of energy levels. In the free atoms, the continuous group and angular momentum theory can be used, and for the electrons of ions in the crystals, one can use the point group theory, which will be discussed in next chapter.

First, we would like to mention the formation of spectral terms, their symbols, and the Hund rule for determining the ground states. By the interaction of electron orbital magnetic moment with spin magnetic moment, the energy levels of each electronic configuration will be split into a series of energy level sets designated by a total spin quantum number S and a total orbital angular momentum quantum number L. Each energy level set contains $(2S + 1)$ $(2L + 1)$ states. Such energy level set is called spectral term using symbol ^{2S+1}L. The interaction between orbital momentum and spin momentum is called the coupling of orbital angular momentum with spin angular momentum.

There are four models of coupling in the atomic spectroscopy theory, that is, ***LS*** coupling (Russell–Saunders coupling), ***jj*** coupling, pair coupling (Racah coupling), and intermediate coupling. These depend on the relative magnitude between the Coulomb and the spin–orbital interactions.

Generally speaking, the spin–orbit interaction has a magnetic origin and will be rapidly intensified as the atomic number Z increased. Calculated by the hydrogen wave function, spin–orbit coupling coefficients can be expressed as [5]

$$\xi_{nl} = \frac{e^2\hbar^2}{2m^2e^2a_1^2} \frac{Z^4}{n^3l(l+\frac{1}{2})(l+1)}$$

The spin–orbit splitting of sodium D line is 17.2 cm^{-1} while that of the cesium atom's main line reaches 554 cm^{-1}. For the rare earth ions, the spin–orbit coupling coefficients of Tm^{3+} ions are more than three times higher than that of the Pr^{3+} ions. On the other hand, the spin–spin interaction is basically independent of Z. This phenomenon is obvious which resulted from the nature of the spin–spin interaction, because it is a direct interaction between two electrons and so is not affected by the nucleus field. Therefore, for the atoms with smaller atomic number Z, the magnitude of spin–spin interaction is larger than that of spin–orbital interaction. In this case, the spin momenta $\boldsymbol{s}_i$ of all the electrons will be coupled by exchange interaction to form a total spin angular momentum $\boldsymbol{S}$. Similarly, the orbital momenta $\boldsymbol{l}_i$ of all the electrons will also be coupled by Coulomb interaction to form a total orbital angular momentum $\boldsymbol{L}$. Finally, the total spin angular momentum $\boldsymbol{S}$ couples with the total orbital angular momentum $\boldsymbol{L}$. This is so-called ***LS*** coupling. However, for the atoms with very large atomic number, and especially for their excited states, the couplings between the spin angular momentum and the orbital angular momentum of the same electron are stronger than the spin–spin couplings between different electrons. Therefore, the orbital angular momentum $\boldsymbol{l}_i$ couples with the spin angular momentum $\boldsymbol{s}_i$ of the same electron to form a total angular momentum $\boldsymbol{j}_i$ of the electron, then $\boldsymbol{j}_i$ of all the electrons will couple to form the total angular momentum $\boldsymbol{J}$ of the whole system. This is so-called ***jj*** coupling.

LS coupling is the most commonly used model. For transition-metal ions, and especially their ground spectral terms, the ***LS*** coupling model can be used. This model can also be used in the first approximation problems of rare earth ion in which the electrostatic interaction between different $4f$ electrons and the spin–orbit interaction have the same order of magnitudes. According to this model, the states can be designated by $^{2S+1}L_J$.

The calculation of the state energies is to solve the eigenvalue problem of Hamiltonian involving all interactions in the atom and so the complicated and tedious work should be done. However, the qualitative rules are helpful to understand the energy level structure. Hund rule is one of the well-known qualitative rules, which has the following guidelines:

(1) The spectral terms (or states) with the largest S values have the lowest energy values among all the spectral terms of an electron configuration. Among these terms, those with the highest L values have the lowest energy values.
(2) For the spectral terms in a configuration with a number of electrons equal to or greater than that of the half-filled shell, the larger the J value, the lower is the energy. On the other hand, if the number of electrons is less than that of the half-filled shell, then the smaller the J values, the lower is the energy.

Hund rule is very convenient to determine the ground state of an electronic configuration. Taking the example of the trivalent rare earth ion Nd^{3+} mentioned above, its electronic configuration ($4f^3$) belongs to the case of an electron number lower than that of a half-filled shell. The total spin of three electrons can be 3/2 in maximum, and the maximum L is 6. According to the second guideline of Hund rule, the energy of state with $J = 9/2$ will be the lowest one. Therefore, the ground states should be $^4I_{9/2}$. The trivalent erbium ions have electronic configuration $4f^{11}$ and have the same spectral terms and states as those for the $4f^3$ configuration. However, its electron number is greater than that of the half-filled shell and so the larger the J, the lower is the energy. Then its ground state should be $^4I_{15/2}$.

In the case of ***LS*** coupling, the energy separation of the energy levels with different total angular quantum number J in the same spectral term ^{2S+1}L has a rule called Landé interval rule. It can be expressed as $\Delta E_J - \Delta E_{J-1} = A(LS)J$, where the constant $A(LS)$ depends on the value of LS and their signs determine whether the electron shell is half-filled or less than half-filled (the electron number in the shell $q \le 2l + 1$, $A(LS) > 0$) or greater than half-filled (the electron number in the shell $q > 2l + 1$, $A(LS) < 0$). This rule can be demonstrated as follows.

Owing to the fact that the splitting of the energy levels with different total angular quantum number J in the same spectral term ^{2S+1}L is the effect of spin–orbit interaction and proportional to $\boldsymbol{L} \cdot \boldsymbol{S}$, this energy separation can be written as $\Delta E_J = C(LS)\boldsymbol{L} \cdot \boldsymbol{S}$. Because of $\boldsymbol{J}=\boldsymbol{L}+\boldsymbol{S}$, $|\boldsymbol{J}|^2 = |\boldsymbol{L}|^2 + |\boldsymbol{S}|^2 + 2\boldsymbol{L} \cdot \boldsymbol{S}$ and so

$$\boldsymbol{L} \cdot \boldsymbol{S} = \frac{1}{2}\left(|\boldsymbol{J}|^2 - |\boldsymbol{L}|^2 - |\boldsymbol{S}|^2\right) = \frac{\hbar^2}{2}[J(J+1) - L(L+1) - S(S+1)]$$

$\Delta E_J(SL) - \Delta E_{J-1}(SL) = \frac{C(LS)\hbar^2}{2}[J(J+1) - J(J-1)] = A(LS)J$ where $A(LS) = C(LS)\hbar^2$.

However, as will be pointed out in the next section, the energy separation of different J in the same spectral term ^{2S+1}L will not follow strictly the Landé interval rule in the case of intermediate-coupling approximation.

It should be pointed out that besides the above-mentioned interaction, the electrons in the atoms also have smaller inter-configuration interaction, Coulomb and spin–orbit-correlated interaction, spin–spin interaction as well as spin–other orbital interaction. We will introduce simply these interactions and their effects in energy level calculation of rare earth ions in Sect. 1.5.

1.3 Energy Levels of Free Transition-Metal Ions

The transition-metal group discussed here is also called first transition group and has partially filled electron shell $3d^q$ ($q < 10$). The LS coupling is suitable for the electrons in this group especially for their ground configuration. The spectral terms in $3d^q$ configuration are listed in Table 1.1. For the $3d^q$ with $q > 3$, two quantum

Table 1.1 Spectral terms of d^q electronic configuration [17]

Configuration	Spectral term
d^1, d^9	$^2D^1$
d^2, d^8	$^1S^1$, $^1D^1$, $^1G^1$, $^3P^1$, $^3F^1$
d^3, d^7	$^2P^1$, $^2D^2$, $^2F^1$, $^2G^1$, $^2H^1$, $^4P^1$, $^4F^1$
d^4, d^6	$^1S^2$, $^1D^2$, $^1F^1$, $^1G^2$, $^1I^1$, $^3P^2$, $^3D^1$, $^3F^2$, $^3G^1$, $^3H^1$, $^5D^1$
d^5	$^2S^1$, $^2P^1$, $^2D^3$, $^2F^2$, $^2G^2$, $^2H^1$, $^2I^1$, $^4P^1$, $^4D^1$, $^4F^1$, $^4G^1$, $^6S^1$

Table 1.2 Classification of the spectral terms in d^3, d^4, and d^5 configuration according to the reducing branching rule of the group chain $U(10) \supset SU(2) \otimes U(5) \supset SU(2) \otimes SO(5) \supset SU(2) \otimes SO(3)$

Configuration	$U(10)$	$SU(2) \otimes U(5)$	$SU(2) \otimes SO(5)$	$SU(2) \otimes SO(3)$
$d^3(d^7)$	$[1^3]$	$^2[2, 1]$	$^2(1, 0)$	^2D
			$^2(2, 1)$	^2P, ^2D, ^2F, ^2G, ^2H
		$^4[1, 1]$	$^4(1, 1)$	^4P, ^4F
$d^4(d^6)$	$[1^4]$	$^1[2, 2]$	$^1(0, 0)$	^1S
			$^1(2, 0)$	^1D, ^1G
			$^1(2, 2)$	^1S, ^1D, ^1F, ^1G, ^1I
		$^3[211]$	$^3(1, 1)$	^3P, ^3F
			$^3(2, 1)$	^3P, ^3D, ^3F, ^3G, ^3H
		$^5[1]$	$^5(1, 0)$	^5D
d^5	$[1^5]$	$^2[2, 2, 1]$	$^2(1, 0)$	^2D
			$^2(2, 1)$	^2P, ^2D, ^2F, ^2G, ^2H
			$^2(2, 2)$	^2S, ^2D, ^2F, ^2G, ^2I
		$^4[2, 1, 1, 1]$	$^4(1, 1)$	^4P, ^4F
			$^4(2, 0)$	^4D, ^4G
			$^6(0, 0)$	^6S

numbers L and S cannot completely discriminate different spectral terms. The right superscript of the spectral term's symbol in Table 1.1 denotes the number of spectral terms with the same quantum numbers L and S. Different spectral terms with the same quantum numbers L and S should be discriminated by different irreducible representation symbol of group chain. By using different irreducible representation symbol in the continuous group chain $U(10) \supset SU(2) \otimes U(5) \supset SU(2) \otimes SO(5) \supset SU(2) \otimes SO(3)$, different spectral terms of this configuration can be completely discriminated. The spectral terms of three $3d^q$ configurations with $q = 3$, 4, and 5 according to irreducible representation branch of above group chain are listed in Table 1.2.

The q values for unfilled electron shell $3d^q$ ($q < 10$) are listed in Table 1.3. Owing to the fact that $3d^q$ is an outer shell, the crystal field interaction is of the same order of magnitude as the electron–electron Coulomb interaction (about 10^4 cm^{-1}). However, it is much stronger than that of the spin–orbital interaction (about 10^2 cm^{-1}). Therefore, energy levels of these ions are strongly affected by the electrostatic field and the vibration of lattice ions. The energy level structure is very different for different host materials. It will be discussed later in a single chapter. Only the main features of the free iron group ions will be introduced here. Slater-Condon parameters F_k were used to describe the electrostatic interaction but now they are usually replaced by Racah parameters A, B, and C. The relations between these two kinds of parameters are:

Table 1.3 Electronic configurations of iron group ions

Ion	q
Ti^{3+}, V^{4+}	1
V^{3+}	2
V^{2+}, Cr^{3+}, Mn^{4+}	3
Mn^{3+}	4
Mn^{2+}, Fe^{3+}	5
Fe^{2+}, Co^{3+}	6
Co^{2+}	7
Ni^{2+}	8
Cu^{2+}	9

$$A = F_0 - 49F_4$$
$$B = F_2 - 5F_4$$
$$C = 35F_4$$

The energies of the spectral terms of the configurations d^2, d^3, d^4, and d^5 can be expressed and some of them are listed in Table 1.4.

Parameter A contributes the same energy to all the spectral terms in the same configuration. Parameters B and C can be calculated by radial wave functions R_{nl} using some kinds of theoretical model or fitted in comparison to experimental energy levels. The crystal field energy is generally higher than that of the spin–orbital interaction for this group of ions and so the spin–orbital splitting should be calculated after the crystal field energy calculation.

Table 1.5 shows the experienced parameters of B, C, B/C, and spin–orbital coupling coefficient.

It can be seen from Table 1.5 that electron Coulomb interaction is indeed larger than that of the spin–orbital interaction for d electrons. Racah parameters for the same configuration increased with atomic number due to the fact that the increasing of nuclear charge will cause decrease of the radius of d electron shell and finally result in a stronger electron Coulomb interaction. On the other hand, the spin–orbital coupling coefficient of Cu^{2+} with atomic number 29 is about five times that of the Ti^{3+} with atomic number 22, and more than that it increases with Z^4 as pointed out in Sect. 1.3.

Table 1.4 Spectral term energy expression of $3d^q$ configuration [17]

d^2	d^3
$^3F = A-8B$	$^4F = 3A-15B$
$^3P = A + 7B$	$^4P = 3A$
$^1G = A + 4B + 2C$	$^2H = {}^2P = 3A-6B + 3C$
$^1D = A - 3B + 2C$	$^2G = 3A - 11B + 3C$
$^1S = A + 14B + 7C$	$^2F = 3A + 9B + 3C$
	$_2D = 3A + 5B + 5C \pm (193B_2 + 8BC + 4C_2)_{1/2}$

(continued)

Table 1.4 (continued)

d^2	d^3
d^4	d^5
^{5}D = $6A - 21B$	^{6}S = $10A - 35B$
^{3}H = $6A - 17B + 4C$	^{4}G = $10A - 25B + 5C$
^{3}G = $6A - 12B + 4C$	^{4}F = $10A - 13B + 7C$
^{3}F = $6A - 5B + 5.5C \pm 1.5(68B_2 + 4BC + C_2)_{1/2}$	^{4}D = $10A - 18B + 5C$
^{3}D = $6A - 5B + 4C$	^{4}P = $10A - 28B + 7C$
$_3$P = $6A - 5B + 5.5C \pm 0.5(912B_2 - 24BC + 9C_2)_{1/2}$	^{2}I = $10A - 24B + 8C$
^{1}I = $6A - 15B + 6C$	^{2}H = $10A - 22B + 10C$
$_1$G = $6A - 5B + 3.5C \pm 0.5(708B_2 - 12BC + 9C_2)_{1/2}$	^{2}G = $10A - 13B + 8C$
$_1$F = $6A + 6C$	^{2}G′ = $10A + 3B + 10C$
$_1$D = $6A + 9B + 3.5C \pm 1.5(144B_2 + 8BC + C_2)_{1/2}$	^{2}F = $10A - 9B + 8C$
$_1$S = $6A + 10B + 10C \pm 2(193B_2 + 8BC + 4C_2)_{1/2}$	^{2}F′ = $10A - 25B + 10C$
	^{2}D′ = $10A - 4B + 10C$
	$_2$D = $10A - 3B + 11C \pm 3(57B_2 + 2BC + C_2)_{1/2}$
	^{2}P = $10A + 20B + 10C$
	^{2}S = $10A - 3B + 8C$

Table 1.5 Racah parameters and spin–orbital coupling coefficient for *d* electron (in unit of cm^{-1}) [18]

Ion (configuration)	B	C	B/C	ξ
$Ti^{3+}(3d^1)$				154
$V^{4+}(3d^1)$				248
$Ti^{2+}(3d^2)$	718	2629	3.66	121
$V^{3+}(3d^2)$	861	4165	4.84	209
$Cr^{4+}(3d^2)$	1039	4238	4.08	327
$V^{2+}(3d^3)$	766	2855	3.73	167
$Cr^{3+}(3d^3)$	918	3850	4.19	273
$Mn^{4+}(3d^3)$	1064			402
$Cr^{2+}(3d^4)$	830	3430	4.13	230
$Mn^{3+}(3d^4)$	965	3675	3.81	352
$Fe^{4+}(3d^4)$	1144	4459	3.90	514
$Mn^{2+}(3d^5)$	960	3325	3.46	347
$Fe^{3+}(3d^5)$	1015	4800	4.73	
$Fe^{2+}(3d^6)$	1058	3901	3.69	410
$Co^{3+}(3d^6)$	1065	5120	4.81	
$Co^{2+}(3d^7)$	971	4366	4.50	533
$Ni^{3+}(3d^7)$	1115	5450	4.89	
$Ni^{2+}(3d^8)$	1041	4831	4.64	649
$Cu^{2+}(3d^9)$	1238	4659	3.76	829

1.4 Energy Levels of Free Rare Earth Ions

Generally, lanthanide elements (from lanthanum $Z = 57$ to lutetium $Z = 71$), scandium ($Z = 21$), and yttrium ($Z = 39$) are referred to as rare earth elements. Except scandium and yttrium, they have similar electron shell structure, that is, the divalent and trivalent ions all have partially filled electron shell structure $4f^q$ (q = 1–13). The $4f$ electrons of these elements are not in the outer shell but are shielded by the electron shell $5s^2 5p^6$, so even in the host materials, their energy level structure and spectra have similar features compared to those of the free ions. From another point of view, it has been shown by photoelectron emission experiments that the position of $4f^q$ states in trivalent rare earth compounds is lower than the top of their valence band by about 5 eV. So, for low $4f^q$ energy levels, the free ion character can be well maintained. This property can also be explained by the energies of different interactions. The order of magnitude of electron Coulomb interaction is 10^4 cm^{-1}, while the spin–orbital and crystal field interactions are in the orders of 10^3 and 10^2 cm^{-1}, respectively. Using the language of quantum mechanics, the wave functions of free rare earth ions are very good zero-order wave functions for the ions in materials. Therefore, having some knowledge about the free ion's wave functions would be helpful for the further study of their spectroscopic properties in laser materials. In a good enough approximation, the principal quantum number n and the orbital angular momentum quantum number l can be still used to describe this multi-electron system, although the radial wave functions of the rare earth ions are different from those of hydrogen atom and the interaction between configurations cannot be neglected.

In Table 1.6, ${}^{2S+1}L^n$ represents that the spectral terms of ${}^{2S+1}L$ appear n times. The J values corresponding to the different possible states are not shown in the table. They can be obtained from L and S as follows: $J = L - S, L - S + 1, \ldots, L + S$. For example, the spectral term ^{4}D has S = 3/2, L = 2, and J can be: 1/2, 3/2, 5/2, 7/2, that is, giving rise to states ${}^4\mathrm{D}_{1/2}, {}^4\mathrm{D}_{3/2}, {}^4\mathrm{D}_{5/2}, {}^4\mathrm{D}_{7/2}$. When the number of electrons is large, quantum number LS are insufficient to classify different spectral terms completely. Spectral terms with the same LS number can appear up to ten times. Therefore, in order to provide a full classification of the states in f^q configuration, the group representation theory should be used.

The continuous group, first used by Racah, was introduced by Judd to classify the states of f^q configuration by a group chain of $SU(2) \otimes U(7) \supset SU(2) \otimes SO(7) \supset SU(2) \otimes G_2 \supset SU(2) \otimes SO(3)$ [4]. It is enough for readers to understand that in Table 1.7, $[\lambda_1, \lambda_2, \ldots]$ is another expression for the total spin quantum number S; on the other hand, (w_1, w_2, w_3) and (u_1, u_2) are quantum numbers in

Table 1.6 Allowed spectral terms for *f* electrons

f^1, f^{13}	$^2F^1$
f^2, f^{12}	$^1S^1$, $^1D^1$, $^1G^1$, $^1I^1$, $^3P^1$, $^3F^1$, $^3H^1$
f^3, f^{11}	$^2P^1$, $^2D^2$, $^2F^2$, $^2G^2$, $^2H^2$, $^2I^1$, $^2K^1$, $^2L^1$, $^4S^1$, $^4D^1$, $^4F^1$, $^4G^1$, $^4I^1$
f^4, f^{10}	$^1S^2$, $^1D^4$, $^1F^1$, $^1G^4$, $^1H^2$, $^1I^3$, $^1K^1$, $^1L^2$, $^1N^1$, $^3P^3$, $^3D^2$, $^3F^4$, $^3G^3$, $^3H^4$, $^3I^2$, $^3K^2$, $^3L^1$, $^3M^1$, $^5S^1$, $^5D^1$, $^5F^1$, $^5G^1$, $^5I^1$
f^5, f^9	$^2P^4$, $^2D^5$, $^2F^7$, $^2G^6$, $^2H^7$, $^2I^5$, $^2K^5$, $^2L^3$, $^2M^2$, $^2N^1$, $^2O^1$, $^4S^1$, $^4P^2$, $^4D^3$, $^4F^4$, $^4G^4$, $^4H^3$, $^4I^3$, $^4K^2$, $^4L^1$, $^4M^1$, $^6P^1$, $^6F^1$, $^6H^1$
f^6, f^8	$^1S^4$, $^1P^1$, $^1D^6$, $^1F^4$, $^1G^8$, $^1H^4$, $^1I^7$, $^1K^3$, $^1L^4$, $^1M^2$, $^1N^2$, $^1Q^1$, $^3P^6$, $^3D^5$, $^3F^9$, $^3G^7$, $^3H^9$, $^3I^6$, $^3K^6$, $^3L^3$, $^3M^3$, $^3N^1$, $^3O^1$, $^5S^1$, $^5P^1$, $^5D^3$, $^5F^2$, $^5G^3$, $^5H^2$, $^5I^2$, $^5K^1$, $^5L^1$, $^7F^1$
f^7	$^2S^2$, $^2P^5$, $^2D^7$, $^2F^{10}$, $^2G^{10}$, $^2H^9$, $^2I^9$, $^2K^7$, $^2L^5$, $^2M^4$, $^2N^2$, $^2O^1$, $^2Q^1$, $^4S^2$, $^4P^2$, $^4D^6$, $^4F^5$, $^4G^7$, $^4H^5$, $^4I^5$, $^4K^3$, $^4L^3$, $^4M^1$, $^4N^1$, $^6P^1$, $^6D^1$, $^6F^1$, $^6G^1$, $^6H^1$, $^6I^1$, $^8S^1$

addition to the *LS*. With these quantum numbers, one can completely differentiate all states with the same *LS* quantum number. For more knowledge one can refer to books on theory of atomic spectroscopy [4, 6]. Table 1.7 contains the example for two rare earth configurations.

Three coupling models have been mentioned in the previous discussion, but not the intermediate-coupling model. However, it should be pointed out that although in the first-order approximation the ***LS*** coupling model can be used to deal with the energy level analysis of rare earth ions, the spin–orbital Hamiltonian cannot be neglected. The wave functions of ***LS*** coupling model were used as the first-order approximation wave functions to calculate eigenvalues and eigenfunctions of the Hamiltonian $H_0 + H_{s0}$. The eigenfunctions obtained in this way are called intermediate-coupling wave functions. Some of these wave functions, expressed as wave functions of ***LS*** coupling model for lower spectral terms in the rare earth ions in laser materials, are listed in Table 1.8.

In this table, $\psi(^{2S+1}L_J)$ refers to intermediate-coupling wave functions and $\varphi(^{2S+1}L_J)$ denotes the ***LS*** coupling wave functions. Some small terms have been neglected.

It can be seen from Table 1.8 that in the intermediate-coupling wave functions the spin quantum number difference between the main wave functions and the mixed wave functions is $\Delta S = 1$. The reason is when the difference of spin quantum number between two wave functions is larger or smaller than 1, the matrix element of the spin–orbital interaction operator between them is equal to zero, so that there is no spin–orbital mixing. The matrix element of the spin–orbital interaction operator can be expressed by the formula (2–106) of Wyboune [7]. In Chap. 4, it will be shown that the transition between the two energy levels with different spin

Table 1.7 Classification of the spectral terms in f^3 and f^4 configuration according to the reducing branching rule of the group chain $SU(2) \otimes U(7) \supset SU(2) \otimes SO(7) \supset SU(2) \otimes G_2 \supset SU(2) \otimes SO(3)$

Configuration	$SU(2) \otimes U(7)$	$SU(2) \otimes SO(7)$	$SU(2) \otimes G_2$	$SU(2) \otimes SO(3)$
$f^3(f^{11})$	$^2[2, 1]$	$^2(1, 0, 0)$	$^2(1, 0)$	^{2}F
	$^2[2, 1]$	$^2(2, 1, 0)$	$^2(1, 1)$	^{2}P, ^{2}H
	$^2[2, 1]$	$^2(2, 1, 0)$	$^2(2, 0)$	^{2}D, ^{2}G, ^{2}I
	$^2[2, 1]$	$^2(2, 1, 0)$	$^2(2, 1)$	^{2}D^2, ^{2}F, ^{2}G, ^{2}H, ^{2}K, ^{2}L
	$^4[1^3]$	$^4(1, 1, 1)$	$^4(0, 0)$	^{4}S
	$^4[1^3]$	$^4(1, 1, 1)$	$^4(1, 0)$	^{4}F
	$^4[1^3]$	$^4(1, 1, 1)$	$^4(2, 0)$	^{4}D, ^{4}G, ^{4}I
$f^4(f^{10})$	$^1[2^2]$	$^1(0, 0, 0)$	$^1(0, 0)$	^{1}S
	$^1[2^2]$	$^1(2, 0, 0)$	$^1(2, 0)$	^{1}D, ^{1}G, ^{1}I
	$^1[2^2]$	$^1(2, 2, 0)$	$^1(2, 0)$	^{1}D, ^{1}G, ^{1}I
	$^1[2^2]$	$^1(2, 2, 0)$	$^1(2, 1)$	^{1}D, ^{1}F, ^{1}G, ^{1}H, ^{1}K, ^{1}L
	$^1[2^2]$	$^1(2, 2, 0)$	$^1(2, 2)$	^{1}S, ^{1}D, ^{1}G, ^{1}H, ^{1}I, ^{1}L, ^{1}N
	$^3[2, 1^2]$	$^3(1, 1, 0)$	$^3(1, 0)$	^{3}F
	$^3[2, 1^2]$	$^3(1, 1, 0)$	$^3(1, 1)$	^{3}P, ^{3}H
	$^3[2, 1^2]$	$^3(2, 1, 1)$	$^3(1, 0)$	^{3}F
	$^3[2, 1^2]$	$^3(2, 1, 1)$	$^3(1, 1)$	^{3}P, ^{3}H
	$^3[2, 1^2]$	$^3(2, 1, 1)$	$^3(2, 0)$	^{3}D, ^{3}G, ^{3}I
	$^3[2, 1^2]$	$^3(2, 1, 1)$	$^3(2, 1)$	^{3}D, ^{3}F, ^{3}G, ^{3}H, ^{3}K, ^{3}L
	$^3[2, 1^2]$	$^3(2, 1, 1)$	$^3(3, 0)$	^{3}P, ^{3}F, ^{3}G, ^{3}H, ^{3}I, ^{3}K, ^{3}M
	$^5[1^4]$	$^5(1, 1, 1)$	$^5(0, 0)$	^{5}S
	$^5[1^4]$	$^5(1, 1, 1)$	$^5(1, 0)$	^{5}F
	$^5[1^4]$	$^5(1, 1, 1)$	$^5(2, 0)$	^{5}D, ^{5}G, ^{5}I

quantum number is forbidden except due to the effect of intermediate coupling, the transition forbidden is partially released. However, for the so-called double-spin flip transition with $\Delta S = 2$, the forbidden rule is strictly established.

Figure 1.1 shows that the radii of 5*s* and 5*p* shells are larger than that of 4*f* shell. However, the quantum defect Δ decreases with orbital angular momentum l, $5 - \Delta(5s) < 5 - \Delta(5p) < 4 - \Delta(4f)$, so that the energy of 4*f* electron is higher than those of 5*s* and 5*p* electrons and the electrons will first fill 5*s* and 5*p* shells and then the 4*f* shell. Therefore, the 4*f* electrons are in the inner shell with outer filled 5*s* and 5*p* shells to shield the interactions of the host materials. So for low $4f^q$ energy levels, the free ion character can be maintained certainly. In a good enough approximation, the principal quantum number n and the orbital angular momentum quantum number l can be still used to describe this multi-electron system in the host materials, although the radial wave functions of the rare earth ions are different from those of hydrogen atom and the interaction between configurations cannot be neglected.

Table 1.8 Some of the intermediate-coupling wave functions of rare earth ions

Ion	Intermediate-coupling wave functions
Nd^{3+} [19]	$\psi(^4I_{9/2}) = 0.9839\varphi(^4I_{9/2}) - 0.1672\varphi\left((210)(21)^2H_{9/2}\right) + 0.0580\varphi\left((210)(11)^2H_{9/2}\right)$
	$\psi(^4I_{11/2}) = 0.9945\varphi(^4I_{11/2}) + 0.0369\varphi\left((210)(11)^2H_{11/2}\right) - 0.0964\varphi\left((210)(21)^2H_{11/2}\right)$
	$\psi(^4I_{13/2}) = 0.9975\varphi(^4I_{13/2}) - 0.066\varphi(^2K_{13/2})$
	$\psi(^4I_{15/2}) = 0.9927\varphi(^4I_{15/2}) + 0.1201\varphi(^2K_{15/2})$
	$\psi(^4F_{3/2}) = 0.9685\varphi(^4F_{3/2}) - 0.0563\varphi(^4S_{3/2}) + 0.2268\varphi\left((210)(20)^2D_{3/2}\right) - 0.0622\varphi(^2P_{3/2}) - 0.0577\varphi\left((210)(21)^2D_{3/2}\right)$
Er^{3+} [20]	$\psi(^4I_{15/2}) = 0.9852\varphi(^4I_{15/2}) - 0.1708\varphi(^2K_{15/2})$
	$\psi(^4I_{13/2}) = -0.9955\varphi(^4I_{13/2}) + 0.0896\varphi(^2K_{13/2})$
	$\psi(^4I_{11/2}) = 0.9125\varphi(^4I_{11/2}) + 0.1094\varphi(^4G_{11/2}) + 0.0631\varphi(^2I_{11/2}) + 0.3740\varphi\left((210)(21)^2H_{11/2}\right) - 0.1073\varphi\left((210)(11)^2H_{11/2}\right)$
	$\psi(^4I_{9/2}) = -0.7322\varphi(^4I_{9/2}) + 0.2765\varphi((210)(20)^2G_{9/2}) - 0.2204\varphi((210)(21)^2G_{9/2}) + 0.1953\varphi\left((210)(11)^2H_{9/2}\right) - 0.4125\varphi\left((210)(21)^2H_{9/2}\right) + 0.3611\varphi(^4F_{9/2})$
	$\psi(^4S_{3/2}) = 0.8371\varphi(^4S_{3/2}) - 0.4196\varphi(^2P_{3/2}) - 0.2666\varphi\left((210)(20)^2D_{3/2}\right) + 0.2237\varphi(^4F_{3/2})$
Ho^{3+} [21]	$\psi(^5I_8) = 0.967\varphi(^5I_8)$
	$\psi(^5I_7) = 0.965\varphi(^5I_7)$
	$\psi(^5I_6) = -0.976\varphi(^5I_6)$
	$\psi(^5I_5) = -0.952\varphi(^5I_5)$
	$\psi(^5I_4) = 0.948\varphi(^5I_4)$
	$\psi(^5S_2) = 0.841\varphi(^5S_2) + 0.377\varphi\left((210)(11)^3P_2\right)$
	$\psi(^5F_5) = 0.896\varphi(^5F_5) - 0.315\varphi\left((211)(21)^3P_2\right)$
	$\psi(^5F_4) = 0.959\varphi(^5F_4)$
	$\psi(^5F_3) = 0.951\varphi(^5F_3)$
	$\psi(^5F_2) = 0.794\varphi(^5F_2) - 0.383\varphi(^5S_2) + 0.345\varphi\left((211)(21)^3D_2\right)$
	$\psi(^5F_1) = -0.892\varphi(^5F_1) - 0.384\varphi\left((211)(21)^3D_1\right)$
Tm^{3+} [20]	$\psi(^3P_0) = 0.9718\varphi(^3P_0) - 0.2354\varphi(^1S_0)$
	$\psi(^3P_1) = \varphi(^3P_1)$
	$\psi(^3P_2) = 0.7693\varphi(^3P_2) - 0.1984\varphi(^3F_2) - 0.607\varphi(^1D_2)$
	$\psi(^1S_0) = 0.9718\varphi(^1S_0) + 0.2354\varphi(^3P_0)$
	$\psi(^3F_2) = 0.8769\varphi(^3F_2) - 0.1374\varphi(^3P_2) - 0.4606\varphi(^1D_2)$
	$\psi(^3F_3) = \varphi(^3F_3)$
	$\psi(^3H_4) = 0.5282\varphi(^3F_4) + 0.7713\varphi(^3H_4) - 0.3549\varphi(^1G_4)$
	$\psi(^3F_4) = 0.787\varphi(^3F_4) - 0.2883\varphi(^3H_4) + 0.5454\varphi(^1G_4)$
	$\psi(^3H_5) = \varphi(^3H_5)$
	$\psi(^3H_6) = 0.9956\varphi(^3H_6) + 0.0931\varphi(^1I_6)$
	$\psi(^1D_2) = 0.6473\varphi(^1D_2) + 0.623\varphi(^3P_2) + 0.4378\varphi(^3F_2)$

(continued)

Table 1.8 (continued)

Ion	Intermediate-coupling wave functions
	$\psi(^1G_4) = 0.7594\varphi(^1G_4) + 0.5674\varphi(^3H_4) - 0.3182\varphi(^3F_4)$
	$\psi(^1I_6) = 0.9956\varphi(^1I_6) - 0.0931\varphi(^2H_6)$
Pr^{3+} [20]	$\psi(^3P_0) = 0.996\varphi(^3P_0) + 0.0876\varphi(^1S_0)$
	$\psi(^3P_1) = \varphi(^3P_1)$
	$\psi(^3P_2) = 0.9592\varphi(^3P_2) + 0.2812\varphi(^1D_2)$
	$\psi(^3F_2) = 0.989\varphi(^3F_2) + 0.1475\varphi(^1D_2)$
	$\psi(^3F_3) = \varphi(^3F_3)$
	$\psi(^3F_4) = 0.8544\varphi(^3F_4) + 0.1035\varphi(^3H_4) - 0.5092\varphi(^1G_4)$
	$\psi(^3H_4) = 0.9878\varphi(^3H_4) + 0.1534\varphi(^1G_4)$
	$\psi(^3H_5) = \varphi(^3H_5)$
	$\psi(^3H_6) = -0.9985\varphi(^3H_6) + 0.0541\varphi(^1I_6)$
	$\psi(^1S_0) = -0.9962\varphi(^1S_0) + 0.0876\varphi(^3P_0)$
	$\psi(^1D_2) = -0.9483\varphi(^1D_2) + 0.2823\varphi(^3P_2) + 0.1452\varphi(^3F_2)$
	$\psi(^1G_4) = 0.8469\varphi(^1G_4) - 0.1167\varphi(^3H_4) + 0.5188\varphi(^3F_4)$
	$\psi(^1I_6) = 0.9956\varphi(^1I_6) - 0.0931\varphi(^3H_6)$

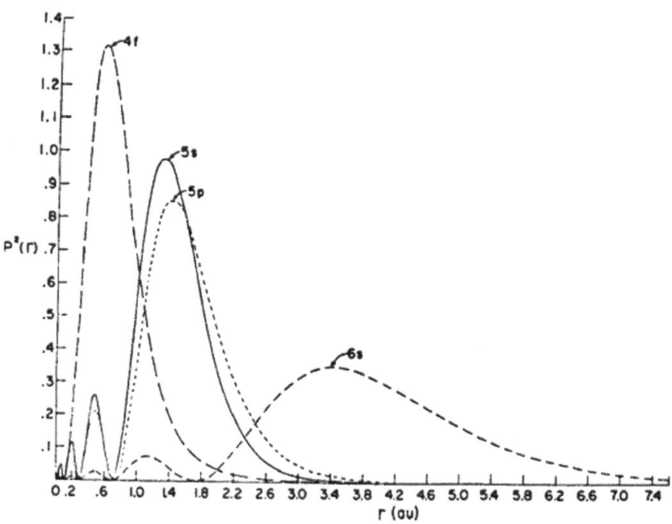

Fig. 1.1 Outer electron shell distribution of rare earth ions

The lower energy levels of $4f^{11}$ configurations of free Er^{3+} ion and this ion in five kinds of host crystals, including LaF_3, are listed in Table 1.9. One can see from this table the differences of a given energy level for free Er^{3+} ion and that this ion in different crystals is generally less than 100 cm^{-1}; only in few cases is about 200 cm^{-1}. Therefore, in the discussion of an energy level position in different hosts, their position in LaF_3 crystal can be seen as a good reference. Figure 1.2 shows the energy levels of 13 rare earth ions. This diagram has been frequently referred to in the literature. It can give us a general idea of the energy level positions for all the rare earth ions and help us in the analysis of an unknown spectrum and determines which energy levels are involved in the spectrum. From this diagram, if one considers the Gd^{3+} ion as a center, it can be seen that the Hund rule is really available and any spectral terms of a rare earth ion in the left hand position can be found in another rare earth ion in the corresponding right hand position. However, for the same spectral term ^{2S+1}L, the lower $^{2S+1}L_J$ levels in the left-hand side correspond to the higher $^{2S+1}L_J$ levels in the corresponding right-hand side. It should also be noted that the energy separations between different spectral terms ^{2S+1}L and different $^{2S+1}L_J$ levels in the same spectral term are larger for the rare earth ions in the right-hand side of Gd^{3+} ion. It is due to the fact that the Coulomb interaction between electrons and the spin–orbital interaction are stronger for the rare earth ions in the right-hand side, because of the effect of lanthanide contraction and the spin–orbital coupling coefficient is greater for ions of larger atomic number. This tendency can be clearly seen from Fig. 1.3 which shows that the Slater parameter F^k and spin–orbital coupling coefficient ζ increase with $4f$ electron number.

In order to investigate the energy levels of rare earth ions, one should also know the separation between ground configuration and excited configurations. This is because the excited configurations have important effects on the ground configuration, which will be discussed in the following. Figure 1.4 is a schematic diagram

Table 1.9 Level energies for the free Er^{3+} ion and that in five host crystals (unit of cm^{-1})

Levels	Free ion [22]	LaF_3 [22]	$LaCl_3$ [23]	$Cs_3Lu_2Cl_9$ [13]	Y_2O_3 [22]	$Y_3Al_5O_{12}$ [24]
$^4I_{15/2}$	0	0	0	0	0	0
$^4I_{13/2}$	6,485	6,480	6,482	6,465	6,458	6,470
$^4I_{11/2}$	10,123	10,123	10,112	10,078	10,073	10,090
$^4I_{9/2}$	12,345	12,350	12,352	12,300	12,287	12,312
$^4F_{9/2}$	15,182	15,235	15,177	15,091	15,071	15,129
$^4S_{3/2}$	18,299	18,353	18,290	18,142	18,072	18,166
$^2H_{11/2}$	19,010	–	–	18,944	18,931	18,980
$^4F_{7/2}$	20,494	20,492	20,407	20,292	20,276	20,348
$^4F_{5/2}$	22,181	22,161	22,066	21,939	21,894	21,992
$^4F_{3/2}$	22,453	22,494	22,407	22,301	22,207	22,370
$^2H_{9/2}$	24,475	24,526	24,453	24,339	24,304	24,368
$^4G_{11/2}$	26,376	26,368	26,257	26,116	26,074	26,166
$^2G_{9/2}$	27,319	27,412	–	27,097	–	–

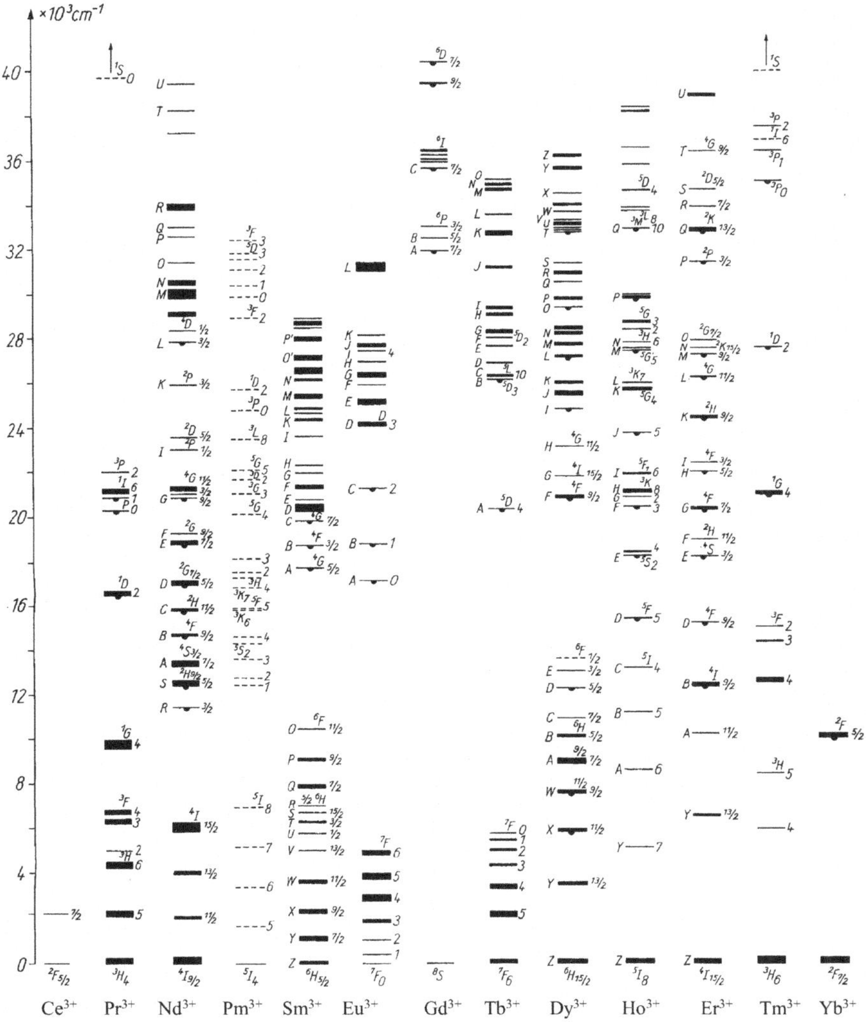

Fig. 1.2 Energy levels of trivalent rare earth ions

for the separation between 4*f* electron and some of the excited shells of divalent and trivalent rare earth ions [8]. Figs. 1.5 and 1.6 are relations between two lowest configurations for the trivalent rare earth ions and divalent rare earth ions, respectively.

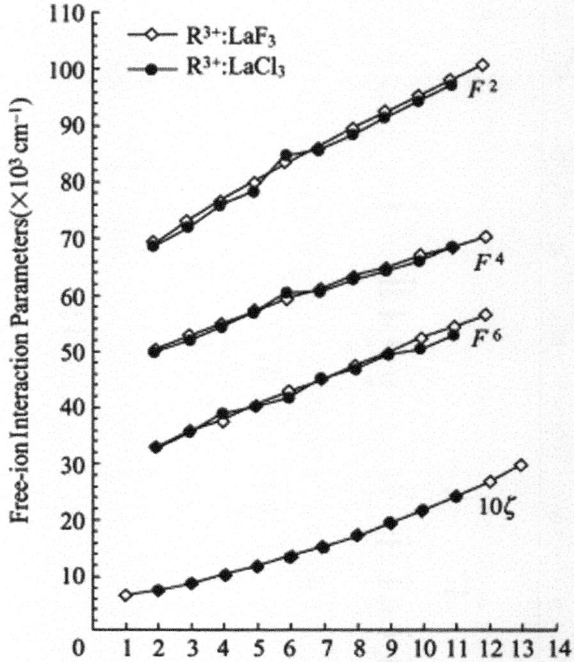

Fig. 1.3 Relation between Slater parameter F^k as well as spin-orbital coupling coefficient ζ and the electron number determined by experiments on crystals of LaF_3 and $LaCl_3$ [16]

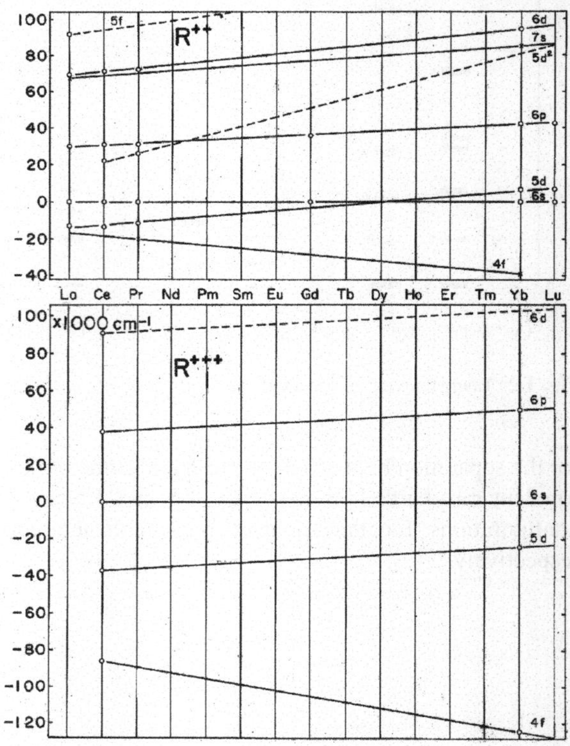

Fig. 1.4 Schematic diagrams for the separation between $4f$ electron and some of the excited shells of the divalent and trivalent rare earth ions [8]

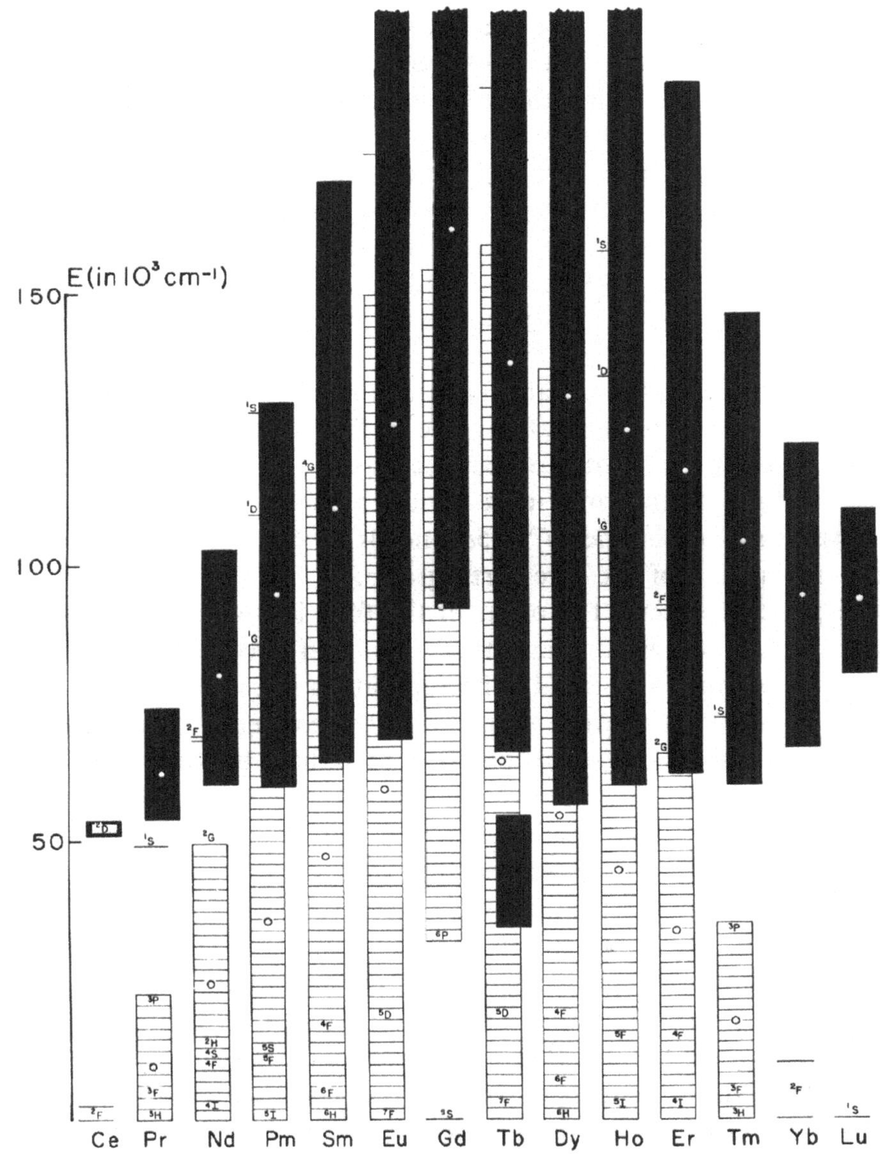

Fig. 1.5 Relation between two lowest configurations of the trivalent rare earth ions [8]

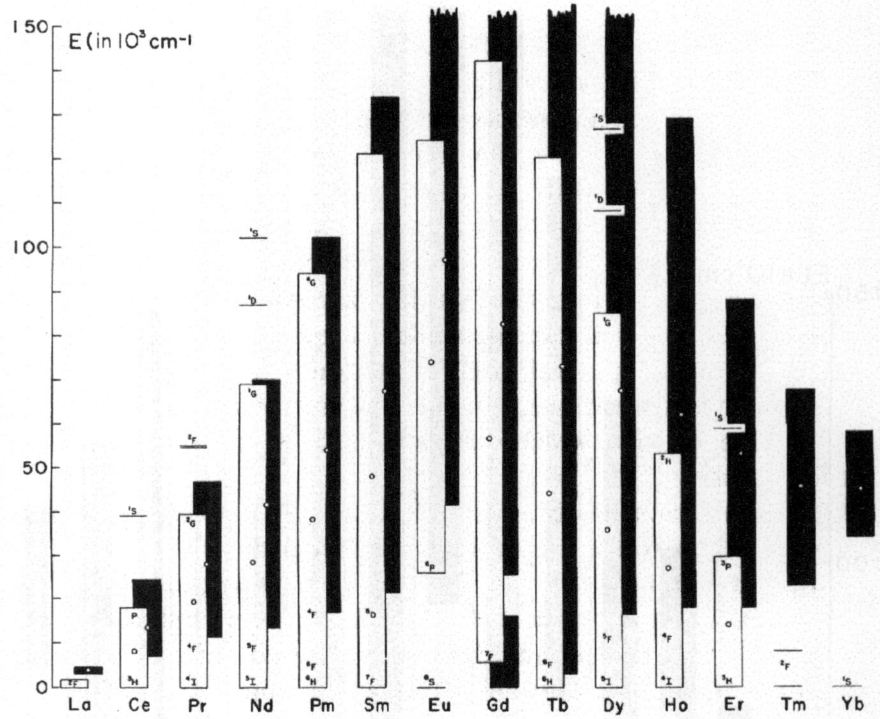

Fig. 1.6 Schematic diagrams for the separation between two lowest configurations of the divalent rare earth ions [8]

1.5 Theory of Interactions in Rare Earth Ions

By the end of this chapter, we would like to return to the subject of a series of interactions in rare earth ions and introduce their effects on the energy level positions. To investigate the energy levels of multi-electron system, a very good example is the application of continuous group in spectroscopy. In this field, Racah in 1940s [9] and Judd in 1950s [6] have performed outstanding works. From the Hamiltonians used to describe the different interactions in the atoms and the wave functions classified by Lie group representations, by means of the tensor operator method established at that time, a series of interaction energies were calculated. We would like to introduce the parameter method that can describe all these interactions but not involving too much mathematics. The number of parameters introduced is the minimum necessary to describe all the interactions. For example, among the parameters t_k used to describe configuration interactions, the effect of t_1 and t_5 can be embodied by the electrostatic Coulomb interaction parameters and so they are omitted.

As presented previously, Coulomb interaction and spin–orbit interaction can be taken into account by constructing intermediate-coupling wave functions. The other method generally adopted is to describe these two interactions by related parameters. The Coulomb interaction in (1.26) is considered individually as

$$H_{\mathrm{ee}} = \sum_{i>j=1}^{n} \frac{e^2}{4\pi\varepsilon_0 r_{i,j}} \tag{1.34}$$

By applying the following expanded expression

$$\frac{1}{r_{ij}} = \frac{4\pi}{2k+1} \sum_{kq} \frac{r_<^k}{r_>^{k+1}} Y_q^{(k)*}(\theta_i, \varphi_i) Y_q^{(k)}(\theta_j, \varphi_j) \tag{1.35}$$

and

$$C_q^{(k)}(\hat{r}) = \left(\frac{4\pi}{2k+1}\right)^{1/2} Y_q^{(k)}(\theta, \phi)$$

for $4f^q$ electronic configuration, H_{ee} becomes

$$H_{\mathrm{ee}} = \sum_{k=0,2,4,6} F^k f_k = \sum_{k=0,2,4,6} F_k f^k \tag{1.36}$$

where

$$f_k = \langle \alpha LS | \sum_{i>j} C^{(k)}(i) \cdot C^{(k)}(j) | \alpha LS \rangle \tag{1.37a}$$

$$F^k = \frac{e^2}{2\pi\varepsilon_0} \int_0^\infty \int_0^\infty R_{4f}^2(r_i) R_{4f}^2(r_j) \frac{r_<^k}{r_>^{k+1}} \mathrm{d}r_i \mathrm{d}r_j \tag{1.37b}$$

where F^k is called Slater parameter, and another parameter introduced by Condon and Shortley is F_k. The relations between them can be written as

$$\frac{F^k}{F_k} = D_k \tag{1.38}$$

where D_k are Condon-Shortley factors, for $4f$ electrons

$$D_0 = 1,\ D_2 = 225,\ D_4 = 1089,\ D_6 = 184041/25$$

Coulomb interaction can also be described by Racah parameters E^0, E^1, E^2, E^3 and expressed as

$$H_{ee} = \sum_{k=0,1,2,3} E^k e_k \tag{1.39}$$

The matrix elements of e_k can be found in the well-known tables provided by Nielson and Koster [10], f^k and e_k have the following relations

$$\begin{aligned}
e_0 &= f^0 = q(q-1) \\
e_1 &= \frac{9}{7}f^0 + \frac{1}{42}f^2 + \frac{1}{77}f^4 + \frac{1}{462}f^6 \\
e_2 &= \frac{143}{42}f^2 - \frac{130}{77}f^4 + \frac{35}{462}f^6 \\
e_3 &= \frac{11}{42}f^2 + \frac{4}{77}f^4 - \frac{7}{462}f^6
\end{aligned} \tag{1.40}$$

where q is the number of electrons in the configuration $4f^q$. The relations between Racah parameters and Slater parameters are

$$\begin{aligned}
E^0 &= F_0 - 10F_2 - 33F_4 - 286F_6 \\
E^1 &= \frac{70F_2 + 231F_4 + 2002F_6}{9} \\
E^2 &= \frac{F_2 - 3F_4 + 7F_6}{9} \\
E^3 &= \frac{5F_2 + 6F_4 - 91F_6}{3}
\end{aligned} \tag{1.41}$$

or

$$\begin{aligned}
F_0 &= \frac{7E^0 + 9E^1}{7} \\
F_2 &= \frac{E^1 + 143E^2 + 11E^3}{42} \\
F_4 &= \frac{E^1 - 130E^2 + 4E^3}{77} \\
F_6 &= \frac{E^1 + 35E^2 - 7E^3}{462}
\end{aligned} \tag{1.42}$$

The spin–orbital interaction of the whole atom system can then be obtained by simply summing (1.2) over all the electrons

$$H_{so} = \sum_{i=1}^{n} \xi(r_i)\boldsymbol{l}_i \cdot \boldsymbol{s}_i \tag{1.43}$$

Integrating over radial variable, it can be expressed as

$$H_{\text{so}} = \zeta_{4f} A_{\text{so}} \tag{1.44}$$

where $\zeta_{4f} = \int_0^{\infty} R_{4f}^2(r)\xi(r)\mathrm{d}r$. A_{so} is the angular operator of spin–orbit interaction and can be expressed as

$$A_{\text{so}}(nl) = (-1)^{L+S'+J}\sqrt{(2l+1)(l+1)l}\,\delta_{JJ'}\delta_{MM'} \times \begin{Bmatrix} L & S & J \\ S' & L' & 1 \end{Bmatrix} \left\langle \tau LS \| V^{(11)} \| \tau' L' S' \right\rangle$$

Reduced matrix elements of tensor operator $\boldsymbol{V}^{(11)}$ can be found in [10].

Considering $4f^q$ configuration, the calculation of energy levels may have errors up to several hundreds of cm^{-1} compared with the experimental results, if only the above two interactions are included. The reason for this error is the Coulomb interaction between ground and excited configurations that has not been taken into account. This interaction can be divided into two kinds, that is, two-body and three-body interactions. Two-body interaction is written as

$$H_{\text{cc1}} = \alpha L(L+1) + \beta G(G_2) + \gamma G(R_7) \tag{1.45}$$

where α, β, γ are fitting parameters, $G(G_2)$ and $G(R_7)$ are Casimir operators for Lie group G_2 and R_7 with eigenvalues to be found from Tables 2.6 and 2.7 in [7].

The three-body interaction should be considered for configurations having three or more electrons and it has the following expression introduced by Judd [11]

$$H_{\text{cc2}} = \sum_i T^i t_i \tag{1.46}$$

where i = 2, 3, 4, 6, 7, 8, T^i and t_i are fitting parameters and operators, respectively. The operators transformed according to the irreducible representations of group G_2 and R_7, also called three-body operators and can be found in [7]. Four-body or many-body interactions can be introduced in the same way, but without substantive effect on the energy level positions.

Besides spin–orbital interaction expressed by ζ_{nl}, the magnet interactions of spin–spin and the spin angular momentum with the orbital angular momentum of other electrons can be expressed as

$$H_{\text{ss+soo}} = \sum_{h=0,2,4} M^h m_h \tag{1.47}$$

where m_h are effective operators and M^h are relative Marvin integral being written as follows

$$M^h = \frac{\alpha^2}{4} \iint_{r_i > r_j} R_{4f}^2(r_i) R_{4f}^2(r_j) \frac{r_i^h}{r_j^{h+3}} \mathrm{d}r_i \mathrm{d}r_j \tag{1.48}$$

In order to improve the fitting of f electron spectroscopic data, it is necessary to introduce three two-body operators to take into account weaker electrostatic and spin–orbit-correlated interactions, expressed as

$$H_{cc3} = \sum_{k=2,4,6} P^k p_k \tag{1.49}$$

P^k are parameters while p_k are two-electron operators depending on the spin angular momentum and can be found in [12].

Considering all the interactions introduced, the total Hamiltonian of the rare earth ions can be written as

$$\begin{aligned} H = & \sum_{k=2,4,6} F^k f_k + \zeta_{4f} A_{so} + \alpha L(L+1) + \beta G(G_2) + \gamma G(R_7) + \sum_{i=2,3,4,6,7,8} T^i t_i \\ & + \sum_{k=2,4,6} P^k p_k + \sum_{h=0,2,4} M^h m_h \end{aligned} \tag{1.50}$$

In this way, the E^0, E^1, E^2, E^3, (F^0, F^2, F^4, F^6); ζ; α, β, γ; T_i (i = 2, 3, 4, 6, 7, 8); p_k (k = 2, 4, 6); and m_k (h = 0, 4, 6), 20 parameters can have a full description of the energy levels of free rare earth ions. It should be pointed out that compared to the electrostatic Coulomb interaction between electrons and the spin–orbit interaction, the configuration and the relativistic magnetic interactions are much weaker. Moreover, we have

$$H_{cc1} > H_{cc2} > H_{cc3} \geq H_{ss+soo}$$

Among all the 20 parameters in (1.50), four parameters should be subjected to the Hartree-Fock restriction [13]

$$\begin{aligned} M^0 &= 1.78M^2 = 2.63M^4 \\ P^2 &= 1.33P^4 = 2P^6 \end{aligned}$$

Therefore, there are only 16 independent parameters. After fitting to the experimental energy level data, the values of these parameters for Er^{3+} ions in three crystal hosts are listed in Table 1.10.

The relative parameters for Nd^{3+} in $NdAl_3(BO_3)_4$ (NAB) crystal can be found in [14]. It can be seen that these parameters are not only different for different host but are different for different sites in the same host. As pointed out by Marshall and Stuart [15], when a free ion placed in crystal (or liquid), its wave function will expand into crystal (or liquid). These kinds of expansion will produce the screen of 4f electron by the overlap electron cloud of 4f electron and its ligands, and finally

Table 1.10 Atomic parameters for Er^{3+} ions in three crystal hosts (units in cm^{-1})

Parameters	$LaCl_3$ [22]	$Cs_3Lu_2Cl_9$ [23]	$Y_3Al_5O_{12}$ [13]
F^2	98,260	97,940	99,333
F^4	69,793	69,850	70,581
F^6	48,114	49,850	49,207
ζ	2362	2366	2368
α	17.4	17.1	17.6
β	−638	−647	−612
γ	2061	1747	1698
T^2	426	299	599
T^3	48	45	38
T^4	22	36	69
T^6	−305	−312	−362
T^7	289	213	323
T^8	353	352	505
P^2	416	539	478
M^0	4.2	4.2	3.6

result in the decrease of effective nuclear charge. Therefore, Slater parameters and spin–orbital coupling coefficient fitting by crystal spectral data will be less than those of free ions. These kinds of wave function expansion obviously will be different for different media so that the parameters fitting by spectral data of different media have some difference. However, the difference of all the 16 parameters for different hosts in general is within a range of 1%.

It is clearly shown in Table 1.10 that all the atomic parameters for rare earth ions in different hosts have nearly the same values. Fitting with so many parameters is complicated but the results after fitting have not substantially improved, so for the practical problems of laser materials in which the gravity center of the spectral terms are non-essential, the atomic parameters can refer to those of other hosts and fitting only the crystal parameters is enough.

References

1. G. Herzberg, *Atomic Spectra and Atomic Structure* (Dover Publication, New York, 1944)
2. E.U. Condon, G.H. Shortley, *The Theory of Atomic Spectra* (Cambridge University Press, New York, 1935)
3. J.C. Slater, *Quantum Theory of Atomic Structure* (McGraw-Hill Book Company, New York, 1960)
4. E.U. Condon, H. Odabasi, *Atomic Structure* (Cambridge University Press, London, 1980)
5. R.C. Powell, *Physics of Solid-State Laser Materials* (Springer, New York Inc, New York, 1998)
6. B.R. Judd, *Operator Techniques in Atomic Spectroscopy* (McGraw-Hill Book Company, New York, 1963)

7. B.G. Wybourne, *Spectroscopic Properties of Rare Earths* (Interscience Publishers, New York, 1965)
8. G.H. Dieke, *Spectra and Energy Levels of Rare Earth Ions in Crystals* (Interscience Publishers, New York, 1968)
9. G. Racah, Phys. Rev., **61**, 186 (1942); **62**, 438 (1942); **63**, 367 (1943); **76**, 1352 (1949)
10. C.W. Nielson, G.F. Koster, *Spectroscopic Coefficients for p^n, d^n and f^n Configurations* (MIT Press, Massachusetts, 1964)
11. B.R. Judd, Phys. Rev. **141**, 4 (1966)
12. B.R. Judd, H.M. Crosswhite, H. Crosswhite, Phys. Rev. **169**, 130 (1968)
13. S.R. Luthi, H.U. Gudel, H.U. Hehlen et al., Phys. Rev. B **57**, 15229 (1998)
14. C. Cascales, C. Zaldo, U. Caldino, Garcia Sole J. Luo Z.D., J. Phys. Condens. Matter **13**, 8071 (2001)
15. W. Marshall, R. Stuart, Phys. Rev. **123**, 2048 (1961)
16. L. Guokui, J. Bernard (eds.), *Spectroscopic of Rare Earths in Optical Materials* (Springer, Berlin Heidelberg New York, 2005)
17. J.S. Griffith, *The Theory of Transition-Metal* (Cambridge University Press, London, 1961)
18. B. Henderson, R.H. Bartram, *Crystal-Field Engineering of Solid-State Laser Materials* (Cambridge University Press, Cambridge, 2000)
19. B.G. Wybourne, J. Chem. Phys. **34**, 279 (1961)
20. R. Pappalardo, J. Lumin. **14**, 159 (1976)
21. K. Rajnak, W.F. Krupke, J. Chem. Phys. **46**, 3532 (1967)
22. S. Hufner, *Optical Spectra of Transparent Rare Earth Compounds* (Academic Press, New York, 1978)
23. K.W. Kramer, H.U. Gudel, R.N. Schwartz, Phys. Rev. B **56**, 13830 (1997)
24. J.B. Gruber, J.R. Quagliano, M.F. Reid et al., Phys. Rev. B **48**, 15561 (1993)

Chapter 2
Group Theory and Quantum Theory

The electron–electron interaction, electron–nucleus interaction as well as the orbit–orbit, orbit–spin, spin–spin couplings, and the other complicate interactions of electrons in free ions have been discussed in Chap. 1. However, the whole system of the free ion containing all the electrons in it can be seen as a spherical symmetry system. This is due to the fact that the space of the free ion is isotropic in a good enough approximation. If the ion is placed in a solid material, this space isotropy will be generally destroyed by the interactions between the ions and the environment. The energy of the electrons in different states will be different because the interactions are different for different relative positions (relative distance or angle) between the electrons of ion studied and the lattice ions. This also means the elimination of energy level degeneracy of the free ions. It can be seen that the symmetry of the interactions plays a very important role in the determination of the electron energy states. Therefore, some knowledge of group theory and related quantum theory should be introduced before a detailed discussion of the electron energy level.

2.1 Mathematical Description of the Symmetry

Let us consider a simple cubic lattice; the ion distribution is shown in Fig. 2.1.

In this special situation, the cube shown can represent the distribution of the ions in space. Therefore, one can find out about the local symmetry of the crystal by investigating only the ion's distribution in this cube. Obviously, if the cube is rotated around an axis passing through the centers of opposite faces, such as $\boldsymbol{a}$ axis, by 90°, 180°, or 270°, the distribution of the ions in the cube and in the whole lattice will not be changed. This is to say that $\boldsymbol{a}$ is one of the symmetry axis. Moreover, because the angle rotated around this axis is an integral multiple of $2\pi/4$, it is called fourth-order axis and denoted by C_4. Similarly, the $\boldsymbol{b}$ and $\boldsymbol{c}$ axes which pass through the centers of the other two sets of opposite faces are also fourth-order

Z. Luo and Y. Huang, *Physics of Solid-State Laser Materials*, Springer Series in Materials Science 289, https://doi.org/10.1007/978-981-32-9668-8_2

Fig. 2.1 Distribution of ions in a simple cubic lattice

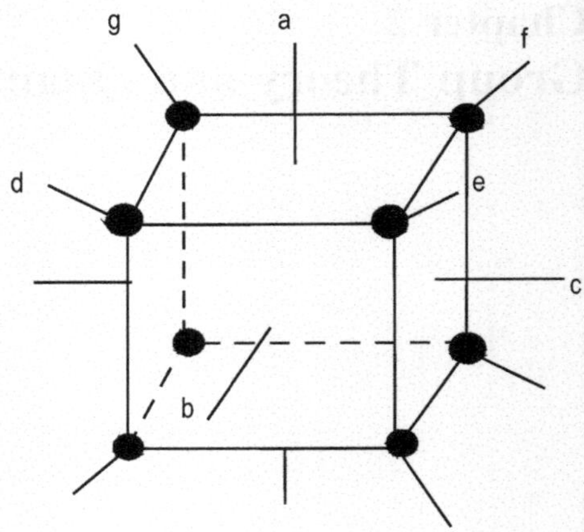

axes. On the other hand, if the cube is rotated 120° or 240° around the axes passing through the opposite corners of the cube, that is, axes ***d***, ***e***, ***f***, and ***g*** then the ion's distribution has no variation. It is to say that ***d***, ***e***, ***f***, and ***g*** are third-order axes and denoted by C_3. In the same way, six axes passing through the midpoints of opposite edges of the cube are second-order axes and denoted by C_2. It means that the configuration will not be changed by the rotation of 180°. Besides, the inversion through the center of the cube and the reflection on the planes σ_h perpendicular to the fourth-order axes or σ_v passing through each pair of opposite edges are also symmetry elements of the cube. It should be noted that most of symmetry axes and symmetry planes include not only one symmetry element. For example, the three fourth-order axes include rotations of 90°, 180°, and 270°, that is, nine symmetry elements $C_4(3)$, $C_4^2(3)$, and $C_4^3(3)$. Similarly, four third-order axes include eight symmetry elements $C_3(4)$ and $C_3^2(4)$ coming from rotations of 120° and 240°, respectively. The figures inside the brackets indicate the number of symmetry elements. On the other hand, the combination (product) of two symmetry elements will generate a new symmetry element. For example, the combination of fourth-order axes and the symmetry planes generate fourth-order rotation–reflection axes $S_4(6)$, and the combination of third-order axes and the symmetry planes generate eight sixth-order rotation–reflection axes $S_6(8)$. Therefore, including the identity element E, there are all together 48 symmetry elements divided into ten classes in this symmetry group:

$$E; C_2(6); C_3, C_3^2(8); C_4, C_4^3(6); C_4^2(3); I; \sigma_v(6); \sigma_h(3); S_6, S_6^5(8); S_4, S_4^3(6)$$

It can be found that a symmetry element in one class can be transformed into a symmetry element in the same class by combining with some symmetry element in other class. For example, eight symmetry elements C_3 and C_3^2 can be transformed into each other by C_4 and C_2, as well as C_4 and C_4^3 can be transformed into each other by C_2. On the other hand, C_4 axes with different orientations can be transformed into each other by C_3 and so on. The symmetry axes in the same class rotate with the same angle; the difference is only the orientation of axes. Therefore, one can say that the symmetry elements in the same class are equivalent. We will see it more clearly later and find that any two successive transformations always can be replaced by one single transformation and any transformation always has a corresponding inverse transformation. In mathematical language, these 48 elements constitute a symmetry group denoted by O_h. The group theory is a mathematical tool to investigate the symmetry of a system and the effects of this symmetry on the characteristics of the system. Almost all the characteristics of the micro-systems have very close relations with its symmetry and so group theory is a powerful tool in dealing with these problems. This is the main reason why we will introduce the basic knowledge of group theory in the following.

2.2 Basic Conception of the Group

By considering every symmetry element as an element of a group, from the example given above, one can see that the elements of a group have the following properties:

(1) The product of any two elements, which is defined as two successive transformations, is also an element of the group.
(2) Each group includes an identity element E which satisfies $(AE) = (EA)$ for any element A in the group.
(3) For any element A in the group, there is an inverse element A^{-1} in the same group satisfying $AA^{-1} = E$.
(4) For the multiplication of elements, the associative law is valid $A(BC) = (AB)C$.

The last property cannot be directly seen from the example given. This is because the multiplication table of those group elements still has not been given. It is complicate to make the multiplication table for such a group of 48 elements. However, we can discuss it by using a simple example related to the symmetry elements of a regular triangle. These symmetry elements constitute a group denoted by C_{3v}. From Fig. 2.2, it can be seen that this group has the following elements: (1) Unit operation (identity transformation) E; (2) Clockwise rotation through 120° around z-axis: C_3; (3) Counter-clockwise rotation through 120° or clockwise rotations through 240° around z-axis: C_3^2; (4) Reflection about plane $Z1$: σ_1; (5) Reflection about plane $Z2$: σ_2; (6) Reflection about plane $Z3$: σ_3.

It is easy to see that these elements satisfy multiplication table (Table 2.1). It can be seen from this table that the association law $(C_3\sigma_1)\sigma_2 = \sigma_3\sigma_2 = C_3^2 = C_3(\sigma_1\sigma_2)$ is satisfied. On the other hand, $\sigma_1\sigma_2 = C_3$, but $\sigma_2\sigma_1 = C_3^2$, so the product is different for different orders of the multiplication. The elements σ_1 and σ_2 are said to be non-commutative. If all the elements of a group are commutative, then the group is called Abelian.

Fig. 2.2 Symmetry elements in a regular triangle

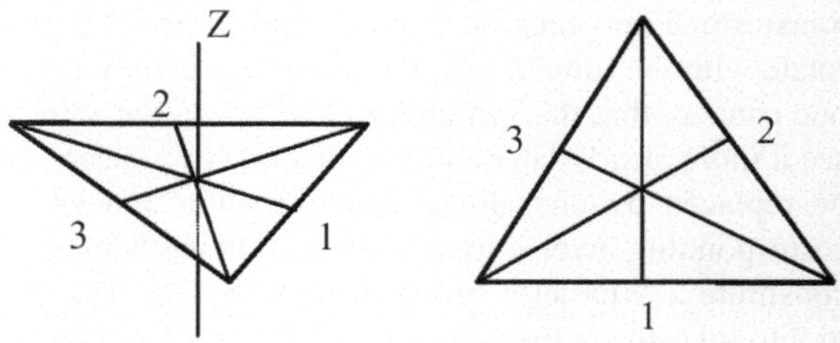

Table 2.1 Multiplication table of C_{3v} group

	E	C_3	C_3^2	σ_1	σ_2	σ_3
E	E	C_3	C_3^2	σ_1	σ_2	σ_3
C_3	C_3	C_3^2	E	σ_3	σ_1	σ_2
C_3^2	C_3^2	E	C_3	σ_2	σ_3	σ_1
σ_1	σ_1	σ_2	σ_3	E	C_3	C_3^2
σ_2	σ_2	σ_3	σ_1	C_3^2	E	C_3
σ_3	σ_3	σ_1	σ_2	C_3	C_3^2	E

One can introduce the concept of the class by investigating this simple group. By applying the above multiplication table, one has:

$$C_3^{-1} = C_3^2, (C_3^2)^{-1} = C_3, \sigma_1^{-1} = \sigma_1, \sigma_2^{-1} = \sigma_2, \sigma_3^{-1} = \sigma_3$$

and so

$$EC_3E = C_3, C_3^{-1}C_3C_3 = C_3, \left(C_3^2\right)^{-1}C_3C_3^2 = C_3, \sigma_1^{-1}C_3\sigma_1 = C_3^2, \sigma_2^{-1}C_3\sigma_2$$
$$= C_3^2, \sigma_3^{-1}C_3\sigma_3 = C_3^2,$$

Similarly, when A in $G^{-1}AG$ is replaced by C_3^2, G being all the elements in the group, the results obtained are still C_3 and C_3^2. $G^{-1}AG$ is called the conjugate of A. According to the definition of group theory, if some elements in a group are conjugate to each other, then these elements form a class in the group. Obviously, C_3 and C_3^2 form a class in the group C_{3v}. It can also be shown that σ_1, σ_2, and σ_3 form another class. If two symmetry operations belong to the same class, then one operation is replaced by the other in a new coordinate system. For example, in the case considered, one can rotate the triangle counter-clockwise of 120°; then the operation σ_1 in the rotated system is equal to the operation σ_2 in the old system. Any group can be decomposed into non-overlapping classes, which cover all the elements of the group. For an Abelian group, $G^{-1}AG = AG^{-1}G = A$, and so each of the group element forms a class.

We would like to introduce here the concept of order. The number of elements in a group is the order of the group. If an element A in the group has relation $A^n = E$, then n is called the order of element A. When B and A are in the same class, then we have:

$$B^n = \left(G^{-1}AG\right)^n = G^{-1}AGG^{-1}AGG^{-1}\ldots G = G^{-1}A^nG = G^{-1}EG = G^{-1}G = E$$

It means that the elements in the same class have the same order.

Another important concept in the group theory is the subgroup. If one set of the elements H of a group G has all the four properties of the group mentioned above, then H is a subgroup of the group G. For example, suppose E, C_3, C_3^2 is a subgroup of group C_{3v}, then all the products and the inverse of these elements are in the subgroup. Note that $C_3^3 = E$ and so all the elements in the subgroup can be expressed as some integer power of one element of this subgroup. If all the elements in a group are integer power of a given element, then this group is cyclic. Therefore, the above-mentioned subgroup is a cyclic group with order 3. Group C_{3v} has five subgroups as follows:

$$\{E\}, \{E, \sigma_1\}, \{E, \sigma_2\}, \{E, \sigma_3\}, \left\{E, C_3, C_3^2\right\}$$

It is useful to point out without proof that if the order of a group is a prime number, then this group is a cyclic group. The order of subgroup $\{E, C_3, C_3^2\}$ is 3, so that it is a cyclic group.

If the subgroup H in the group G is self-conjugated to any element g in group G, that is, $g^{-1}H g = H$, then subgroup H is called an invariant subgroup of the group G. The subgroup $\{E, C_3, C_3^2\}$ satisfies this condition and is an invariant subgroup of the group C_{3v}.

Factor group is another important concept in the group theory. In order to introduce this concept, it is necessary to mention about coset. If H is a subgroup of a group G, then from an element g in G but not in H, one can form a set of elements as follows

$$gH = \{gh_a | h_a \in H\}$$

It is defined as the left coset of H. The right coset of H can be formed similarly

$$Hg = \{h_a g | h_a \in H\}$$

Obviously, if H is an invariant subgroup, then the left coset should be equal to the right coset and it is not necessary to distinguish between the left and the right coset. In this case, if one considers H as an element of the group, then all the cosets of H are also the elements of the group. The new group formed in this way is called a factor group and denoted by G/H. For example, for invariant subgroup $H = \{E, C_3, C_3^2\}$ in C_{3v}, H and $\{\sigma_1, \sigma_2, \sigma_3\}$ constitute a factor group of order two.

2.3 Theory of Group Representations

The representation of a group is a method to embody the concept of the group elements. It relates the abstract group elements to the concrete symmetry operations and is an important way for the application of the group theory to spectroscopy.

Abstractly, if any set of elements can be in correspondence with the elements of the group and if the same multiplication table can be applied, then this set of elements is a representation of the group. There are two kinds of correspondence in group theory, one is homomorphism, and the other is isomorphism. If the set of elements G' and the set of elements G (a group) have the same multiplication table and the following correspondence relation:

$$A \to A'_1, A'_2 \ldots A'_i; B \to B'_1, B'_2 \ldots B'_i; AB \to A'_1B'_1, A'_2B'_2, \ldots, A'_iB'_i$$

where A, B,... are elements of G and $A'_1, B'_1, A'_2, B'_2, \ldots$ are elements of G', then G is said to be homomorphism to G'. If there is a one-to-one correspondence, then G is said to be isomorphism to G'. Obviously, isomorphism is also homomorphism, but homomorphism is not necessarily isomorphism. Two isomorphic groups have completely the same structure and have not any difference from the point of view of group theory. The aforementioned symmetry elements of a regular triangle and the following matrices are two isomorphic groups.

$$E \Leftrightarrow \begin{pmatrix} 1 & 0 \\ 0 & 1 \end{pmatrix}, \sigma_1 \Leftrightarrow \begin{pmatrix} 1 & 0 \\ 0 & -1 \end{pmatrix}, \sigma_2 = \begin{pmatrix} -\frac{1}{2} & \frac{\sqrt{3}}{2} \\ \frac{\sqrt{3}}{2} & \frac{1}{2} \end{pmatrix},$$

$$\sigma_3 = \begin{pmatrix} -\frac{1}{2} & -\frac{\sqrt{3}}{2} \\ -\frac{\sqrt{3}}{2} & \frac{1}{2} \end{pmatrix}, C_3 = \begin{pmatrix} -\frac{1}{2} & \frac{\sqrt{3}}{2} \\ -\frac{\sqrt{3}}{2} & -\frac{1}{2} \end{pmatrix}, C_3^2 = \begin{pmatrix} -\frac{1}{2} & -\frac{\sqrt{3}}{2} \\ \frac{\sqrt{3}}{2} & -\frac{1}{2} \end{pmatrix}$$

It is a general way to use matrices in the linear space to represent a group. The transformations in the n-dimensional space Vn form a group designated by G. If one chooses a set of base functions for V_n, then each transformation will be represented by an $n \times n$ matrix. The set of matrices representing the linear transformation will form a matrix representation of G and the base functions are said to form the bases of the representation. The matrix representation of G: $M(A)$, $M(B)$, $M(C)$, … have a one-to-one correspondence with elements A, B, C, … of G and they are multiplied according to the same multiplication table

$$AB = C \to M(A)M(B) = M(C), \quad AA^{-1} = E \to M(A)M(A)^{-1} = M(E)$$

The representation matrices depend, in general, on the bases chosen for V_n. From linear algebra one knows that the matrices are transformed according to $M'(B) = S_{-1}M(B)SM'(A) = S^{-1}M(A)S$, and so on, when the bases transform by S. These kinds of transformations are defined as similarity transformation. It is easy to show that M' also forms a representation of the group G

$$M'(A)M'(B) = S^{-1}M(A)SS^{-1}M(B)S = S^{-1}M(A)M(B)S = S^{-1}M(AB)S = M'(AB)$$

Two representations which differ only by a similarity transformation are said to be equivalent.

It should be pointed out that many linear transformations in physics keep the inner product of two vectors invariant. It is to say that for any two vectors x and y, the following equation is always valid:

$$(Mx, My) = (x, y)$$

This transformation M is a unitary transformation and the transformation matrix of the unitary matrix satisfies

$$M^+M = MM^+ = 1$$

where M^+ is transposed complex conjugate of matrix M. It can be shown [1] that for a finite group, every representation is equivalent to a unitary representation.

If a matrix M can be expressed as

$$M(R) = \left(\begin{array}{c|c} M_1(R) & 0 \\ \hline 0 & M_2(R) \end{array}\right) \tag{2.1}$$

and it is also written as

$$M(R) = M_1(R) \oplus M_2(R) \tag{2.2}$$

then the matrix M is said to be reducible representation. In some cases, $M_1(R)$ and $M_2(R)$ can still be written into above direct summation further. When the matrix M cannot be reduced, then it is said to be an irreducible representation. A series of properties of the irreducible representation are very important in the application of group theory. Some orthogonal relations are given here without proof.

If $M_\alpha(R)$ and $M_\beta(R)$ are two irreducible representations for the group G, then their matrix elements have the following relation

$$\sum_R M_\alpha(R)^*_{ik} M_\beta(R)_{lj} = \frac{g_o}{n_\alpha} \delta_{\alpha\beta}\delta_{il}\delta_{kj} \tag{2.3}$$

where g_o is the order of group G and n_α is the dimension of irreducible representation. If irreducible representation α is non-equivalent to irreducible representation β, then $\delta_{\alpha\beta} = 0$. Equation (2.3) can be introduced by Shur lemma [1, 2].

Matrices of an irreducible representation are dependent on the choice of basis. However, the summation of diagonal matrix elements—the trace of the matrix (called the character of irreducible representation in group theory)—is independent of the basis chosen. It is easy to show in mathematics that the character $\chi(R) = \sum_i M(R)_{ii}$ does not change in the transformation of the coordinate. If

$$M'(R) = S^{-1}M(R)S$$

then

$$\chi'(R) = \sum_i M'(R)_{ii} = \sum_i (S^{-1}MS)_{ii} = \sum_{imn} (S^{-1})_{im} M_{mn} S_{ni} = \sum_{mn} (SS^{-1})_{nm} M_{mn}$$
$$= \sum_{mn} \delta_{mn} M_{mn} = \sum_m M_{mm} = \chi(R)$$

Obviously, any two elements A and B in the same class can be written as $A = S^{-1}BS$, and so the group elements in the same class have the same character. Two non-equivalent irreducible representations have different characters. Let $k = i$ and $l = j$, then (2.3) becomes

$$\sum_R M_\alpha(R)^*_{ii} M_\beta(R)_{jj} = \frac{g_0}{n_\alpha} \delta_{\alpha\beta}\delta_{ij} \tag{2.4}$$

Summing over i and j in the above equation, then

$$\sum_R \chi_\alpha(R)^* \chi_\beta(R) = g_0 \delta_{\alpha\beta} \tag{2.5}$$

Owing to the fact that the characters of the elements belonging to the same class are the same, if there are hi elements in a class C_i, then the summation over the elements of the group can be changed to the summation over the classes of the group. If there are r classes in the group, then

$$\sum_{i=1}^{r} h_i \chi_\alpha(C_i)^* \chi_\beta(C_i) = g_o \delta_{\alpha\beta} \tag{2.6}$$

By using orthogonal relation (2.5), one can obtain the number of times l_i that a particular irreducible representation appears in any reducible representation. One has

$$\chi(R) = \sum_i l_i \chi_i(R)$$

and

$$l_i = \frac{1}{g_0} \sum_R \chi(R)^* \chi_i(R) \tag{2.7}$$

In addition, if a representation is irreducible, then

$$\sum_R \chi(R)^* \chi(R) = g_o$$

because

$$\sum_R \chi(R)^* \chi(R) = \sum_{Rij} l_i \chi_i(R) l_j \chi_j(R) = g_o \sum_{ij} l_i l_j \delta_{ij} = g_o \sum_i l_i^2$$

If the representation is irreducible, then $\chi(R)$ cannot be decomposed into a sum of several terms, that is, the only solution is $l_1 = 1$, and so $\sum_i l_i^2 = 1$.

By means of (2.3) and (2.6), the following two important properties of irreducible representation can also be introduced:

(a) The sum of the squares of the dimensions of all irreducible representations is equal to the order of the group.
(b) The number of different irreducible representations is equal to the number of classes in the group.

The first property can be shown by (2.3). Suppose $M_\alpha(R)_{ik}$ forms a g_o-dimensional vector (the group element R has g_o elements, it constitutes a g_o-dimensional space). This vector is orthogonal, according to (2.3), to all other g_o-dimensional vectors formed by $M_\alpha(R)_{lj}$. There are n_α^2 such orthogonal vectors. Besides, there are n_β^2 and n_γ^2 vectors formed by $M_\beta(R)$ and $M_\gamma(R)$, respectively, and all these vectors are orthogonal to each other. The total number of orthogonal vectors is $\sum_i n_i^2$ and it is obviously equal to g_0, and so

$$\sum_i n_i^2 = g_0$$

The second property can be demonstrated by rewriting (2.6) as

$$\sum_{i=1}^{r} \chi_\alpha(C_i)^* \left(\frac{h_i}{g_0}\right)^{1/2} \chi_\beta(C_i) \left(\frac{h_i}{g_0}\right)^{1/2} = \delta_{\alpha\beta} \tag{2.8}$$

where $\chi_\alpha(C_i)^* \left(\frac{h_i}{g_0}\right)^{1/2}$ can be considered as an r-dimensional vector (there are r classes in the group). Equation (2.8) means that the vectors formed by non-equivalent irreducible representations are orthogonal. If the number of non-equivalent irreducible representations is n, then $n = r$, because the orthogonal vector in an r-dimensional space is r.

These two properties are very useful in determining the character table of irreducible representations.

2.4 Direct Product Group and Direct Product Representation

One way to construct a complicate group from simple groups is to form a product group. If one has two groups G_1 and G_2

$$G_1 = E_1, A_1, B_1, C_1 \ldots$$
$$G_2 = E_2, A_2, B_2, C_2 \ldots$$

then all the products R_1R_2 between the elements of the two groups are commutative

$$R_1 R_2 = R_2 R_1$$

If R_1 and R_2 are represented by matrix $M(R_1)$ and matrix $M(R_2)$, respectively, then the direct product $G_1 \otimes G_2$ is represented by matrices $M(R_1) \otimes M(R_2)$ with $g_1g_2 \times g_1g_2$ elements. Obviously, their characters have the following relation

$$\chi(G_1 \otimes G_2) = \chi(G_1) \times \chi(G_2) \tag{2.9}$$

As an example to construct a complex group from simple groups, we will discuss the construction of D_{3h} from D_3 and C_{1h}. The group C_{1h} contains only a unit element and a symmetry plane, while the group D_3 includes a unit element, a third-order axis, and a second-order axis perpendicular to the third-order axis. Another two second-order axes will be generated by the third-order axis. Their characters tables are Tables 2.2, 2.3.

The characters table of $D_{3h} = D_3 \otimes C_{1h}$ is (Table 2.4).
Similarly, other groups can also be formed by direct product, such as $C_{4h} = C_4 \otimes C_i, C_{6h} = C_6 \otimes C_i$, $S_6 = C_3 \otimes C_i, D_{3d} = D_3 \otimes C_i$, $D_{2h} = D_2 \otimes C_i$, $D_{4h} = D_4 \otimes C_i$, $D_{6h} = D_6 \otimes C_i, O_h = O \otimes C_i, T_h = T \otimes C_i$.

The direct product representation for a group can be formed from two different irreducible representations of the same group. The character of the direct product representation can be easily shown as the product of two characters of the related irreducible representations, that is

$$\chi(\gamma) = \chi(\alpha \otimes \beta) = \chi(\alpha) \times \chi(\beta) \tag{2.10}$$

This direct product representation is generally reducible and the numbers of irreducible representation contained can be calculated by the above-mentioned

Table 2.2 Characters table of C_{1h} group

C_{1h}	E	σ_h
A'	1	1
A''	1	−1

Table 2.3 Characters table of D_3 group

D3	E	$2C_3$	$3C_2'$
A_1	1	1	1
A_2	1	1	−1
E	2	−1	0

Table 2.4 Characters table of D_{3h} group

$D_{3h} = D_3 \otimes C_{1h}$	E	σ_h	$2C_3$	$2S_3\ (\sigma_h)C_3$	$3C_2'$	$3\sigma_v(\sigma_h C_2')$
$A_1'(A' \otimes A_1)$	1	1	1	1	1	1
$A_2'(A' \otimes A_2)$	1	1	1	1	−1	−1
$A_1''(A'' \otimes A_1)$	1	−1	1	−1	1	−1
$A_2''(A'' \otimes A_2)$	1	−1	1	−1	−1	1
$E'(A' \otimes E)$	2	2	−1	−1	0	0
$E''(A'' \otimes E)$	2	−2	−1	1	0	0

orthogonal relations. If one uses $C_{\alpha\beta\gamma}$ to express the number of γ irreducible representation contained in the direct product representation $\alpha \otimes \beta$, then

$$C_{\alpha\beta\gamma} = \frac{1}{g_0}\sum_R \chi_\gamma(R)^* \chi_\alpha(R)\chi_\beta(R) \tag{2.11}$$

To investigate the properties of a quantum system, the wave functions of the system are always expressed as products of those of the subsystems. For example, electronic wave function is a product of orbital wave functions and spin wave functions, and multi-electron wave function is a product of single-electron wave functions. A representation for a system can be expressed as a product of those of its subsystems. If the first function belongs to representation α and the second one belongs to representation β, then the product function belongs to the representation $\alpha \otimes \beta$. In the application of group theory, we often express the direct product wave function as a linear combination of wave functions belonging to different irreducible representations. For this purpose we use the Clebsch-Gordan coefficients (C-G coefficients) as the combination coefficients. They will be discussed in detail later.

2.5 Sketches of the Group in Spectroscopy

2.5.1 *Finite Group*

Finite group is the group with finite number of elements; among this, the point group and the space group are most frequently used in physics.

1. Point group and space group

Point groups have been mentioned many times in the above discussion, which is a symmetry group of finite bodies having at least one point fixed. In crystal, point group is the set of symmetry elements of lattice points and will be discussed in a separate section. Space group is the symmetry group of configuration in space. Crystallographic space group includes all the symmetry elements in crystal and can describe the symmetry of whole crystal lattice. Symmetry elements of space group can be expressed by $\{\alpha|\boldsymbol{n}\}$, where $\{\varepsilon|0\}$ is identity element and $\{\alpha|0\}$ is symmetry operation for some fixed point, that is, the symmetry operation of point group while $\{\varepsilon|\boldsymbol{n}\}$ is symmetry operation of pure translation group. Multiplication is defined as $\{\alpha|\boldsymbol{n}\}\{\beta|\boldsymbol{n}'\} = \{\alpha\beta|\alpha\boldsymbol{n}' + \boldsymbol{n}\}$, and the inverse element is $\{\alpha|\boldsymbol{n}\}^{-1}\{\alpha^{-1}| - \alpha^{-1}\boldsymbol{n}\}$, owing to the fact that $\{\alpha|\boldsymbol{n}\}\{\alpha|\boldsymbol{n}\}^{-1} = \{\alpha|\boldsymbol{n}\}\{\alpha^{-1}| - \alpha^{-1}\boldsymbol{n}\} = \{\alpha\alpha^{-1}| - \alpha\alpha^{-1}\boldsymbol{n} + \boldsymbol{n}\} = \{\varepsilon|0\}$. It can be easily demonstrated that the associative law of multiplication is also valid and the four criteria of groups are all satisfied.

Because of the limit of lattice translation symmetry, the point group $\{\alpha|0\}$ in crystal can only be one of 32 kinds of point group, and conversely, by the limit of $\{\alpha|0\}$, $\{\varepsilon|\boldsymbol{n}\}$ should belong to one of 14 kinds of Bravias lattice. The 32 point groups combine with 14 Bravias lattices and form 230 space groups, which include 73 symmorphic space groups and 157 non-symmorphic space groups and can be divided into seven crystallographic systems. The symmorphic space groups are composed of all the symmetry elements of one translation group and one point group, while the non-symmorphic space groups include screw axis (a rotation through an angle φ about symmetry axis followed by a fractional lattice vector translation along this axis) and glide reflection plane (a plane reflection followed by a fractional lattice vector translation parallel to this plane). The translation lattice vector may not necessarily be a Bravias lattice vector. The point groups, Bravias lattices, and the number of space group for seven crystallographic systems are listed in Table 2.5. It should be noted that a crystal belonging to a certain space group of a certain crystallographic system does not mean that all the lattice points have the point symmetry belonging to that crystallographic systems. For example, the laser crystal YAG ($Y_3Al_5O_{12}$) belongs to space group I*a*3*d* (O_h^{10}), but the trivalent Y^{3+} ions occupy site A with point symmetry of D_2, and the trivalent Al^{3+} ions occupy two kinds of sites. Among these the site B has point symmetry of S_6 and the site C has point symmetry of S_4, but the point symmetry of divalent anion O^{2-} is S_2. The possible lattice point symmetry of the entire space group can be found in books or data compilation of crystallography.

Space group is an important theoretical tool for crystallography and solid-state physics, especially the band theory of the solid and lattice vibration spectroscopy. This book is mainly concerned with the energy levels of local optical centers and transition between these energy levels; therefore, the knowledge of space group has only a brief introduction.

Table 2.5 Point groups, Bravias lattices, and number of space group for seven crystallographic systems

Crystal system	Point group	Bravias lattice	Number of space group
Triclinic	C_1, S_2	Simple triclinic	Symmorphic 2
Monoclinic	C_2, C_{1h}, C_{2h}	Simple and end-centered monoclinic	Symmorphic 6, non-symmorphic 7
Orthorhombic	D_2, C_{2v}, D_{2h}	Simple, end-centered, body-centered, and face-centered orthorhombic	Symmorphic 13, non-symmorphic 46
Tetragonal	C_4, S_4, C_{4h}, D_4, C_{4v}, D_{2d}, D_{4h}	Simple and body-centered tetragonal	Symmorphic 16, non-symmorphic 52
Trigonal	C_3, S_6, D_3, C_{3v}, D_{3d}	Simple trigonal	Symmorphic 13, non-symmorphic 12
Hexagonal	C_6, C_{3h}, C_{6h}, D_6, C_{6v}, D_{3h}, D_{6h}	Simple hexagonal	Symmorphic 8, non-symmorphic 19
Cubic	T, T_h, O, T_d, O_h	Simple, body-centered and face-centered cubic	Symmorphic 15, non-symmorphic 21

2.5.2 Permutation Group

Permutation symmetry is the symmetry of quantum system consisting of identical particles. For example, in the $3d^q$ and $4f^q$ configuration of transition-metal ion and rare earth ion, the q electrons in these systems have permutation symmetry. Therefore, the permutation group has important application in the spectroscopy of these ions.

Permutation group with n elements is usually denoted by S_n. In order to differentiate from symbol of relate point group, the symbol of permutation group is adopted as follows

$$P_n = \begin{pmatrix} 1 & 2 & \dots & n \\ a_1 & a_2 & \dots & a_n \end{pmatrix}$$

where P_n denotes the permutation of 1 by a_1, 2 by $a_2 \ldots$, n by a_n. The order of two permutation product PP' is defined from the right to the left, for example

$$\begin{pmatrix} 1 & 2 & 3 \\ 2 & 1 & 3 \end{pmatrix}\begin{pmatrix} 1 & 2 & 3 \\ 2 & 3 & 1 \end{pmatrix} = \begin{pmatrix} 1 & 2 & 3 \\ 1 & 3 & 2 \end{pmatrix}$$

and the product of $P'P$ is

$$\begin{pmatrix} 1 & 2 & 3 \\ 2 & 3 & 1 \end{pmatrix}\begin{pmatrix} 1 & 2 & 3 \\ 2 & 1 & 3 \end{pmatrix} = \begin{pmatrix} 1 & 2 & 3 \\ 3 & 2 & 1 \end{pmatrix}$$

This result shows that the permutation does not satisfy the commutative law but it is easy to show that the associative law is valid and other requirements of group are all satisfied. The identity element is

$$P_n^{(0)} = \begin{pmatrix} a_1 & a_2 & \dots & a_n \\ a_1 & a_2 & \dots & a_n \end{pmatrix}$$

If

$$P_n = \begin{pmatrix} b_1 & b_2 & \dots & b_n \\ a_1 & a_2 & \dots & a_n \end{pmatrix}$$

then the inverse element will be

$$P_n^{-1} = \begin{pmatrix} a_1 & a_2 & \dots & a_n \\ b_1 & b_2 & \dots & b_n \end{pmatrix}$$

because the P_n^{-1} satisfies the following expression

$$\begin{aligned} P_n P_n^{-1} &= \begin{pmatrix} b_1 & b_2 & \dots & b_n \\ a_1 & a_2 & \dots & a_n \end{pmatrix}\begin{pmatrix} a_1 & a_2 & \dots & a_n \\ b_1 & b_2 & \dots & b_n \end{pmatrix} = \begin{pmatrix} a_1 & a_2 & \dots & a_n \\ a_1 & a_2 & \dots & a_n \end{pmatrix} \\ &= P_n^{(0)} \end{aligned}$$

The group element number of permutation P_n is $n!$, that is, the order of group P_n is $n!$ Any permutation can be resolved into cycles which affect only a smaller fixed number of elements; for example, a P_8 can be resolved as follows

$$\begin{pmatrix} 1 & 2 & 3 & 4 & 5 & 6 & 7 & 8 \\ 2 & 3 & 1 & 5 & 4 & 7 & 6 & 8 \end{pmatrix} = (123)(45)(67)(8)$$

The cycles in the right-hand side of the above expression have no common symbols. Permutations with same cycle structure belong to the same class of the group. Cycle structure can be written as

$$(v) = (1^{v_1}, 2^{v_2}, \dots, n^{v_n})$$

It expresses that the cycle structure of permutation has v_1 1-cycle (unchanged), v_2 2-cycles (transposition), …, v_n n-cycles. The length of all the cycles satisfies

$$\sum_{i=1}^{n} i v_i = n$$

The number of permutation group P_n with cycle structure (v) can be expressed as

$$N_{(v)} = \frac{n!}{1^{v_1} v_1! 2^{v_2} v_2! \ldots n^{v_n} v_n!}$$

For example, the 4-symbol permutation group P_4 can be divided into five classes and have altogether 4! = 24 elements:

1. Identity element $P_4^{(0)} = (1)(2)(3)(4)$—$N_{(v)} = \frac{4!}{1^4 4!} = 1$
2. (123)(4), (132)(4), (124)(3), (142)(3), (134)(2), (143)(2), (234)(1), (243)(1) —$N_{(v)} = \frac{4!}{1^1 1! 3^1 1!} = 8$
3. (12)(34), (13)(24), (14)(23)—$N_{(v)} = \frac{4!}{2^2 2!} = 3$
4. (12)(3)(4), (13)(2)(4), (14)(2)(3), (23)(1)(4), (24)(1)(3), (34)(1)(2) —$N_{(v)} = \frac{4!}{1^2 1 2^1 2!} = 6$
5. (1234), (1243), (1324), (1324), (1423), (1432)—$N_{(v)} = \frac{4!}{4^1 1!} = 6$

The class of P_n can generally be expressed by using partition $[\lambda] = [\lambda_1 \lambda_1 \ldots \lambda_n]$, where

$$\begin{aligned} \lambda_1 &= v_1 + v_2 + \cdots + v_n \\ \lambda_2 &= v_2 + v_3 + \cdots + v_n \\ &\cdots\cdots\cdots\cdots \\ \lambda_n &= v_n \end{aligned}$$

Because v_i is a positive number, so $\lambda_1 \geq \lambda_2 \cdots \geq \lambda_n \geq 0$, and also $\lambda_1 + \lambda_2 + \cdots + \lambda_n = n$. The possible class number of P_n is determined by the number of partition. Group theory uses Young diagrams to express partition $[\lambda] = [\lambda_1 \lambda_1 \ldots \lambda_n]$, where there are λ_1 cells in the first row and λ_2 cells in the second row, and so on. For example, for S_4, its first class can be written as $[1111] = [1^4]$, the second class [31], the third class $[22] \equiv [2^2]$, the fourth class $[211] \equiv [21^2]$, and the fifth class [4]. Their Young diagrams shown follow from left to right for the first to fifth class. It should be pointed out that there is a requirement that no lower row can be longer than upper row.

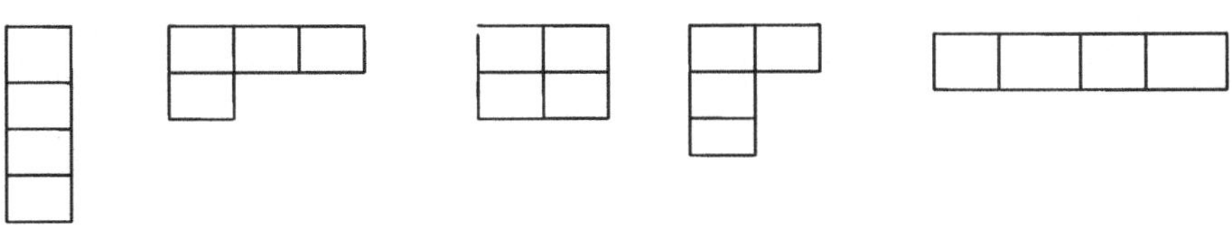

When the symbols of 1, 2, 3, 4 are placed into the Young diagram, one can obtain a table called Young tableau. If the numbers in a Young tableau are arranged so that they increase from left to right and from top to bottom, the tableau is called a standard Young tableau. For example, the partition $[21^2]$ has three standard Young tableaux as follows

1	2
3	
4	

1	3
2	
4	

1	4
2	
3	

It can be demonstrated that the number of standard Young tableau can be generated from a given Young diagram which is equal to the dimensions of the corresponding irreducible representation. For example, the dimension of partition $[21^2]$ is 3 due to the fact that it has the above three standard Young tableaux. The method of using Young tableau to introduce the characters of irreducible representation can be referred from the books of group theory [1].

By Cayley theorem, any n order group isomorphism with one subgroup of permutation group P_n and the irreducible representation of unitary group $U(n)$ and special unitary group $SU(n)$, which have important application in physics, can be obtained by using irreducible representation of P_n. Therefore, permutation group is very important not only in mathematics but also in physics.

2.5.3 Continuous Groups

The groups in which elements can be labeled by continuously varying parameters are called as continuous groups. They have infinite elements. It can be found in linear algebra that n dimension vector can be transformed by the following relation

$$\boldsymbol{y} = \boldsymbol{A}\boldsymbol{x}, \quad \boldsymbol{z} = \boldsymbol{B}\boldsymbol{y} \quad \Rightarrow \boldsymbol{z} = \boldsymbol{C}\boldsymbol{x}, \quad \boldsymbol{C} = \boldsymbol{B}\boldsymbol{A}$$

where $\boldsymbol{A}$, $\boldsymbol{B}$, and $\boldsymbol{C}$ are matrices of n line and n row. It is easy to show that the transformation matrices form a group due to the fact that the unit matrix $\boldsymbol{I}$ corresponds to the identity element, the product of two transformation matrices is in this matrix set, and the transformation of vector $\boldsymbol{y}$ into vector $\boldsymbol{x}$ satisfies the inverse transformation relation $\boldsymbol{A}^{-1}\boldsymbol{A} = \boldsymbol{I}$, as well as the associative law of multiplication is valid. Rotation group $SO(3)$ which transforms one vector into another vector in three-dimensional space is the simplest example, in which the rotation angle is expressed as Euler angle. α, β, γ obviously can be continuously varied, that is, it has infinite number. If the angle parameters of matrix $\boldsymbol{A}$ are α, β, γ, matrix $\boldsymbol{B}$ are α_1, β_1, γ_1, and matrix $\boldsymbol{C}$ are α_2, β_2, γ_2, then the angle parameters of matrix $\boldsymbol{C}$ can be expressed by those of matrices $\boldsymbol{A}$ and $\boldsymbol{B}$; for example,

$$\sin\beta_2 = \sin\beta\cos\alpha_1\cos\beta_1 + \cos\beta(\cos\gamma\sin\beta_1 - \sin\gamma\sin\alpha_1\cos\beta_1)$$

Continuous groups commonly used in spectroscopy are:
Groups of $U(n)$ and $SU(n)$—$n \times n$ unitary matrix set constitutes the n dimension unitary group $U(n)$, where the condition of a matrix $\boldsymbol{A}$ to be a unitary will be

$$\boldsymbol{A}^+\boldsymbol{A}=\boldsymbol{I},\ \text{i.e.}(\boldsymbol{A}^+\boldsymbol{A})_{ij}= \sum_k (A^+)_{ik}A_{kj} = \sum_k A^*_{ki}A_{kj} = \delta_{ij}$$

(Matrix $\boldsymbol{A}^+$ is the conjugate transposed matrix of matrix $\boldsymbol{A}$)

It has n^2 independent real parameters. On the other hand, all the $n \times n$ unitary matrices with determinant equal to +1 constitute special unitary group $SU(n)$ having $n^2 - 1$ independent real parameters.

Group $O(n)$ and $SO(n)$—$n \times n$ real orthogonal matrices constitute an n dimension real orthogonal group $O(n)$, where the matrix $\boldsymbol{A}$ satisfies

$$\boldsymbol{A}^t\boldsymbol{A} = I, \text{i.e.}(\boldsymbol{A}^t\boldsymbol{A})_{ij}= \sum_k (A^t)_{ik}A_{kj} = \sum_k A_{ki}A_{kj} = \delta_{ij}$$

(Matrix $\boldsymbol{A}^t$ is the transposed matrix of matrix $\boldsymbol{A}$)

Its group elements are described by $n(n-1)/2$ independent real parameters. Among the $O(n)$ groups, the $n\times n$ real orthogonal matrices with their determinants equal to +1 constitute the n dimension special real orthogonal group $SO(n)$. Group $O(n)$ is defined in real number field but group $U(n)$ is defined in complex number field so that group $O(n)$ is the subgroup of group $U(n)$. The group $SO(n)$ is a special group of the group $O(n)$, and it is certainly a subgroup of group $O(n)$. On the other hand, group $O(n)$ is a subgroup of the group $U(n)$ and $SU(n)$ is a subgroup of $U(n)$ group with determinant equal to +1; therefore, the group $SO(n)$ is the common subgroup of groups $O(n)$ and $SU(n)$. One example of the application of continuous group in spectroscopy is that mentioned in Chap. 1 about the classification of the spectral terms of nl^q configuration. By introducing high-order continuous group $U(4l + 1)$ (group $SO(3)$ is its subgroup), one can use different irreducible representation of following group chain

$$U(4l+1) \supset SU(2l+1) \otimes SU(2) \supset SO(2l+1) \otimes SU(2) \supset SO(3) \otimes SU(2)$$

to distinguish spectral different terms which have the same quantum number LS. The properties of permutation group and Young tableau can be used in the reduction of irreducible representation along the group chain because there is close connection between permutation group $P(n)$ and unitary group $U(n)$. The theory of continuous group and its application in atomic spectroscopy can be referred to the books of [1, 3].

2.6 Point Group and Their Representation

The symmetry properties of the energy levels and the energy level splitting of the ions (rare earth or transition-metal ions) in crystal are determined by symmetry of the surrounding lattice ions. This is the point symmetry around the positions where the active ions are located. These points remain fixed when any symmetry transformation of the crystal is applied and so are the intersections of all the rotation axes and all the reflection planes, because two successive rotations of the body about two non-intersecting axes or two reflections in two planes without a common point result in a translation of the body. Symmetry operations in fact can be divided into three different fundamental kinds: rotation around an axis, reflection in a plane, and translations in certain direction. The third operation has not relation to the problems discussed in this book, and so we are not concerned with it. Before going to enumerate all the point symmetry groups, it is necessary to discuss the common features of the symmetry operations of rotations around an axis:

(1) If a group includes element R_g which takes the axis a to axis b, then the symmetry elements rotating around the axis a and axis b through the same angle are conjugated, that is

$$R_b = R_g R_a R_g^{-1} \tag{2.12}$$

We can explain this formula in the following way: the operation $R_g R_a R_g^{-1}$ takes the axis b to axis a followed by rotating around the axis a through an angle φ and then rotating from axis a to axis b at last. The final result of this operation is a rotation around axis b through an angle φ. Therefore, the operations which rotate around different axes through the same angle are conjugate elements belonging to the same class of the group.

(2) If there is a 2-order axis perpendicular to the n-order axis C_n in a group, then C_n^k and C_n^{-k} will belong to the same class. If the axis C_n lies on a plane σ_v, then $C_n^{-k} = \sigma_v C_n^k (\sigma_v)^{-1}$, because the symmetry plane σ_v changed the direction of rotation and so C_n^k and C_n^{-k} belong to the same class. However, if there is a plane σ_h perpendicular to the C_n axis, which changes the rotation direction and at the same time reverses the direction of the axis, then C_n^k and C_n^{-k} belong to different classes.

(3) Many point groups are direct products of simple point group G (does not include inversion i) and the group of symmetry center I containing only unit element e and inversion i. One half of the direct product $G \otimes I$ is the original group G and the other elements are the elements of G multiplied by inversion element i. The inversion element i commutes with the elements of G, so the direct product contains two times the elements of group G. Therefore, the number of class of direct product group $G \otimes i$ is two times that of the group G;

and each class C of the group G corresponds to two classes C and $C \otimes i$. The new irreducible representations produced have the same number and dimension as those of the group G. This property makes it very simple to construct the character table of direct product group.

There are two kinds of point groups: proper and improper rotations. The so-called improper rotation means a rotation about some axis followed by a reflection through a plane or followed by an inversion through the origin.

First kind of point group

(1) C_n Groups

Cyclic groups of n order. It is a group of rotations about an axis of n order. Each one of its n elements forms a class itself and can be generated by any element in the group. Group C_1 contains only the identical transformation E and without any symmetry. All the irreducible representations are of one-dimension. In a crystal there are only five kinds of rotation groups: C_1, C_2, C_3, C_4, and C_6.

(2) D_n Groups

These groups are produced by adding a 2-order axis perpendicular to n-order axis C_n. By the action of C_n rotation, further $n - 1$ 2-order axes are generated. Including n horizontal 2-order axis intersecting at angles π/n, there are altogether $2n$ elements. The number of classes is different for n is odd and n is even. If n is odd ($n = 2p + 1$), all the 2-order axes can be transformed to each other by C_n, and so they are in the same class. The rotations of clockwise C_n and those of counter-clockwise C_n are conjugate, because of the existence of perpendicular 2-order axes. Therefore, besides the identity element $E = C_n^n$, $2p$ rotations form p classes and so there are $p + 2$ classes. If n is even ($n = 2p$), all the 2-order axes can not be arrived separately by C_n and so they are divided into two classes. The other elements in this group form $p + 1$ classes, that is, the identity element E forms one class and C_{2p} is divided into p classes, because rotations of clockwise and those of counter-clockwise are conjugate. In this case, there are $p + 3$ classes. For example, there are three classes in D_3 and five classes in D_4. According to the properties of the irreducible representation illustrated in Sect. 2.4, because $2^2 + 1^2 + 1^2 = 6$, there are two one-dimensional irreducible representations and one two-dimensional irreducible representation in group D_3. For group D_4, owing to the fact that $2^2 + 1^2 + 1^2 + 1^2 + 1^2 = 8$, there are four one-dimensional irreducible representations and one two-dimensional irreducible representation. By means of orthogonal relations, it is easy to obtain their characters tables.

(3) Tetrahedral Group T

This group is formed by all the symmetry axes of a tetrahedron, that is, three 2-order axes passing through two midpoints of opposite edges and four 3-order axes passing through the vertices of the tetrahedron. Three 2-order axes are equivalent by rotating the 3-order axes, and four 3-order axes are also equivalent by rotating the

2-order axes. Therefore, there are 12 elements divided into four classes: E; three 2-order axes C_2, four 3-order axes C_3, four 3-order axes C_2^3. Considering that $3^2 + 1^2 + 1^2 + 1^2 = 12$, there are one three-dimensional irreducible representation and three one-dimensional irreducible representations.

(4) **Octahedron Group *O***

The symmetry axes of a octahedron contain three 4-order axes passing through the centers of opposite faces, four 3-order axes which are body diagonal, and six 2-order axes passing through the midpoints of opposite edges. All the axes of the same order are equivalent. There are 24 elements divided into following five classes: E; 6 rotations C_4 and C_4^3; 3 rotations C_4^2; 8 rotations C_3 and C_3^2; and 6 rotations C_2. Because $3^2 + 3^2 + 2^2 + 1^2 + 1^2 = 24$, there are two three-dimensional irreducible representations, one two-dimensional irreducible representation and two one-dimensional irreducible representations.

Second kind of point group

(1) **C_{nv} Groups**

The structure of groups C_{nv} is very similar to that of the groups D_n, so long as one replaces the 2-order axes perpendicular to C_n to a vertical plane σ_v passing through C_n. It is also a $2n$ order group containing n rotations about the axis of n order and n vertical reflections plane σ_v. When n is odd ($n = 2p + 1$), it has $p + 2$ classes, but has $p + 3$ classes when n is even ($n = 2p$). Therefore, the irreducible representation number is the same as that of the group D_n and its character table is listed together with that of group D_n.

(2) **C_{nh} Groups**

These groups are obtained by adding a perpendicular plane σ_h to C_n and have $2n$ elements: n rotations C_n^k and n rotation–reflections $C_n^k\sigma_h$, with $k = 1, 2, 3, \ldots, n$. All the elements are commutative and so is a Abelian group. The number of class is equal to the number of elements. If n is even, it includes symmetry inversion: $C_{2p}^p\sigma_h = C_2\sigma_h = i$. The simplest group $C_{1h} \equiv Cs$ contains only elements E and σ_h.

(3) **S_{2n} Groups**

These are $2n$ order rotation–reflection cyclic groups. $S_2 \equiv C_i$ contains only identity and inversion elements. If $2n = 4p + 2$, owing to the fact that $\left(S_{4p+2}\right)^{2p+1} = C_2\sigma_h = C_i$, it includes symmetry center. These kind of groups are S_2 and S_6, and can be written as $S_{4p+2} = C_{2p+1} \otimes C_i$. The number of classes is equal to the order of the group and all their irreducible representations are of one-dimension. Representation E of S_4 is not a real two-dimensional representation but two one-dimensional representations (Γ_3 and Γ_4) merged into one.

(4) D_{nh} Groups

The $4n$ elements in these groups are obtained by adding to D_n a horizontal reflection plane σ_h perpendicular to the n-order symmetry axis. A 2-order axis and the plane σ_h passing through it will produce vertical plane σ_v perpendicular to the reflection plane σ_h. Besides the original $2n$ elements of group D_n, there are n reflection planes σ_v and n rotation–reflection elements $C_n^k\sigma_h$. Owing to the fact that σ_h is commutative with all the other group elements, group D_{nh} can be expressed as $D_{nh} = D_n \otimes C_s$, where C_s is a group that contains only two elements E and σ_h. If n is even, then D_{nh} includes symmetry center and $D_{2ph} = D_{2p} \otimes C_i$. The number of classes for groups D_{nh} is twice the number in the group D_n, while half of its classes are the same as those of D_n. The other classes can be obtained by multiplying these by σ_h. When n is odd, then all the σ_v planes belong to one class. When n is even, the σ_v planes belong to two classes. The rotation–reflections $C_n^k\sigma_h$ and $C_n^{-k}\sigma_h$ are conjugated in pair. The number of irreducible representations is twice the number in the group D_n, so the number of one-dimension and two-dimension representations is also twice the number in the group D_n.

(5) D_{nd} Groups

The other way to add the reflection plane to D_n is to draw vertical planes through the axis of the n-order axis and midway between each adjacent pair of horizontal 2-order axes. Another $n - 1$ similar planes are produced by the rotation of the n-order axis. Therefore, these groups also contain $4n$ elements. Besides the original $2n$ elements of group D_n, there are n reflection planes σ_d, n rotation–reflection axes, like $S_{2n}^{2k+1}, k = 1, 2, 3, \ldots, n - 1$. It can be shown that

$$\sigma_d S_{2n}^{2k+1}\sigma_d^{-1} = \sigma_d\sigma_h C_{2n}^{2k+1}\sigma_d^{-1} = \sigma_h\sigma_d C_{2n}^{2k+1}\sigma_d^{-1} = \sigma_h C_{2n}^{-(2k+1)} = S_{2n}^{-(2k+1)},$$

It means that the rotation–reflection axes are conjugated in pairs. When n is even $n = 2p$. The identity element E forms the first class and p times the rotation C_{2p} around n-order axis $C_{2p}^p = C_2$ forms the second class. The $2p$ second-order rotations around horizontal axis, which are equivalent by the n-order axis rotation, constitute the third class, and the $2p$ reflection planes σ_d constitute the fourth class. The rotation around n-order axis (for $k = 1, 2, 3, \ldots, p-1$) in which C_{2p}^k and C_{2p}^{-k} are conjugated constitute p-1 classes in pairs. Finally, the rotation–reflection axes S_{2n}^{2k+1} and $S_{2n}^{-(2k+1)}$ constitute p classes in pairs. Therefore, when n is even $n = 2p$, there are $2p + 3$ classes. When n is odd, $n = 2p + 1$, there are horizontal 2-order axes perpendicular to the vertical plane and so a symmetry center C_i is produced. Therefore, $D_{2p+1,d} = D_{2p+1} \otimes C_i$ and $D_{2p+1,d}$ has $2p + 4$ classes, because D_{2p+1} has $p + 2$ classes and the class number of direct product of $D_{2p+1} \otimes C_i$ is twice the number in the group D_{2p+1}. Their irreducible representations can be deduced by those of the groups D_{2p+1}.

(6) Group O_h

It is the cubic group introduced previously.

(7) **Group T_d**

This group is formed by all the symmetry elements of a tetrahedron. It can be obtained by adding to the group T a reflection plane passing through one 2-order and two 3-order axes. In this way, the 2-order axes become 4-order rotation–reflection axes. All the 24 elements are divided into five classes: E, 8 rotations C_3 and C_3^2, 6 reflection planes, 6 rotation–reflection axes, and 3 rotations. It has the same irreducible representations as that of the group O.

(8) **Group T_h**

This group is obtained by adding to the group T a symmetry center i: $T_h = T \otimes C_i$. The appearance of i produces three reflection planes mutually perpendicular to each other, because $i = C_2\sigma_h$. On the other hand, by means of $S_6 = C_3 \otimes C_i$, 6-order rotation–reflection axis is presented. Thus, 24 elements of this group are divided into eight classes, which has the number double that of the group T, and so the number of irreducible representations.

Noteworthily, owing to the restriction of the translations, axes of 5-order and those higher than 6-order cannot exist and hence there are only 32 point groups in the crystal. They are:

$C_n, C_{nh} (n = 1, 2, 3, 4, 6)$	10 groups
$C_{nv}, D_n, D_{nh} (n = 2, 3, 4, 6)$	12 groups
$S_{2n} (n = 1, 2, 3), D_{nd} (n = 2, 3)$	5 groups
O, O_h, T, T_h, T_d	5 groups

Among these 32 point groups, there are 11 point groups, which include symmetry center:

$$S_2 \equiv C_i, S_6, C_{2h}, C_{4h}, C_{6h}, D_{3d}, D_{2h}, D_{4h}, D_{6h}, O_h, T_h$$

The irreducible representations of these 11 groups are divided into even representations g and odd representations u.

2.7 Symmetry and Quantum Theory of the Ions in Solids

We will turn to the topic of the energy levels of the active ions in solids. Owing to the fact that this problem is greatly dependent on the symmetry around the ions, we should deal with it by associating the group theory with quantum mechanics.

Let us consider the relation between Schrödinger equation and the representation of symmetry transformation groups. If the ion is in a symmetry position represented by a group $G = \{G_i\}$, then under the action of an group element G_i, the wave functions of the system, that is, the eigenfunctions of the Hamiltonian operator should not have any change. Consider the Schrödinger equation

$$H\psi = E\psi \tag{2.13}$$

In order to express the transformation of the wave function, the following two equations can be used

$$G_i^{-1}\psi(x) = \psi(x') = \psi(G_i x) \tag{2.14}$$

$$G_i\psi(x') = \psi(x) = \psi\left(G_i^{-1}x'\right) \tag{2.15}$$

These two equations show that the transformation of the system by the operation G_i is the same as to keep the system invariant but to transform the coordinate system by G_i^{-1}. On the other hand, the linear algebra tells us that a real orthogonal transformation, which keeps the distance of two points invariant, is a unitary transformation (remember that the eigenfunctions of the Hamiltonian operator in quantum mechanics are orthogonal and normalized). Transforming $\varphi(x) = H(x)\psi(x)$ by means of (2.14) and (2.15), the following can be obtained

$$G_i^{-1}\varphi(x) = H(G_i x)\psi(G_i x) = H(G_i x)G_i^{-1}\psi(x)$$

Multiplying both sides of above equation from the left by G_i, one can obtain

$$\varphi(x) = G_i H(G_i x)G_i^{-1}\psi(x)(G_i G_i^{-1} = 1\,\text{has been used}).$$

Comparing with the original formula $\varphi(x) = H(x)\psi(x)$, the following result is obvious

$$H(x) = G_i H(G_i x)G_i^{-1}.$$

Note that the Hamiltonian should be invariant because the system is invariant under the transformation, that is, $H(x) = H(G_i x)$; therefore

$$H(x)G_i = G_i H(x) \tag{2.16}$$

Operating G_i on both sides of (2.13) and using (2.16), then

$$H(x)G_i\psi(x) = G_i H(x)\psi(x) = G_i E\psi(x) = E(G_i\psi(x)) \tag{2.17}$$

It shows that the wave functions belong to the same energy eigenvalue before and after the transformation.

If the energy eigenvalue E has degeneracy g, that is, there are g wave functions ψ_j belonging to the same energy eigenvalue E, $G_i\psi_j$ should be a linear combination of g wave functions ψ_j according to (2.17)

$$G_i\psi_j = \sum_{l=1}^{g} M(G_i)_{lj}\psi_l \tag{2.18}$$

Operating with another group element, one can obtain a similar result

$$G_k\psi_j = \sum_{l=1}^{g} M(G_k)_{lj}\psi_l$$

Each group element in $\{G_i\}$ has a corresponding matrix $M\{G_i\}$, and for the joint operation of product G_iG_k, one has

$$G_iG_k\psi_j = \sum_{l=1}^{g} M(G_k)_{lj}G_i\psi_l = \sum_{l,m=1}^{g} M(G_k)_{lj}M(G_i)_{ml}\psi_m = \sum_{m=1}^{g} M(G_iG_k)_{mj}\psi_m$$

Formula $M(G_iG_k)_{mj} = \sum_{l=1}^{g} M(G_i)_{ml}M(G_k)_{lj}$ shows that the matrix for the group element product is equal to the product of the matrices of group elements. The existence of identity element and inverse element can also be demonstrated. Therefore, these matrices form a representation of the group G, and the related wave functions form the bases. It can be shown that the representation formed by the degeneracy bases is irreducible and its dimension is equal to g. Certainly, the accidental degeneracy and Kramers degeneracy should be excluded.

After establishing the relation between irreducible representation and the degeneracy of energy levels, one can go back to the problem of free ions. Obviously, the free ions possess spherical symmetry: the symmetry group is three-dimensional rotation group R_3 ($SO(3)$). The representation of the R_3 group, as will be discussed in the following, can be reduced to a series of irreducible representations labeled by angular momentum quantum numbers J:

$$D_{J_1}, D_{J_2}, \ldots, D_{J_i}, \ldots$$

The dimension of D_{J_i} is $2J_i + 1$, the degeneracy of the spectral term with angular momentum J. The eigen wave functions of free ions (the wave functions of the spectral terms) constitute the bases of the irreducible representations.

The active ions in crystals have interactions with the surrounding ions and a lower symmetry perturbation potential should be added to the Hamiltonian with spherical symmetry. Then, the crystal field potential has symmetry group G which is a subgroup of R_3. The representations originally irreducible in R_3 are now reducible and they can be split into two or more irreducible representations. For example,

$$D_{J_i} = \Gamma^{(1)}(g_1) + \Gamma^{(2)}(g_2) \tag{2.19}$$

then, it is said that the spectral term J_i is split into two levels with degeneracy g_1 and g_2.

Having this general understanding, the next step is to realize a detailed method for the calculation of energy level splitting. In order to achieve this goal, one should have more knowledge of group theory and the theory of angular momentum as well as the application of the group theory to the calculation of matrix elements.

2.8 Full Rotation Group and Angular Momentum Theory

Full rotation group, that is, spherical symmetry group, is the symmetry group of free ions. It can be seen from previous analysis that the free ion's wave functions constitute the bases of the irreducible representations of this group. On the other hand, these functions are also eigenfunctions of angular momentum. The essence of this relation is that the angular momentum operator is in fact a rotation operator. This can be clearly seen from the following analysis.

Consider the transformation of a wave function $\psi(x, y, z)$ after rotation around z axis

$$\psi(x', y', z') = R\psi(x, y, z)$$

If one rotates around z axis through an angle α, then according to the familiar formula

$$\begin{aligned} x' &= x\cos\alpha + y\sin\alpha \\ y' &= -x\sin\alpha + y\cos\alpha \\ z' &= z \end{aligned}$$

Substituting $\alpha = d\varphi$ into this equation, then

$$\begin{aligned} x' &= x + yd\varphi \\ y' &= -xd\varphi + y \\ z' &= z \end{aligned}$$

Because $d\varphi$ is a very small angle, the following relations can be used:

$$\cos(d\varphi) = 1, \sin(d\varphi) = d\varphi.$$

Therefore

$$\psi(x', y', z') = \psi(x + yd\varphi, y - xd\varphi, z)$$

Expanding this formula into Taylor series

$$\psi(x', y', z') = \psi(x, y, z) + d\varphi\left(y\frac{\partial}{\partial x} - x\frac{\partial}{\partial y}\right)\psi(x, y, z)$$

and using the angular momentum operator in quantum mechanics

$$J_z = -i\hbar\left(x\frac{\partial}{\partial y} - y\frac{\partial}{\partial x}\right) = -i\hbar\frac{\partial}{\partial \varphi}$$

We have

$$R(d\phi)\psi(x, y, z) = (1 - id\phi J_z/\hbar)\psi(x, y, z) = \exp(-id\phi J_z/\hbar)\psi(x, y, z)$$

Rotating z axis through a finite angle φ, then

$$R(\phi)\psi(x, y, z) = \exp(-i\phi J_z/\hbar)\psi(x, y, z)$$

The electronic eigenfuctions of free ion constitute the bases of the irreducible representations of group R_3. It can be seen from Chap. 1 that they are spherical harmonic functions

$$Y_{jm}(\theta, \varphi) \equiv |jm\rangle$$

We also have

$$J_z|jm\rangle = m\hbar|jm\rangle \tag{2.20}$$

Therefore, the $2j + 1$ functions with $m = -j, -j + 1, -j + 2, \ldots, j - 2, j - 1, j$ under the symmetry operation of rotation through an angle φ around z axis is a diagonal matrix with diagonal matrix elements

$$e^{ij\varphi}, e^{i(j-1)\varphi}, e^{i(j-2)\varphi}, \ldots, e^{-i(j-2)\varphi}, e^{-i(j-1)\varphi}, e^{-ij\varphi}$$

Obviously, the character of the operation matrix (irreducible representations of group R_3) is

$$\chi(\varphi) = \sum_{m=-j}^{m=j} e^{im\varphi} = \frac{\sin(j + 1/2)\varphi}{\sin(\varphi/2)} \tag{2.21}$$

When $\varphi = 0$ or 2π, it should be calculated by the ratio of differential quotient and the result is $\chi(0) = 2j + 1$, $\chi(2\pi) = 2j + 1$ for j is an integer and $\chi(2\pi) = -(2j + 1)$ for j is a half-integer.

Equation (2.21) will be used frequently to solve the problem of energy level splitting by crystal field. It should be noted that when j is a half-integer, so $j + 1/2$ is an integer, then by (2.21)

$$\chi(\varphi+2\pi)=\frac{\sin(j+1/2)\varphi}{\sin(\varphi/2)}=\frac{\sin(j+1/2)\varphi}{\sin(\varphi/2+\pi)}=-\frac{\sin(j+1/2)\varphi}{\sin(\varphi/2)}=-\chi(\varphi) \quad (2.22)$$

Therefore, for the half-integer j we have a double-valued character. In order to cope with this situation, a new group element $\overline{E}$ will be introduced, which represents a rotation of 2π and now the identity element E is the rotation of 4π.

The augmented group contains twice as many elements as the original one and is call the double group. Besides the original elements G_1, G_2, …, G_n, there are new elements $\overline{G}_1, \overline{G}_2, \ldots, \overline{G}_n$ in the double group, which have the relation of $\overline{G}_i = \overline{E}G_i$. The irreducible representation of the double group can be divided into two parts, one is called single-valued representation, in which G_i and $\overline{G}_i$ have the same representation, and the other one is called double-valued representation, in which $\chi(\overline{G}_i) = -\chi(G_i)$.

It should be noted that by using (2.21) and (2.22) to reduce the irreducible representation from higher symmetry to lower symmetry, the angle φ for symmetry center and reflection mirror is taken to be π. For rotation–reflection axis S_3, φ = π/3, and for rotation–reflection axis S_4 and S_6, φ should be π/2 and π/3, respectively.

Now let us consider the coupling between two angular momenta, which can be different momenta of the same electron (e.g. one spin momentum and one orbital momentum) or the same kind of momentum for different electrons. In quantum mechanics, the zero-order wave function of an interaction system is the product of the wave functions of its subsystems. By using the language of group representation, the basis space describing the interaction system is the direct product of the basis spaces describing the subsystems. The direct product space is generally reducible, that is, it can be reduced to the sum of a series of subspaces. By a linear transformation, one can express the direct product space as the direct sum of a series of subspace orthogonal to each other. If the wave functions of two angular momenta are $|j_1m_1\rangle$ and $|j_2m_2\rangle$, respectively, their direct product space can be transformed to irreducible space of the basis $|j_1j_2jm\rangle$ by the following linear transformation

$$|j_1j_2jm\rangle=\sum_{m_1m_2}|j_1m_1j_2m_2\rangle\langle j_1m_1j_2m_2|j_1j_2jm\rangle \quad (1.23)$$

The coefficients $\langle j_1m_1j_2m_2|j_1j_2jm\rangle$ are called vector-coupling coefficients. In order to write it in a more symmetric form, Wigner introduced the so-called 3-j symbols defined as

$$\begin{pmatrix} j_1 & j_2 & j_3 \\ m_1 & m_2 & m_3 \end{pmatrix}=(-1)^{j_1-j_2-m_3}(2j_3+1)^{-1/2}\langle j_1m_1j_2m_2 \mid j_1j_2j_3-m_3\rangle \quad (2.24)$$

They have the following symmetric properties:
The even permutation of the columns is symmetry

$$\begin{pmatrix} j_1 & j_2 & j_3 \\ m_1 & m_2 & m_3 \end{pmatrix} = \begin{pmatrix} j_2 & j_3 & j_1 \\ m_2 & m_3 & m_1 \end{pmatrix} = \begin{pmatrix} j_3 & j_1 & j_2 \\ m_3 & m_1 & m_2 \end{pmatrix} \tag{2.25}$$

The odd permutation of the columns involves a multiplying factor of $(-1)^{j_1+j_2+j_3}$

$$\begin{pmatrix} j_3 & j_2 & j_1 \\ m_3 & m_2 & m_1 \end{pmatrix} = \begin{pmatrix} j_2 & j_1 & j_3 \\ m_2 & m_1 & m_3 \end{pmatrix} = \begin{pmatrix} j_1 & j_3 & j_2 \\ m_1 & m_3 & m_2 \end{pmatrix} = (-1)^{j_1+j_2+j_3}\begin{pmatrix} j_1 & j_2 & j_3 \\ m_1 & m_2 & m_3 \end{pmatrix} \tag{2.26}$$

and

$$\begin{pmatrix} j_1 & j_2 & j_3 \\ m_1 & m_2 & m_3 \end{pmatrix} = (-1)^{j_1+j_2+j_3}\begin{pmatrix} j_1 & j_2 & j_3 \\ -m_1 & -m_2 & -m_3 \end{pmatrix} \tag{2.27}$$

The 3-j symbols also have the following orthogonal properties:

$$\sum_{j_3,m_3}(2j_3+1)\begin{pmatrix} j_1 & j_2 & j_3 \\ m_1 & m_2 & m_3 \end{pmatrix}\begin{pmatrix} j_1 & j_2 & j_3 \\ m_1' & m_2' & m_3 \end{pmatrix} = \delta(m_1m_1')\delta(m_2m_2') \tag{2.28}$$

$$\sum_{m_1m_2}\begin{pmatrix} j_1 & j_2 & j_3 \\ m_1 & m_2 & m_3 \end{pmatrix}\begin{pmatrix} j_1 & j_2 & j_3' \\ m_1 & m_2 & m_3' \end{pmatrix} = (2j_3+1)^{-1}\delta(j_3j_3')\delta(m_3m_3') \tag{2.29}$$

Introducing by (2.29), it can be shown

$$\sum_{m_1m_2m_3}\begin{pmatrix} j_1 & j_2 & j_3 \\ m_1 & m_2 & m_3 \end{pmatrix}^2 = 1$$

The values of 3-j symbols can be calculated by formula (e.g. Equations (2.1–2.20) provided by Judd in reference [3]), and they can also be found from tables (e.g. Rotenberg et al. "The 3-j and 6-j symbols" [4]) or calculated by means of computer programs [5]. Two useful simple formulas are

$$\begin{pmatrix} j & j & 1 \\ m & -m & 0 \end{pmatrix} = (-1)^{j-m}\frac{m}{[(2j+1)(j+1)j]^{1/2}} \tag{2.30}$$

$$\begin{pmatrix} j_1 & j_2 & j_3 \\ 0 & 0 & 0 \end{pmatrix} = (-1)^{J/2}\left[\frac{(J-2j_1)!(J-2j_2)!(J-2j_3)!}{(J+1)!}\right]^{1/2}\frac{(\frac{J}{2})!}{(\frac{J}{2}-j_1)!(\frac{J}{2}-j_2)!(\frac{J}{2}-j_3)!} \tag{2.31}$$

When $2j + 1$ is even, (2.30) is not valid. In (2.31) $J = j_1 + j_2 + j_3$, when J is odd, this 3-j symbol is equal to zero. It is the result obtained by (2.27). Some more formulas for the calculation of 3-j symbols in special cases can be found from the book of Rotenberg et al. [4] or directly calculated by a computer program [5]. There is an important property of the 3-j symbols that should be mentioned, that is, the triangle condition of 3-j symbols. It means that 3-j symbols will be zero unless three j are satisfied by the triangle condition $j_1 + j_2 \geq j_3 \geq |j_1 - j_2|$. On the other hand, the 3-j symbols vanish unless $m1 + m2 + m3 = 0$.

We should also mention briefly some of the properties of 6-j and 9-j symbols, which are used in the wave functions calculations for the coupling of three and four angular momenta. For the coupling of three angular momenta, the ways for j_1, j_2, j_3 coupling in pair can be different. For example, it can be $(j_1j_2)j_{12}j_3J$ or $j_1(j_2j_3)j_{23}J$, that is, j_1j_2 couple at first and then couple with j_3, or j_2j_3 couple at first and then couple with j_1. The wave functions of reducible space obtained by different coupling ways can be transformed by

$$|(j_1j_2)j_{12}j_3JM\rangle = \sum_{j_{23}} \langle (j_1j_2)j_{12}j_3J \mid j_1(j_2j_3)j_{23}J\rangle |j_1(j_2j_3)j_{23}JM\rangle \tag{2.32}$$

and the 6-j symbols are defined as

$$\begin{Bmatrix} j_1 & j_2 & j_{12} \\ j_3 & J & j_{23} \end{Bmatrix} = (-1)^{j_1+j_2+j_3+J}[(2j_{12}+1)(2j_{23}+1)]^{-1/2}\langle (j_1j_2)j_{12}j_3J \mid j_1(j_2j_3)j_{23}J\rangle \tag{2.33}$$

The values of 6-j symbols can also be found from tables [4]. However, in these tables, only the values of some 6-j symbols are listed. Values of other 6-j symbols can be found by their symmetry properties. Each 6-j symbol can be seen as a regular tetrahedron with 6j vectors as its six edges and so it has the symmetry of a regular tetrahedron. Its symmetry properties can be expressed as: the 6-j symbol is invariant under any permutation of the columns and is also invariant under an interchange of the upper and lower arguments in any two columns or exchange of any two columns:

$$\begin{Bmatrix} j_1 & j_2 & j_3 \\ j_4 & j_5 & j_6 \end{Bmatrix} = \begin{Bmatrix} j_4 & j_5 & j_3 \\ j_1 & j_2 & j_6 \end{Bmatrix} = \begin{Bmatrix} j_2 & j_1 & j_3 \\ j_5 & j_4 & j_6 \end{Bmatrix} \tag{2.34}$$

The aforementioned four planes of the regular tetrahedron are triangles and the value of 6-j symbol will be zero unless the following triangle conditions are satisfied

$$\Delta(j_1j_2j_3), \Delta(j_1j_5j_6), \Delta(j_4j_5j_3)\Delta(j_2j_4j_6)$$

That means, for example,

$$j_1 + j_2 \geq j_3 \geq |j_1 - j_2|$$

In addition, the 6-j symbols have the following orthogonal relation:

$$\sum_{j_6} (2j_3+1)(2j_6+1)\begin{Bmatrix} j_1 & j_2 & j_3 \\ j_4 & j_5 & j_6 \end{Bmatrix}\begin{Bmatrix} j_1 & j_2 & j'_3 \\ j_4 & j_5 & j_6 \end{Bmatrix} = \delta(j_3j'_3) \tag{2.35}$$

One can find the values of 6-j symbols and many useful calculation formulas in the book written by Rotenberg et al. [4].

The 9-j symbols are used to correlate different coupling ways for four angular momenta. For example, in the relation between $|(j_1j_2)j_{12}(j_3j_4)j_{34}J\rangle$ and $|(j_1j_3)j_{13}(j_2j_4)j_{24}J\rangle$, the following coefficients can be used in the transformation

$$|(j_1j_2)j_{12}(j_3j_4)j_{34}JM\rangle = \sum_{j_{13}j_{24}} \langle(j_1j_2)j_{12}(j_3j_4)j_{34}J \mid (j_1j_3)j_{13}(j_2j_4)j_{24}J\rangle|(j_1j_3)j_{13}(j_2j_4)j_{24}JM\rangle \tag{2.36}$$

The above coefficients can be expressed as 9-j symbols by

$$\begin{aligned} &\langle(j_1j_2)j_{12}(j_3j_4)j_{34}J \mid (j_1j_3)j_{13}(j_2j_4)j_{24}J\rangle \\ &= [(2j_{12}+1)(2j_{34}+1)(2j_{13}+1)(2j_{24}+1)]^{1/2}\begin{Bmatrix} j_1 & j_2 & j_{12} \\ j_3 & j_4 & j_{34} \\ j_{13} & j_{24} & J \end{Bmatrix} \end{aligned} \tag{2.37}$$

We will not use 9-j symbols directly in this book and so would like to mention only two formulas, which are useful in the calculation of matrix elements

$$\begin{aligned} \begin{Bmatrix} j_{11} & j_{12} & j_{13} \\ j_{21} & j_{22} & j_{23} \\ j_{31} & j_{32} & j_{33} \end{Bmatrix} = \sum_{all\ m} &\begin{pmatrix} j_{11} & j_{12} & j_{13} \\ m_{11} & m_{12} & m_{13} \end{pmatrix}\begin{pmatrix} j_{21} & j_{22} & j_{23} \\ m_{21} & m_{22} & m_{23} \end{pmatrix}\begin{pmatrix} j_{31} & j_{32} & j_{33} \\ m_{31} & m_{32} & m_{33} \end{pmatrix} \\ \times &\begin{pmatrix} j_{11} & j_{21} & j_{31} \\ m_{11} & m_{21} & m_{31} \end{pmatrix}\begin{pmatrix} j_{12} & j_{22} & j_{32} \\ m_{12} & m_{22} & m_{32} \end{pmatrix}\begin{pmatrix} j_{13} & j_{23} & j_{33} \\ m_{13} & m_{23} & m_{33} \end{pmatrix} \end{aligned} \tag{2.38}$$

$$\begin{Bmatrix} j_1 & j_2 & j \\ j_3 & j_4 & j \\ j' & j' & 0 \end{Bmatrix} = (-1)^{j_2+j_3+j+j'}[(2j+1)(2j'+1)]^{-1/2}\begin{Bmatrix} j_1 & j_2 & j \\ j_4 & j_3 & j' \end{Bmatrix} \tag{2.39}$$

2.9 Irreducible Tensor Operators and the Calculation of Matrix Elements

The calculation of matrix elements and energy levels are the most basis calculations in the theory of crystal spectroscopy. It is an important problem to use group theory and the theory of angular momentum in these calculations. This problem will be discussed in this section.

The transformation characteristics of the angular momentum eigenfunctions under rotation groups have been investigated in the last section. The wave function spaces of the interacting systems have also been correlated with the representation spaces of the direct product of the rotation group. Obviously, if the operators are classified according to the transformation properties of the angular momentum eigenfunctions, that is to say, let the operators constitute the bases of the irreducible representation space of the rotational group; undoubtedly, we can make full use of the symmetry properties of the system and a series of formulas developed by group theory to simplify the computation of various matrix elements in spectroscopy, which is the reason for the introduction of irreducible tensor operators. By means of these transformation properties, one can define irreducible tensor operators: irreducible tensor operators are operators forming the bases of the irreducible representation of the rotation group. They satisfy the following relation in mathematics

$$R(\alpha, \beta, \gamma) T_q^{(k)} R^{-1}(\alpha, \beta, \gamma) = \sum_{q'} T_{q'}^{(k)} R_{q'q}^{(k)}(\alpha, \beta, \gamma) \tag{2.40}$$

Practically, (2.40) is the same as the transformation of angular momentum eigenfunctions under rotation expressed as follows

$$R(\alpha, \beta, \gamma)|kq\rangle = \sum_{q'} |kq'\rangle R_{q'q}^{(k)}(\alpha, \beta, \gamma)$$

To view it at a different angle, irreducible tensor operators $T_q^{(k)}$ have the same transformation properties under operation of rotation group as those of angular momentum eigenfunctions $|kq\rangle$. Therefore, under the action of angular momentum —rotation operator, $T_q^{(k)}$ and $|kq\rangle$ should have the same transformation, that is

$$J_z|k, q\rangle = q|k, q\rangle$$

$$J_{\pm}|k, q\rangle = [k(k+1) - q(q \pm 1)]^{1/2}|k, q\rangle \tag{2.41a}$$

$$\left[J_z, T_q^{(k)}\right] = qT_q^{(k)}$$

$$\left[J_\pm, T_q^{(k)}\right] = [k(k+1) - q(q\pm 1)]^{1/2} T_{q\pm 1}^{(k)} \tag{2.41b}$$

Equations (2.40) and (2.41a, 2.41b) can be seen as the mathematical definition of irreducible tensor operator, in which J_z is the z component of angular momentum and $J_\pm$ are the shift operators which raise and lower the component q of angular momentum k, respectively.

Having introduced the irreducible tensor operators, the calculation of matrix elements becomes very easy, because the transformation property of the product of operators and the wave functions can be easily obtained. In order to calculate matrix elements, it is necessary to know Wigner-Eckart theorem first.

The physics idea for the introduction of this theorem has been mentioned above. If one would like to calculate matrix elements $\langle\alpha' j' m'|T_q^{(k)}|\alpha jm\rangle$, the first step is to investigate the transformation property of $T_q^{(k)}|\alpha jm\rangle$. According to the result discussed previously, it is transformed as the vectors in the direct product space $R^{(k)} \otimes R^{(j)}$, or one can say that $T_q^{(k)}|\alpha jm\rangle$ with all the different k and j are the bases of the direct product space $R^{(k)} \otimes R^{(j)}$. Reducing the reducible space, the basis can be written as $|\beta JM\rangle$. By means of the angular momentum coupling method discussed previously, these vectors can be written as a linear combination of the bases in the direct product space $R^{(k)} \otimes R^{(j)}$ (remember that the transformation property of the operators and the bases of wave function space is the same)

$$|\beta JM\rangle = \sum_{qm} \langle kqjm \mid kjJM\rangle T_q^{(k)}|\alpha jm\rangle \tag{2.42}$$

One can make an inverse transformation by using the orthogonal property of the coefficients

$$T_q^{(k)}|\alpha jm\rangle = \sum_{JM} \langle kqjm \mid kjJM\rangle |\beta JM\rangle$$

Multiplying both sides of this equation by $\langle a' j' m'|$, then

$$\left\langle \alpha' j' m' \middle| T_q^{(k)} \middle| \alpha jm \right\rangle = \sum_{JM} \langle kqjm \mid kjJM\rangle \langle \alpha' j' m' \mid \beta JM\rangle$$

By using orthogonal property of angular momentum eigenfunctions, it can be shown that $\langle\alpha' j' m'|\beta JM\rangle$ is zero except $J = j'$, $M = m'$ and its values is independent of m'. It is denoted as $\left\langle\alpha' j'\left\|T^{(k)}\right\|\alpha j\right\rangle$ and called irreducible matrix element, that is

$$\langle \alpha' j' m' \mid \beta JM\rangle = \left\langle \alpha' j' \left\| T^{(k)} \right\| \alpha j \right\rangle \delta_{Jj'} \delta_{Mm'}$$

then

$$\left\langle \alpha' j' m' \middle| T_q^{(k)} \middle| \alpha j m \right\rangle = \langle kqjm \mid kjj'm' \rangle \left\langle \alpha' j' \middle\| T^{(k)} \middle\| \alpha j \right\rangle$$

This is so-called Wigner-Eckart theorem (W-E theorem). Writing the coefficients in the form of 3-j symbols, then

$$\left\langle \alpha' j' m' \middle| T_q^{(k)} \middle| \alpha j m \right\rangle = (-1)^{j'-m'} \begin{pmatrix} j' & k & j \\ -m' & q & m \end{pmatrix} \left\langle \alpha' j' \middle\| T^{(k)} \middle\| \alpha j \right\rangle \tag{2.43a}$$

According to the definition (2.24) of 3-j symbol, the expression (2.43a) should be

$$\left\langle \alpha' j' m' \middle| T_q^{(k)} \middle| \alpha j m \right\rangle = (-1)^{k-j+m'} (2j\prime + 1)^{1/2} \begin{pmatrix} j' & k & j \\ -m' & q & m \end{pmatrix} \left\langle \alpha' j' \middle\| T^{(k)} \middle\| \alpha j \right\rangle \tag{2.43b}$$

However, many authors adopt (2.43a) as the expression of W-E theorem. The reduced matrix elements of the two different formulas are different. On the other hand, the W-E theorem for point group can be expressed as

$$\langle \Gamma_1 \gamma_1 | O_\gamma^{(\Gamma)} | \Gamma_2 \gamma_2 \rangle = (\Gamma)^{-1/2} \langle \Gamma_1 \gamma_1 | \Gamma_2 \gamma_2 \Gamma \gamma \rangle \langle \Gamma_1 || O^{(\Gamma)} || \Gamma_2 \rangle \tag{2.43c}$$

where $\langle \Gamma_1 \gamma_1 | \Gamma_2 \gamma_2 \Gamma \gamma \rangle$ is Clebsch-Gordan coefficients (C-G coefficients) of point group, Γ and γ denote irreducible representation of point group and the component of that irreducible representation, respectively, and (Γ) is the dimension of irreducible representation Γ. C-G coefficients of triangle basis can be found in Appendix F.

Using this theorem, we can save a lot of time in the calculation of matrix elements, because with only one of the reduced matrix element $\left\langle \alpha' j' \middle\| T^{(k)} \middle\| \alpha j \right\rangle$, all the $(2j + 1)(2k + 1)(2j' + 1)$ matrix elements can be calculated. Furthermore, by the properties of the 3-j symbols, some of the matrix elements are equal to zero and without calculation. We will use it frequently in the calculation of the energy level and transition probability.

As an example, we would like to prove a useful formula for the calculation of energy level and transition probability. Many spectroscopic problems involved orbital and spin variables. If these two kinds of angular momentum are denoted by j_1 and j_2, respectively, by using (2.21), a general coupling state (basis $\langle JM|$ of direct product space) can be expressed as a linear combination of the direct products of two subspace bases $\langle j_1 m_1|$ and $\langle j_2 m_2|$ as follows

$$\langle j_1 j_2 JM| = \sum_{m_1 m_2} \langle j_1 m_1 j_2 m_2| \begin{pmatrix} j_1 & j_2 & J \\ m_1 & m_2 & -M \end{pmatrix} (2J+1)^{1/2} (-1)^{j_1 - j_2 + M}$$

Similarly, a tensor operator $T_Q^{(K)}(1,2)$ of a direct product space $\mathfrak{R}(KQ)$ can be expressed as the following linear combination of tensor operators $T_{q_1}^{(k_1)}(1)$ and $T_{q_2}^{(k_2)}(2)$ of subspaces $\mathfrak{R}(k_1 q_1)$ and $\mathfrak{R}(k_2 q_2)$

$$T_Q^{(K)}(1,2) = \sum_{q_1 q_2} T_{q_1}^{(k_1)}(1) T_{q_2}^{(k_2)}(2) \begin{pmatrix} k_1 & k_2 & K \\ q_1 & q_2 & -Q \end{pmatrix} (2K+1)^{1/2} (-1)^{k_1 - k_2 + Q} \tag{2.44a}$$

The formula for a point group can be expressed as follows

$$T_\gamma^{(\Gamma)} = \left[T_{\gamma_1}^{(\Gamma_1)} \otimes T_{\gamma_2}^{(\Gamma_2)} \right] = \sum_{\gamma_1 \gamma_2} T_{\gamma_1}^{(\Gamma_1)} T_{\gamma_2}^{(\Gamma_2)} \langle \Gamma_1 \Gamma_2 \gamma_1 \gamma_2 | \Gamma \gamma \rangle \tag{2.44b}$$

Equation (2.44b) is practically a general formula for the direct product of two irreducible tensors. If a point group irreducible representation Γ replaced by an angular momentum quantum number K, a component γ of Γ replaced by Q and the C-G coefficients of the point group replaced by 3-j symbol then formula (2.44a) can be obtained.

The reduced matrix element expression of (2.44a) can be obtained as follows:

The left side of (2.44a) can be expressed as

$$\langle j_1 j_2 JM | T_Q^{(K)}(1,2) | j_1' j_2' J' M' \rangle = (-1)^{J-M} \begin{pmatrix} J & K & J' \\ -M & Q & M' \end{pmatrix} \left\langle j_1 j_2 J \| T^{(K)}(1,2) \| j_1' j_2' J' \right\rangle \tag{2.44c}$$

By using the detailed expansion of the wave functions $\langle j_1 j_2 JM|$ and $|j_1' j_2' J' M'\rangle$, the summation of the right side of (2.44a) can be expressed as

$$\begin{aligned} &\sum_{m_1 m_2} \sum_{m_1' m_2'} \sum_{q_1 q_2} (-1)^{k_1 - k_2 + Q + 2j_1 + M + j_1' - j_2' + M' - m_1 - m_2} \begin{pmatrix} j_1 & j_2 & J \\ m_1 & m_2 & -M \end{pmatrix} \begin{pmatrix} j_1' & j_2' & J' \\ m_1' & m_2' & -M' \end{pmatrix} \begin{pmatrix} k_1 & k_2 & K \\ q_1 & q_2 & -Q \end{pmatrix} \\ &\times \begin{pmatrix} j_1 & k_1 & j_1' \\ -m_1 & q_1 & m_1' \end{pmatrix} \begin{pmatrix} j_2 & k_2 & j_2' \\ -m_2 & q_2 & m_2' \end{pmatrix} [(2J+1)(2J'+1)(2K+1)]^{1/2} \left\langle j_1 \| T^{(k_1)}(1) \| j_1' \right\rangle \langle j_2 \| T^{(k_2)}(2) \| j_2' \rangle \end{aligned} \tag{2.44d}$$

Multiply the right side of (2.44a) and (2.44d) by $(-1)^{J-M} \begin{pmatrix} J & K & J' \\ -M & Q & M' \end{pmatrix}$ and sum over $\sum_{MM'Q}$. Take the orthogonal property of the 3-j symbol into account and change the order of the column and row in the 3-j symbol. Eliminate the phase factors by means of Eqs. (2.26), (2.28), and (2.38), and then comparing both sides of the obtained equation, it can be shown that

$$\left\langle j_1 j_2 J \| T^{(K)}(1,2) \| j'_1 j'_2 J' \right\rangle$$
$$= [(2J+1)(2J'+1)(2K+1)]^{1/2} \begin{Bmatrix} j_1 & j'_1 & k_1 \\ j_2 & j'_2 & k_2 \\ J & J' & K \end{Bmatrix} \left\langle j_1 \| T^{(k_1)}(1) \| j'_1 \right\rangle \left\langle j_2 \| T^{(k_2)}(2) \| j'_2 \right\rangle \tag{2.45}$$

Relation (2.45) is a very useful formula in spectroscopy.

If the discussion only concerns the matrix elements of "the first part" operator $T^{(k_1)}(1)$ of the state $|j_1 j_2 J\rangle$, this formula can be further simplified. Let $T^{(k_2)}(2) = I$, that is, unit operator, then $k_2 = 0$, $j_2 = j'_2$, and $K = k_1$, using (2.39) and noting that in this case one has

$$\left\langle j_2 \| T^{(k_2)}(2) \| j'_{\in} \right\rangle = (2j_2+1)^{1/2} \delta_{j_2 j'_{\in}}$$

the following expression can be obtained

$$\left\langle j_1 j_2 J \| T^{(k_1)}(1) \| j'_{\infty} j'_{\in} J' \right\rangle = (-1)^{j_1+j_2+J'+k_1} [(2J+1)(2J'+1)]^{1/2} \begin{Bmatrix} J & J' & k_1 \\ j'_1 & j_1 & j_2 \end{Bmatrix} \left\langle j_1 \| T^{(k_1)}(1) \| j'_1 \right\rangle \delta_{j_2 j'_{\in}} \tag{2.46}$$

If we are concerned about the matrix elements of "the second part" operator $T^{(k_2)}(2)$, then let $T^{(k_1)}(1) = I$, and $k_1 = 0$, $j_1 = j'_1$, $K = k_2$. By using (2.39) once again, the following expression can be obtained similarly

$$\left\langle j_1 j_2 J \| T^{(k_2)}(2) \| j'_1 j'_2 J' \right\rangle = (-1)^{j_1+j'_2+J+k_2} [(2J+1)(2J'+1)]^{1/2} \begin{Bmatrix} J & J' & k_2 \\ j'_2 & j_2 & j_1 \end{Bmatrix} \left\langle j_2 \| T^{(k_2)}(2) \| j'_2 \right\rangle \delta_{j_1 j'_1} \tag{2.47}$$

Although (2.45) can be used to calculate the matrix elements of a two-electron system, the previous discussion is about the matrix elements of single electron. Most of the problems in spectroscopy are concerned with multi-electron, and so the calculation of multi-electron matrix elements should be introduced. It should be pointed out that to construct the wave functions of a multi-electron system and calculate their matrix elements, the coefficients of fractional parentage should be introduced. We would not go so far in mathematics but only make use of the unit operator of multi-electron, which has appropriate direct reading tables.

Two operators $A_q^{(k)}$ and $O_q^{(k)}$ with the same transformation property obviously have the relation

$$\left\langle \alpha' j' m' \middle| T_q^{(k)} \middle| \alpha j m \right\rangle = (-1)^{j'-m'} \begin{pmatrix} j' & k & j \\ -m' & q & m \end{pmatrix} \left\langle \alpha' j' \| T^{(k)} \| \alpha j \right\rangle$$

$T_q^{(k)}$ can be $A_q^{(k)}$ or $O_q^{(k)}$. In this way

$$\left\langle \alpha' j' m' \middle| A_q^{(k)} \middle| \alpha j m \right\rangle = \left\langle \alpha' j' m' \middle| O_q^{(k)} \middle| \alpha j m \right\rangle \times \frac{\langle \alpha' j' \| A^{(k)} \| \alpha j \rangle}{\langle \alpha' j' \| O^{(k)} \| \alpha j \rangle} \tag{2.48}$$

In case $O_q^{(k)}$ is a unit operator, (2.48) shows that any operator can be expressed as a unit operator multiplied by a certain value. The unit operator for a single electron is denoted by $u_q^{(k)}(i)$, then its reduced matrix element will be

$$\left\langle \alpha' j' \middle\| u^{(k)} \middle\| \alpha j \right\rangle = \delta_{\alpha\alpha'} \delta_{jj'}$$

and for multi-electron system $U_q^{(k)} = \sum_i u_q^{(k)}(i)$, the summation being over all the electrons. The related 6-j symbols and the coefficients of fractional parentage are used to determine their irreducible matrix elements. Nielson and Koster [6] have given the tables for the irreducible matrix elements of the unit operator for the configurations of p^n, d^n, f^n. We show now an example of the irreducible matrix elements of unit operators.

The crystal field potential energy of rare earth ions in crystal can be expressed as

$$V = \sum_{k,q} B_{kq} C_q^{(k)} \tag{2.49}$$

where multi-electron operators $C_q^{(k)} = \sum_i C_q^{(k)}(\theta_i, \varphi_i)$ are related to the spherical harmonics functions by $C_q^{(k)}(\theta_i, \varphi_i) = \left(\frac{4\pi}{2k+1}\right)^{1/2} Y_q^{(k)}(\theta_i, \varphi_i)$ and B_{kq} are the so-called crystal field parameters. By means of a relation like (2.48), it is obvious that

$$\langle f^n SLJM | V | f^n S'L'J'M' \rangle = \sum_{kq} B_{kq} \left\langle f^n SLJM \middle| U_q^{(k)} \middle| f^n S'L'J'M' \right\rangle \left\langle f \middle\| C^{(k)} \middle\| f \right\rangle \tag{2.50}$$

According to the formulas given above, one has

$$\left\langle f^n SLJM \middle| U_q^{(k)} \middle| f^n S'L'J'M' \right\rangle = (-1)^{J-M} \begin{pmatrix} J & k & J' \\ -M & q & M' \end{pmatrix} \left\langle f^n SLJ \middle\| U^{(k)} \middle\| f^n S'L'J' \right\rangle \tag{2.51}$$

$\langle f^n SLJ \| U^{(k)} \| f^n S'L'J' \rangle$ can be obtained by using (2.46). Remember that the crystal field potential is an orbital operator

$$\left\langle f^n SLJ \| U^{(k)} \| f^n S'L'J' \right\rangle = (-1)^{S+L'+J+k}[(2J+1)(2J'+1)]^{1/2} \begin{Bmatrix} J & J' & k \\ L' & L & S \end{Bmatrix} \left\langle f^n SL \| U^{(k)} \| f^n SL' \right\rangle \delta_{SS'} \tag{2.52}$$

where $\langle f^n SLJ \| U^{(k)} \| f^n SL'J' \rangle$ can be found in the tables published by Kamiskii [7], which involves the data of all high-lying $^{2S+1}L_J$ states (part of these data can be found in Appendix D of this book). On the other hand, we have

$$\langle l_1 \| C^{(k)} \| l_2 \rangle = (-1)^{l_1} \times [(2l_1+1)(2l_2+1)]^{1/2} \begin{pmatrix} l_1 & k & l_2 \\ 0 & 0 & 0 \end{pmatrix} \tag{2.53}$$

For f electron it is

$$\left\langle 3 \| C^{(k)} \| 3 \right\rangle = (-1)^3 \times 7 \times \begin{pmatrix} 3 & k & 3 \\ 0 & 0 & 0 \end{pmatrix}$$

The 3-j symbols in (2.53) are easy to be calculated by (2.31).

References

1. M. Hamermesh, *Group Theory and Its Application to Physical Problems* (Addison—Wesley Publishing Company, INC, London, 1962)
2. M. Tinkham, *Group Theory and Quantum Mechanics* (McGraw-Hill Book Company, New York, 1964)
3. B.R. Judd, *Operator Techniques in Atomic Spectroscopy* (McGraw-Hill Book Company, New York, 1963)
4. M. Rotenberg, R. Bivins, N. Metropolis, J.K. Wooton, *The 3-j and 6-j Symbols, Technology Press, MIT* (Mass, Cambridge, 1959)
5. D.J. Newman, N.G. Betty, *Crystal Field Handbook* (Cambridge, University Press, Cambridge, 2000)
6. C.W. Nielson, G.F. Koster, *Spectroscopic Coefficients for the p^n, d^n, and f^n Configuration, MIT Press* (Mass, Cambridge, 1963)
7. A.A. Kaminskii, *Crystalline Lasers: Physical Processes and Operating Schemes* (CRC Press, Boca Raton, 1996)

Chapter 3
Rare Earth Ions in Materials

3.1 Crystal Field on the Active Ions

The crystal field potential matrix element has been briefly mentioned in the last chapter. We should study it in much more detail, because it is a central problem in the energy level calculation of solid-state laser materials.

First, it is necessary to have a general idea about the physical mechanisms of crystal field. In free ions, there are electron–nucleus Coulomb interaction, electron–electron Coulomb interaction, orbital–spin interaction, and electron–electromagnetic field interaction. Besides, the electrons in the active ions in a medium have complex interactions with the lattice ions in the environment that can be described by crystal fields. It includes at least the following interaction mechanisms [1, 2]: (1) Overlap and exchange interaction of electric charge distribution; (2) Covalence interaction; (3) Point charge Coulomb interaction; (4) Dispersing charge Coulomb interaction; (5) Dipole polarization interaction; (6) Quadrupolar polarization interaction; (7) Charge penetration interaction. In the case of rare earth ions, some authors claim that the first two interactions make dominant contribution to the crystal field [2]. Nevertheless, if the molecular dynamics simulation (MDS) method is used to determine the actual position of ligands, the calculation results of energy levels and electric-dipole transition intensities are satisfactory even only the point charge contribution is taken into account [3].

Thus the Hamiltonian for the electrons of active ions in crystals can be written as

$$H = H_{\mathrm{en}} + H_{\mathrm{ee}} + H_{\mathrm{so}} + H_{\mathrm{c}} + H_{\mathrm{ext}} \tag{3.1}$$

where H_{en} is the electron–nucleus interaction Hamiltonian, H_{so} denotes the spin–orbit interaction, H_{c} represents the crystal field interaction Hamiltonian, and H_{ext} is the interaction Hamiltonian with external field. The last interaction will not be discussed in this book and the first three interactions have been dealt with in Chap. 1. The main objective of this chapter is to discuss the crystal field interaction

Z. Luo and Y. Huang, *Physics of Solid-State Laser Materials*, Springer Series in Materials Science 289, https://doi.org/10.1007/978-981-32-9668-8_3

represented by H_c. The property of the system, especially the energy level structure, depends on the magnitude of the perturbation H_c. If H_c is smaller than $H_{en} + H_{ee}$ as well as H_{so}, then the energy levels are similar to those of the free ions. Thus we can use eigenfunctions of the free ions as zero-order wave functions to find the crystal field energy level splitting of the state $^{2S+1}L_J$. If the magnitude of H_c is the same as or even larger than that of H_{so}, then one should consider the interaction of H_c before that of the H_{so}. The characteristics of the energy level are reflected by the symmetry of the crystal field and its point group irreducible representation. We now show the orders of magnitude of various interactions for transition metal ions and rare earth ions (in unit of wave number cm^{-1}).

	H_{en}	H_{ee}	H_{so}	H_c
Iron group	10^5	10^4	10^2	10^4
Rare earth	10^5	10^4	10^3	10^2

In order to solve the problem of the energy level splitting for the active ions in the materials, besides to treat crystal field interaction as a perturbation, the following two approximations should be assumed:

(1) The interactions between the electrons of neighbor active ions can be neglected;
(2) The interactions between the electrons of the same active ions can be neglected.

The first approximation is correct for the active ions in the insulated laser materials. Solid laser materials (excluding semiconductors) doped with a small amount of rare earth or transition ions as the luminescent centers are in this category. The second approximation, that is, single-electron approximation, has been proved to provide good enough calculation results for energy levels. Certainly, there are a series of published papers on the correlative crystal field [4–6]. The readers interested in the effects of the correlative crystal field can refer to those papers.
Indeed, the effects of lattice vibration on the energy levels cannot be neglected. However, we will discuss the effect of the crystal field in a static lattice first and then deal with the effects of lattice vibration.

On the other hand, for a general purpose, we will not calculate the crystal field splitting of different mechanisms separately, but use a phenomenological parameter method to summarize all the splitting effects caused by different mechanisms. Equation (2.48) is a general expression and valid not only for rare earth ions but also for transition-metal ions, because spherical harmonic functions $C_q^{(k)}$ are orthogonal basis functions in a completed space and any function can be expanded by these functions. The expanded coefficients B_{kq} are the so-called crystal field parameters in general crystal field theory. By using electrostatic model, one can obtain an expression of parameter B_{kq}. Classically, it is supposed that the charge distribution among lattice ions in the environment will not overlap with that of the electrons in the active ions and so can comply with Laplace equation. Thus the crystal field potential can be written as

$$H_c = -\sum_i \int d\tau \frac{e\rho(\boldsymbol{R})}{|\boldsymbol{R} - \boldsymbol{r}_i|}$$

where the summation is over the electrons at different position $\boldsymbol{r}_i$, and $\rho(\boldsymbol{R})$ is the external charge distribution density. The related potential should be integrated over all the crystal. On the other hand, $|\boldsymbol{R} - \boldsymbol{r}_i|^{-1}$ can be expanded as

$$|\boldsymbol{R} - \boldsymbol{r}_i|^{-1} = \sum_{k=0}^{\infty} \frac{r_{\langle}^k}{r_{\rangle}^{k+1}} P_k(\cos\omega_i)$$

where ω_i is the angle between the radius vector of external charge described by spherical coordinates $\boldsymbol{R}$, Θ, Φ and the radius vector of the electron described by spherical coordinates $\boldsymbol{r}_i$, θ_i, ϕ_i in the active ion. $r_<$ and $r_>$ is the shorter and longer between $\boldsymbol{R}$ and $\boldsymbol{r}_i$, respectively. On the other hand,

$$\begin{aligned} P_k(\cos\omega_i) &= \sum_{q=-k}^{k} (-1)^q \frac{4\pi}{2k+1} Y_{k,-q}(\Theta, \Phi) Y_{kq}(\theta_i, \varphi_i) k \\ &= \sum_{q=-k}^{k} (-1)^q C_{-q}^{(k)}(\Theta, \Phi) C_q^{(k)}(\theta_i, \varphi_i) \end{aligned}$$

If H_c is written as

$$H_c = \sum_{kq} B_{kq} \sum_i C_q^{(k)}(\theta_i, \phi_i) \tag{3.2}$$

then

$$B_q^k = -e \int (-1)^q \rho(\boldsymbol{R}) C_{-q}^k(\Theta, \Phi) \frac{r_{\langle}^k}{r_{\rangle}^{k+1}} d\tau \tag{3.3}$$

Neglecting the electron overlapping effect as well as the exchanging and covalent ones between electrons in adjacent ions, one always has $|\boldsymbol{R}|\rangle|\boldsymbol{r}_i|$, then

$$B_{kq} = -e \int (-1)^q \rho(\boldsymbol{R}) C_{-q}^{(k)}(\Theta, \Phi) \frac{r_i^k}{R^{k+1}} d\tau = \left[-e \int d\tau \frac{(-1)^q \rho(\boldsymbol{R}) C_{-q}^{(k)}(\Theta, \Phi)}{R^{k+1}} \right] r_i^k \tag{3.4}$$

In the calculation of energy level splitting caused by the crystal field, the crystal field Hamiltonian has been integrated over electronic wave functions so that the crystal field parameter can be expressed as

$$B_{kq} = -e \int d\tau \frac{(-1)^q \rho(\boldsymbol{R}) C_{-q}^{(k)}(\Theta, \Phi)}{R^{k+1}} \langle r^k \rangle \tag{3.5}$$

where

$$\langle r^k \rangle = \int_0^\infty dr R_{4f}^2(r) r^2 \tag{3.6}$$

In (3.6), R_{4f} is electronic radial wave function. It should be noted that both (3.2) and (2.49) are universally valid, but the detailed expression of B_{kq}, that is, (3.5) is only valid under a special assumption of electrostatic model.
There is a factor $\langle l \| C^{(k)} \| l \rangle$ in the matrix elements of crystal field potential. Therefore, according to (2.49) and the triangle condition of 3-j symbol or the triangle rule of the angular momentum coupling, for transition-metal ions $l = 2$, one need to consider only $0 \leq k \leq 4$, and for rare earth ions $l = 3$, then $0 \leq k \leq 6$ should be satisfied. For the crystal field splitting, the energy levels concerned all have the same parity, and k can only be even. The symmetry group of the active ions will also impose restrictions on the number of crystal field parameters.

The energy level structure has a big difference between the rare earth ions and the transition metal ions and so we would like to discuss only the former case in this chapter. The energy level structure of the transition metal ions will be discussed in Chap. 7.

3.2 Energy Level Splitting of the Rare Earth Ions

It was mentioned in the previous section that the crystal field action on the rare earth ions in materials could be seen as a perturbation on the free ions. Hence, in order to calculate the energy level splitting of the rare earth ions, one can use the wave functions (more accurately, it should be intermediate-coupling wave functions) of free ions as the zero-order approximate wave functions. According to the crystal field, wave functions are expanded as the bases of irreducible representation of rotation group or those of point group, and the method of crystal field can be divided into two classes. In the first class, that is the traditional crystal field method, the basis functions used are the bases of rotation group irreducible representation. This traditional crystal field method can be further divided into two kinds. The first kind uses the Hamiltonian expansion (3.2) or (2.49) and, by means of the calculation method introduced in the last chapter, we can calculate matrix elements between energy levels and then solve the secular equation. The second kind of traditional crystal field method is different from the first only in the bases used. They are the eigenfunctions of equivalent angular momentum operators and so this method is called equivalent operator method. If H_c is expressed as

$$H_c = \sum_{km} A_{km} \sum_i P_{km}(r_i) \tag{3.7}$$

The following replacement is usually carried out

$$\sum_i P_{2m}(r_i) \rightarrow \alpha_J \langle r^2 \rangle O_{2m}, \sum_i P_{4m}(r_i) \rightarrow \beta_J \langle r^4 \rangle O_{4m}, \sum_i P_{6m}(r_i) \rightarrow \gamma_J \langle r^6 \rangle O_{6m} \tag{3.8}$$

The equivalent operators O_{km} are obtained by replacing x, y, z in $P_{km}(r)$ (k = 2, 4, 6) by J_x, J_y, J_z. The matrix elements of the polynomial $P_{km}(r)$ and equivalent operator O_{km} can be found in Hufner's book (1978) [7]. In order to correctly calculate the equivalent operator factors α_J, β_J, γ_J, the intermediate-coupling approach should be considered [8, 9]. For most of the rare earth ions, these factors can also be found directly from tables [9]. This method has the advantage of direct and simple, but it cannot take into account of the mixing effect of free ion state's functions belonging to different J values (J mixing effect). If the crystal field is very strong, this effect can introduce a big error in the calculations of energy levels.

By using the above two methods, the number of crystal field parameters should be determined by the restrictions of the point symmetry at the site of active ions, that is, by the symmetry conditions and some of the parameters will be zero. In the following we will introduce the method for obtaining the crystal field parameters in different point symmetry conditions.

In the case of rare earth ions, as having been pointed out, $0 \le k \le 6$, (3.2) can be written as

$$H_c = \sum_{k=0}^{6} \sum_{q=0}^{k} \left(B_{kq} C_q^{(k)} + B_{k,-q} C_{-q}^{(k)} \right) \tag{3.9}$$

Note that writing in this form the terms with $q = 0$ have been repeatedly calculated and so the extra term should be eliminated. B_{kq} is usually a complex number and can be written as $B_{kq} = RB_{kq} + iIB_{kq}$, with real part RB_{kq} and imaginary part IB_{kq}. In order to ensure that the crystal field potential is a real number, the real and imaginary parts of the crystal field parameters should satisfy the following relations:

$$RB_{kq} = (-1)^q \, RB_{k-q}, \; IB_{k-q} = (-1)^{q+1} IB_{kq} \tag{3.10}$$

In the introduction of (3.10), the following relations have been used

$$C_q^{(k)} = \left(\frac{4\pi}{2k+1}\right) Y_{kq}(\theta,\varphi) = (-1)^q \left[\frac{(k-q)!}{(k+q)!}\right]^{1/2} r^k P_{kq}(\cos\theta) e^{iq\varphi}$$

It can be shown[1]

---

[1]The associated Legendre polynomial is defined as

$$P_{kq}(\cos\theta) \equiv P_{kq}(x) = \frac{(1-x^2)^{q/2}}{2^k k!} \frac{d^{k+q}}{dx^{k+q}} (1-x^2)^k \text{ where } x = \cos\theta$$

in which

$$\frac{d^{k+q}}{dx^{k+q}} (x^2-1)^k = \sum_{r=0}^{k+q} \binom{k+q}{r} \frac{d^r}{dx^r} (x-1)^k \frac{d^{k+q-r}}{dx^{k+q-r}} (x+1)^k$$

where $\binom{k+q}{r} \equiv \frac{(k+q)!}{r!(k+q-r)!}$

Because $\frac{d^{k+q-r}}{dx^{k+q-r}} (x+1)^k = 0$ when $r < q$, after differential we have

$$\frac{d^{k+q}}{dx^{k+q}} (x^2-1)^k = \sum_{r=q}^{k} \binom{k+q}{r} \frac{k!}{(k-r)!} (x-1)^{k-r} \frac{k!}{(r-q)!} (x+1)^{r-q}$$

r in the above expressions is a summing index, and one can make a replacement $r = r' + q$, then the expression changes into

$$\begin{aligned}
\frac{d^{k+q}}{dx^{k+q}} (x^2-1)^k &= \sum_{r'=0}^{k-q} \binom{k+q}{r'+q} \frac{k!}{(k-q-r')!} (x-1)^{k-q-r'} \frac{k!}{r'!} (x+1)^{r'} \\
&= \frac{(k+q)!}{(k-q)!} (x^2-1)^{-q} \sum_{r'=0}^{k-q} \binom{k-q}{r'} \frac{k!}{(k-r')!} (x-1)^{k-r'} \frac{k!}{(r'+q)!} (x+1)^{r'+q} \\
&= (-1)^q \frac{(k+q)!}{(k-q)!} (x^2-1)^{-q} \frac{d^{k-q}}{dx^{k-q}} (x^2-1)^k
\end{aligned}$$

$$\begin{aligned}
P_{kq}(x) &= \frac{(1-x^2)^{q/2}}{2^k k!} \frac{d^{k+q}}{dx^{k+q}} (1-x^2)^k = \frac{(1-x^2)^{q/2}}{2^k k!} \times (-1)^q \frac{(k+q)!}{(k-q)!} (x^2-1)^{-q} \frac{d^{k-q}}{dx^{k-q}} (x^2-1)^k \\
&= (-1)^q \frac{(k+q)!}{(k-q)!} \frac{(1-x^2)^{-q/2}}{2^k k!} \frac{d^{k-q}}{dx^{k-q}} (1-x^2) = (-1)^q \frac{(k+q)!}{(k-q)!} P_{k-q}(x)
\end{aligned}$$

Therefore

$$P_{kq}(\cos\theta) = (-1)^q \frac{(k+q)!}{(k-q)!} P_{k-q}(\cos\theta)$$

$$P_{kq}(\cos\theta) = (-1)^{-q}\frac{(k+q)!}{(k-q)!}P_{k-q}(\cos\theta)$$

so that

$$\left(C_q^{(k)}\right)^* = (-1)^q C_{-q}^{(k)}$$

Therefore, we have the following expression

$$\begin{aligned} H_c &= \sum_{k=0}^{6}\sum_{q=0}^{k}\left\{RB_{kq}\left[C_q^{(k)} + (-1)^q C_{-q}^{(k)}\right] + iIB_{k,-q}\left[C_q^{(k)} + (-1)^{q+1}C_{-q}^{(k)}\right]\right\} \\ &\equiv \sum_{k=0}^{6}\sum_{q=0}^{k}\left(V_{kq} + \bar{V}_{kq}\right) \end{aligned} \tag{3.11}$$

We name the coefficient N_{kq} as

$$\begin{aligned} N_{kq} &= (-1)^{2q}\tfrac{(k+q)!}{(k-q)!} \\ V_{kq} &= N_{kq}^{-1/2} r^k P_{kq}(\cos\theta)\cos q\phi RB_{kq} \\ \bar{V}_{kq} &= N_{kq}^{-1/2} r^k P_{kq}(\cos\theta)\sin q\phi IB_{kq} \end{aligned}$$

From the above expression of $P_{kq}(\cos\theta)$, we have

$$P_{kq}[\cos(\pi-\theta)] = P_{kq}(-\cos\theta) = (-1)^{k+q}P_{kq}(\cos\theta)$$

and

$$e^{2\pi iq/n} = \left(e^{\pi i}\right)^{2q/n} = (-1)^{2q/n}$$

In this way we can introduce the restriction condition of crystal parameters under the symmetry operation as shown in Table 3.1 given by Sachs [10].

As two examples, we will investigate the cases of point groups D_3 and D_{3d}. The point group D_3 consists of three classes, that is, E, C_3, and U_2 (C_2 axis which is perpendicular to the principal axis). We first study the real crystal field components. It can be seen from Table 3.1 that the invariance requirements are: (I) $q = 3m$; (II) $k + 2q/3 + q = 2m$. By requirement (I), it is necessary that q is an integer multiple of 3 and can only be 0, 3, 6 and so on. For even k, in order to satisfy the requirement (II), $2q/3 + q$ should be even. It is easy to see that in this case q should be

Table 3.1 Invariance requirements for the crystal field components under typical point group operations ($m = 0, 1 \ldots$) [10]

	Symmetry operation						
	E	C_n	σ_h	σ_v	$\sigma_d(n)$[a]	$U_2(n)$[a]	S_n
$(\theta, \phi) \rightarrow$	(θ, ϕ)	$(\theta, \phi + 2\pi/n)$	$(\pi - \theta, \phi)$	$(\theta, -\phi)$	$(\theta, -\phi + \pi/n)$	$(\pi-\theta, -\phi + 2\pi/n)$	$(\pi - \theta, \phi + 2\pi/n)$
V_{kq} invariance requirement	V_{kq} None	$(-1)^{\frac{2q}{n}} V_{kq}$ $q = nm$	$(-1)^{k+q} V_{kq}$ $k + q = 2m$	V_{kq} None	$(-1)^{\frac{q}{n}} V_{kq}$ $q = 2nm$	$(-1)^{k+\frac{2q}{n}+q} V_{kq}$ $k + 2q/n + q = 2m$	$(-1)^{k+\frac{2q}{n}+q} V_{kq}$ $k + 2q/n + q = 2m$
$\bar{V}_{kq}$ invariance requirement	$\bar{V}_{kq}$ None	$(-1)^{\frac{2q}{n}} \bar{V}_{kq}$ $q = nm$	$(-1)^{k+q} \bar{V}_{kq}$ $k + q = 2m$	$-\bar{V}_{kq}$ $\bar{V}_{kq} = 0$	$(-1)^{\frac{q}{n}+1} \bar{V}_{kq}$ $q = n(2m - 1)$	$(-1)^{k+\frac{2q}{n}+1+q} \bar{V}_{kq}$ $k + 2q/n + q = 2m - 1$	$(-1)^{k+\frac{2q}{n}+q} \bar{V}_{kq}$ $k + 2q/n + q = 2m$

[a] n is the order of rotation and m is an integer, 0, 1, 2, …

an integer multiple of 6 and can be written as $q = 6m$, which satisfy the requirement (I) as well; for odd k, in order to satisfy the requirement (II), $2q/3 + q$ should be odd and can be 3, 9, 15, …, which also satisfy requirement (I) and can be written as $q = 6m - 3$.

Secondly, for the imaginary part of crystal field parameters the invariance requirements are: (I) $q = 3m$; (II) $k + 2q/3 + q + 1 = 2m$. It is easy to find the conditions for the non-zero crystal field components: for even k, $q = 6m - 3$, for odd k, $q = 6m$.

In this chapter, we will only concern with energy level splitting and discuss only the effects of even components of crystal field. For the rare earth ions, as has been pointed out, even k can only be 2, 4, and 6. Therefore, in case of D_3 point group symmetry, the real crystal field components will be V_{k0} ($k = 2, 4, 6$) and V_{66} corresponding to crystal field parameters B_{k0} ($k = 2, 4, 6$) and B_{66}. On the other hand, the imaginary components of crystal field parameters are: IB_{43} and IB_{63}. In the following table it is simply written as B_{43} and B_{63}.

The point group D_{3d} is the direct product of D_3 and $C_i \equiv S_2$. The crystal field invariance requirements for D_{3d} are those for D_3 plus that for S_2, which is $k + 2q = 2m$. It practically requires that k should be an even number. Therefore, as far as the even crystal field is concerned, the crystal field parameters are the same for the cases D_{3d} and D_3. The even crystal field parameters for 32 point groups are shown in Table 3.2.

Table 3.2 Even crystal field parameters for 32 point groups

Point group	Crystal field parameters	Point group	Crystal field parameters
C_1, C_i	All B_{kq} except B_{k0} ($k = 2, 4, 6$) are real, others are complex	C_3, $S_6(C_{3i})$	B_{k0} ($k = 2, 4, 6$) are real, B_{43}, B_{63}, and B_{66} are complex
C_2, $C_s(C_{1h})$, C_{2h}	B_{k0} ($k = 2, 4, 6$) are real, B_{22}, B_{4q}($q = 2, 4$), B_{6q} ($q = 2, 4, 6$) are complex	D_2, C_{2v}, D_{2h}	B_{k0} ($k = 2, 4, 6$), B_{22}, B_{4q} ($q = 2, 4$) and B_{6q} ($q = 2, 4, 6$) are all real
C_4, S_4 C_{4h}	B_{k0} ($k = 2, 4, 6$) are real, B_{44} and B_{64} are complex	C_6, C_{3h} C_{6h}	B_{k0} ($k = 2, 4, 6$) are real, B_{66} is complex
D_4, C_{4v} D_{2d}, D_{4h}	B_{k0} ($k = 2, 4, 6$), B_{44}, B_{64} are all real	C_{6v}, D_{3h}, D_{6h} D_6	B_{k0} ($k = 2, 4, 6$), B_{66} are all real
T, O T_d, O_h, T_h	B_{k0} ($k = 4, 6$), B_{44}, and B_{64} are all real, and only two independent parameters. In this case: $B_{44} = \frac{5}{\sqrt{70}} B_{40}$, $B_{64} = -\sqrt{\frac{2}{7}} B_{60}$	D_3 D_{3d}	B_{k0} ($k = 2, 4, 6$) and B_{66} are all real, B_{43}, B_{63} are imaginary
		C_{3v}	B_{k0} ($k = 2, 4, 6$), B_{43}, B_{63}, B_{66} are all real

Note All the complex numbers have both imaginary and real parts, and they include two parameters

Table 3.3 Ratio of crystal field parameters used in the two traditional crystal field methods

k	Q	$B_{kq}/A_{kq}\langle r^k\rangle$
2	0	2
2	2	$\sqrt{6}/3$
4	0	8
4	2	$2\sqrt{10}/5$
4	3	$-2\sqrt{35}/35$
4	4	$4\sqrt{70}/35$
6	0	16
6	2	$16\sqrt{105}/105$
6	3	$-8\sqrt{105}/105$
6	4	$8\sqrt{14}/21$
6	6	$16\sqrt{231}/231$

The basic functions having the same k and q values in the two traditional crystal field methods are different by a constant factor and so do their crystal field parameters. The ratio between their crystal field parameters is listed in Table 3.3.

The energy level splitting of rare earth ions in a crystal field can be obtained by using the method introduced in the last chapter, and the results are listed in Table 3.4. It can be seen that the energy level of even number electronic system has no Kramers degeneracy. When the site symmetry of rare earth ions belongs to 8-point groups of triclinic, monoclinic, and rhombic crystal systems, the energy level of even number electronic system has no any degeneracy; but if their site symmetry is higher, the degeneracy of some energy levels is completely removed while the others remain twofold or threefold degeneracy, as shown in the following table. For the odd number electronic system, the crystal field energy level number are all equal to J + 1/2, and each crystal field energy level has only twofold Kramers degeneracy, except for the site symmetry of rare earth ion belongs to 5 cubic point groups. All the crystal field energy levels of rare earth ion site in the 5 cubic symmetric point group position still have twofold Kramers degeneracy while some of the crystal field energy levels have both the twofold crystal field degeneracy and twofold Kramers degeneracy, thus form the fourfold degeneracy energy levels. The values of x and y in Table 3.4 are obtained by the relations $J = x + 6\lambda$ and $J = y + 3\lambda$ (λ = 0, 1, 2, …). The square bracket in this table indicates that in case an integer number is obtained this integer should be taken, but if a non-integer number is obtained, then the next smaller integer number should be taken.

x	0 1 2 3 4 5	y	1/2 3/2 5/2
$\varphi_1(x)$	1 0 0 1 1 0	$\varphi(y)$	1 0 1
$\varphi_2(x)$	0 0 1 0 1 1		

Table 3.4 The relationship between the energy level splitting and the symmetry of crystal field [7]

Symmetry/Total angular momentum	J: 0, 1, 2, 3, 4, 5, ...	P
Triclinic C_1, C_i Monoclinic C_s, C_2, C_{2h} Rhombic C_{2v}, D_2, D_{2h}	1, 3, 5, 7, 9, 11, ..., $2J + 1$	1
Trigonal C_3, S_6, C_{3v}, D_3, D_{3d} Hexagonal C_{3h}, C_6, C_{6h}, D_{3h}, C_{6v}, D_6, D_{6h}	1, 1, 1, 3, 3, 3, ..., $2[J/3] + 1$ 0, 1, 2, 2, 3, 4, ..., $[(2J - 1)/3] + 1$	1 2
Tetragonal S_4, C_4, C_{4h}, D_{2d}, C_{4v}, D_4, D_{4h}	1, 1, 3, 3, 5, 5, ..., $2[J/2] + 1$ 0, 1, 1, 2, 2, 3, ..., $[(J + 1)/2]$	1 2
Cubic T, T_h, T_d, O, O_h	1, 0, 0, 1, 1, 0, ..., $\varphi_1(x) + [J/6]$ 0, 0, 1, 0, 1, 1, ..., $\varphi_2(x) + [J/6]$ 0, 1, 1, 2, 2, 3, ..., $[(J + 1)/2]$	1 2 3
Symmetry/Total angular momentum	J: 1/2, 3/2, 5/2, 7/2, 9/2, 11/2, ..., $2J + 1$	P
27 point group except cubic	1, 2, 3, 4, 5, 6, ..., $J + 1/2$	2
Cubic T, T_h, T_d, O, O_h	1, 0, 1, 2, 1, 2, ..., $\varphi(y) + [J/3]$ 0, 1, 1, 1, 2, 2, ..., $[(J + (3/2))/3]$	2 4

Note p is the degeneracy of the level and functions $\varphi(x_1)$, $\varphi(x_2)$, and $\varphi(y)$ are given as follows

The research work on the crystal field of rare earth ions has two approaches. The first is by calculating the crystal field parameters through fitting values of energy level splitting obtained from spectroscopy experiments to the eigenvalues of crystal field Hamiltonian. The second is by calculating the crystal field parameters using some kind of theoretical models. The second approach is not really a successful example, although many scientists have done a lot of work and proposed several kinds of theoretical models. This is because the crystal field effect is a complicated effect involving many mechanisms. However, the calculation results of the simple point charge model considering the shielding effect can be used as staring values for crystal field fitting. If the crystal field parameters of the point charge model are denoted by B'_{kq}, then the reasonable starting values for the crystal field fitting can be expressed as B_{kq} and $B_{kq} = (1 - \sigma_k)B'_{kq}$. The value of σ_2 is about 0.8, σ_4 and σ_6 are about 0.1 [11], and some recent calculation even shows that σ_2 approaches 0.9 [2].

The crystal field fitting is generally divided into the following steps: 1. Estimate initial values of the crystal field parameters; 2. Construct the energy matrix using estimated or referenced crystal field parameters; 3. The estimation eigenvalues of the energy level are obtained by diagonalization of the matrix; 4. Establish a one-to-one correspondence between the experimental and calculated level positions; 5. Keep the eigenvector of the energy matrix fixed and determine the crystal field parameter values to minimize the mean square deviation between the experiment and the calculation energy levels; 6. Substitute the crystal field parameters obtained in the fifth step to the second step, and repeat second to fifth step until the difference of calculated and experimental energy levels reaches the minimum.

By using fitting method, crystal field analyses have been carried out for many crystals doped with rare earth ions and there is a large number of data of energy

Table 3.5 Crystal field parameters of trivalent rare earth ions in yttrium aluminum garnet crystals [12]

R^{3+}	B_{20}	B_{21}	B_{40}	B_{42}	B_{44}	B_{60}	B_{62}	B_{64}	B_{66}
Nd	514	129	−363	−2005	−950	−1702	−763	837	−718
Nd	424	101	−191	−2108	−1082	−1784	−681	868	−641
Sm	502	95	800	−1542	−1003	−1386	−650	917	−542
Sm	599	−356	394	−1464	−212	−315	−481	684	−633
Eu	627	56	−344	−1611	−817	−1437	−608	600	−632
Tb	461	165	−169	−1720	−900	−1324	−621	599	−561
Dy	492	89	−60	−1743	−852	−1140	−434	557	−462
Dy	504	72	−71	−1737	−895	−1171	−454	576	−425
Dy	502	73	−77	−1771	−906	−1164	−446	552	−432
Er	424	82	−288	−1522	−899	−1122	−303	492	−322
Er	401	53	−107	−1591	−717	−1021	−349	538	−404

Table 3.6 Crystal field parameters of trivalent rare earth ions in yttrium lithium fluoride [12]

R^{3+}	B_{20}	B_{40}	B_{44}	B_{60}	RB_{64}	IB_{64}
Pr	489	−1043	1242	−42	1213	22.5
Pr	485	−1061	1296	−57.5	1186	0
Nd	441	−906	1115	−26.3	1073	20.6
Nd	480	−973	1119	−60.6	1051	49.0
Nd	502	−962	1105	−27.4	1019	35
Nd	401	−1088	1230	30	1074	0
Dy	330	−704	937	−70.4	609	92.7
Ho	410	−615	819	−27.9	677	32.8
Er	400	−692	925	−21.3	610	149
Er	314	−625	982	−32.4	584	171
Er	190	−1184	858	−44.8	295	0
Er	380	−640	975	−36.8	599	99.8
Tm	359	−608	844	−173	629	0
Tm	333	−648	867	−141	623	3
Yb	281	−556	569	−106	840	953

level and crystal field parameter in the literature. For example, in the Morrison and Leavitt's work [12], the energy level and crystal field parameter data for rare earth ions in 26 crystals are reported. Many data of a series of rare earth ions in different laser crystals can be found in Kaminskii's books [13, 14]. In Tables 3.5 and 3.6, data for two typical laser crystals are listed. It can be seen from these tables that different rare earth ions have nearly the same crystal field parameters.

It should be pointed out that the two methods mentioned above have a common disadvantage, that is, their crystal field Hamiltonian and related eigenfunctions are expanded in terms of basis functions of the rotation group and so the site symmetry properties of the electronic wave function cannot be directly determined. This increases the indefiniteness and the amount of work in identification of the spectral lines. There are two methods to remedy this limitation. One is to introduce crystal field quantum number which is related to irreducible representation of point group. By using the fixed relationship between the crystal field quantum number and the magnetic quantum number, the irreducible representation of the electronic wave function can be determined. The second way is to expand the crystal field Hamiltonian and related eigenfunctions in terms of the basis functions of point group. By the second method, the crystal field Hamiltonian and the eigenfunctions are designated by a series of irreducible representations of the group-chain with the symmetry reduced step by step. This method is hereby called group-chain scheme of crystal field analysis.

3.3 Crystal Field Quantum Number

In the study of crystal electronic energy level, Hellwege in 1948 [15] introduced crystal field quantum number according to the following fundamental expression:

$$M = \mu(\bmod q) \Rightarrow M = \mu + kq, \quad k = 0, \pm 1, \pm 2, \ldots$$

The value of q is the order of highest symmetry axis of international point group symbol or the crystal system, that is, triclinic $q = 1$; monoclinic and rhombic $q = 2$; trigonal $q = 3$; tetragonal $q = 4$; hexagonal $q = 6$, for example

$C_2(2)$, $C_s(\mathrm{m})$, $D_2(222)$, $C_{2v}(mm2)$, $D_{2h}(2/m)$—$q = 2$; $C_{3v}(3\mathrm{m})$, $D_3(32)$—$q = 3$; $D_4(422)$, $C_{4v}(4\mathrm{mm})$, $S_4(\bar{4})$, $D_{2d}(\bar{4}\mathrm{m}2)$—$q = 4$; $D_6(622)$, $C_{6v}(6\mathrm{mm})$, $D_{3h}(\bar{6}\mathrm{m}2)$, $C_{3h}(\bar{6})$—$q = 6$.

For an odd number electronic system, from the M composition of the wave function, the crystal field quantum number μ can be determined and then the point group irreducible representation can be found, because in this case one μ value corresponds to one irreducible representation. Therefore, by means of crystal field quantum number, one can judge the irreducible representation character of energy level after their wave function is calculated by usual crystal field analysis. However, for an even number electronic system, it happens for many point groups, only one crystal field quantum number μ cannot completely determine the irreducible representation of the wave function, that is, one μ value corresponds to two different irreducible representations and the wave function should be completely characterized by crystal field quantum number μ and v. According to the definition of Hellwege [15], the wave function transforms as $\psi(-x, y, -z) = e^{i\pi v}\psi(x, y, z)$ by the rotation around y axis through an angle π, on the other hand, the reflection in the plane xz equal to the rotation around y axis through an angle π plus the inversion through the symmetry center, that is, one has the relation $\psi(x, -y, z) = e^{i\pi v}I\psi(x, y, z)$, where I is the parity of the wave function. For the even electronic system $I = +1$ so that one also has $\psi(x, -y, z) = e^{i\pi v}\psi(x, y, z)$. Therefore, for C_{nv} point group and D_n point group, the same crystal field quantum numbers can be used to describe the relation between the wave function and the irreducible representation. In this case, $S = e^{i\pi v}$ in [15]. On the other hand, because $D_{2h} = D_2 \otimes C_i$, the relation between the quantum number of wave function and the irreducible representation for point group D_{2h} is similar to that of the point group D_2; the only difference is that the group D_{2h} has both even irreducible representation Γ_i^+ and odd irreducible representation Γ_i^-. The crystal field quantum number μ for an even number electronic system and the crystal field quantum number μ, v for an odd number electronic system with wave functions of $J_Z \equiv M$ are listed in the following to supplement the deficiency of many books in this field.

1. **C_2, C_s point groups $q = 2$.**
   See Tables 3.7 and 3.8.

Table 3.7 Correspondence of crystal field quantum numbers and irreducible representations with magnetic quantum number M of wave function and energy level number N for odd number electronic system

$\mu = \pm 1/2$ (Γ_3, Γ_4 conjugate forming a twofold degeneracy energy level)		
J	M	N
1/2	±1/2	1
3/2	±1/2, ∓3/2	2
5/2	±5/2, ±1/2, ∓3/2	3
7/2	±5/2, ±1/2, ∓3/2, ∓7/2	4
9/2	±9/2, ±5/2, ±1/2, ∓3/2, ∓7/2	5
11/2	±9/2, ±5/2, ±1/2, ∓3/2, ∓7/2, ∓11/2	6
13/2	±13/2, ±9/2, ±5/2, ±1/2, ∓3/2, ∓7/2, ∓11/2	7
15/2	±13/2, ±9/2, ±5/2, ±1/2, ∓3/2, ∓7/2, ∓11/2, ∓15/2	8

Table 3.8 Correspondence of crystal field quantum numbers and irreducible representations with magnetic quantum number M of wave function and energy level number N for even number electronic system

Γ_1—$\mu = 0$; Γ_2—$\mu = 1$							
J	μ	M	N	J	μ	M	N
0	0	0	1	5	0	−4, −2, 0, +2, +4	5
1	0	0	1		1	−5, −3, −1, +1, +3, +5	6
	1	−1, +1	2	6	0	−6, −4, −2, 0, +2, +4, +6	7
2	0	−2, 0, +2	3		1	−5, −3, −1, +1, +3, +5	6
	1	−1, +1	2	7	0	−6, −4, −2, 0, +2, +4, +6	7
3	0	−2, 0, +2	3		1	−7, −5, −3, −1, +1, +3, +5, +7	8
	1	−3, −1, +1, 3	4	8	0	−8, −6, −4, −2, 0, +2, +4, +6, +8	9
4	0	−4, −2, 0, +2, +4	5		1	−7, −5, −3, −1, +1, +3, +5, +7	8
	1	−3, −1, +1, 3	4				

2. **D_2, C_{2v} point groups q = 2.**
   See Tables 3.9 and 3.10.

Table 3.9 Correspondence of crystal field quantum numbers and irreducible representations with magnetic quantum number M of wave function and energy level number N for odd number electronic system

$\mu = \pm 1/2\ (\Gamma_5)$		
J	M	N
1/2	±1/2	1
3/2	±1/2, ∓3/2	2
5/2	±5/2, ±1/2, ∓ 3/2	3
7/2	±5/2, ±1/2, ∓3/2, ∓7/2	4
9/2	±9/2, ±5/2, ±1/2, ∓3/2, ∓7/2	5
11/2	±9/2, ±5/2, ±1/2, ∓3/2, ∓7/2, ∓11/2	6
13/2	±13/2, ±9/2, ±5/2, ±1/2, ∓3/2, ∓7/2, ∓11/2	7
15/2	±13/2, ±9/2, ±5/2, ±1/2, ∓3/2, ∓7/2, ∓11/2, ∓15/2	8

Table 3.10 Correspondence of crystal field quantum numbers and irreducible representations with magnetic quantum number M of wave function and energy level number N for even number electronic system

Γ_1—$\mu = 0$, $v = 0$; Γ_3—$\mu = 0$, $v = 1$; Γ_2—$\mu = 1$, $v = 0$; Γ_4—$\mu = 1$, $v = 1$									
J	μ	v	M	N	J	μ	v	M	N
0	0	0	0	1	5	0	0	−4, −2, +2, +4	2
1	0	1	0	1		0	1	−4, −2, 0, +2, +4	3
	1	0	−1, +1	1		1	0	−5, −3, −1, +1, +3, +5	3
	1	1	−1, +1	1		1	1	−5, −3, −1, +1, +3, +5	3
2	0	0	−2, 0, 2	2	6	0	0	−6, −4, −2, 0, +2, +4, +6	4
	0	1	−2, +2	1		0	1	−6, −4, −2, +2, +4, +6	3
	1	0	−1, +1	1		1	0	−5, −3, −1, +1, +3, +5	3
	1	1	−1, +1	1		1	1	−5, −3, −1, +1, +3, +5	3
3	0	0	−2, +2	1	7	0	0	−6, −4, −2, +2, +4, +6	3
	0	1	−2, 0, +2	2		0	1	−6, −4, −2, 0, +2, +4, +6	4
	1	0	−3, −1, +1, 3	2		1	0	−7, −5, −3, −1, +1, +3, +5, +7	4
	1	1	−3, −1, +1, +3	2		1	1	−7, −5, −3, −1, +1, +3, +5, +7	4
4	0	0	−4, −2, 0, +2, +4	3	8	0	0	−8, −6, −4, −2, 0, +2, +4, +6, +8	5
	0	1	−4, −2, +2, +4	2		0	1	−8, −6, −4, −2, +2, +4, +6, +8	4
	1	0	−3, −1, +1, +3	2		1	0	−7, −5, −3, −1, +1, +3, +5, +7	4
	1	1	−3, −1, +1, +3	2		1	1	−7, −5, −3, −1, +1, +3, +5, +7	4

In Table 3.10 there are some cases where two different μ, ν sets (corresponding to two wave functions with two different irreducible representations) have the same M component set. For example, if $J = 2$, two wave functions constituted by the same $M = -1$ and $M = +1$ component set corresponds to two different irreducible representations $\Gamma_2(\mu = 1, \nu = 0)$ and $\Gamma_4(\mu = 1, \nu = 1)$. It means that this component set forms two linear combinations of $\varphi(2, -1)$ and $\varphi(2, 1)$ corresponding to two irreducible representations Γ_2 and Γ_4. One can find out their irreducible representation according to their transformation character. For the D_2 point group, the wave function $\varphi(J, M)$ transformed by U_2 is $U_2\varphi(J, M) = (-1)^{J+M}\varphi(J, -M)$ (because $U_2 Y_{MM}(\theta, \varphi) = Y_{JM}(\pi - \theta, \pi - \varphi)$). In this example one has

$$\psi_1 = \frac{1}{\sqrt{2}}\varphi(2, 1) + \frac{1}{\sqrt{2}}\varphi(2, -1), \psi_2 = \frac{1}{\sqrt{2}}\varphi(2, 1) - \frac{1}{\sqrt{2}}\varphi(2, -1)$$

In this case $U_2\varphi(2, 1) = -\varphi(2, -1)$, $U_2\varphi(2, -1) = -\varphi(2, 1)$; therefore, $U_2\psi_1 = -\psi_1$ and $U_2\psi_2 = \psi_2$. By the character table D_2 point group in Appendix B, it can be decided that ψ_1 has irreducible representation Γ_4, that is $\mu = 1$, $\nu = 1$, on the other hand, ψ_2 has irreducible representation Γ_2, that is $\mu = 1$, $\nu = 0$.

3. **C_{3v}, D_3 point groups q = 3.**
   See Tables 3.11 and 3.12.

Table 3.11 Correspondence of crystal field quantum numbers and irreducible representations with magnetic quantum number M of wave function and energy level number N for odd number electronic system

J	$\mu = \pm 1/2$ (Γ_4) M	$\mu = \pm 3/2$ (Γ_5,Γ_6) M	N
1/2	±1/2		1
3/2	±1/2	±3/2	2
5/2	∓5/2, ±1/2	±3/2	3
7/2	∓5/2, ±1/2, ±7/2	±3/2	4
9/2	∓5/2, ±1/2, ±7/2	±3/2, ±9/2	5
11/2	∓11/2, ∓5/2, ±1/2, ±7/2	±3/2, ±9/2	6
13/2	∓11/2, ∓5/2, ±1/2, ±7/2, ±13/2	±3/2, ±9/2	7
15/2	∓11/2, ∓5/2, ±1/2, ±7/2, ±13/2	±3/2, ±9/2, ±15/2	8

Table 3.12 Correspondence of crystal field quantum numbers and irreducible representations with magnetic quantum number M of wave function and energy level number N for even number electronic system

Γ_1—$\mu = 0$, $\nu = 0$; Γ_2—$\mu = 0$, $\nu = 1$; Γ_3—$\mu = \pm 1$

J	μ	ν	M	N	J	μ	ν	M	N
0	0	0	0	1	5	0	0	−3, +3	1
1	0	1	0	1		0	1	−3, 0, +3	2
	±1		±1	1		±1		∓5, ∓2, ±1, ±4	4
2	0	0	0	1	6	0	0	−6, −3, 0, +3, +6	3
	±1		∓2, ±1	2		0	1	−6, −3, +3, +6	2
3	0	0	−3, +3	1		±1		∓5, ∓2, ±1, ±4	4
	0	1	−3, 0, +3	2	7	0	0	−6, −3, +3, +6	2
	±1		∓2, ±1	2		0	1	−6, −3, 0, +3, +6	3
4	0	0	−3, 0, +3	2		±1		∓5, ∓2, ±1, ±4, ±7	5
		1	−3, +3	1	8	0	0	−6, −3, 0, +3+6	3
	±1		∓2, ±1, ±4	3		0	1	−6, −3, +3+6	2
						±1		∓8, ∓5, ∓2, ±1, ±4, ±7	6

4. **D_4, C_{4v}, D_{2d}, S_4 point groups $q = 4$.**
   See Tables 3.13 and 3.14.

Table 3.13 Correspondence of crystal field quantum numbers and irreducible representations with magnetic quantum number M of wave function and energy level number N for odd number electronic system

J	$\mu = \pm 1/2$ (D_4, C_{4v}, $D_{2d} - \Gamma_6$; $S_4 - \Gamma_5$, Γ_6)) M	$\mu = \pm 3/2$ (D_4, C_{4v}, $D_{2d} - \Gamma_7$; S_4–Γ_7, Γ_8) M	N
1/2	±1/2		1
3/2	±1/2	±3/2	2
5/2	±1/2	±3/2, ∓5/2	3
7/2	±1/2, ∓7/2	±3/2, ∓5/2	4
9/2	±9/2, ±1/2, ∓7/2	±3/2, ∓5/2	5
11/2	±9/2, ±1/2, ∓7/2	±11/2, ±3/2, ∓5/2	6
13/2	±9/2, ±1/2, ∓7/2	±11/2, ±3/2, ∓5/2, ∓13/2	7
15/2	±9/2, ±1/2, ∓7/2, ∓15/2	±11/2, ±3/2, ∓5/2, ∓13/2	8

Table 3.14 Correspondence of crystal field quantum numbers and irreducible representations with magnetic quantum number M of wave function and energy level number N for even number electronic system

Γ_1—$\mu = 0$, $\nu = 0$; Γ_2—$\mu = 0$, $\nu = 1$; Γ_3—$\mu = 2$, $\nu = 0$; Γ_4—$\mu = 2$, $\nu = 1$; Γ_5—$\mu = \pm 1$
For point group S_4, only one crystal field quantum numberμ is enough. In this case $\nu = 0$ and $\nu = 1$ has no difference, Γ_1—$\mu = 0$; Γ_2—$\mu = 2$; Γ_3—$\mu = 1$; Γ_4—$\mu = -1$

J	μ	ν	M	N	J	μ	ν	M	N
0	0	0	0	1	5	0	0	−4, +4	1
1	0	1	0	1		0	1	−4, 0, +4	2
	±1		±1	1		±1		∓3, ±1, ±5	3
2	0	0	0	1		2	0	−2, +2	1
	±1		±1	1		2	1	−2, +2	1
	2	0	−2, +2	1	6	0	0	−4, 0, +4	2
	2	1	−2, +2	1		0	1	−4, +4	1
3	0	1	0	1		±1		∓3, ±1, ±5	3
	±1		∓3, ±1	2		2	0	−6, −2, +2, +6	2
	2	0	−2, +2	1		2	1	−6, −2, +2, +6	2
	2	1	−2, +2	1	7	0	0	−4, +4	1
						0	1	−4, 0, +4	2
4	0	0	−4, 0, +4	2		±1		∓7, ∓3, ±1, ±5	4
	0	1	−4, +4	1		2	0	−6, −2, +2, +6	2
	±1		∓3, ±1	2		2	1	−6, −2, +2, +6	2
	2	0	−2, +2	1	8	0	0	−8, −4, 0, +4, +8	3
	2	1	−2, +2	1		0	1	−8, −4, +4, +8	2
						±1		∓7, ∓3, ±1, ±5	4
						2	0	−6, −2, +2, +6	2
						2	1	−6, −2, +2, +6	2

Note The energy level number N in the above table is for point group D_4, C_{4v} and D_{2d}. Γ_1 energy level of D_{2d} corresponds to Γ_1 energy level of S_4, Γ_5 energy level of D_{2d} split into Γ_3 and Γ_4 energy levels of S_4. Γ_3 and Γ_4 energy levels of D_{2d} correspond to Γ_2 of S_4. The determination of the irreducible representation for the wave function having same M component set can be done by transformation property

5. **D_6, C_{6v}, D_{3h}, C_3 point groups $q = 6$.**
   See Tables 3.15 and 3.16.

Table 3.15 Correspondence of crystal field quantum numbers and irreducible representations with magnetic quantum number M of wave function and energy level number N for odd number electronic system

J	$\mu = \pm 1/2$ (D_6, C_{6v}, $D_{3h} - \Gamma_7$; $C_{3h} - \Gamma_7, \Gamma_8$) M	$\mu = \pm 3/2$ (D_6, C_{6v}, $D_{3h} - \Gamma_9$; $C_{3h} - \Gamma_{11}, \Gamma_{12}$) M	$\mu = \pm 5/2$ (D_6, C_{6v}, $D_{3h} - \Gamma_8$; $C_{3h} - \Gamma_9, \Gamma_{10}$) M	N
1/2	±1/2			1
3/2	±1/2	±3/2		2
5/2	±1/2	±3/2	±5/2	3
7/2	±1/2	±3/2	±5/2, ∓7/2	4
9/2	±1/2	±3/2, ∓9/2	±5/2, ∓7/2	5

(continued)

Table 3.15 (continued)

J	$\mu = \pm 1/2$ (D_6, C_{6v}, $D_{3h} - \Gamma_7$; $C_{3h} - \Gamma_7, \Gamma_8$) M	$\mu = \pm 3/2$ (D_6, C_{6v}, $D_{3h} - \Gamma_9$; $C_{3h} - \Gamma_{11}, \Gamma_{12}$) M	$\mu = \pm 5/2$ (D_6, C_{6v}, $D_{3h} - \Gamma_8$; $C_{3h} - \Gamma_9, \Gamma_{10}$) M	N
11/2	±1/2, ∓11/2	±3/2, ∓9/2	±5/2, ∓7/2	6
13/2	±1/2, ∓11/2, ±13/2	±3/2, ∓9/2	±5/2, ∓7/2	7
15/2	±1/2, ∓11/2±13/2	±15/2, ±3/2, ∓9/2	±5/2, ∓7/2	8

Note In literature [6] the μ value of first column should be exchanged with that of its second column. In literature [16], $\mu = \pm 3/2$ and $\mu = \pm 5/2$ were mistakenly corresponded to Γ_8 and Γ_9 of D_{3h}, respectively

Table 3.16 Correspondence of crystal field quantum numbers and irreducible representations with magnetic quantum number M of wave function and energy level number N for even number electronic system

D_6, C_{6v}, D_{3h}: $\Gamma_1 \to \mu = 0, \nu = 0$; $\Gamma_2 \to \mu = 0, \nu = 1$; $\Gamma_3 \to \mu = 3, \nu = 1$; $\Gamma_4 \to \mu = 3, \nu = 0$
$\Gamma_5 \to \mu = \pm 1$ twofold degeneracy level; $\Gamma_6 \to \mu = \pm 2$ twofold degeneracy level
C_{3h}; $\Gamma_1 \to \mu = 0$, ν is not necessary; Γ_5 and $\Gamma_6 \to \mu = \pm 1$ are two conjugate levels, Γ_3 and $\Gamma_2 \to \mu = \pm 2$ are also two conjugate levels; $\Gamma_4 \to \mu = 3$, ν is not necessary

J	μ	ν	M	N	J	μ	ν	M	N
0	0	0	0	1	5	0	1	0	1
1	0	1	0	1		±1		∓5, ±1	2
	±1		±1	1		±2		±2, ∓4	2
2	0	0	0	1		3	1	−3, +3	1
	±1		±1	1		3	0	−3, +3	1
	±2		±2	1	6	0	0	−6, 0, +6	2
3	0	1	0	1		0	1	−6, +6	1
	±1		±1	1		±1		∓5, ±1	2
	±2		±2	1		±2		∓4, ±2	2
	3	1	−3, +3	1		3	1	−3, +3	1
	3	0	−3, +3	1		3	0	−3, +3	1
4	0	0	0	1	7	0	0	−6, +6	1
	±1		±1	1		0	1	+6, 0, −6	2
	±2		∓4, ±2	2		±1		∓5, ±1, ±7	3
	3	1	−3, +3	1		±2		∓4, ±2	2
	3	0	−3, +3	1		3	1	−3, +3	1
						3	0	−3, +3	1
					8	0	0	−6, 0, +6	2
						0	1	+6−6	1
						±1		∓5, ±1, ±7	3
						±2		∓4, ±2, ±8	3
						3	1	−3, +3	1
						3	0	−3, +3	1

The determination of the irreducible representation for the wave functions having same M component sets corresponding to different μ and ν can be made by the transformation properties under the symmetry operation

The crystal field quantum number and the Butler symbol [17] of irreducible representation have a corresponding relationship. For the example of point group C_{3v} and D_3, in the case of an even electronic system, μ can be 0 and ±1. The energy levels with crystal field quantum number $\mu = 0$, $v = 0$ and 1 belong to representations Γ_1 and Γ_2. They are two one-dimensional irreducible representations corresponding to two single energy levels and both have the same quantum number $\mu = 0$ but different v. Butler used "0" and '$\tilde{0}$' to discriminate them. The energy level of $\mu = \pm1$ belongs to representation Γ_3 and is denoted as "1" by Butler to indicate its twofold degeneracy property. In the case of an odd electronic system, μ can only be $\pm1/2$ and $\pm1/3$. The energy levels of $\mu = \pm1/2$ belong to representation Γ_4. A single number "1/2" was used by Butler to not only correspond crystal field quantum number but also indicate its twofold degeneracy property. On the other hand, for the energy levels Γ_5 and Γ_6 of $\mu = \pm3/2$, Butler used "3/2" and "−3/2" to indicate that these original two single energy levels are accidental degeneracy to become a twofold degeneracy level.

By the typical crystal field analysis method, the introduction of crystal field quantum number has two advantages: First, by means of crystal field quantum number, the order of secular equation can be reduced in the solution of energy level and wave function, because different crystal field quantum numbers correspond to different irreducible representations between which the matrix elements of crystal field Hamiltonian are zero. For example, if the site symmetry is D_3 then to solve the secular equation for energy level with $J = 11/2$, by means of crystal field quantum number in Table 3.11, it can be divided into one 4-order equation with $M = \mp11/2$, $\mp5/2$, $\pm1/2$, $\pm7/2$ and two 2-order equation with $M = \pm3/2$, $\pm9/2$, while their degeneracy is clear. Secondly, by means of crystal field quantum number, the irreducible representation character of wave functions solved in the crystal field analysis can be determined and so it is easy to obtain the information data of polarization properties of the emission and absorption spectra between different energy levels. Here is an example of $NdAl_3(BO_3)_4$ (NAB) crystal. By the typical crystal field analysis method, the wave functions of two crystal field energy levels of multiplet $^4F_{3/2}$ and five crystal field energy levels of multiplet $^4I_{9/2}$ are as follows [18] (all the energy levels are twofold degeneracy):

$$
\begin{aligned}
{}^4\mathrm{F}_{3/2}(1) - \psi_1 &= |1/2\rangle \\
\psi_2 &= -|1/2\rangle \\
{}^4\mathrm{F}_{3/2}(2) - \psi_1 &= 0.7071i|3/2\rangle - 0.7071|-3/2\rangle \\
\psi_2 &= 0.7071|3/2\rangle - 0.7071i|-3/2\rangle \\
{}^4\mathrm{I}_{9/2}(1) - \psi_1 &= -0.6578|7/2\rangle - 0.5706|1/2\rangle - 0.4916|-5/2\rangle \\
\psi_2 &= 0.6578|-7/2\rangle - 0.5706|-1/2\rangle + 0.4916|5/2\rangle \\
{}^4\mathrm{I}_{9/2}(2) - \psi_1 &= (-0.1575 - 0.5796i)|9/2\rangle + (-0.1575i + 0.5796)|-9|2\rangle \\
&\quad + (-0.3576 - 0.1070i)|3/2\rangle + (+0.3576i - 0.1070)|-3/2\rangle \\
\psi_2 &= (-0.1575 - 0.5796i)|-9/2\rangle + (-0.1575i + 0.5796)|9|2\rangle \\
&\quad + (-0.3576 - 0.1070i) - 3/2\rangle + (+0.3576i - 0.1070)|3/2\rangle
\end{aligned}
$$

$$
\begin{aligned}
{}^4\mathrm{I}_{9/2}(3) - \psi_1 &= (0.3694 - 0.0536i)|9/2\rangle + (0.3694i + 0.0536)|-9/2\rangle \\
&\quad + (-0.2404 - 0.5504i)|3/2\rangle + (0.2404i + 0.5504)|-3/2\rangle \\
\psi_2 &= (0.3694 - 0.0536i)|-9/2\rangle + (0.3694i + 0.0536)|9/2\rangle \\
&\quad + (-0.2404 - 0.5504i) - 3/2\rangle + (0.2404i + 0.5504)|3/2\rangle \\
{}^4\mathrm{I}_{9/2}(4) - \psi_1 &= -0.9390|7/2\rangle - 0.1755|1/2\rangle - 0.2957|-5/2\rangle \\
\psi_2 &= -0.9390|-7/2\rangle - 0.1755|-1/2\rangle - 0.2957|5/2\rangle \\
{}^4\mathrm{I}_{9/2}(5) - \psi_1 &= 0.1746|7/2\rangle - 0.9842|1/2\rangle + 0.0295|-5/2\rangle \\
\psi_2 &= 0.1746|-7/2\rangle - 0.9842|-1/2\rangle + 0.0295|5/2\rangle
\end{aligned}
$$

Above are the linear compositions of wave functions with the same total angular momentum quantum number J but different magnetic quantum number M (quantum number J is omitted). From Table 3.11, it can be determined that the crystal field quantum number of ${}^4F_{3/2}(1)$ is $\mu = \pm 1/2$, corresponding to irreducible representation Γ_4; the crystal field quantum number of ${}^4F_{3/2}(2)$ is $\mu = \pm 3/2$, corresponding to irreducible representations Γ_5, Γ_6; on the other hand, crystal field quantum number of energy levels ${}^4I_{9/2}(1)$, ${}^4I_{9/2}(4)$, and ${}^4I_{9/2}(5)$ is $\mu = \pm 1/2$ corresponding to irreducible representation Γ_4, while the crystal field quantum number of ${}^4I_{9/2}(2)$ and ${}^4I_{9/2}(3)$ is $\mu = \pm 3/2$ corresponding to irreducible representations Γ_5, Γ_6. By the irreducible representation correlation of $SO_3 \rightarrow O \rightarrow D_3$ in Appendix B, it can be seen that multiplet $J = 3/2$ is decomposed into one energy level of irreducible representation Γ_4 and one energy level of irreducible representations Γ_5, Γ_6, while multiplet $J = 9/2$ decomposed into three energy levels of irreducible representation Γ_4 and two energy levels of irreducible representations Γ_5, Γ_6. This is the same as the result obtained from the wave functions of literature [19] by using crystal field quantum number method. (The precise point symmetry of Nd^{3+} ions in NAB crystal is C_2, but in a good enough approximation, it can be described by point group D_3 [19].)

In the NAB crystal, the wave functions and crystal field quantum numbers as well as their irreducible representations for two crystals field energy levels of $^4I_{11/2}(2)$ and $^4I_{11/2}(3)$ among six energy levels of multiplet $^4I_{11/2}$ are listed as follows [18]

$$\begin{aligned} {}^4I_{11/2}(2) - \psi_1 &= 0.8434|7/2\rangle + 0.0965|1/2\rangle - 0.4182|-5/2\rangle - 0.3234|-11/2\rangle \\ \psi_2 &= 0.8434|-7/2\rangle + 0.0965|-1/2\rangle - 0.4182|5/2\rangle - 0.3234|11/2\rangle \\ \mu &= \pm 1/2 - \Gamma_4 \end{aligned}$$

$$\begin{aligned} {}^4I_{11/2}(3) - \psi_1 &= (-0.0413 - 0.6617i)|9/2\rangle + (-0.6617 + 0.0413i)|-9/2\rangle \\ &\quad + (0.1651 - 0.1822i)|3/2\rangle + (0.1822 + 0.1651i)|-3/2\rangle \\ \psi_2 &= (-0.0413 - 0.6617i) - 9/2 > + (-0.6617 + 0.0413i)|9/2\rangle \\ &\quad + (0.1651 - 0.1822i)|-3/2\rangle + (0.1822 + 0.1651i)|3/2\rangle \\ &= \pm 3/2 - \Gamma_5, \Gamma_6 \end{aligned}$$

By using the relation between the crystal field quantum number and the magnetic quantum number M of wave function component, the following results can be obtained from the literature [18]: energy levels $^4I_{11/2}(1)$ and $^4I_{11/2}(4)$ both belong to irreducible representation Γ_4; energy level $^4I_{11/2}(5)$ belongs to irreducible representations Γ_5, Γ_6, and energy level $^4I_{11/2}(6)$ also belongs to Γ_4. In Chap. 4, this information will be used to determine the polarization of emission line.

It should be pointed out that the irreducible representations of above energy levels are mostly the same as those of Cascales [20] and Jaque [21] obtained by the analysis of NAB crystal spectral lines. The only difference is in energy levels $^4I_{9/2}(3)$, $^4I_{9/2}(4)$, $^4I_{11/2}(2)$, and $^4I_{11/2}(3)$. One possible reason for this inconsistency is that the D_3 point group symmetry of Nd^{3+} ions in NAB crystal is only an approximation. There is C_2 symmetry component in its crystal field [18]. In this situation, σ and π polarization transitions are all permitted. This will affect the results of energy level irreducible representations determined by the polarization character of spectral lines. It is for this reason, the irreducible representations of energy levels obtained by spectral data of Nd^{3+}:$YAl_3(BO_3)_4$ (NYAB) crystal, in which Nd^{3+} ions occupy precisely the D_3 symmetry positions [22] has some difference with that of NAB crystal [21].

3.4 Group Chain Scheme Method in Crystal Field Analysis

The main idea of this method is to use a series of irreducible representations of a group–subgroup chain to designate the crystal field wave functions and potential Hamiltonian. If the active ions occupy a G group symmetry position, then from SO_3

(or O_3) to G, the group chain will be SO_3 (or O_3) $\supset H_1 \supset H_2 \supset H_3 \supset G$. According to the difference in the group G, one can choose the first group as SO_3 or O_3. For example, if G is group D_3 then one should select SO_3 as the first group of the group chain, because group D_3 is a subgroup of group O and group O is a subgroup of group SO_3, but if G is group D_{2d} then one should select O_3 as the first group of the group chain, because group D_{2d} is a subgroup of group T_d, group T_d is a subgroup of group O_h, and group O_h is a subgroup of group O_3.

In the case of crystal Nd^{3+}:$YAl_3(BO_3)_4$, Nd^{3+} ions occupy D_3 symmetry positions then the group chain will be $SO_3 \supset O \supset D_3$. One can use $\psi(J, \mu, v, \lambda, \xi)$ to describe the energy levels of a multiplet with angular momentum J split by the crystal field, where μ denotes the irreducible representation (irresp.) of first subgroup H_1 in the first symmetry sub-chain $SO_3 \supset H_1$ and v denotes the irresp. of the subgroup H_2 in the second symmetry sub-chain $H_1 \supset H_2$ reduced from μ irreducible representation of H_1. λ and ξ have similar meanings. The crystal field potential Hamiltonian can be written as

$$H_c = \sum_{k,\mu,v,\lambda,\xi} V^k_{\mu,v,\lambda,\xi} = \sum_{k,\mu,v,\lambda,\xi} C^k_{\mu,v,\lambda,\xi} b^{(k)}_{\mu,v,\lambda,\xi} \tag{3.10}$$

where $C^k_{\mu,v,\lambda,\xi}$ denote the crystal field parameters in the group chain scheme and $b^{(k)}_{\mu,v,\lambda,\xi}$ are the basic vectors.

For rare earth ions, $k = 2, 4, 6$. All the $V^k_{\mu,v,\lambda,\xi}$ terms should belong to the identity irresp. of group G, because the crystal field potential must be invariant under the transformation of rare earth position point group. The method for writing down (3.10) for rare earth ions is to reduce $k = 2, 4$ and 6 along the group chain $SO_3 \supset H_1 \supset H_2 \supset H_3 \supset G$ and to find all the channels for reaching the identity representations of group G (decomposition of group O_3 to point group, the angular momentum quantum number divided into even parity J^+ and odd parity J^-, for the crystal field energy level calculations, only involve the even parity J^+. It should be pointed out that the group chain can have different length for different symmetry condition; the higher the site symmetry of the active ion, the shorter the length of group chain. There are a series of tables in Butler's book [17] which provide a powerful tool to carry out this work. A simple example is given here to show how to use the tables for the purpose of writing down crystal field potential Hamiltonian and their eigenfunctions. According to Butler [17], the notations of irreducible representations are different from those generally used. The irreducible representations of the wave functions, crystal field parameters, and coupling coefficient as well as those in Appendices 2, 3, and 5 will be denoted by the subscripts of related Beth's notations in this book. Here, for example, we show the comparison between them in groups O and D_3:

Group O	Mulliken	A_1	A_2	E	T_1	T_2	E'	E''	U'
	Bethe	Γ_1	Γ_2	Γ_3	Γ_4	Γ_5	Γ_6	Γ_7	Γ_8
	Butler	0	$\tilde{0}$	2	1	$\tilde{1}$	$\frac{1}{2}$	$\frac{\tilde{1}}{2}$	$\frac{3}{2}$
	This book	1	2	3	4	5	6	7	8

Group D_3	Mulliken	A_1	A_2	E	E'	E''	
	Bethe	Γ_1	Γ_2	Γ_3	Γ_4	Γ_5	Γ_6
	Butler	0	$\tilde{0}$	1	$\frac{1}{2}$	$\frac{3}{2}$	$-\frac{3}{2}$
	This book	1	2	3	4	5	6

Let's consider the group chain scheme analysis of crystal field energy levels of the ground multiplet $^4I_{9/2}$ of Nd^{3+} ions occupying the D_3 symmetry positions. The first step is to investigate the first section $SO_3 \supset O$ of the group chain. From the table of branching rule of $SO_3 \rightarrow O$ section (Chap. 12 in Butler's book [17] or Appendix B of this book), one can find that the representation of 9/2 of SO_3 group can be reduced into one $6(\Gamma_6)$ and two $8(\Gamma_8)$ representations. By using the branching rule for the $O \rightarrow D_3$ section or related irreducible representation reduction tables, it can be found that the $6(\Gamma_6)$ representation of group O corresponds to the $4(\Gamma_4)$ representation of group D_3. On the other hand, the $8(\Gamma_8)$ representation of group O can be reduced to $4(\Gamma_4) \oplus 5(\Gamma_5) \oplus 6(\Gamma_6)$ representations of group D_3. In this way, the resolution of the multiplet $^4I_{9/2}$ can be written as follows (note that in order to distinguish the difference between two $8(\Gamma_8)$ representations of group O, the subscripts 0 and 1 are used to write it as $8_0(\Gamma_{80})$ and $8_1(\Gamma_{81})$)

$$
\begin{aligned}
|^4I_{9/2}\rangle \rightarrow |^4I_{9/2}64\rangle \oplus |^4I_{9/2}8_04\rangle \oplus |^4I_{9/2}8_05\rangle \oplus |^4I_{9/2}8_06\rangle \oplus |^4I_{9/2}8_14\rangle \\
\oplus |^4I_{9/2}8_15\rangle \oplus |^4I_{9/2}8_16\rangle
\end{aligned}
\tag{3.11}
$$

That is, it splits into seven states. Among these states those corresponding to the representation 4 of group D_3 have twofold degeneracy and those corresponding to representations 5 and 6 of group D_3 have also twofold degeneracy, although each of them has only one-dimensional representations, but these two irreducible representations are conjugated and constitute a twofold degeneracy energy level. Therefore, the multiplet $^4I_{9/2}$ will be split into five sub-states by the crystal field with D_3 point group symmetry. The wave functions of other energy states can also be introduced in the same way. The results are given in the paper published by the authors [18]. Generally, the crystal field wave function can be expressed as

$$
\Psi = \sum_{a,a_1,a_2,a_3,a_4} C^a_{a_1a_2a_3a_4} |aa_1a_2a_3a_4\rangle \tag{3.12}
$$

Let's consider the expression for the crystal field Hamiltonian. It can be shown by the branching rule or by the related irresp. reduction tables that the irresp. for $k = 2$ in SO_3 group will be reduced to two irresps. $3 \oplus 5$ of group O, but irresp. for

$k = 4$ will be reduced to four irresps $1 \oplus 2 \oplus 3 \oplus 5$ of group O. On the other hand, the irresp. for $k = 6$ of SO_3 group will be reduced to six irresps. $1 \oplus 2 \oplus 3 \oplus 4 \oplus 5_0 \oplus 5_1$ of group O. It can be found that only irresp. 1 and 5 of group O can be reduced to the identity irresp. 1 of group D_3. The possible ways leading to identity irresp. 1 of group D_3 can be shown as follows

$$2 \to 5 \to 1, 4 \to 1, 4 \to 5 \to 1, 6 \to 1 \to 1, 6 \to 5_0 \to 1, 6 \to 5_1 \to 1$$

It must be noted that $k = 6$ has two representations 5 of group O, and so two subscripts 0 and 1 should be used to distinguish them. In this way, the crystal field Hamiltonian can be written as

$$H_c = C_5^2 b_5^{(2)} + C_1^4 b_1^{(4)} + C_5^4 b_5^{(4)} + C_1^6 b_1^{(6)} + C_{5_0}^6 b_{5_0}^{(6)} + C_{5_1}^6 b_{5_1}^{(6)} \tag{3.13}$$

where $b_\mu^{(k)}$ denotes the related unit vectors. In (3.13) the subscript 1 used to denote identity representation of D_3 has been omitted. On comparing with Table 3.2, it can be seen that the number of parameters in (3.13) is the same as that for the traditional crystal field theory, that is, there are six parameters. By observing the branching rule, it can be seen, for the even crystal field potential, that the even representations of groups O_h and T_d have one-to-one correspondence and $O_h \to D_{3d}$, $T_d \to C_{3v}$, and $O \to D_3$ all have the same table of branching rule. Therefore, one can write the six parameter expression of the crystal field potential for point group of D_{3d} and C_{3v} similar to that of D_3. The same method can be used to perform the analysis for other point groups. The crystal field parameters for the 32 point symmetry groups are listed in Table 3.17.

The numbers of above crystal field parameters in the group chain scheme are the same as those of Table 3.2 for traditional crystal field scheme. In the case of cubic system O, O_h, and T_d, among the four parameters of the traditional crystal field scheme, there are two relations shown previously, and so only two independent parameters, and the same is shown in Table 3.17.

Note that in the above table, owing to the fact that the crystal field potential with even parity is discussed, the crystal field parameters concerned only with even parity and so the superscript + used to denote even parity is omitted.

Having the expression for the crystal field Hamiltonian and its eigenfunctions, the following work is to calculate the matrix elements. In order to do this calculation, the Wigner–Eckart theorem within the group chain scheme should be used. A simple introduction of this theorem is given here. For a more detailed discussion, one can refer to the book published by Butler [17] and Piepho and Schatz [23]. The Wigner–Eckart theorem has been given in (2.43). Expression of this theorem used in group chain scheme analysis will be introduced in the following. Similar to (2.43c), the expression of Wigner–Eckart theorem for point group can be expressed as

Table 3.17 Crystal field parameters in the group chain scheme

Crystalline system	Group	Crystal field parameters	Subscript point group
Triclinic	C_1	$C^2_{3111}, C^2_{5431}, C^2_{3311}, C^2_{5522}, C^2_{5542}, C^4_{1111}, C^4_{4231}$ $C^4_{3111}, C^4_{3311}, C^4_{5431}, C^4_{5522}, C^4_{5542}, C^4_{4522}, C^4_{4542}$ $C^6_{1111}, C^6_{3111}, C^6_{3311}, C^6_{4231}, C^6_{5_0431}, C^6_{5_1431}, C^6_{2311}$ $C^6_{5_0522}, C^6_{5_1522}, C^6_{5_0542}, C^6_{5_1542}, C^6_{4522}, C^6_{4542}$	$O \to D_4 \to D_2 \to C_2$
	C_1	The same as above	$O_h \to D_{4h} \to D_{2h} \to C_{2h}$
Monoclinic	C_2	$C^2_{311}, C^2_{543}, C^2_{331}, C^4_{111}, C^4_{423}, C^4_{311}, C^4_{331}$ $C^4_{543}, C^6_{111}, C^6_{311}, C^6_{331}, C^6_{423}, C^6_{5_043}, C^6_{5_143}$ C^6_{231}	$O \to D_4 \to D_2$
	C_{2h}	The same as above	$O_h \to D_{4h} \to D_{2h}$
	C_s	$C^2_{3111}, C^2_{5431}, C^2_{3311}, C^4_{1111}, C^4_{4231}, C^4_{3111},$ $C^4_{3311}, C^4_{5431}, C^6_{1111}, C^6_{3111}, C^6_{3311}, C^6_{4231},$ $C^6_{5_0431}, C^6_{5_1431}, C^6_{2311}$	$O_h \to D_{4h} \to D_{2h} \to C_{2h}$
Orthorhombic	D_2	$C^2_{31}, C^2_{33}, C^4_{11}, C^4_{31}, C^4_{33}, C^6_{11}, C^6_{31}, C^6_{33}, C^6_{23}$	$O \to D_4$
	D_{2h}	The same as above	$O_h \to T_h$
	C_{2v}	$C^2_{331}, C^2_{333}, C^4_{111}, C^4_{331}, C^4_{333}, C^6_{111}, C^6_{331}, C^6_{333}, C^6_{223},$	$O_h \to T_d \to D_{2d}$
Triangular	D_3	$C^2_5, C^4_1, C^4_5, C^6_1, C^6_{5_0}, C^6_{5_1}$	O
	D_{3d}	The same as above	O_h
	C_{3v}	$C^2_{55}, C^4_{11}, C^4_{55}, C^6_{11}, C^6_{5_05}, C^6_{5_15}$	$O_h \to T_d$
	C_3	$C^2_{51}, C^4_{11}, C^4_{51}, C^4_{42} C^6_{11}, C^6_{5_01}, C^6_{5_11}, C^6_{21}, C^6_{42}$	$O \to D_3$
	C_{3i}	$C^2_{54}, C^4_{11}, C^4_{54}, C^4_{44} C^6_{11}, C^6_{5_04}, C^6_{5_14}, C^6_{21}, C^6_{44}$	$O_h \to T_h$
Tetragonal	D_4	$C^2_{33}, C^4_{11}, C^4_{33}, C^6_{11}, C^6_{33},$	O
	D_{4h}	$C^2_3, C^4_1, C^4_3, C^6_1, C^6_3,$	O_h
	D_{2d}	The same as above	$O_h \to T_d$
	C_{4v}	$C^2_{31}, C^4_{11}, C^4_{31}, C^6_{11}, C^6_{31},$	$O_h \to D_{4h}$
	C_4	$C^2_{31}, C^4_{11}, C^4_{31}, C^4_{42}, C^6_{11}, C^6_{31}, C^6_{42}$	$O \to D_4$
	C_{4h}	The same as above	$O_h \to D_{4h}$
	S_4	$C^2_{331}, C^4_{111}, C^4_{331}, C^4_{442}, C^6_{111}, C^6_{331}, C^6_{442}$	$O_h \to T_d \to D_{2d}$
Hexagonal	D_6	$C^2_1, C^4_1, C^6_1, C^6_6$	D_∞
	D_{6h}	The same as above	$D_{\infty h}$
	C_{6v}, D_{3h}	$C^2_{11}, C^4_{11}, C^6_{11}, C^6_{61}$	$D_{\infty h} \to D_{6h}$
	C_6	$C^2_{11}, C^4_{11}, C^6_{11}, C^6_{61}, C^6_{62}$	$D_\infty \to D_6$
	C_{6h}	The same as above	$D_{\infty h} \to D_{6h}$
	C_{3h}	$C^2_{111}, C^4_{111}, C^6_{111}, C^6_{611}, C^6_{622}$	$D_{\infty h} \to D_{6h} \to D_{3h}$
Cubic	T	C^4_1, C^6_1, C^6_2	O
	T_h	The same as above	O_h
	O	C^4_1, C^6_1	O
	T_d, O_h	The same as above	O_h

$$\left\langle a\alpha \middle| T_{\phi}^{(f)} \middle| b\beta \right\rangle = \begin{pmatrix} a \\ \alpha \end{pmatrix} \begin{pmatrix} a^{*} & f & b \\ \alpha^{*} & \phi & \beta \end{pmatrix} \left\langle a \middle\| T^{(f)} \middle\| b \right\rangle \tag{3.14}$$

For the cases in which the same irreducible representation appears r times, there are r different coefficients, we have

$$\left\langle a\alpha \middle| T_{\phi}^{(f)} \middle| b\beta \right\rangle = \sum_{r} \begin{pmatrix} a \\ \alpha \end{pmatrix} \begin{pmatrix} a^{*} & f & b \\ \alpha^{*} & \phi & \beta \end{pmatrix}^{r} \left\langle a \middle\| T^{(f)} \middle\| b \right\rangle \tag{3.15}$$

where $\begin{pmatrix} a & f & b \\ \alpha & \phi & \beta \end{pmatrix}$ refers to the $3jm$ coefficients corresponding to the $3j$ coefficients of the rotation group and $\begin{pmatrix} a \\ \alpha \end{pmatrix}$ called $2jm$ coefficient corresponding to the coefficient of $(-1)^{j-m} \equiv \begin{pmatrix} j \\ m \end{pmatrix}$ of the rotation group. It can be shown [17] that for the group chain $G \supset G_1 \supset G_2 \supset C_n$, the $3jm$ coefficients can be reduced as following product

$$\begin{pmatrix} a & b & c \\ a_1 & b_1 & c_1 \\ a_2 & b_2 & c_2 \\ a_3 & b_3 & c_3 \end{pmatrix} \begin{matrix} rG \\ G_1 \\ G_2 \\ C_n \end{matrix} = \begin{pmatrix} a & b & c \\ a_1 & b_1 & c_1 \end{pmatrix}_{G_1}^{rG} \begin{pmatrix} a_1 & b_1 & c_1 \\ a_2 & b_2 & c_2 \end{pmatrix}_{G_2}^{G_1} \begin{pmatrix} a_2 & b_2 & c_2 \\ a_3 & b_3 & c_3 \end{pmatrix}_{C_n}^{G_2} \tag{3.16}$$

where for the group chain $G_2 \supset C_n$, $3jm$ factor $\begin{pmatrix} a_2 & b_2 & c_2 \\ a_3 & b_3 & c_3 \end{pmatrix}_{C_n}^{G_2}$ also is the $3jm$ coefficients, because the Abelian groups C_n are all one dimension. Therefore, the $3jm$ coefficients for a group chain are equal to the product of $3jm$ factors of all the sections of the chain, no matter how long is the group chain. On the other hand, there is a factorization relation for the $2jm$ coefficients, and so one has

$$\begin{aligned} &\left\langle aa_1a_2a_3 \middle| T_{f_1f_2f_3}^{(k)} \middle| bb_1b_2b_3 \right\rangle \\ &= \begin{pmatrix} a \\ a_1 \end{pmatrix}_{G_1}^{G} \begin{pmatrix} a_1 \\ a_2 \end{pmatrix}_{G_2}^{G_1} \begin{pmatrix} a_2 \\ a_3 \end{pmatrix}_{G_3}^{G_2} \sum_{r} \begin{pmatrix} a^{*} & f & b \\ a_1^{*} & f_1 & b_1 \end{pmatrix}_{r_1G_1}^{rG} \begin{pmatrix} a_1^{*} & f_1 & b_1 \\ a_2^{*} & f_2 & b_2 \end{pmatrix}_{r_2G_2}^{r_1G_1} \begin{pmatrix} a_2^{*} & f_2 & b_2 \\ a_3^{*} & f_3 & b_3 \end{pmatrix}_{G_3}^{r_2G_2} \left\langle a \middle\| T^{(k)} \middle\| b \right\rangle_{r}^{G} \end{aligned} \tag{3.17}$$

Applying this theorem to the calculation of the matrix elements of the crystal field Hamiltonian, and substituting into (3.10), we can obtain

$$\begin{aligned}&\langle aa_1a_2a_3a_4\|H_c\|bb_1b_2b_3b_4\rangle\\&=\begin{pmatrix}a\\a_1\end{pmatrix}^{G}_{G_1}\begin{pmatrix}a_1\\a_2\end{pmatrix}^{G_1}_{G_2}\begin{pmatrix}a_2\\a_3\end{pmatrix}^{G_2}_{G_3}\begin{pmatrix}a_3\\a_4\end{pmatrix}^{G_3}\sum_{C_n\,krr_1r_2r_3}C^k_{\mu\nu\xi}\begin{pmatrix}a^* & k & b\\a_1^* & \mu & b_1\end{pmatrix}^{rG}_{r_1G_1}\\&\times\begin{pmatrix}a_1^* & \mu & b_1\\a_2^* & \nu & b_2\end{pmatrix}^{r_1G_1}_{r_2G_2}\begin{pmatrix}a_2^* & \nu & b_2\\a_3^* & \xi & b_3\end{pmatrix}^{r_2G_2}_{r_3G_3}\begin{pmatrix}a_3^* & \xi & b_3\\a_4^* & 0 & b_4\end{pmatrix}^{r_3G_3}\left\langle a\|b^{(k)}\|b\right\rangle\end{aligned}\tag{3.18}$$

where $\langle a\|b^k\|b\rangle$ is an irreducible matrix element of basis vector in the representation space of rotation group for a multi-electronic system. According to the previous description of the method of multi-electron unit vector (referred to the discussion of (2.48) to (2.49)), one has

$$\left\langle a\|b^{(k)}\|b\right\rangle=\left\langle f^nSLa\|U^{(k)}\|f^nSL'b\right\rangle\left\langle 3\|C^{(k)}\|3\right\rangle\tag{3.19}$$

All the matrix elements can be obtained by using (3.18) and (3.19) as well as the tables in the books of Butler [17] or Piepho and Schatz [23].

As an example, it can be used to the calculation of the energy levels of Nd^{3+} ions in crystal $YAl_3(BO_3)_4$, which have symmetry of point group D_3.

Let's consider the multiplet $^4F_{3/2}$. From the branching rule of $SO_3 \rightarrow O$ reduction, it can be seen that $J = 3/2$ corresponds to the irresp. 8 of group O. By the branching rule of $O \rightarrow D_3$ reduction, the irresp. 8 of group O corresponds to the irresps. $4 \oplus 5 \oplus 6$ of group D_3. Therefore, the multiplet $^4F_{3/2}$ will be reduced to three states with wave functions $|3/2, 8, 4\rangle$, $|3/2, 8, 5\rangle$, and $|3/2, 8, 6\rangle$. One can omit the common first label and write them as $|8, 4\rangle$, $|8, 5\rangle$, and $|8, 6\rangle$. Equation (3.18) becomes

$$\langle 3/2\,a_1a_2|H_c|3/2\,b_1b_2\rangle=\frac{1}{\sqrt{|a_2|}}\begin{pmatrix}3/2\\a_1\end{pmatrix}^{SO_3}_{O}\begin{pmatrix}a_1\\a_2\end{pmatrix}^{O}_{D_3}\sum_k\begin{pmatrix}3/2 & k & 3/2\\a_1^* & \mu & b_1\end{pmatrix}^{SO_3}_{O}\begin{pmatrix}a_1^* & \mu & b_1\\a_2^* & 0 & b_2\end{pmatrix}^{O}_{D_3}C^k_\mu\left\langle 3/2\|b^{(k)}\|3/2\right\rangle\tag{3.20}$$

The dimension factor $|a_2|$ introduced in (3.20) is due to the fact that the group D_3 includes irreducible representation higher than one dimension. Actually, if the calculation is to go down to the component of related irresps. (i.e. to multiply a $3jm$ coefficients or a $3jm$ factor of $D \rightarrow C_3$), the result will be the same.

It can be seen by the branching rule of $SO_3 \rightarrow O$ and $O \rightarrow D_3$ reductions that $2jm$ factors in (3.20) are all equal to +1. Obviously, the $3jm$ factor table for $SO_3 \rightarrow O$ reduction shows that only when $k = 2$ the first $3jm$ factor in (3.20) has non-zero value. On the other hand, by (3.13), one knows that the only term remain is that including the factor of $C_5^{(2)}$

$$\left\langle {}^3\!/\!_2 84 \middle| H_c \middle| {}^3\!/\!_2 84 \right\rangle = \frac{1}{\sqrt{2}} \begin{pmatrix} {}^3\!/\!_2 & 2 & {}^3\!/\!_2 \\ 8 & 5 & 8 \end{pmatrix}_O^{SO_3} \begin{pmatrix} 8 & 5 & 8 \\ 4 & 1 & 4 \end{pmatrix}_{D_3}^{O} C_5^2 \left\langle {}^3\!/\!_2 \| b^{(2)} \| {}^3\!/\!_2 \right\rangle \tag{3.21}$$

Note the irresps. Γ_4 of D_3 group is a two-dimensional representation.

In the tables of the $3jm$ factors for $SO_3 \to O$ reduction, one can see the following line (can be found in Appendix E or table of Butler [17]):

$$\begin{matrix} 2 & {}^3\!/\!_2 & {}^3\!/\!_2 \\ 5 & 8 & 8 \end{matrix} \; 0000 + \; -\sqrt{3}\big/\sqrt{5}$$

Behind the $3jm$ symbol, there are four "0" in the same line. The first number indicates the repeat number of the subgroup, and number 0 means there is no repetition of the subgroup. The number in the second, third, and fourth positions indicate the repeat number of the branches, and number 0 means there is no repetition of the branches. The symbol + indicates that the $3jm$ factor will not change sign under the exchange of columns while the symbol − indicates that the $3jm$ factor will change sign under the exchange of columns. Therefore

$$\begin{pmatrix} {}^3\!/\!_2 & 2 & {}^3\!/\!_2 \\ 8 & 5 & 8 \end{pmatrix}_O^{SO_3} = \begin{pmatrix} 2 & {}^3\!/\!_2 & {}^3\!/\!_2 \\ 5 & 8 & 8 \end{pmatrix}_O^{SO_3} = -\sqrt{3}\big/\sqrt{5}$$

Similarly, from the table of $3jm$ factors for $O \to D_3$ reduction, it can be found that

$$\begin{pmatrix} 8 & 5 & 8 \\ 4 & 1 & 4 \end{pmatrix}_{D_3}^{O} = \begin{pmatrix} 5 & 8 & 8 \\ 1 & 4 & 4 \end{pmatrix}_{D_3}^{O} = -1\big/\sqrt{2\times 3}$$

Substituting into (3.21), the following can be obtained

$$\left\langle {}^3\!/\!_2 84 \middle| H_c \middle| {}^3\!/\!_2 84 \right\rangle = \frac{1}{\sqrt{2}} \times \frac{1}{\sqrt{10}} C_5^2 \left\langle {}^3\!/\!_2 \| b^{(2)} \| {}^3\!/\!_2 \right\rangle = \frac{1}{2\sqrt{5}} \times C_2 U^{(2)} C_5^{(2)} \tag{3.22}$$

Similarly

$$\begin{aligned} \left\langle {}^3\!/\!_2 85 \middle| H_c \middle| {}^3\!/\!_2 85 \right\rangle &= \frac{1}{\sqrt{2}} \begin{pmatrix} {}^3\!/\!_2 & 2 & {}^3\!/\!_2 \\ 8 & 5 & 8 \end{pmatrix}_O^{SO_3} \begin{pmatrix} 8 & 5 & 8 \\ 6 & 1 & 5 \end{pmatrix}_{D_3}^{O} C_5^{(2)} \left\langle {}^3\!/\!_2 \| b^{(2)} \| {}^3\!/\!_2 \right\rangle \\ &= \left(-\frac{\sqrt{3}}{\sqrt{5}}\right) \times \left(\frac{1}{2\sqrt{3}}\right) C_2 U^{(2)} C_5^{(2)} = -\frac{1}{2\sqrt{5}} \times C_2 U^{(2)} C_5^{(2)} \end{aligned} \tag{3.23}$$

$$
\begin{aligned}
\left\langle {}^{3}\!/\!_{2}86\middle|H_c\middle|{}^{3}\!/\!_{2}86\right\rangle &= \frac{1}{\sqrt{2}}\begin{pmatrix} 3/2 & 2 & 3/2 \\ 8 & 5 & 8 \end{pmatrix}_{O}^{SO_3}\begin{pmatrix} 8 & 5 & 8 \\ 5 & 1 & 6 \end{pmatrix}_{D_3}^{O} C_5^{(2)}\left\langle {}^{3}\!/\!_{2}\middle\|b^{(2)}\middle\|{}^{3}\!/\!_{2}\right\rangle \\
&= \left(-\frac{\sqrt{3}}{\sqrt{5}}\right)\times\left(\frac{1}{2\sqrt{3}}\right)C_2U^{(2)}C_5^{(2)} = -\frac{1}{2\sqrt{5}}\times C_2U^{(2)}C_5^{(2)}
\end{aligned}
\tag{3.24}
$$

In the above equation the fact that in D_2 group irresp. 5 and irresp. 6 are conjugated to each other has been taken into account and the formulas for C_k and $U^{(k)}$ (in the above equation $k = 2$, $J = 3/2$) are

$$
C_k \equiv \left\langle 4f\middle\|C^{(k)}\middle\|4f\right\rangle;\ U^{(k)} \equiv \left\langle 4f,{}^4F_J\middle\|U^{(k)}\middle\|4f,{}^4F_J\right\rangle
$$

The table of the $3jm$ factors for $O \rightarrow D_3$ reduction shows that all the non-diagonal matrix elements between different crystal field states of this multiplet are zero.

More calculation should be done for the states with higher angular momentum numbers but the method is the same. Let's look at the example of $J = 9/2$. Their energy levels have been given in (3.11) and as a calculation example one of the off-diagonal matrix elements is calculated in the following

$$
\begin{aligned}
\left\langle {}^{9}\!/\!_{2}64\middle|H_c\middle|{}^{9}\!/\!_{2}8_04\right\rangle &= \frac{1}{\sqrt{2}}\begin{pmatrix} 9/2 & 2 & 9/2 \\ 6 & 5 & 8_0 \end{pmatrix}_{O}^{SO_3}\begin{pmatrix} 6 & 5 & 8_0 \\ 4 & 1 & 4 \end{pmatrix}_{D_3}^{O} C_5^2\left\langle {}^{9}\!/\!_{2}\middle\|b^{(2)}\middle\|{}^{9}\!/\!_{2}\right\rangle \\
&+ \frac{1}{\sqrt{2}}\begin{pmatrix} 9/2 & 4 & 9/2 \\ 6 & 5 & 8_0 \end{pmatrix}_{O}^{SO_3}\begin{pmatrix} 6 & 5 & 8_0 \\ 4 & 1 & 4 \end{pmatrix}_{D_3}^{O} C_5^4\left\langle {}^{9}\!/\!_{2}\middle\|b^{(4)}\middle\|{}^{9}\!/\!_{2}\right\rangle \\
&+ \frac{1}{\sqrt{2}}\begin{pmatrix} 9/2 & 6 & 9/2 \\ 6 & 5_0 & 8_0 \end{pmatrix}_{O}^{SO_3}\begin{pmatrix} 6 & 5_0 & 8_0 \\ 4 & 1 & 4 \end{pmatrix}_{D_3}^{O} C_{5_0}^6\left\langle {}^{9}\!/\!_{2}\middle\|b^{(6)}\middle\|{}^{9}\!/\!_{2}\right\rangle \\
&+ \frac{1}{\sqrt{2}}\begin{pmatrix} 9/2 & 6 & 9/2 \\ 6 & 5_1 & 8_0 \end{pmatrix}_{O}^{SO_3}\begin{pmatrix} 6 & 5_1 & 8_0 \\ 4 & 1 & 4 \end{pmatrix}_{D_3}^{O} C_{5_1}^6\left\langle {}^{9}\!/\!_{2}\middle\|b^{(6)}\middle\|{}^{9}\!/\!_{2}\right\rangle \\
&= \frac{1}{\sqrt{2}}\times\left(-\frac{\sqrt{2}}{\sqrt{3\times 11}}\right)\times\frac{1}{\sqrt{3}}C_5^2\left\langle {}^{9}\!/\!_{2}\middle\|b^{(2)}\middle\|{}^{9}\!/\!_{2}\right\rangle + \frac{1}{\sqrt{2}}\times\frac{19}{\sqrt{2\times 3\times 5\times 11\times 13}}\times\frac{1}{\sqrt{3}}C_5^4\left\langle {}^{9}\!/\!_{2}\middle\|b^{(4)}\middle\|{}^{9}\!/\!_{2}\right\rangle \\
&+ \frac{1}{\sqrt{2}}\times\left(-\frac{\sqrt{2}}{\sqrt{3\times 5\times 13}}\right)\times\frac{1}{\sqrt{3}}C_{5_0}^6\left\langle {}^{9}\!/\!_{2}\middle\|b^{(6)}\middle\|{}^{9}\!/\!_{2}\right\rangle + \frac{1}{\sqrt{2}}\times\left(-\frac{2\sqrt{6}}{5\sqrt{11\times 13}}\right)\times\frac{1}{\sqrt{3}}C_{5_1}^6\left\langle {}^{9}\!/\!_{2}\middle\|b^{(6)}\middle\|{}^{9}\!/\!_{2}\right\rangle \\
&= -\frac{1}{3\sqrt{11}}C_2U^{(2)}C_5^2 + \frac{19}{6\sqrt{715}}C_4U^{(4)}C_5^4 - \frac{1}{3\sqrt{65}}C_6U^{(6)}C_{5_0}^6 - \frac{2}{5\sqrt{143}}C_6U^{(6)}C_{5_1}^6
\end{aligned}
\tag{3.25}
$$

The matrix elements for other J values can be found in the literature [18]. The crystal field and energy level analysis for Nd^{3+}:YVO_4 crystal (in which Nd^{3+} ions occupy D_{2d} point group position) by group chain method can be found in the author's paper [24].

After obtaining the matrix elements, one can use exactly the same method as that used in the traditional crystal field scheme to solve the secular equations. However, it is necessary to use the point charge model to calculate the starting values of crystal field parameters in the traditional crystal field scheme and then obtain those parameters in the group chain scheme. The relations between these two sets of crystal field parameters can be introduced as follows.

The crystal field parameters in the traditional crystal field scheme are the expanded coefficients of the crystal field Hamiltonian in terms of the irreducible representation basis functions of the full-rotation group. At the same time, the crystal field parameters in the group-chain scheme are those expanded in terms of the irreducible representation basis functions of the point symmetry group. Therefore, it is easy to obtain their relations by the relation between the two kinds of basis functions

$$H_c = \sum_{k,\mu,\nu,\lambda,\xi} C^{(k)}_{\mu,\nu,\lambda,\xi} b^{(k)}_{\mu,\nu,\lambda,\xi} = \sum_{kq} B_{kq} C^{(k)}_q$$
$$C^{(k)}_q = \left(\frac{4\pi}{2k+1}\right)^{1/2} Y_{kq} \tag{3.26}$$

Using the following relation between spherical harmonic function and the basis function of SO_3 group [17]

$$Y_{kq} = \left(\frac{2k+1}{4\pi}\right)^{1/2} r^{-k} \sqrt{\frac{(2k-1)(2k-3)(2k-5)\cdots}{k!}} |kq\rangle$$
$$C^{(k)}_q = \left(\frac{4\pi}{2k+1}\right)^{1/2} Y_{kq}$$

that is

$$C^{(k)}_q = \left(\frac{4\pi}{2k+1}\right)^{1/2} Y_{kq} = r^{-k} \sqrt{\frac{(2k-1)(2k-3)(2k-5)\cdots}{k!}} |kq\rangle$$

and

$$b^{(k)}_{\mu\nu\lambda\xi} = \sqrt{\frac{(2k-1)(2k-3)(2k-5)\cdots}{k!}} |k\mu\nu\lambda\xi\rangle$$

We obtain

$$\sum_{k,\mu\nu\lambda\xi} C^{k}_{\mu\nu\lambda\xi} b^{(k)}_{\mu\nu\lambda\xi} = \sum_{k,\mu\nu\lambda\xi} C^{k}_{\mu\nu\lambda\xi} \sqrt{\frac{(2k-1)(2k-3)(2k-5)\cdots}{k!}} |k\mu\nu\lambda\xi\rangle$$
$$= \sqrt{\frac{(2k-1)(2k-3)(2k-5)\cdots}{k!}} \sum_{kq} B_{kq} |kq\rangle \tag{3.27}$$

Hence the relation between the two basis functions can be written as

$$|k'\mu'\nu'\xi'\rangle = \sum_{k'q'} a_{k'q'} |k'q'\rangle$$

The conjugate of the above formula will be

$$\langle k'\mu'\nu'\xi'| = \sum_{k'q'} a^{+}_{k'q'} \langle k'q'| \tag{3.28}$$

Multiply the left-hand side and right-hand side of the second equation of (3.27) with left side and right side of (3.28), respectively, and apply the orthogonal and normalized relation, then the following can be shown

$$C^{k}_{\mu\nu\lambda\xi} = \sum_{kq} a^{*}_{kq} B_{kq} \tag{3.29}$$

The following relation is satisfied by the two sets of crystal parameters

$$\sum_{k\mu\nu\lambda\xi} (\hat{C}^{k}_{\mu\nu\lambda\xi})^2 = \sum_{kq} (B_{kq})^2 \tag{3.30}$$

because the two sets of basis functions are transformed by orthogonal transformation and the length of the vector is invariable by this transformation.

Equation (3.29) is the crystal field parameters in the group chain scheme. By using this equation the relation between the two kinds of crystal parameters for the crystal field with D_3 point group symmetry will be

$$\begin{aligned}
C^{2}_{5} &= -B_{20}\\
C^{4}_{1} &= -\frac{1}{9}\sqrt{21}B_{40} - \frac{2}{9}\sqrt{30}B_{43}\\
C^{4}_{5} &= -\frac{2}{9}\sqrt{15}B_{40} + \frac{1}{9}\sqrt{42}B_{43}\\
C^{6}_{1} &= -\frac{4}{9}\sqrt{2}B_{60} + \frac{2}{27}\sqrt{105}B_{63} - \frac{1}{27}\sqrt{462}B_{66}\\
C^{6}_{5_0} &= \frac{1}{21}\sqrt{462}B_{63} + 2\sqrt{\frac{5}{21}}B_{66}\\
C^{6}_{5_1} &= \frac{7}{9}B_{60} + \frac{8}{189}\sqrt{210}B_{63} - \frac{8}{189}\sqrt{231}B_{66}
\end{aligned} \tag{3.31}$$

Noting that in D_3 point group, $B_{kq} = B_{k,-q}$, it is easy to show that (3.31) satisfied by the invariant relation (3.30).

By using the group chain scheme, a series of problems in crystal spectroscopy, including relative intensities of emission spectra in crystal, J mixing effect and up-conversion emission have been studied by the authors [18, 25–32]. The calculated results are in good agreement with the experimental data. In Chap. 12, this method is used to deal with spectroscopy problems of the rare earth ions in glasses; hence one can see that an incisive knowledge on their spectroscopic properties can be obtained.

References

1. D.J. Newman, Adv. Phys. **20**, 197 (1971)
2. D.J. Newman, B. Ng, *Crystal Field Handbook* (Cambridge University Press, 2000)
3. M. Klintenberg, S. Edvardsson, J.O. Thomas, Phys., Rev. B **55**, 10369 (1997)
4. B.R. Judd, Phys. Rev. Let. **39**, 242 (1977)
5. B. Ng, D.J. Newman, J. Chem. Phys. **87**, 7086, 7110 (1987)
6. E. Rukmini, C.K. Jayasankar, M.F. Reid, J. Phys. Condens. Matter **6**, 5919 (1994)
7. S. Hufner, *Optical Spectra of Transparent Rare Earth Compounds* (Academic Press, New York, 1978)
8. B.R. Judd, Proc. Phys. Soc. (London) **A227**, 552 (1955)
9. G.H. Dieke, *Spectra and Energy Levels of Rare Earth Ions in Crystals* (Interscience, New York, 1968)
10. M. Sachs, *Solid State Theory* (McGraw-Hill Book Company, New York, 1963)
11. D. Sengupta, J.O. Artman, Phys. Rev. B **1**, 2986 (1970)
12. C.A. Morrison, R.P. Leavitt, *Handbook on the Physics and Chemistry of Rare Earths*, vol. 5, ed. by K.A. Gschneidner Jr., L. Eyring (North-Holland Publishing Company, 1982)
13. A.A. Kaminskii, *Laser Crystals, their Physics and Properties* (Springer, Heidelberg, 1981 and 1990)
14. A.A. Kaminskii, *Crystalline Lasers: Physical Processes and Operating Schemes* (CRC Press, Boca Raton, 1996)
15. K.H. Hellwege, Ann. Phys. **4**, 95 (1948)
16. G.K. Liu, B. Jacquier, *Spectroscopic Properties of Rare Earths in Optical Materials* (Tsinghua University Press and Springer, Berlin Heidelberg, 2005)
17. P.H. Butler, *Point Group Symmetry Application: Method and Tables* (Plenum, New York, 1981)
18. Z.D. Luo, Y.D. Huang, J. Phys. Condens. Matter **5**, 6949 (1993)
19. A. Benayas, D. Jaque, S.J. Gacia et al., J. Phys. Condens. Matter **19**, 24640 (2007)
20. C. Cascales, C. Zaldo, U.J. Caldino, J. Phys. Condens. Matter **13**, 8071 (2001)
21. D. Jaque, O. Enguitia, U.J. Caldino, J. Appl. Phys. **90**, 561 (2001)
22. D. Jaque, J. Capmany, Z.D. Luo, J. Phys. Condens. Matter **9**, 9715 (1997)
23. S.B. Piepho, P.N. Schatz, *Group Theory in Spectroscopy with Applications to Magnetic Circular Dichroism* (Wiley, New York, 1983)
24. Y.D. Huang, Z.D. Luo, J. Phys. Condens. Matter **6**, 3737 (1994)
25. X.Y. Chen, Z.D. Luo, J. Phys. Condens. Matter **8**, 2571 (1996)
26. X.Y. Chen, Z.D. Luo, Acta Physica Sinica (in Chinese) **46**, 181 (1997)
27. X.Y. Chen, Z.D. Luo, J. Phys. Condens. Matter **9**, 4197 (1997)
28. X.Y. Chen, Z.D. Luo, Chinese J. Struct. Chem. **6**, 358 (1997)
29. X.Y. Chen, Z.D. Luo, J. Phys. Condens. Matter **9**, 7981 (1997)
30. X.Y. Chen, Z.D. Luo, J. Phys. Condens. Matter **10**, 5147 (1998)
31. X.Y. Chen, Z.D. Luo, Acta Physica Sinica (in Chinese) **7**, 773 (1998)
32. X.Y. Chen, Z.D. Luo, Acta Physica Sinica (in Chinese) **8**, 607 (1999)

Chapter 4
Theory of Radiative Transition

Light emission and absorption are the main processes to be discussed in spectroscopy and laser physics, while absorption coefficient, emission and absorption cross-sections, and oscillator strength are fundamental in the investigation of solid-state laser materials. Therefore, in order to have a deep understanding of these processes and calculate correctly all the spectral and laser parameters, it is necessary to study and systematically analyze these processes from the first principles of electrodynamics and quantum mechanics as well as to discuss the selection rule and polarization character of radiative transition by symmetry property of the energy levels.

4.1 Interactions Between Active Ions and Radiation

The active ions in solid materials have interactions with electromagnetic field, besides those with lattice ions and lattice vibrations. These interactions are the physical mechanisms of various performances of solid-state laser. In this chapter, we will discuss the interaction of active ions with radiation field and introduce the formulas for calculating the emission and absorption probabilities and other important spectral parameters.

In order to deal with both stimulated and spontaneous emissions, quantum theory should be used, because spontaneous emission cannot be accounted for by the classical or semi-classical theory. As it is well known from quantum theory, the photon creation operators $a_{\boldsymbol{k}\alpha}^{+}$ and photon annihilation operators $a_{\boldsymbol{k}\alpha}$ are used to describe radiation field. These operators satisfy the following commutation relations [1]

$$\left[a_{\boldsymbol{k}\alpha}, a_{\boldsymbol{k}'\alpha'}^{+}\right] = \delta_{\boldsymbol{k},\boldsymbol{k}'}\delta_{\alpha\alpha'},\ [a_{\boldsymbol{k}\alpha}, a_{\boldsymbol{k}'\alpha'}] = 0,\ \left[a_{\boldsymbol{k}\alpha}^{+}, a_{\boldsymbol{k}'\alpha'}^{+}\right] = 0 \tag{4.1}$$

Z. Luo and Y. Huang, *Physics of Solid-State Laser Materials*, Springer Series in Materials Science 289, https://doi.org/10.1007/978-981-32-9668-8_4

where α denotes the polarization state of the photon with energy $\hbar\omega_k$ and momentum $\hbar\boldsymbol{k}$. The creation and annihilation operators have the following eigenvalues in the photon-number representation $|n_{k\alpha}\rangle$

$$a_{k\alpha}^{+}|n_{k\alpha}\rangle = \sqrt{n_{k\alpha}+1}|n_{k\alpha}+1\rangle_{\dot{\alpha}_\alpha},\ a_{k\alpha}|n_{k\alpha}\rangle = \sqrt{n_{k\alpha}}|n_{k\alpha}-1\rangle,\ a_{k\alpha}^{+}a_{k\alpha}|n_{k\alpha}\rangle = n_{k\alpha}|n_{k\alpha} \tag{4.2}$$

In this representation, if the vector potential $\boldsymbol{A}$ of the field is expressed as the following expression

$$\boldsymbol{A} = \sum_{k\alpha}\left(\frac{2\pi\hbar c^2}{\omega_k V\varepsilon}\right)^{1/2}\boldsymbol{e}_\alpha(\boldsymbol{k})\left[a_{k\alpha}(t)e^{i\boldsymbol{k}\cdot\boldsymbol{r}} + a_{k\alpha}^{+}(t)e^{-i\boldsymbol{k}\cdot\boldsymbol{r}}\right] \tag{4.3}$$

then the field energy can be expressed as the sum of all the photons, where

$$a_{k\alpha}(t) = a_{k\alpha}e^{-i\omega_k t},\ a_{k\alpha}^{+}(t) = a_{k\alpha}^{+}e^{i\omega_k t} \tag{4.4}$$

where $\boldsymbol{e}_\alpha(\boldsymbol{k})$, in which $\alpha = 1, 2$, is the unit vector used to represent the polarization direction of the photons, c is the velocity of the photons in vacuum, ε refers to the dielectric constant of the solid material, and V denotes the volume occupied by the radiation field. According to electromagnetic dynamics [2]

$$\boldsymbol{E} = -\frac{1}{c}\frac{\partial\boldsymbol{A}}{\partial t} \tag{4.5a}$$

and then one can obtain

$$\boldsymbol{E} = -\frac{1}{c}\frac{\partial\boldsymbol{A}}{\partial t} = i\sum_{k\alpha}\left(\frac{2\pi\hbar\omega_{\mathbf{k}}}{V\varepsilon}\right)^{1/2}\boldsymbol{e}_\alpha(\boldsymbol{k})\left[a_{k\alpha}(t)e^{i\boldsymbol{k}\cdot\boldsymbol{r}} - a_{k\alpha}^{+}(t)e^{-i\boldsymbol{k}\cdot\boldsymbol{r}}\right] \tag{4.5b}$$

By using the following relations

$$\int e^{i(\boldsymbol{k}-\boldsymbol{k}')\cdot\boldsymbol{r}}d\tau = V\delta_{\boldsymbol{k}\boldsymbol{k}'}$$

and

$$\boldsymbol{e}_\alpha(\boldsymbol{k})\cdot\boldsymbol{e}_{\alpha'}(\boldsymbol{k}) = \delta_{\alpha\alpha'}$$

Noting that the polarization vectors are perpendicular to vector $\boldsymbol{k}$ representing the direction of the light wave, that is, $\boldsymbol{e}_1(\boldsymbol{k}) \cdot \boldsymbol{k} = \boldsymbol{e}_2(\boldsymbol{k}) \cdot \boldsymbol{k} = 0$, then the energy Hamiltonian of the field can be written as

$$H_{\mathrm{ph}} = \int \left(\frac{\varepsilon}{4\pi}\right) \boldsymbol{E}^2 d\tau = \sum_{k\alpha} \hbar\omega_{\boldsymbol{k}} \left(a_{k\alpha}^{+} a_{k\alpha} + \frac{1}{2} \right) \tag{4.6}$$

where $a_{k\alpha}^{+} a_{k\alpha}$ represents the operator of the photon number. Equation (4.6) shows that the energy of each photon mode is equal to the number of photon multiplied by the energy of a single photon plus zero field energy, and the total energy of the whole field is the sum of the energy of all the photons. This is a well-known physical reality. Obviously, (4.3) can be used to obtain a correct result and so it is justified.

Next, the interaction Hamiltonian between the radiation field and the electron without spin will be introduced. For the case studied, the electric potential (scalar potential) can be assumed to be zero. It can be shown that if the interaction Hamiltonian is written in the form:

$$H = \frac{1}{2m} \left(\boldsymbol{p} - \frac{e\boldsymbol{A}}{c} \right)^2 \tag{4.7}$$

where $\boldsymbol{p}$ is the momentum of the electron with mass m, then the Newtonian equations for the electron in the field can be introduced correctly. According to Hamiltonian equations in mechanics, one has [3]

$$\frac{d}{dt}\boldsymbol{r} = \frac{\partial H}{\partial \boldsymbol{p}}, \cdot \frac{d}{dt}\boldsymbol{p} = -\frac{\partial H}{\partial \boldsymbol{r}}$$

where $\boldsymbol{r}$ is the vector of electronic coordinate. There are six equations here, from which the equations for the component x are:

$$\frac{d}{dt}x = \frac{\partial H}{\partial p_x}, \frac{d}{dt}p_x = -\frac{\partial H}{\partial x}$$

By substituting (4.7) in the above equation, we obtain

$$\frac{d}{dt}x = \frac{1}{m}\left(p_x - \frac{e}{c} A_x \right) \tag{4.8a}$$

Obviously, its vector form is

$$\frac{d}{dt}\boldsymbol{r} = \frac{1}{m}\left(\boldsymbol{p} - \frac{e}{c}\boldsymbol{A} \right) \tag{4.8b}$$

On the other hand, by (4.7) and x component of the second Hamiltonian equation

$$\frac{d}{dt}p_x = -\frac{1}{m}\left(\boldsymbol{p} - \frac{e}{c}\boldsymbol{A}\right) \cdot \left(-\frac{e}{c}\frac{\partial \boldsymbol{A}}{\partial x}\right) \tag{4.9}$$

Differentiating (4.8a) and using (4.9) and (4.8b), one obtains

$$m\frac{d^2x}{dt^2} = \frac{d}{dt}p_x - \frac{e}{c}\frac{d}{dt}A_x = \frac{d\boldsymbol{r}}{dt} \cdot \frac{e}{c}\frac{\partial \boldsymbol{A}}{\partial x} - \frac{e}{c}\frac{d}{dt}A_x \tag{4.10}$$

Owing to the fact that $A_x = A_x(x, y, z, t)$, the total differential of the last term in the above equation can be expressed as

$$\frac{dA_x}{dt} = \frac{\partial A_x}{\partial x}\frac{dx}{dt} + \frac{\partial A_x}{\partial y}\frac{dy}{dt} + \frac{\partial A_x}{\partial z}\frac{dz}{dt} + \frac{\partial A_x}{\partial t}$$

Therefore, (4.10) becomes

$$m\frac{d^2x}{dt^2} = \frac{e}{c}\left[\frac{dy}{dt}\left(\frac{\partial A_y}{\partial x} - \frac{\partial A_x}{\partial y}\right) - \frac{dz}{dt}\left(\frac{\partial A_x}{\partial z} - \frac{\partial A_z}{\partial x}\right)\right] - \frac{e}{c}\frac{\partial}{\partial t}A_x \tag{4.11}$$

By means of the well-known electromagnetic dynamics equation [2]

$$\boldsymbol{B} = \nabla \times \boldsymbol{A} \quad \boldsymbol{E} = -\frac{1}{c}\frac{\partial \boldsymbol{A}}{\partial t}$$

It is easy to show, by (4.11), that

$$m\frac{d^2x}{dt^2} = eE_x + \frac{e}{c}(v \times B)_x \tag{4.12}$$

The right-hand side of the above equation is the x component of the Lorentz force, and so the formula (4.12) corresponds to the x component of the Newtonian equation for the electron in the radiation field. Therefore, the interaction Hamiltonian of (4.7) is justified.

Expanding formula (4.7), one has

$$\frac{1}{2m}\left(\boldsymbol{p} - \frac{e\boldsymbol{A}}{c}\right)^2 = \frac{1}{2m}\boldsymbol{p}^2 - \frac{e}{2mc}(\boldsymbol{p} \cdot \boldsymbol{A} + \boldsymbol{A} \cdot \boldsymbol{p}) + \frac{e^2}{2mc^2}\boldsymbol{A}^2$$

If the term having the factor $\boldsymbol{A}^2$ is neglected, then the interaction Hamiltonian can be written as

$$H_i = \frac{e}{2mc}(\boldsymbol{p}\cdot\boldsymbol{A} + \boldsymbol{A}\cdot\boldsymbol{p})$$

Expressing $\boldsymbol{p}$ as $-i\hbar\nabla$ and remembering that $\nabla\cdot\boldsymbol{A} = 0$, it can be shown that

$$\boldsymbol{p}\cdot\boldsymbol{A} = \boldsymbol{A}\cdot\boldsymbol{p} - i\hbar\nabla\cdot\boldsymbol{A} = \boldsymbol{A}\cdot\boldsymbol{p}$$

Therefore

$$H_i = \frac{e}{mc}\boldsymbol{p}\cdot\boldsymbol{A} = \frac{e}{m}\sum_{k\alpha}\left(\frac{2\pi\hbar}{V\omega_k\varepsilon}\right)^{1/2} e_\alpha(\boldsymbol{k})\cdot\boldsymbol{p}\left[a_{k\alpha}(t)e^{ik\cdot r} + a^{+}_{k\alpha}(t)\mathrm{e}^{-ik\cdot r}\right] \tag{4.13}$$

where $\boldsymbol{p}$ is an operator acting on the electronic wave functions while the creation and annihilation operators are photon operators acting on photon wave functions.

4.2 Probability of Emission and Absorption Processes

In the last section, the mechanism of the interaction between electron and radiation field is discussed and the formula for the interaction Hamiltonian has been introduced. The emission and absorption processes and the formulas of their transition probability will be discussed and derived in this section.

The Hamiltonian H_i, a time-dependent perturbation operator, causes transitions between different electronic energy levels. According to time-dependent perturbation theory [4], transition probability can be calculated by using the following formula

$$W_{if} = \frac{2\pi}{\hbar^2}\left|\langle i|H_i|f\rangle\right|^2\rho(\omega_{\mathbf{k}}) \equiv \frac{2\pi}{\hbar^2}M^2_{if}\rho(\omega_{\boldsymbol{k}}) \tag{4.14}$$

where the final state of the radiation field has state density $\rho(\omega_{\boldsymbol{k}})$.

Let us suppose that the radiation field is limited in a cube with volume $V = L^3$, then the $\boldsymbol{k}$ vector in (4.13) should satisfy the following equations obtained by period conditions

$$k_x = \frac{2\pi n_x}{L},\ k_y = \frac{2\pi n_y}{L},\ k_z = \frac{2\pi n_z}{L}$$

where each set of n_x, n_y, n_z represents a state. Thus the number of states between $\boldsymbol{k}$ and $\boldsymbol{k} + d\boldsymbol{k}$ will be

$$\rho(\boldsymbol{k})d\boldsymbol{k} = dn_x dn_y dn_z = \frac{V}{8\pi^3}dk_x dk_y dk_z = \frac{V}{8\pi^3}k^2 dk d\Omega_k$$

where $d\Omega_{\boldsymbol{k}}$ is the infinitesimal solid angle in the direction $\boldsymbol{k}$. By using $k = \omega/c$, the above expression can be written as

$$\rho(\omega_k)d\omega_k = \frac{V}{8\pi^3 c^3}\omega_k^2 d\omega_k d\Omega_k$$

Therefore

$$\rho(\omega_k) = \frac{V}{8\pi^3 c^3}\omega_k^2 d\Omega_k \tag{4.15}$$

where c is the velocity of light; it should be replaced by c/n in solid materials.

In the process of absorbing a photon and jumping to the final state, the radiation field has one photon less than that in the initial state. Expressing the wave function of initial state as $|i> = |\varphi_i^e > |n_k >$, then that of the final state is $|f> = |\varphi_f^e > |n_k - 1 >$. If the ion emits a photon, the wave function of initial state will be $|i> = |\varphi_i^e > |n_k >$ and that of the final state can be expressed as $f> = |\varphi_f^e > |n_k + 1$, where φ_i^e and φ_f^e are electronic wave functions for the initial and final states, respectively. If one calculates the matrix elements of photon operators at first, then the electronic transition matrix element for the emission process will be

$$\boldsymbol{M}_{if} = \frac{e}{m}\left[\frac{2\pi\hbar}{V\omega_k\varepsilon}(n_k+1)\right]^{1/2}\left\langle \varphi_f^e \left| \sum_i \mathbf{e}_\alpha(\boldsymbol{k}) \cdot \boldsymbol{p}_i e^{-i\boldsymbol{k}\cdot\boldsymbol{r}_i} \right| \varphi_i^e \right\rangle \tag{4.16}$$

For the absorption process it is

$$\boldsymbol{M}_{if} = \frac{e}{m}\left[\frac{2\pi\hbar}{V\omega_k\varepsilon}n_k\right]^{1/2}\left\langle \varphi_f^e \left| \sum_i \mathbf{e}_\alpha(\boldsymbol{k}) \cdot \boldsymbol{p}_i e^{-i\boldsymbol{k}\cdot\boldsymbol{r}_i} \right| \varphi_i^e \right\rangle \tag{4.17}$$

In these two expressions, the summations are over all the involved electrons. It can be seen from expression (4.16) that even if there is no radiation field present before emission, that is, $n_k = 0$, the transition matrix elements for emission are different from zero. It gives rise to spontaneous emission. The other term proportional to n_k presenting the photon number in the radiation field describes the stimulated emission.

In order to carry out a detailed calculation, it is necessary to expand $e^{i\boldsymbol{k}\cdot\boldsymbol{r}_i}$. One can estimate the order of magnitude of $\boldsymbol{k}\cdot\boldsymbol{r}_i$ at first. The electronic coordinate $\boldsymbol{r}_i$ is in the range of an ion radius of 10^{-8} cm, and the light wavelength λ is in the range of 10^{-4}–10^{-5} cm, so we have

$$k \cdot r_i = \frac{2\pi a}{\lambda} \sim 10^{-3}.$$

Thus it is enough to keep only the first and second terms in the expansion of $e^{i\boldsymbol{k}\cdot\boldsymbol{r}_i}$

$$e^{i\boldsymbol{k}\cdot\boldsymbol{r}_i} = 1 + i\boldsymbol{k}\cdot\boldsymbol{r}_i,\ e^{-i\boldsymbol{k}\cdot\boldsymbol{r}_i} = 1 - i\boldsymbol{k}\cdot\boldsymbol{r}_i \tag{4.18}$$

The first term corresponds to the electric-dipole transition. The transition probability per unit time for the stimulated absorption and emission of a photon with frequency ω_k can be expressed as

$$P_{\mathbf{k}}(\text{abs.}) = P_{\mathbf{k}}(\text{st.em.}) = \frac{\omega_k^3}{2\pi\hbar c^3 \varepsilon} n_k \sum_{\alpha=1,2} \int d\Omega_k \left| \left\langle \varphi_f^{\text{e}} \middle| \mathbf{e}_\alpha(\boldsymbol{k}) \cdot \sum_i e\boldsymbol{r}_i \middle| \varphi_i^{\text{e}} \right\rangle \right|^2 \tag{4.19}$$

The transition probability for spontaneous emission can be expressed as

$$P_{\boldsymbol{k}}(\text{sp.em.}) = \frac{\omega_k^3}{2\pi\hbar c^3 \varepsilon} \sum_{\alpha=1,2} \int d\Omega_k \left| \left\langle \varphi_f^{\text{e}} \middle| \mathbf{e}_\alpha(\boldsymbol{k}) \cdot \sum_i e\boldsymbol{r}_i \middle| \varphi_i^{\text{e}} \right\rangle \right|^2 \tag{4.20}$$

In the deduction of expressions (4.19) and (4.20), the Poisson equation in quantum mechanics $\dot{\boldsymbol{r}} = \frac{i}{\hbar}[H, \boldsymbol{r}]$ [1] has been used; therefore

$$\boldsymbol{p} = m\dot{\boldsymbol{r}} = \frac{im}{\hbar}[H, \boldsymbol{r}], \left\langle \varphi_f^e | \boldsymbol{p} | \varphi_i^e \right\rangle = \frac{im}{\hbar}\left(E_f^e - E_i^e\right)\left\langle \varphi_f^e | \boldsymbol{r} | \varphi_i^e \right\rangle = im\omega_k \left\langle \varphi_f^e | r | \varphi_i^e \right\rangle$$

In expressions (4.19) and (4.20), $\sum_i e\boldsymbol{r}_i$ is electric-dipole of the electronic system and so the corresponding transition is referred to as the electric-dipole transition.

The unit polarization vectors $\boldsymbol{e}_1(\boldsymbol{k})$ and $\boldsymbol{e}_2(\boldsymbol{k})$ are perpendicular to each other and perpendicular to the photon propagate direction $\boldsymbol{k}$ and so one can discuss the problem in a coordinate system constituting by these three vectors. If the angle between radius $\boldsymbol{r}_i$ of the electron i and the photon propagating direction $\boldsymbol{k}$ is θ_i while the angle between the projection of radius $\boldsymbol{r}_i$ on the plane $\boldsymbol{e}_1(\boldsymbol{k})$–$\boldsymbol{e}_2(\boldsymbol{k})$ and the polarization vector $\boldsymbol{e}_1(\boldsymbol{k})$ is φ_i, then we have

$$\sum_{\alpha=1,2} \boldsymbol{e}_\alpha \cdot \sum_i e\mathbf{r}_i = \sum_i er_i(\sin\theta_i\cos\phi_i + \sin\theta_i\sin\phi_i)$$

Owing to the fact that the angles between the photon propagating direction $\boldsymbol{k}$ and the radius $\boldsymbol{r}_i$ of different electron positions are independent of each other and can assume any different values, then the expressions (4.19) and (4.20) should be integrated over $d\Omega = \sin\theta d\theta d\phi$. The integral of the term of $\cos\varphi_i\sin\varphi_i$ over φ_i is zero. Therefore, the probabilities of stimulated and spontaneous electric-dipole emissions can be written, respectively, as

$$P_{\mathbf{k}}(\text{st.em.}) = \frac{\omega_{\mathbf{k}}^3}{2\pi\hbar c^3 \varepsilon} n_k \sum_i \int \sin^2\theta_i d\Omega \left| \left\langle \varphi_f^e \middle| r_i \middle| \varphi_i^e \right\rangle \right|^2$$

$$P_{\mathbf{k}}(\text{sp.em.}) = \frac{\omega_{\mathbf{k}}^3}{2\pi\hbar c^3 \varepsilon} \sum_i \int \sin^2\theta_i d\Omega \left| \left\langle \varphi_f^e \middle| \boldsymbol{r}_i \middle| \varphi_i^e \right\rangle \right|^2$$

In the processes of emission and absorption, the electrons with different subscript i are independent and the infinitesimal solid angle all have the same expression $d\Omega = \sin\theta d\theta d\varphi$, and so for the spontaneous electric-dipole emission we have

$$\begin{aligned} P_{\mathbf{k}}(\text{sp.em.}) &= \frac{\omega_{\mathbf{k}}^3}{2\pi\hbar c^3 \varepsilon} \sum_i \int_0^{2\pi} \int_0^{\pi} \sin^3\theta d\theta d\varphi \left| \left\langle \varphi_f^e \middle| r_i \middle| \varphi_i^e \right\rangle \right|^2 \\ &= \frac{4\omega_{\mathbf{k}}^3}{3\hbar c^3 \varepsilon} \left| \sum_i e \left\langle \varphi_f^e \middle| r_i \middle| \varphi_i^e \right\rangle \right|^2 \end{aligned} \tag{4.21}$$

In the case of isotropic space, we have

$$\left\langle \varphi_f^e | x_i | \varphi_i^e \right\rangle = \left\langle \varphi_f^e | y_i | \varphi_i^e \right\rangle = \left\langle \varphi_f^e | z_i | \varphi_i^e \right\rangle = \left\langle \varphi_f^e | r_i | \varphi_i^e \right\rangle$$

The spontaneous transition probability for emission light polarized in any direction will be the same. In the general cases, the spontaneous transition probability for emission light polarized in any direction should be calculated according to the *Theorem on the Total Probability* [5]. By this theorem, if any event B occurs by the ways of A_1, A_2, A_3,…, A_n having probability of $p(A_1)$, $p(A_2)$, $p(A_3)$, …, $p(A_n)$, respectively, then the probability of event B to take place by using any of the above-mentioned ways can be calculated by:

$$P(\text{B}) = \sum_i p(\text{A}_i) p(\text{A}_i | \text{B})$$

where $p(A_i|B)$ is the probability of event B occurred by the way of A_i, and $p(A_i)$ is the probability of the way of A_i. The above summation should be changed to an integral in the case of continuous variables. Applying it to the isotropic solid material, considering that the probability of polarization direction in $d\Omega_p$ is $d\Omega_p/4\pi$, the probability of spontaneous electric-dipole emission with polarization in any direction becomes:

$$P_k = \frac{1}{4\pi} \int \frac{4\omega_k^3}{3\hbar c^3 \varepsilon} \left| \sum_i e \left\langle \varphi_f^e | r_i | \varphi_i^e \right\rangle \right|^2 d\Omega_p \tag{4.22}$$

The result is $\frac{4\omega_k^3}{3\hbar c^3\varepsilon}\left|\sum_i e\left\langle \varphi_f^e | r_i | \varphi_i^e \right\rangle\right|^2$, because the space is isotropic. It can be shown that in this case

$$P_k = \frac{4\omega_k^3}{3\hbar c^3\varepsilon}\left|\sum_i e\left\langle \varphi_f^e | x_i | \varphi_i^e \right\rangle\right|^2 = \frac{4\omega_k^3}{3\hbar c^3\varepsilon}\left|\sum_i e\left\langle \varphi_f^e | y_i | \varphi_i^e \right\rangle\right|^2$$
$$= \frac{4\omega_k^3}{3\hbar c^3\varepsilon}\left|\sum_i e\left\langle \varphi_f^e | z_i | \varphi_i^e \right\rangle\right|^2$$

This result has been obtained previously. It leads to the conclusion that in isotropic solid materials, the electric-dipole spontaneous emission with polarization in any direction has the same transition probability as that with polarization along x, y, or z axis. This conclusion seems to be obvious; it is in fact the result of the theorem on the total probability. Understanding of this point is very important for the calculation of the transition probabilities in the case of anisotropic crystal.

In the case of anisotropic crystal, the spontaneous emission probability for the transition emitting the light polarized in the direction of X, Y, and Z of crystallo physics axes can be written, respectively, as $P^{(X)}(\text{sp.em.})$, $P^{(Y)}(\text{sp.em.})$, and $P^{(Z)}(\text{sp.em.})$. For the sake of convenience, $S(i, f)$ is used to represent the sum of the electric-dipole moment matrix element square in the above formula. The probability of electric-dipole transition polarizing along a certain direction q is written as

$$P^{(q)}(\text{sp.em.}) = \frac{4\omega_{\mathbf{k}}^3 e^2}{3\hbar c^3\varepsilon} S^{(q)}(i,f) \tag{4.23}$$

The theorem on the total probability gives the total spontaneous emission probability

$$\begin{aligned} P(\text{sp.em.}) &= p(X)P^{(X)}(\text{sp.em.}) + p(Y)P^{(Y)}(\text{sp.em.}) + p(Z)P^{(Z)}(\text{sp.em.}) \\ &= \frac{1}{3}\left[P^{(X)}(\text{sp.em.}) + P^{(Y)}(\text{sp.em.}) + P^{(Z)}(\text{sp.em.})\right] \end{aligned} \tag{4.24a}$$

For an uniaxial crystal

$$\begin{aligned} P(\text{sp.em.}) &= p(\pi)P^{(\pi)}(\text{sp.em.}) + p(\sigma)P^{(\sigma)}(\text{sp.em.}) \\ &= \frac{1}{3}P^{(\pi)}(\text{sp.em.}) + \frac{2}{3}P^{(\sigma)}(\text{sp.em.}) \end{aligned} \tag{4.24b}$$

Note: Dealing with the transition in the solid material, the correction factor introduced by the refractive index should be included, and this will be discussed later.

Obviously, by using (4.24a, 4.24b), the error of over estimation in the transition probability as well as the Judd–Ofelt parameters can be avoided.

The related expressions for magnetic-dipole and electric-quadrupole transitions will be briefly discussed now. Suppose the related electric-dipole matrix element is equal to zero, that is, $\left\langle \varphi_f^e | \boldsymbol{r} | \varphi_i^e \right\rangle = 0$, assume the single electron approximation at the same time. Referring to formula (4.16), the expression of matrix element in (4.18) will be

$$\left\langle \varphi_f^e | \boldsymbol{e} \cdot \boldsymbol{p} e^{-i\boldsymbol{k}\cdot\boldsymbol{r}} | \varphi_i^e \right\rangle = -i \left\langle \varphi_f^e | (\boldsymbol{k} \cdot \boldsymbol{r})(\boldsymbol{e} \cdot \boldsymbol{p}) | \varphi_i^e \right\rangle \tag{4.25}$$

where $\boldsymbol{e}_\alpha(\boldsymbol{k})$ has been represented by $\boldsymbol{e}$, for the sake of convenience. In order to understand their transition mechanisms $(\boldsymbol{k}\cdot\boldsymbol{r})$ $(\boldsymbol{e}\cdot\boldsymbol{p})$ is separated into two parts

$$\begin{aligned}(\boldsymbol{k} \cdot \boldsymbol{r})(\boldsymbol{e} \cdot \boldsymbol{p}) &= \sum_{i,j} k_i r_i e_j p_j = \sum_{i,j} k_i e_j r_i p_j \\ &= \frac{1}{2} \sum_{i,j} k_i e_j \left(r_i p_j - r_j p_i \right) + \frac{1}{2} \sum_{i,j} k_i e_j \left(r_i p_j + r_j p_i \right) \\ &= \frac{1}{2} (\boldsymbol{k} \times \boldsymbol{e}) \cdot \boldsymbol{l} + \frac{1}{2} \sum_{i,j} k_i e_j \left(r_i p_j + r_j p_i \right) \\ &= -\frac{1}{2} \mathbf{e} \cdot (\boldsymbol{k} \times \boldsymbol{l}) + \frac{1}{2} \sum_{i,j} k_i e_j \left(r_i p_j + r_j p_i \right)\end{aligned} \tag{4.26}$$

where i and j are used to denote three components of related vectors and $\boldsymbol{l} = \boldsymbol{r} \times \boldsymbol{p}$ is the orbital angular momentum operator. Similar to the transformation of the matrix elements of $\boldsymbol{p}$ to that of $\boldsymbol{r}$ in the introduction of formulas (4.19) and (4.20), the matrix element of the second term in (4.26) can be transformed into

$$\left\langle \varphi_f^e | r_i p_j + r_j p_i | \varphi_i^e \right\rangle = im\omega_{\boldsymbol{k}} \left\langle \varphi_f^e | r_i r_j | \varphi_i^e \right\rangle \tag{4.27}$$

Therefore

$$\left\langle \varphi_f^e | (\boldsymbol{k} \cdot \boldsymbol{r})(\boldsymbol{e} \cdot \boldsymbol{p}) | \varphi_i^e \right\rangle = -\frac{1}{2} \left\langle \varphi_f^e | \mathbf{e} \cdot (\boldsymbol{k} \times \boldsymbol{l}) | \varphi_i^e \right\rangle + \frac{im\omega_{\mathbf{k}}}{2} \left\langle \varphi_f^e | (\boldsymbol{e} \cdot \boldsymbol{r})(\boldsymbol{k} \cdot \boldsymbol{r}) | \varphi_i^e \right\rangle \tag{4.28}$$

Substituting (4.28) into (4.16), the spontaneous emission matrix element in single electron approximation can be expressed as

$$M_{if} = \frac{e}{m} \left(\frac{2\pi\hbar}{V\omega_k \varepsilon} \right)^{1/2} \left[\frac{i}{2} \left\langle \varphi_f^e | \mathbf{e} \cdot (\boldsymbol{k} \times \boldsymbol{l}) | \varphi_i^e \right\rangle + \frac{m\omega_k}{2} \left\langle \varphi_f^e | (\boldsymbol{e} \cdot \boldsymbol{r})(\boldsymbol{k} \cdot \boldsymbol{r}) | \varphi_i^e \right\rangle \right] \tag{4.29}$$

The first and the second terms in (4.29) correspond to the magnetic-dipole transition and the electric-quadrupole transition, respectively. Based on the above relations, the more common Hamiltonians of the magnetic-dipole and the electric-quadrupole transitions can be obtained by using (4.13).

$$H_{\mathrm{md}} = -i\sum_{\boldsymbol{k}\alpha}\left(\frac{2\pi\hbar\omega_{\mathbf{k}}}{V\varepsilon}\right)^{1/2} a_{\boldsymbol{k}\alpha}\boldsymbol{e}_{\alpha}(\boldsymbol{k})\cdot(\boldsymbol{i}_{k}\times\boldsymbol{M}_{\mathrm{md}}) + i\sum_{\boldsymbol{k}\alpha}\left(\frac{2\pi\hbar\omega_{\boldsymbol{k}}}{V\varepsilon}\right)^{1/2} a^{+}_{\boldsymbol{k}\alpha}\boldsymbol{e}_{\alpha}(\boldsymbol{k})\cdot(\boldsymbol{i}_{k}\times\boldsymbol{M}_{\mathrm{md}}) \tag{4.30}$$

and

$$\begin{aligned} H_{\mathrm{eq}} = &-\sum_{\boldsymbol{k}\alpha}\left(\frac{\pi\hbar\omega_{\mathbf{k}}^{3}}{2c^{2}V\varepsilon}\right)^{1/2} a_{\boldsymbol{k}\alpha}(\boldsymbol{e}_{\alpha}(\boldsymbol{k})\cdot e\boldsymbol{r})(\boldsymbol{i}_{\boldsymbol{k}}\cdot\boldsymbol{r}) \\ &+\sum_{\boldsymbol{k}\alpha}\left(\frac{\pi\hbar\omega_{\boldsymbol{k}}^{3}}{2c^{2}V\varepsilon}\right)^{1/2} a^{+}_{\boldsymbol{k}\alpha}(\boldsymbol{e}_{\alpha}(\boldsymbol{k})\cdot e\boldsymbol{r})(\boldsymbol{i}_{\boldsymbol{k}}\cdot\boldsymbol{r}) \end{aligned} \tag{4.31}$$

While introducing the above equations, unit vector $\boldsymbol{i}_k$ in $\boldsymbol{k}$ direction and $\boldsymbol{k} = \frac{\omega_k}{c}\boldsymbol{i}_{\boldsymbol{k}}$ has been used and the magnetic-dipole of electron is expressed as follows by including spin magnetic-dipole moment of electrons $\boldsymbol{M}_{\mathrm{sp}} = (e/mc)\boldsymbol{s}$

$$\boldsymbol{M}_{\mathrm{md}} = \frac{e\hbar}{2mc}(\boldsymbol{L}+2\boldsymbol{S}).$$

By using formulas (4.30) and (4.31), the probabilities of magnetic-dipole transition and electric-quadrupole transition can be obtained, respectively. The probability of spontaneous magnetic-dipole transition is given by

$$P^{\mathrm{md}}_{\mathrm{sp.em.}} = \frac{4\omega_{\boldsymbol{k}}^{3}}{3\hbar c^{3}\varepsilon}\left|\langle\boldsymbol{M}_{\mathrm{md}}\rangle_{if}\right|^{2} \tag{4.32a}$$

while the probabilities of stimulated emission and absorption of magnetic-dipole transition are given, respectively, by

$$P^{\mathrm{md}}_{\mathrm{st.em.}} = P^{\mathrm{md}}_{\mathrm{ab.}} = \frac{4\omega_{\boldsymbol{k}}^{3}n_{k}}{3\hbar c^{3}\varepsilon}\left|\langle\boldsymbol{M}_{\mathrm{md}}\rangle_{if}\right|^{2} \tag{4.32b}$$

Obviously, the magnetic-dipole and electric-quadrupole transition probabilities are much lower than that of the electric-dipole transition, when all the transitions are allowed by selection rules. A simple comparison of the order of magnitude can give us a general idea. For the free ions and atoms, the magnetic-dipole transition probability will be about 10^{-5}–10^{-6} times of that of electric-dipole transitions, while the corresponding figures for electric-quadrupole transition will be only 10^{-7}–10^{-8}.

The situation in solid-state laser materials is different. Owing to the fact that many transitions in the materials are observed within the same electronic configuration, the electric-dipole transitions are forbidden in zero-order approximation and allowed only if there is a perturbation caused by odd-parity crystal field or lattice vibration which mix some wave function components with different parity into the wave functions of the initial and/or final states of the transitions. Therefore, the transition probability is determined by the odd-parity perturbation. Without this kind of perturbation, the transitions between states with the same parity can only be magnetic-dipole or electric-quadrupole type.

The relationship between the probability of the electric-dipole transition and the symmetry of the active ions can be divided into three kinds [6]:

I. The active ions occupy position with symmetry of the following 11 point groups:

$$S_2 \equiv C_{\mathrm{i}}, S_6, C_{2\mathrm{h}}, C_{4\mathrm{h}}, C_{6\mathrm{h}}, D_{3\mathrm{d}}, D_{2\mathrm{h}}, D_{4\mathrm{h}}, D_{6\mathrm{h}}, O_{\mathrm{h}}, T_{\mathrm{h}}$$

In these cases, the electric-dipole transition is forbidden. Weak electric-dipole transition can still be introduced by odd-parity lattice vibration mode. The intensity of phonon-assisted forced electric-dipole transition is proportional to the reduced temperature T/T_m (T_m is the melting temperature of the crystal in Kelvin scale). At the same temperature T, the lower the melting temperature T_m, the higher is the transition intensity. By this reason, at room temperature, rather strong fluorescence can be observed in some crystals with low melting temperature, but in the crystals with high melting temperature any fluorescence cannot be seen. The associated spectral line is called the phonon sideband, and at the location of one phonon energy away from the zero phonon line.

II. The active ions occupy position with symmetry of the following 18 point groups:

$$C_1, C_2, C_{1\mathrm{h}}, D_2, C_{2v}, C_4, S_4, D_4, C_{4v}, D_{2d}, C_{3v}, C_6, D_6, C_{6v}, D_{3\mathrm{h}}, T, T_d, O$$

In these cases, some pairs of nearest neighbor ion still have or nearly have inversion symmetry centered on active ions, although the electric-dipole transitions induced by crystal field and forced by lattice vibration are all allowed, and the wave function admixture with different parity generally is not large enough and the electric-dipole transition is generally not strong enough, except in the case where some pairs of nearest neighbor ion have a large deviation from the inversion symmetry.

III. The active ions occupy position with symmetry of the following three point groups:

$$C_3, D_3, C_{3\mathrm{h}}$$

In these cases, the wave function with different parity has larger admixture and so has higher electric-dipole transition rate.

It should be noted that for the active ion, the intensity of electric-dipole transition depends not only on its site symmetry but also on the number of nearest neighbor anions deviating from the inversion symmetry as well as their deviating magnitude.

The selection rules for the transition will be given in the next section.

4.3 Selection Rules for Radiative Transition

In principle, if the related matrix element for a transition is non-zero, then the transition can take place. According to this principle, the selection rules for electric-dipole, magnetic-dipole, and electric-quadrupole transitions will be introduced. The results can be easily obtained by using the group theory. The discussion in this chapter does not involve complex coupling coefficients. The notation of the point group irreducible representation still uses the Bethe's notation.

4.3.1 Selection Rules for Radiative Transition of Free Ions and Atoms

According to the group theory, the transition matrix element is a physical quantity measuring the probability of transition so that it is invariant in a symmetry transformation; therefore, it must belong to the identity representation of the symmetry group. The total angular momentum quantum number J is a good quantum number for atoms and free ions. By Chap. 2 we know that the initial state i and the final state f belong to the representations D_{J_i} and D_{J_f} of the full-rotation group, respectively.

In the case of an electric-dipole transition, the electric-dipole is a vector belonging to D_1 [7]. The non-zero transition matrix element requires that the product of $D_{J_i} \otimes D_1$ includes D_{J_f}. It is easy to obtain that $D_{J_i} \otimes D_1 = D_{J_i+1} \oplus D_{J_i} \oplus D_{J_i-1}$. Therefore, the selection rule for the electric-dipole transitions is $\Delta J = 0, \pm 1$. It should be pointed out that when $J_i = 0$, J_f can only be equal to 1 and so the $J_i = 0 \rightarrow J_f = 0$ transition is forbidden. On the other hand, the parity of final state should be different from that of the initial state, because the parity of electric-dipole is odd. In the case of weak spin–orbit coupling, Russell-Saunders approximation is good enough and so L and S are good quantum numbers. In this case, the electric-dipole moment operator belongs to representation D_1 ($L = 1$, $S = 0$), and the initial state belongs to $D_{L_i} \otimes D_S$ but the final state belongs to $D_{L_f} \otimes D_S$. Obviously, we have the following selection rules.

$\Delta S = 0$, $\Delta L = 0, \pm 1$, the parity should be changed in the transition but $L_i = 0 \rightarrow L_f = 0$ is forbidden.

For the magnetic-dipole transitions, remembering that magnetic-dipole moment is an axial vector, which belongs to representation D_{1g} (having even parity), one can obtain their selection rules by using the same method:

$\Delta J = 0, \pm 1$; the parity should not be changed in the transition while $J_i = 0 \rightarrow J_f = 0$ is forbidden.

When Russell-Saunders approximation is reasonable, we have the following selection rules:

$\Delta S = 0$, $\Delta L = 0, \pm 1$ but $L_i = 0 \rightarrow L_f = 0$ is forbidden while it is relaxed by intermediate coupling; the parity should not be changed in the transition.

Finally, for electric-quadrupole transition, the electric-quadrupole moment is a second-order tensor belonging to representation D_{2g}. According to tables of representation product, one has:

When $J_i \geq 2$

$$D_{J_i} \otimes D_2 = D_{J_i+2} \oplus D_{J_i+1} \oplus D_{J_i} \oplus D_{J_i-1} \oplus D_{J_i-2}$$

When $J_i = 1$

$$D_{J_i} \otimes D_2 = D_3 \oplus D_2 \oplus D_1$$

When $J_i = 0$

$$D_{J_i} \otimes D_2 = D_2$$

Therefore, the selection rules for an electric-quadrupole transition are:

$\Delta J = 0, \pm 1, \pm 2$; the parity should not be changed in the transition but $J_i = 0 \rightarrow J_f = 0, 1$ and $J_i = 1 \rightarrow J_f = 0$ are forbidden.

For Russell-Saunders coupling, $\Delta S = 0$, $\Delta L = 0, \pm 1, \pm 2$; the parity should not be changed in the transition but $L_i = 0 \rightarrow L_f = 0, 1$ and $L_i = 1 \rightarrow L_f = 0$ are forbidden.

4.3.2 Selection Rules for Radiative Transition of Ions in Materials

In materials, quantum numbers J, L, and S of active ions are not really good quantum numbers due to the action of crystal field. The good quantum numbers are the irreducible representations of active ion's site symmetry group. Therefore, transition selection rules should be those involving their irreducible representations.

Certainly, when the crystal field strength is weak enough, the selection rules for free ions still have weak limitation. As it has been pointed out previously, the crystal field interaction of transition-metal ions is stronger than that of the spin–orbit interaction, and so the energy levels of transition-metal ions should be denoted first by the related irreducible representation but not the atomic quantum number J, L, and S. Remembering that the electric-dipole has the same irreducible representations as that of a polar vector (the radius of the electron), but the magnetic-dipole moment can be expressed as a vector product of two polar vectors, and the transition selection rules can be introduced easily by applying the tables of the irreducible representation product.

Although the crystal field interaction of rare earth ions in a material is weaker than their spin–orbit coupling and the quantum numbers J, L, and S can be still used to describe their energy levels, the odd parity crystal field interaction has its effect on the electric-dipole transition. Therefore, in this case the selection rules for free ions will not be valid except that of $\Delta S = 0$. It can be shown that the following selection rules are valid:

$\Delta l = \pm 1$, $\Delta S = 0$, $\Delta L \leq 2l$, $\Delta J \leq 2l$ (for rare earth ions $l = 3$), $\Delta J = 2, 4, 6$, if $J = 0_i$ or $J_f = 0$. The forbidden rule of $J_i = 0 \rightarrow J_f = 0$ transition is generally valid except the special cases in which Judd–Ofelt parameter $\Omega_0 \neq 0$. This will be demonstrated in Chap. 5.
Obviously, the electric-dipole transitions inside $4f^n$ configuration ($\Delta l = 0$) are forbidden in the strict sense but this forbidden rule is partly broken by the odd parity perturbation. When the spin–orbit coupling is strong, selection rules on L and S cannot be used. However, if J is still a good quantum number, the selection rule of $\Delta J \leq 2l$ is still valid. For example, in most of solids and liquids, it is difficult to observe the $^6H_{15/2} \rightarrow {}^6F_{1/2}$ transition of Dy^{3+} ions, although it is easy to observe the $^6H_{15/2} \rightarrow {}^6F_{3/2}$ transitions.

Nevertheless, the selection rules related to the irreducible representations of the site symmetry of active ions are strictly valid. The irreducible representation D_{1u} of the electric-dipole moment of full-rotational group can be reduced to the irreducible representations of the site symmetry of active ions as:

$$D_{1u} = \sum_j \Pi_j$$

If the final state is denoted by $|JS\Gamma_f\rangle$ and the initial state by $|JS\Gamma_i\rangle$, then the selection rule is: $\Gamma_i \otimes \Gamma_f^* \supseteq \sum_j \Pi_j$, that is, the product of the irreducible representation of the initial state Γ_i with the complex conjugate irreducible representation of final state Γ_f^* should include the irreducible representation of the dipole moment operator $\sum_j \Pi_j$.

For a uniaxial crystal, the π spectra have selection rules different from those of the σ spectra, because the irreducible representations of π and σ dipole moments are different. Let us take Nd^{3+}:YVO_4 as an example to explain how to use group theory to introduce the selection rule. Nd^{3+} ions in this crystal occupy D_{2d} point group positions; in this group all the irreducible representations are real and so $\Gamma_f^* = \Gamma_f$. The electric-dipole moment can be reduced to the following irreducible representations according to branching rule by Butler's book [8] as well as the table in Appendix B.

$$\Pi_{1u}(SO_3) = \Gamma_5(D_{2d}) \oplus \Gamma_4(D_{2d})$$

Among these representations, it can be seen by Appendix A that the one-dimensional $\Gamma_4(D_{2d})$ corresponds to the z component of the electric-dipole moment and the two-dimensional $\Gamma_5(D_{2d})$ corresponds to the x, y components of the electric-dipole moment. If one has $\Gamma_i \otimes \Gamma_4 = \Gamma_f \oplus \cdots$, then the polarization direction of emission or absorption spectral lines will be along the z direction, that is, the π polarization transition is allowed. If one has $\Gamma_i \otimes \Gamma_5 = \Gamma_f \oplus \cdots$, then the polarization direction of emission or absorption spectral lines will be along the x, y directions, that is, the σ polarization transition will be allowed. For the crystal Nd^{3+}:YVO_4, the crystal field wave functions with D_{2d} point symmetry were given in reference [9]. For the fluorescence spectra corresponding to the transition $^4F_{3/2} \rightarrow {}^4I_{11/2}$, one of the spectra denoted by $b \rightarrow a$ means transition from b level of $^4F_{3/2}$ to a level of $^4I_{11/2}$. These two levels belong to the irreducible representation Γ_7. By using the table of irreducible representation product (Appendix C or reference [10]), it is easy to obtain

$$\Gamma_7 \otimes \Gamma_5 = \Gamma_7 \oplus \Gamma_6, \quad \Gamma_7 \otimes \Gamma_4 = \Gamma_6$$

Obviously, the second product does not include irreducible representation Γ_7 of the final state. Therefore, the emission spectral line can only be σ polarization. For some other transitions, the above two direct products of irreducible representations can include irreducible representations of the final state. Thus, the emission or absorption spectral line can be both σ and π polarizations. What kind of polarization can be observed in a particular situation should be determined by their relative strengths and this problem will be introduced later. The selection rules for transitions of active ions in materials can be tabulated as in the work of Wybourne [11] for the case of electric-dipole transitions in the D_{3h} symmetry system which has even number of electrons (Table 4.1).

For a system with odd number of electrons, similar selection rules can be obtained. However, double-value representations were introduced to match the

Table 4.1 Selection rule of electric-dipole transition in the D_{3h} symmetry system having even number of electrons

Irresp. of the level	Γ_1	Γ_2	Γ_3	Γ_4	Γ_5	Γ_6
Γ_1	–	–	–	π	–	σ
Γ_2	–	–	π	–	–	σ
Γ_3	–	π	–	–	σ	–
Γ_4	π	–	–	–	σ	–
Γ_5	–	–	σ	σ	σ	π
Γ_6	σ	σ	–	–	π	σ

situations of the half-integer J. In this case, electric-dipole moment belongs to Γ_4 (π-polarization) and Γ_6 (σ-polarization) and all the energy levels belong to Γ_7, Γ_8, and Γ_9 representations. According to the table of representation product, it is easy to see that:

$$\Gamma_7 \otimes \Gamma_4 = \Gamma_8, \Gamma_8 \otimes \Gamma_4 = \Gamma_7, \Gamma_9 \otimes \Gamma_4 = \Gamma_9, \Gamma_9 \otimes \Gamma_7 = \Gamma_9 \otimes \Gamma_8 = \Gamma_5 \oplus \Gamma_6$$
$$\Gamma_7 \otimes \Gamma_6 = \Gamma_8 \oplus \Gamma_9, \Gamma_8 \otimes \Gamma_6 = \Gamma_7 \oplus \Gamma_9, \Gamma_9 \otimes \Gamma_6 = \Gamma_7 \oplus \Gamma_8, \Gamma_9 \otimes \Gamma_9$$
$$= \Gamma_1 \oplus \Gamma_2 \oplus \Gamma_3 \oplus \Gamma_4$$

Therefore, one has the Table 4.2.

Selection rules for other point group symmetry systems can be obtained in a similar way. We show here some examples for laser crystals. The rare earth ions occupy D_{2d}, S_4, and D_3 positions in YVO_4, $LiYF_4$, and $YAl_3(BO_3)_4$ crystals, respectively. It should be noted that some of the irreducible representations of the point groups D_3 and S_4 are complex representations, and to determine the selection rule the conjugate irreducible representation of final state must be used. For example, the electric-dipole transition $\Gamma_3 \rightarrow \Gamma_4$ of S_4 position because the z component of the electric-dipole moment belongs to irreducible representation Γ_2 and $\Gamma_3 \otimes (\Gamma_4)^* = \Gamma_2$ and that the electric-dipole transition $\Gamma_3 \rightarrow \Gamma_4$ is π polarized. According to the same method, the electrical-dipole transition rules of even and odd electronic systems are listed in Table 4.3 for the positions with point symmetries of D_{2d}, S_4, and D_3.

From the wave functions, crystal field quantum numbers, and their corresponding irreducible representations of point group D_3 for Nd^{3+}:NAB crystal quoted in Chap. 3, it can be seen that the polarization character of 22 spectral lines produced by the transitions from ${}^4F_{3/2}(1, 2)$ to ${}^4I_{9/2}(1, 2 \ldots 5)$ and ${}^4I_{11/2}(1, 2 \ldots 6)$ are all the same as those calculated in the literature [12] by relating intensity of emission spectra (see Tables 2–5 of that paper). Especially, two transition lines which have been observed by Winzer [13] in his laser experiment: irreducible representations of energy levels ${}^4F_{3/2}(1)$ and ${}^4I_{11/2}(2)$ are Γ_4 and that for energy level ${}^4I_{11/2}(3)$ is $\Gamma_{5,6}$. According to Table 4.3, the light emitted by the transition of ${}^4F_{3/2}(1) \rightarrow {}^4I_{11/2}(2)$ can be σ or π polarization, but the light emitted by the transition of ${}^4F_{3/2}(1) \rightarrow {}^4I_{11/2}(3)$ can only be σ polarization. These are in agreement with the calculation results of Table 2 of paper [12] and fully confirmed by the laser experiment of Winzer [13]. Winzer used two Nd^{3+}:NAB crystal plates with two different orientations. The surface of first crystal plate was perpendicular to the optic axis while the direction of pumping light and laser light were parallel to the optical axis. On the other hand, the surface of second crystal plate was parallel to the

Table 4.2 Selection rule of electric-dipole transition in the D_{3h} symmetry system having odd number of electrons

Irresp. of the level	Γ_7	Γ_8	Γ_9
Γ_7	–	σ π	σ
Γ_8	σ π	–	σ
Γ_9	σ	σ	π

Table 4.3 Selection rule of electric-dipole transition in the D_{2d}, S_4, and D_3 symmetry systems

D_{2d}	Γ_1	Γ_2	Γ_3	Γ_4	Γ_5	Γ_6	Γ_7
Γ_1	–	–	–	π	σ		
Γ_2	–	–	π	–	σ		
Γ_3	–	π	–	–	σ		
Γ_4	π	–	–	–	σ		
Γ_5	σ	σ	σ	σ	π		
Γ_6						σ	σ π
Γ_7						σ π	σ

S_4	Γ_1	Γ_2	Γ_3	Γ_4	Γ_5	Γ_6	Γ_7	Γ_8
Γ_1	–	π	σ	σ				
Γ_2	π	–	σ	σ				
Γ_3	σ	σ	–	π				
Γ_4	σ	σ	π	–				
Γ_5					–	σ	π	σ
Γ_6					σ	–	π	σ
Γ_7					π	σ	–	σ
Γ_8					σ	π	σ	–

D_3	Γ_1	Γ_2	Γ_3	Γ_4	Γ_5	Γ_6
Γ_1	–	π	σ			
Γ_2	π	–	σ			
Γ_3	σ	σ	σ π			
Γ_4				σ π	σ	σ
Γ_5				σ	–	π
Γ_6				σ	π	–

optical axis while the direction of pumping light and laser light were perpendicular to the optic axis. From the first crystal plate, σ polarization laser emission at wavelength 1065.5 nm corresponding to the transition of $^4F_{3/2}(1) \rightarrow {}^4I_{11/2}(3)$ was observed, while from the second crystal plate, σ polarization as well as π polarization laser emission at wavelength 1063.8 nm corresponding to the transition of $^4F_{3/2}(1) \rightarrow {}^4I_{11/2}(2)$ also were observed.

Finally, some problems about magnetic-dipole selection rules should be mentioned. This kind of transition is allowed within the same configuration. The probability of magnetic-dipole transition is proportional to the following square of matrix element

$$|\langle\psi_i|L+2S|\psi_f\rangle|^2$$

In the case that the related quantum numbers are approximately good quantum numbers, the selection rules for free ions are mention previously.

However, the total transition probability between multiplets is distributed into different crystal field levels by the interaction of crystal field and their selection rule should be introduced by the group theory method.
For example in the uniaxial crystal, in case of magnetic-dipole transition, it is generally necessary to measure three kinds of polarization spectra: α spectrum, the light transmits along optic axis $\boldsymbol{c}$ with $\boldsymbol{E}\perp\boldsymbol{c}$, $\boldsymbol{H}\perp\boldsymbol{c}$; σ spectrum, the light transmits perpendicular to optic axis $\boldsymbol{c}$ with $\boldsymbol{E}\perp\boldsymbol{c}$, $\boldsymbol{H}//\boldsymbol{c}$, and π spectrum, the light transmits perpendicular to optic axis $\boldsymbol{c}$ with $\boldsymbol{E}//\boldsymbol{c}$, $\boldsymbol{H}\perp\boldsymbol{c}$. $\boldsymbol{L} = \boldsymbol{r} \times \boldsymbol{p}$ is an axial vector and so the transformation property of L_x, L_y, and L_z is the same as that of R_x, R_y, and R_z. In case of $\boldsymbol{E}\perp\boldsymbol{c}$ or $\boldsymbol{H}\perp\boldsymbol{c}$, the condition of electric-dipole transition and magnetic-dipole transition is

$$\Gamma_i \otimes \Gamma_{R_{x,y}} \otimes \Gamma_f^* = \Gamma_i \otimes \Gamma_{x,y} \otimes \Gamma_f^* \supseteq \Gamma_1$$

In case of $\boldsymbol{E}//\boldsymbol{c}$ or $\boldsymbol{H}//\boldsymbol{c}$, the condition of electric-dipole transition and magnetic-dipole transition is

$$\Gamma_i \otimes \Gamma_{R_z} \otimes \Gamma_f^* = \Gamma_i \otimes \Gamma_z \otimes \Gamma_f^* \supseteq \Gamma_1$$

If the rare earth ions occupy position with D_3 point group, by the irreducible representation character table of group D_3 in Appendix A, it can be seen that R_x, R_y and x, y belong to irreducible representation Γ_3 while R_z and z belong irreducible representation Γ_2. Therefore, in case of $\boldsymbol{E}\perp\boldsymbol{c}$ or $\boldsymbol{H}\perp\boldsymbol{c}$ the condition of electric-dipole transition and magnetic-dipole transition is

$$\Gamma_i \otimes \Gamma_3 \otimes \Gamma_f^* \supseteq \Gamma_1$$

On the other hand, in case of $\boldsymbol{E}//\boldsymbol{c}$ or $\boldsymbol{H}//\boldsymbol{c}$, the condition of electric-dipole transition and magnetic-dipole transition is

$$\Gamma_i \otimes \Gamma_2 \otimes \Gamma_f^* \supseteq \Gamma_1$$

First, to see transition $\Gamma_4 \Leftrightarrow \Gamma_4$, it can be found from Appendix C that the related irreducible representation product is (note that Γ_4 is a real irreducible representation)

$$\Gamma_4 \otimes \Gamma_3 \otimes \Gamma_4 \supseteq \Gamma_1 \quad \Gamma_4 \otimes \Gamma_2 \otimes \Gamma_4 \supseteq \Gamma_1$$

Therefore, no matter $\boldsymbol{E}//\boldsymbol{c}$ or $\boldsymbol{E}\perp\boldsymbol{c}$ and $\boldsymbol{H}//\boldsymbol{c}$ or $\boldsymbol{H}\perp\boldsymbol{c}$, $\Gamma_4 \Leftrightarrow \Gamma_4$ can have electric-dipole and magnetic-dipole transitions. It means that the α,σ, and π polarized spectra between different energy Γ_4 levels can be observed and these three kinds of spectra all have electric-dipole and magnetic-dipole transition components.

Next to see the transition of $\Gamma_4 \Leftrightarrow \Gamma_{5,6}$. Here Γ_5 and Γ_5 are conjugate to each other and constitute a doubly degenerate level and so written as $\Gamma_{5,6}$

$$\Gamma_4 \otimes \Gamma_3 \otimes \Gamma_{5,6} \supseteq \Gamma_1, \ \Gamma_4 \otimes \Gamma_2 \otimes \Gamma_{5,6} \notin \Gamma_1$$

Therefore, between Γ_4 and $\Gamma_{5,6}$ we can have electric-dipole transition with $\boldsymbol{E} \perp \boldsymbol{c}$ and magnetic-dipole transition with $\boldsymbol{H} \perp \boldsymbol{c}$. It means that between these two kinds of energy levels, the α, σ, and π polarized spectra can be observed but the σ polarized spectrum only has the component of electric-dipole transition and the π polarized spectrum only has the component of magnetic-dipole transition.

Last to see the transition of $\Gamma_{5,6} \Leftrightarrow \Gamma_{5,6}$, one has

$$\Gamma_{5,6} \otimes \Gamma_2 \otimes \Gamma_{5,6} = \Gamma_1, \ \Gamma_{5,6} \otimes \Gamma_3 \otimes \Gamma_{5,6} \notin \Gamma_1$$

By using the above discussion method, it can be shown that between two energy levels $\Gamma_{5,6}$, the π and σ polarized spectra can be observed, but π polarized spectrum only has the component of electric-dipole transition and σ polarized spectrum only has the component of magnetic-dipole transition. The selection rule can be summarized in the Table 4.4.

When rare earth ion occupies a position with other point group, their transition selection rules can be introduced by the similar method. It should be noted that Table 4.4 only shows the polarization character between crystal field levels. The precondition for magnetic-dipole transition is the related transition should be allowed by magnetic-dipole selection rule between related multiplets.

Because the intensity of electric-quadrupole transition is very weak, we would not discuss their transition selection rule. Actually, it is easy to use similar method to introduce their transition selection rule so long as one knows that the components of quadrupole moment transform as those of x^2, y^2, z^2, xy, yz, and zx, then find out the irreducible representation of these functions from character table of irreducible representation.

Table 4.4 Selection rule for rare earth ion occupy D_3 point group position

Irresp. of the energy level	Γ_4	$\Gamma_{5,6}$
Γ_4	α(ed.,md.), σ(ed.,md.), π(ed., md.)	α(ed., md.), σ(ed.), π(md.)
$\Gamma_{5,6}$	α(ed., md.), σ(ed.), π(md.)	π(ed.), σ(md.)

References

1. A.S. Davydov, *Quantum Mechanics*, 2nd edn. (Pergamon Press, Oxford, 1976)
2. L.D. Landau, E.M. Lifshitz, *Field Theory* (Pergamon Press, Oxford, 1958)
3. H. Goldstein, *Classical Mechanics* (Addison-Wesley Publishing Company, Inc., Third Printing, 1953)
4. W. Heitler, *The Quantum Theory of Radiation* (Clarendon Press, Oxford, 1954)
5. I.N. Broushtein, K.A. Semendyayev, *Handbook of Mathematics* (Van Nostrand Reinhold Company, New York, 1985)
6. P. Hong, S.R. Shinn, Matter. Res. Bull. **11**, 461 (1976)
7. M. Tinkham, *Group Theory and Quantum Mechanics* (McGraw-Hill Inc., New York, 1964)
8. P.H. Butler, *Point Group Symmetry Application Method & Tables* (Plenum Press, New York, 1981)
9. Z.D. Luo, Y.D. Huang, J. Phys, Condens. Matter. **6**, 3737 (1994)
10. G.F. Koster, J.O. Dimmock, R.G. Wheeler, et al. *Properties of Thirty-Two Point Group* (M.I. T. Press, Massachusetts, 1963)
11. B.G. Wybourne, *Spectroscopic Properties of Rare Earths* (Wiley, New York, 1965)
12. Y.D. Huang, Z.D. Luo, J. Phys.: Condens. Matter **5**, 1581 (1993)
13. G. Winzer, P.C. Mocke, W.W. Krcehler, IEEE J. Quant. Elect. QE, **14**(840) (1978)

Chapter 5
Spectroscopic Parameter and Their Calculation

5.1 Absorption Coefficient, Absorption (Emission) Cross-Section, and Oscillator Strength

In the studies of spectroscopic properties of crystals and non-crystalline solid materials, the physical parameters normally used are absorption coefficient, absorption (emission) cross-section, and oscillator strength but not the transition probability. We would like to discuss their physical meaning and calculation formulas while only deal with particular calculation problems of trivalent rare earth ion in this chapter. The calculation of spectroscopic parameters for transition metal ion is put into Chap. 7.

In the previous discussion on transition probability, it was assumed that the electric field acting on the electrons was the electric field of external radiation field. This is in fact only valid in vacuum. In solid materials, if radiation electric field is denoted by $\boldsymbol{E}$, and the local field of electron is denoted by $\boldsymbol{E}_{\text{loc}}$, then the transition matrix elements, which involve the electric field acting on the electrons, should be multiplied by a factor of $\boldsymbol{E}_{\text{loc}}/\boldsymbol{E}$ and the transition probabilities should be multiplied by a factor of $(\boldsymbol{E}_{\text{loc}}/\boldsymbol{E})^2$. This factor has different expressions for different local symmetries. In cases of high symmetry, it can be shown that its form is $[(n^2 + 2)/3]^2$ (where n is refractive index of the solid material). This correction factor is also used in low symmetry situations now. On the other hand, the velocity of light in the materiel should be c/n. Therefore, from (4.19) and (4.20), the electric-dipole absorption (stimulated emission) and spontaneous emission probabilities will become:

$$P_k^{\text{ed}}(\text{abs.}) = P_k(\text{st.em.}) = \frac{4\omega_{\mathbf{k}}^3 n}{3\hbar c^3}\left(\frac{n^2+2}{3}\right)^2 n_{\mathbf{k}}|\boldsymbol{\mu}|^2 \tag{5.1}$$

Z. Luo and Y. Huang, *Physics of Solid-State Laser Materials*, Springer Series in Materials Science 289, https://doi.org/10.1007/978-981-32-9668-8_5

$$P_k^{\text{ed}}(\text{sp.em.}) = \frac{4\omega_{\mathbf{k}}^3 n}{3\hbar c^3}\left(\frac{n^2+2}{3}\right)^2 |\boldsymbol{\mu}|^2 \tag{5.2}$$

where $\varepsilon = n^2$ has been used in obtaining the above formulas. The spontaneous emission probability is usually denoted by A and given by

$$A^{\text{ed}} = \frac{4\omega_{\boldsymbol{k}}^3 n}{3\hbar c^3}\left(\frac{n^2+2}{3}\right)^2 |\boldsymbol{\mu}|^2 \tag{5.2a}$$

By similar introduction, replacing the refractive index correction factor $(n^2 + 2)^2/9n$ by n, it can be shown that the probability of spontaneous magnetic-dipole emission is

$$A^{\text{md}} = \frac{4\omega_{\mathbf{k}}^3 n^3}{3\hbar c^3}|\boldsymbol{M}^{\text{md}}|^2 \tag{5.2b}$$

Integrations over the solid angle have been made in obtaining these formulas and $\boldsymbol{\mu}$ is used to denote the electric-dipole matrix element. $\boldsymbol{M}^{\text{md}}$ is used to denote the magnetic-dipole matrix element.

In order to express the transition probabilities of absorption and stimulated emission, photon number n_k is usually replaced by photon energy density $\text{d}(\omega_k)$ or light intensity $I(\omega_k)$. One can obtain their relation in the following discussion.

The energy of photon field per unit volume is photon energy density $\text{d}(\omega_k)$ integrated over frequency ω_k. It can also be expressed as the k integral of two times the product of photon energy, photon number, and wave vector density (there are two polarization directions belonging to each particular direction of $\boldsymbol{k}$). Therefore one has

$$\int \text{d}(\omega_k)\text{d}\omega_k = \frac{2}{V}\int \hbar\omega_k\left(n_k + \frac{1}{2}\right)\rho(k)\text{d}k$$

By using

$$\rho(k)\text{d}k = \frac{V}{8\pi^3}4\pi k^2\text{d}k = \frac{V\omega_k^2 n^3}{2\pi^2 c^3}\text{d}\omega_k$$

we finally have

$$\text{d}(\omega_k) = \frac{\hbar\omega_k^3 n^3}{\pi^2 c^3}n_k$$

and

$$n_k = \frac{\pi^2 c^3}{\hbar \omega_k^3 n^3} \mathrm{d}(\omega_k) \tag{5.3}$$

Substituting (5.3) into (5.1), the following can be obtained

$$P_k^{\mathrm{ed}}(\mathrm{abs.}) = P_k^{\mathrm{ed}}(\mathrm{st.em.}) = \frac{4\pi^2}{3\hbar^2 n^2}\left(\frac{n^2+2}{3}\right)^2 \mathrm{d}(\omega_k)|\boldsymbol{\mu}|^2 \tag{5.4}$$

By using $I(\omega_k) = \frac{c}{n}\mathrm{d}(\omega_k)$, formula (5.4) changes into

$$P_k^{\mathrm{ed}}(\mathrm{abs.}) = P_k^{\mathrm{ed}}(\mathrm{st.em.}) = \frac{4\pi^2}{3\hbar^2 nc}\left(\frac{n^2+2}{3}\right)^2 |\boldsymbol{\mu}|^2 I(\omega_k) \tag{5.5a}$$

Similarly

$$P_k^{\mathrm{md}}(\mathrm{abs.}) = P_k^{\mathrm{md}}(\mathrm{st.em.}) = \frac{4\pi^2 n}{3\hbar^2 c}\left|\boldsymbol{M}^{\mathrm{md}}\right|^2 I(\omega_k) \tag{5.5b}$$

Let us consider a light beam passing through a system with two energy levels E_1 and E_2 along x direction (E_2 and E_1 satisfying $\omega_k = \frac{E_2 - E_1}{\hbar}$). N_2 and N_1 are the particle densities of the high and low energy levels having degeneracy g_1 and g_2, respectively. The electron in the low energy level absorbs light and go up to the high energy level; at the same time, the electron in the high energy level emits light and go down to the low energy level. Therefore, in the calculation of light absorption, the contributions from two kinds of transition must be considered. $P_k(1_n, 2_m)$ is used to denote the transition probability from the nth degenerate state in level 1 to the mth degenerate state in level 2 and the transition probability from the mth degenerate state in level 2 to the nth degenerate state in level 1 is expressed by $P_k(2_m, 1_n)$. m and n are used to denote the degenerate states of high and low levels, respectively. Therefore, the total transition probability of the level 1 into level 2 should be averaged over all the initial degenerate states of level 1 and summed over all the final degenerate states of level 2. In this calculation, the spontaneous emission has been neglected, because it is very weak compared to the stimulated emission. Thus

$$W_{12} = \frac{1}{g_1}\sum_{1_n, 2_m} P_k(1_n, 2_m) \tag{5.6}$$

Similarly, the total transition probability from level 2 to level 1 is

$$W_{21} = \frac{1}{g_2}\sum_{1_n, 2_m} P_k(2_m, 1_n) \tag{5.7}$$

According to (5.1) or (5.5a, 5.5b), $P_k\,(2_m, 1_n) = P_k\,(1_n, 2_m)$, therefore

$$g_1 W_{12} = g_2 W_{21} \tag{5.8}$$

By means of the results obtained, it can be seen that because of the absorption, the light passing through a solid material will be attenuated. Its intensity change in a distance dx can be expressed as

$$\begin{aligned} \mathrm{d}I(\omega_k) &= -(N_1 W_{12} - N_2 W_{21})\hbar\omega_k \mathrm{d}x = -\left(N_1 - \frac{g_2}{g_1} N_2\right) W_{12}\hbar\omega_k \mathrm{d}x \\ &= -\left[\frac{4\pi^2}{3\hbar c}\frac{(n^2+2)^2 N_1}{9ng_1}\sum_{1_n,2_m}\left|\boldsymbol{\mu}_{1_n 2_m}\right|^2 \omega_k g(\omega_\mathrm{k})\right] I(\omega_k)\mathrm{d}x \\ &\equiv -\kappa^{ed}(\omega_k) I(\omega_k)\mathrm{d}x \end{aligned} \tag{5.9}$$

where $g(\omega_k)$ is introduced to take into account the absorption line shape and the value of g_1N_2/g_2 has been neglected, because particle number in the excited state is much smaller than that in the ground state. Hence the intensity variation by passing through the media for distance x can be expressed as

$$I(\omega_\mathrm{k}) = I_0(\omega_k)\mathrm{e}^{-\kappa(\omega_k)x} \tag{5.10}$$

where $\kappa(\omega_k)$ is the absorption coefficient defined as follows

$$\kappa^{\mathrm{ed}}(\omega_k) = \frac{4\pi^2}{3\hbar c}\frac{(n^2+2)^2 N_1}{9ng_1}\sum_{1_n,2_m}\left|\boldsymbol{\mu}_{1_n 2_m}\right|^2 \omega_k g(\omega_\mathrm{k}) \tag{5.11a}$$

Similarly

$$\kappa^{\mathrm{md}}(\omega_k) = \frac{4\pi^2 n}{3\hbar c}\frac{N_1}{g_1}\sum_{1_n,2_m}\left|\boldsymbol{M}_{12}^{\mathrm{md}}\right|^2 \omega_k g(\omega_\mathrm{k}) \tag{5.11b}$$

The line shape factor $g(\omega_k)$ is normalized according to

$$\int g(\omega_\mathrm{k})\mathrm{d}\omega_\mathrm{k} = 1$$

If one uses wavelength as variable, it should be noted that

$$\int g(\omega_\mathrm{k})\mathrm{d}\omega_\mathrm{k} = \int \frac{2\pi c}{\lambda_k^2 n} g(\omega_k)\mathrm{d}\lambda_k = \int g(\lambda_k)\mathrm{d}\lambda_k = 1$$

so that

$$g(\omega_k) = \frac{\lambda_k^2 n}{2\pi c} g(\lambda_k)$$

The result of $\int \kappa(\lambda_k)\mathrm{d}\lambda_k$ should be

$$\begin{aligned}\int \kappa^{\mathrm{ed}}(\lambda_\mathrm{k})\mathrm{d}\lambda_\mathrm{k} &= \int \frac{\lambda_k^2 n}{2\pi c}\frac{4\pi^2 n}{3\hbar c}\frac{(n^2+2)^2 N_1}{9ng_1}\left(\sum_{1_n 2_m}|\boldsymbol{\mu}_{1_n 2_m}|^2\right)\frac{2\pi c}{\lambda_k n} g(\lambda_k)\mathrm{d}\lambda_k \\ &= \frac{8\pi^3\bar{\lambda}}{3hc}\frac{(n^2+2)^2 N_1}{9ng_1}\sum_{1_n,2_m}|\boldsymbol{\mu}_{1_n 2_m}|^2\end{aligned} \tag{5.12a}$$

Similarly

$$\begin{aligned}\int \kappa^{\mathrm{md}}(\lambda_\mathrm{k})\mathrm{d}\lambda_\mathrm{k} &= \int \frac{\lambda_k^2 n}{2\pi c}\frac{4\pi^2 n}{3\hbar c}\frac{N_1}{g_1}\left(\sum_{1_n 2_m}|(\boldsymbol{M}_{12}^{\mathrm{md}})|^2\right)\frac{2\pi c}{\lambda_k n} g(\lambda_k)d\lambda_k \\ &= \frac{8\pi^3\bar{\lambda}_k n N_1}{3hcg_1}\ {}_1\sum_{1_n,2_m}|(\boldsymbol{M}_{12}^{\mathrm{md}})|^2\end{aligned} \tag{5.12b}$$

where the average wavelength can be expressed as $\bar{\lambda} = \int \lambda_k g(\lambda_k)\mathrm{d}\lambda_k$. Equations (5.12a) and (5.12b) are widely applied in spectroscopic calculations. $\chi = (n^2 + 2)^2/9n$ was introduced as a correction factor by refractive index of the solid material in the literature (e.g. [1]). Therefore

$$\int \kappa^{\mathrm{ed}}(\lambda_k)\mathrm{d}\lambda_k = \left(\frac{N_1}{g_1}\frac{8\pi^3\bar{\lambda}}{3hc}\chi\sum_{1_n,2_m}|\boldsymbol{\mu}_{1_n 2_m}|^2\right) \tag{5.13}$$

The absorption cross-section and emission cross-section are often used to express absorption rate and emission rate in spectroscopy. Absorption cross-section is defined as the quotient of the absorption coefficient divided by the number of absorption centers per unit volume. Therefore, the absorption cross-section $\sigma_{12}(\omega_\mathrm{k})$ and $\sigma_{12}(\lambda_\mathrm{k})$ as the functions of ω_k and λ_k, respectively, can be expressed as

$$\sigma_{12}^{\mathrm{ed}}(\omega_k) = \frac{4\pi^2\omega_k}{3\hbar c g_1}\chi\sum_{1_n,2_m}|\boldsymbol{\mu}_{1_n 2_m}|^2 g(\omega_\mathrm{k}) \tag{5.14a}$$

$$\sigma_{12}^{\mathrm{md}}(\omega_k) = \frac{4\pi^2\omega_k n}{3\hbar c g_1}\sum_{1_n,2_m}|(\boldsymbol{M}_{12}^{\mathrm{md}})|^2 g(\omega_\mathrm{k}) \tag{5.14b}$$

$$\int \sigma_{12}^{\mathrm{ed}}(\lambda_k)\mathrm{d}\lambda_k = \left(\frac{8\pi^3\bar{\lambda}}{3hcg_1}\chi\sum_{1_n 2_m}|\boldsymbol{\mu}_{1_n 2_m}|^2\right) \tag{5.15a}$$

$$\int \sigma_{12}^{\mathrm{md}}(\lambda_k)\mathrm{d}\lambda_k = \left(\frac{8\pi^3\bar{\lambda}n}{3hcg_1}\sum_{1_n 2_m}\left|\left(\boldsymbol{M}_{12}^{\mathrm{md}}\right)\right|^2\right) \tag{5.15b}$$

The oscillator strength of certain transition in spectroscopy can be defined as the ratio of the power emitted or absorbed by the transition to the power emitted or absorbed by a classical oscillator. The power emitted or absorbed by a classical oscillator can be introduced from the following discussion.

Let us discuss the oscillator strength of electric-dipole transition first. By using classical radiation theory, the movement equation for an electric-dipole oscillator can be written as

$$m\ddot{x} + R\dot{x} + Gx = e\boldsymbol{E}_\omega e^{i\omega t} \tag{5.16}$$

The second and the third terms in the above equation are the resistant force and the restoring force of the free oscillator, respectively. The term on the right-hand side of the equation represents the driving force of the radiation field, where $\dot{x}$ denotes the first-order derivative $\mathrm{d}\boldsymbol{x}/\mathrm{d}t$ and $\ddot{x}$ is the second-order derivative $\mathrm{d}^2\boldsymbol{x}/\mathrm{d}t^2$. Assuming that without the driving force of external electric field, the vibrational frequency of the free oscillator is ω_0. The solution of (5.16) should be $x_0 e^{-(\gamma/2)t}\cos(\omega_0 t - \beta)$ (the general solution of the homogeneous equation) plus $\frac{eE_\omega \cos(\omega t - \alpha)}{m} / \left((\omega_0^2 - \omega^2)^2 + \gamma^2\omega^2\right)^{1/2}$ ($\gamma = R/m$) (the special solution of the inhomogeneous equation). The work done by the radiation field driving force $\boldsymbol{F}(\omega) = e\boldsymbol{E}_\omega e^{i\omega t}$ moving a distant $\mathrm{d}\boldsymbol{x}$ is $\mathrm{d}w = \boldsymbol{F}(\omega)\cdot d\boldsymbol{x} = \boldsymbol{F}(\omega)(\mathrm{d}\boldsymbol{x}/\mathrm{d}t)\mathrm{d}t$. If $\boldsymbol{F}(\omega)$ is calculated by the left side of (5.16), then the power done by radiation field driving force for all the frequency ω will be

$$P = \int_0^\infty \left(\frac{\mathrm{d}w}{\mathrm{d}t}\right)\mathrm{d}\omega = \int_0^\infty (m\ddot{x} + R\dot{x} + Gx)\dot{x}\mathrm{d}\omega$$

Substituting the solution of $\boldsymbol{x}$ into above equation and using the approximate relations: $\omega_0 \approx \omega$, $\omega^2 - \omega_0^2 \approx 2\omega(\omega_0 - \omega)$, and noting that the time averages of both $\ddot{x}\dot{x}$ and $\dot{x}x$ are zero and the free oscillation component of $\boldsymbol{x}$ decays to zero quickly, one can obtain

$$\overline{P} = \int_0^\infty R\overline{\dot{x}^2}\mathrm{d}\omega = \frac{e^2E_\omega^2\gamma}{8m}\int_0^\infty \frac{\mathrm{d}\omega}{(\omega_0 - \omega)^2 + \gamma^2/4} = \frac{2\pi e^2 E_\omega^2}{8m}$$

where $\gamma = R/m$ has been taken in the above equation. Taking into account the time-averaged energy density of the radiation field $\rho_\omega = E_\omega^2/8\pi$ and the radiation intensity $I(\omega) = c\rho_\omega$, the following relationship between the average absorption power of a classical oscillator in the vacuum and the intensity of the radiation field can be obtained

$$\overline{P} = \frac{2\pi^2 e^2}{mc} I(\omega) \tag{5.17}$$

According to the definition, the power absorbed by the transition of an electron $\hbar\omega W_{12}$ must be equal to the product of absorption oscillator strength f_{12} multiplied by $\overline{P}$ and the factor of line shape $g_L(\omega_k)$, then for the electric-dipole absorption process

$$\begin{aligned}\frac{2\pi^2 e^2}{mc} I(\omega_k) f_{12}^{ed} g_L(\omega_k) = \hbar\omega W_{12} &= \frac{\hbar\omega_k \sum_{n,k} P_k(1_n, 2_m)}{g_1} \\ &= \left[\frac{4\pi^2}{3\hbar c}\frac{(n^2+2)^2}{9ng_1}\sum_{1_n,2_m}\left|\boldsymbol{\mu}_{1_n2_m}\right|^2 \omega_k g(\omega_k)\right] I(\omega_k)\end{aligned}$$

Equation (5.5a) has been used in the derivation of above equation. Therefore, the absorption oscillator strength f_{12}^{ed} of the electric-dipole is expressed as

$$f_{12}^{ed} = \frac{2}{3\hbar e^2}\frac{m}{g_1}\omega_k \frac{(n^2+2)^2}{9n}\sum_{1_n,2_m}\left|\boldsymbol{\mu}_{1_n2_m}\right|^2 \tag{5.18a}$$

Similarly, the emission oscillator strength f_{21}^{ed} of the electric-dipole transition is expressed as

$$f_{21}^{\mathrm{ed}} = \frac{2}{3\hbar e^2}\frac{m}{g_2}\omega_k \frac{(n^2+2)^2}{9n}\sum_{1_n,2_m}\left|\boldsymbol{\mu}_{2_n1_m}\right|^2 \tag{5.18b}$$

According to the same discussion method, remember that the refractive index correction factor $(n^2 + 2)^2/9n$ should be replaced by n and the electric-dipole matrix element replaced by the magnetic coupling matrix element. The absorption oscillator strength f_{12}^{md} of the magnetic-dipole transition is expressed as

$$f_{12}^{\mathrm{md}} = \frac{2}{3\hbar e^2}\frac{m}{g_1}\omega_k n \sum_{1_n,2_m}\left|\boldsymbol{M}_{12}^{\mathrm{md}}\right|^2 \tag{5.18c}$$

The corresponding emission oscillator strength f_{21}^{md} of the magnetic-dipole transition is expressed as

$$f_{21}^{\mathrm{md}} = \frac{2}{3\hbar e^2}\frac{m}{g_2}\omega_k n \sum_{2,1_m}\left|\boldsymbol{M}_{21}^{\mathrm{md}}\right|^2 \tag{5.18d}$$

The absorption oscillator strength f_{12}^{ed} of electric-dipole transition for the trivalent rare earth ions in solids is in the order of magnitude 10^{-5}–10^{-6} and for the divalent rare earth ions, it can reach the order of magnitude 10^{-2}–10^{-3}, nearly the same as that for the transition in $3d^n$ configurations of transition-metal ions.

One can see another definition of the absorption oscillator strength in the literature as follows

$$f_{12}^{ed} = \frac{2}{3\hbar e^2} \frac{m}{g_1} \omega_k \sum_{1_n, 2_m} \left| \boldsymbol{\mu}_{1_n 2_m} \right|^2 \tag{5.19}$$

It gives the ratio of the power of the absorption to that of the classical oscillator, if the transition concerned is taking place in the vacuum. Obviously, the definition what we adopted can describe the actual situation for the transition occur in material and is much more reasonable. It is also the same as that given by Judd [2] and other classical literature [3].

5.2 Analysis of the Absorption Coefficients of Anisotropic Crystal

Consider the general situation of biaxial crystals. If the absorption coefficients of the crystal for the light polarized in the direction of crystal physical axes *X*, *Y*, and *Z* are denoted by κ_X, κ_Y, and κ_Z, respectively, the absorption coefficients for the light polarized in any direction can be calculated by the following way.

Suppose the light is polarized in the direction with angle variables θ and φ in the spherical coordinate system constituted by crystal physical axes *X*, *Y*, and *Z*, then the electric field $\boldsymbol{E}$ of the light has projections of $\boldsymbol{E}\sin\theta\cos\varphi$, $\boldsymbol{E}\sin\theta\sin\varphi$, and $\boldsymbol{E}\cos\theta$ in the directions of coordinate axes *X*, *Y*, and *Z*, respectively. Obviously, the light intensity of the component polarized in the directions of *X*, *Y*, and *Z* will be $I_0\sin^2\theta\cos^2\varphi$, $I_0\sin^2\theta\sin^2\varphi$, and $I_0\cos^2\theta$, respectively. After the light pass through a crystal with thickness *L*, the intensity *I* of the transmitted light will be

$$I(\theta,\varphi) = I_0 \sin^2\theta\cos^2\varphi \exp\left(-\kappa^{(x)}L\right) + I_0 \sin^2\theta\sin^2\varphi \exp\left(-\kappa^{(y)}L\right) + I_0\cos^2\theta\exp\left(-\kappa^{(z)}L\right)$$

$$I(\theta,\varphi)/I_0 = \sin^2\theta\cos^2\varphi \exp\left(-\kappa^{(x)}L\right) + \sin^2\theta\sin^2\varphi \exp\left(-\kappa^{(y)}L\right) + \cos^2\theta\exp\left(-\kappa^{(z)}L\right)$$

where $\kappa^{(x)}$, $\kappa^{(y)}$, $\kappa^{(z)}$ are the absorption coefficients of the crystal for the light polarized in the direction of the axes *X*, *Y*, and *Z*, respectively. It can be demonstrated that the absorption coefficients of the crystal for the light with wavelength λ can be calculated by

$$\begin{aligned}\kappa(\lambda,\theta,\varphi) &= \lim_{\kappa L\to 0}\left(-\frac{1}{L}\ln\left(\frac{I(\theta,\varphi)}{I_0}\right)\right) \\ &= \kappa^{(x)}(\lambda)\sin^2\theta\cos^2\varphi + \kappa^{(y)}(\lambda)\sin^2\theta\sin^2\varphi + \kappa^{(z)}(\lambda)\cos^2\theta \end{aligned} \tag{5.20}$$

In a uniaxial crystal, the absorption coefficients for the light polarized in X and Y directions are equal and denoted by κ_σ and the absorption coefficient for the light polarized in Z direction is denoted by κ_π, then

$$\kappa(\lambda,\theta,\varphi) = \kappa^{(\sigma)}(\lambda)\sin^2\theta + \kappa^{(\pi)}(\lambda)\cos^2\theta \tag{5.21}$$

If the unit vector of the electric field $\boldsymbol{E}$ is denoted by $\boldsymbol{e}$, and the unit vectors of X, Y, and Z axes are denoted by $\boldsymbol{x}$, $\boldsymbol{y}$, and $\boldsymbol{z}$, respectively, then (5.20) can be expressed as

$$\kappa(\boldsymbol{e},\lambda) \approx \kappa^{(x)}(\lambda)(\boldsymbol{e}\cdot\boldsymbol{x})^2 + \kappa^{(y)}(\lambda)(\boldsymbol{e}\cdot\boldsymbol{y})^2 + \kappa^{(z)}(\lambda)(\boldsymbol{e}\cdot\boldsymbol{z})^2 \tag{5.20a}$$

It is known from Born and Wolf [4] that an unpolarized light transmitting in the direction of θ', φ' with intensity I_0 can be considered as two lights polarized in the direction perpendicular to the direction of θ', φ' with intensity $I_0/2$. If these two polarized light beams pass through a uniaxial crystal with thickness L, the intensity of the transmitted light can be expressed as

$$I_t(\lambda,\theta',\varphi') = \frac{1}{2}\sin^2\theta' I_0\exp\left[-\kappa^{(\pi)}L\right] + \frac{1}{2}(1+\cos^2\theta')I_0\exp\left[-\kappa^{(\sigma)}L\right]$$

where θ' is the angle between the incident light I_0 (i.e. the normal direction of the crystal) and the optic c axis (i.e. the direction of the crystal physical Z axis). The absorption coefficient can be expressed as

$$\kappa(\lambda,\theta',\varphi') = \lim_{\kappa L\to 0} -\frac{1}{L}\ln\left\{\frac{1}{2}\sin^2\theta'\exp\left[-\kappa^{(\pi)}(\lambda)L\right] + \frac{1}{2}(1+\cos^2\theta')\exp\left[-\kappa^{(\sigma)}L\right]\right\} \tag{5.22}$$

This expression is the same as that obtained by Lomheim and Deshazer [11]. Physically, $\kappa(\lambda, \theta',\varphi')$ should be obtained in the condition of $\kappa L \ll 1$. Satisfying this condition then one has

$$\kappa(\lambda,\theta',\varphi') \approx \frac{1}{2}\kappa^{(\pi)}\sin^2\theta' + \frac{1}{2}\kappa^{(\sigma)}(1+\cos^2\theta') \tag{5.23}$$

Note that the angle parameters θ', φ' are different from θ, φ.

The expression (5.23) can also be obtained by (5.21): According to the method mentioned above, an unpolarized light with intensity I_0 and transmitting in the direction of θ', φ' can be decomposed into two lights of intensity $I_0/2$ with polarization direction $\theta_1 = \theta' + \pi/2$, $\varphi_1 = \varphi'$ and $\theta_2 = \pi/2$, $\varphi_2 = \varphi' + \pi/2$, respectively. Then by (5.21) one can obtain

$$\begin{aligned}\kappa(\lambda,\theta',\varphi') &= \frac{1}{2}\left(\kappa^{(\sigma)}(\lambda)\sin^2(\theta'+\pi/2)+\kappa^{(\pi)}(\lambda)\cos^2(\theta'+\pi/2)\right)\\ &\quad+\frac{1}{2}\left(\kappa^{(\sigma)}(\lambda)\sin^2(\pi/2)+\kappa^{(\pi)}(\lambda)\cos^2(\pi/2)\right)\\ &= \frac{1}{2}\left(\kappa^{(\sigma)}(\lambda)\cos^2\theta'+\kappa^{(\pi)}(\lambda)\sin^2\theta'\right)+\frac{\kappa^{(\sigma)}(\lambda)}{2}\\ &= \frac{1}{2}\kappa^{(\pi)}(\lambda)\sin^2\theta'+\frac{1}{2}\kappa^{(\sigma)}(\lambda)(1+\cos^2\theta')\end{aligned}$$

Clearly, for a biaxial crystal, calculated by (5.20), we can show

$$\begin{aligned}\frac{1}{4\pi}\int\Gamma\mathrm{d}\Omega &= \int\left[\frac{1}{4\pi}\int\kappa(\lambda,\theta,\varphi)\mathrm{d}\Omega\right]\mathrm{d}\lambda = \frac{1}{3}\int\kappa^{(x)}\mathrm{d}\lambda+\frac{1}{3}\int\kappa^{(y)}\mathrm{d}\lambda+\frac{1}{3}\int\kappa^{(z)}\mathrm{d}\lambda\\ &= \frac{1}{3}\left(\Gamma^{(X)}+\Gamma^{(Y)}+\Gamma^{(Z)}\right)\end{aligned} \tag{5.24}$$

where Γ is the integral of the absorption coefficient over the wavelength range and is usually called absorbance. For uniaxial crystal, one has

$$\begin{aligned}\frac{1}{4\pi}\int\Gamma\mathrm{d}\Omega &= \int\left[\frac{1}{4\pi}\int\kappa(\lambda,\theta,\varphi)\mathrm{d}\Omega\right]\mathrm{d}\lambda = \frac{1}{3}\int\kappa^{(\pi)}\mathrm{d}\lambda+\frac{2}{3}\int\kappa^{(\sigma)}\mathrm{d}\lambda\\ &= \frac{1}{3}\Gamma^{(\pi)}+\frac{2}{3}\Gamma^{(\sigma)}\end{aligned} \tag{5.25}$$

The probability of spontaneous transition polarized in q direction can be expressed as

$$P^{(q)}(\text{sp.em.}) = \frac{8\pi cn^2}{3\lambda^4 N_1}\Gamma^{(q)} \tag{5.26}$$

$$P(\text{sp.em.}) = \frac{8\pi cn^2}{3\lambda^4 N_1}\left(\Gamma^{(X)}+\Gamma^{(Y)}+\Gamma^{(Z)}\right) \tag{5.27}$$

$$P(\text{sp.em.}) = \frac{8\pi cn^2}{\lambda^4 N_1}\left(\frac{1}{3}\Gamma^{(\pi)}+\frac{2}{3}\Gamma^{(\sigma)}\right) \tag{5.28}$$

One of the author's paper [5] has shown that if one cannot measure polarized absorbance data but would like to obtain effective J–O parameters, then the total integrated absorbance resulting from unpolarized integrated absorbance data measured by three thin crystal slice mutual perpendicularly cutting from an arbitrary orientation crystal cube is a good approach to the space integrated averaged absorbance required by (5.24) and (5.25). This can also be generally demonstrated by (5.20a). Suppose three unpolarized light beams are input in three mutually perpendicular directions of $\boldsymbol{e}_1$, $\boldsymbol{e}_2$, and $\boldsymbol{e}_3$, respectively, then by using the method mentioned above, an unpolarized light of intensity I_0 can be decomposed into two polarized lights of intensity $I_0/2$ with polarization directions perpendicular to each other, in the condition of $\kappa L \ll 1$. According to (5.20a), the absorption coefficients in these three directions can be calculated by

$$\kappa^1 \equiv \kappa(\boldsymbol{e}_1) \approx \frac{1}{2}\left((\boldsymbol{e}_2 \cdot \boldsymbol{x})^2\kappa^{(x)} + (\boldsymbol{e}_2 \cdot \boldsymbol{y})^2\kappa^{(y)} + (\boldsymbol{e}_2 \cdot \boldsymbol{z})^2\kappa_z\right) + \frac{1}{2}\left((\boldsymbol{e}_3 \cdot \boldsymbol{x})^2\kappa^{(x)} + (\boldsymbol{e}_3 \cdot \boldsymbol{y})^2\kappa^{(y)} + (\boldsymbol{e}_3 \cdot \boldsymbol{z})^2\kappa_z\right)$$

$$\kappa^2 \equiv \kappa(\boldsymbol{e}_2) \approx \frac{1}{2}\left((\boldsymbol{e}_1 \cdot \boldsymbol{x})^2\kappa^{(x)} + (\boldsymbol{e}_1 \cdot \boldsymbol{y})^2\kappa^{(y)} + (\boldsymbol{e}_1 \cdot \boldsymbol{z})^2\kappa^{(z)}\right) + \frac{1}{2}\left((\boldsymbol{e}_3 \cdot \boldsymbol{x})^2\kappa^{(x)} + (\boldsymbol{e}_3 \cdot \boldsymbol{y})^2\kappa^{(y)} + (\boldsymbol{e}_3 \cdot \boldsymbol{z})^2\kappa^{(z)}\right)$$

$$\kappa^3 \equiv \kappa(\boldsymbol{e}_3) \approx \frac{1}{2}\left((\boldsymbol{e}_1 \cdot \boldsymbol{x})^2\kappa^{(x)} + (\boldsymbol{e}_1 \cdot \boldsymbol{y})^2\kappa^{(y)} + (\boldsymbol{e}_1 \cdot \boldsymbol{z})^2\kappa^{(z)}\right) + \frac{1}{2}\left((\boldsymbol{e}_2 \cdot \boldsymbol{x})^2\kappa^{(x)} + (\boldsymbol{e}_2 \cdot \boldsymbol{y})^2\kappa^{(y)} + (\boldsymbol{e}_2 \cdot \boldsymbol{z})^2\kappa^{(z)}\right)$$

Taking $\boldsymbol{e}_1$, $\boldsymbol{e}_2$, and $\boldsymbol{e}_3$ as three axes of the coordinate system, the transformation of the *XYZ* coordinate system into the $\boldsymbol{e}_1\boldsymbol{e}_2\boldsymbol{e}_3$ coordinate system is a real orthogonal transformation with constant vector length. Remember that $\boldsymbol{x}, \boldsymbol{y}, \boldsymbol{z}$ are unit vectors in the *XYZ* coordinate system, so that

$$\sum_{i=1,2,3} (\boldsymbol{e}_i \cdot \boldsymbol{x})^2 = |\boldsymbol{x}|^2 = 1, \quad \sum_{i=1,2,3} (\boldsymbol{e}_i \cdot \boldsymbol{y})^2 = |\boldsymbol{y}|^2 = 1, \quad \sum_{i=1,2,3} (\boldsymbol{e}_i \cdot \boldsymbol{z})^2 = |\boldsymbol{z}|^2 = 1$$

Apply these three relations, then one can obtain

$$\kappa^{(1)} + \kappa^{(2)} + \kappa^{(3)} \approx \kappa^{(x)} + \kappa^{(y)} + \kappa^{(z)}$$

On integration over the wavelength, it is shown that

$$\Gamma^{(1)} + \Gamma^{(2)} + \Gamma^{(Z)} \approx \Gamma^{(X)} + \Gamma^{(Y)} + \Gamma^{(Z)} \tag{5.29}$$

Obviously, when one cannot measure the polarized absorption spectra of anisotropic crystals, it is possible to use the above method and formulas (5.27) and (5.28) to calculate the probability of spontaneous emission and radiative lifetime. It should be emphasized once again that in order to obtain a precise result the condition $\kappa L \ll 1$ should be satisfied.

5.3 Judd–Ofelt Approximation and Related Parameter

In the studies of spectroscopic properties of rare earth ions, the intensity parameter method introduced by Judd [2] and Ofelt [6] has become a popular and useful method, especially the Judd–Ofelt three-parameter method, although it involves many approximations. In this section, the Judd-Ofelt three-parameter method will be introduced. By using this method, radiative lifetime, quantum efficiency, emission cross-section, and absorption cross-section of laser and luminescent materials can be calculated (see, e.g., [7]).

From the discussion previously given, it can be seen that the spontaneous electric-dipole emission polarized in the direction $\boldsymbol{e}_q$ from an energy level with degeneracy g_2 has a probability of

$$P_{\rm sp}^{{\rm ed}(q)} = \frac{64\pi^4 v^3 n}{3hc^3 g_2}\left(\frac{n^2+2}{3}\right)^2 \left|\langle\varphi_{\rm f}|\boldsymbol{e}_q\cdot\boldsymbol{P}|\varphi_{\rm i}\rangle\right|^2 \tag{5.30}$$

where $\boldsymbol{P}$ denotes the electric-dipole moment

$$\boldsymbol{P} = -e\sum_i \boldsymbol{r}_i \tag{5.31}$$

If one uses spherical tensors C_q^1 to express electronic radius $\boldsymbol{r}$, then

$$\boldsymbol{e}_q\cdot\boldsymbol{P} = -e\sum_i r_i\left(C_q^1\right)_i \equiv D_q^{(1)} \tag{5.32}$$

where q can be −1, 0, and +1, for uniaxial crystal, $q = 0$ corresponds to π polarization and $q = \pm 1$ corresponds to σ polarization.

Similar to (5.14a), the integral of the absorption coefficient can be written as

$$\int \kappa^{{\rm ed}(q)}(\omega){\rm d}\omega = \frac{4\pi^2\bar{\omega}N_1}{3\hbar c g_1}\chi\left|\langle\varphi_f|\boldsymbol{e}_q\cdot\boldsymbol{P}|\varphi_i\rangle\right|^2 = \frac{4\pi^2 e^2\bar{\omega}N_1}{3\hbar c g_1}\chi\left|\langle\varphi_f|D_q^{(1)}|\varphi_i\rangle\right|^2 \tag{5.33}$$

The problem now is how to calculate the above matrix elements. For the rare earth ions, electric-dipole transition between levels of the same $4f^n$ configuration is forbidden originally. The method of Judd and Ofelt is based on the consideration that the wave function of another configuration with different parity can mix into the transition involving initial and final state wave functions and then the electric-dipole transition matrix element will be non-zero. This admixture is produced by the interactions of odd parity, namely odd crystal field potential and odd crystal vibrational mode. If one denotes the odd crystal field potential as $V_{\rm od}$, then the perturbed wave functions expanded to first-order terms of $V_{\rm od}$ are given by

$$\varphi_i = \varphi(f^n J_i M_i) - \sum_{\varphi(nl)} \frac{\langle f^n J_i M_i | V_{\mathrm{od}} | \varphi(nl) \rangle}{E(f^n J_i M_i) - E(\varphi(nl))} \varphi(nl) \tag{5.34}$$

where $\varphi(nl)$ are the wave functions with parity different from those of the $4f^{\,n}$ configuration (the most important is the $4f^{\,n-1}5d$ configuration). An expression for φ_f can be written in the same way. The odd parity crystal field potential V_{od} can be expressed as

$$V_{\mathrm{od}} = \sum_{t,p} B_{tp} D_p^{(t)},\ D_p^{(t)} = \sum_i r_i^t \left(C_p^{(t)} \right)_i \quad (t \text{ is an odd number})$$

In this way, the matrix element of the electric-dipole transition in a particular polarization direction q can be expressed as

$$\begin{aligned} \left\langle \varphi_f \middle| D_q^{(1)} \middle| \varphi_i \right\rangle = & - \sum_{\varphi(nl)} \frac{\langle f^n J_i M_i | V_{\mathrm{od}} | \varphi(nl) \rangle \left\langle \varphi(nl) \middle| D_q^{(1)} \middle| f^n J_f M_f \right\rangle}{E(f^n J_i M_i) - E(\varphi(nl))} \\ & - \sum_{\varphi(nl)} \frac{\left\langle f^n J_i M_i \middle| D_q^{(1)} \middle| \varphi(nl) \right\rangle \langle \varphi(nl) | V_{\mathrm{od}} | f^n J_f M_f \rangle}{E(f^n J_f M_f) - E(\varphi(nl))} \end{aligned} \tag{5.35}$$

It can be shown that [2]

$$\begin{aligned} & \sum_{\varphi(nl)} \left\langle f^n J_i M_i \middle| D_q^{(1)} \middle| \varphi(nl) \right\rangle \left\langle \varphi(nl) \middle| D_p^{(t)} \middle| f^n J_f M_f \right\rangle \\ & = \sum_k (-1)^{p+q} B_{tp} (2k+1) \begin{pmatrix} 1 & k & t \\ q & -p-q & p \end{pmatrix} \begin{Bmatrix} 1 & k & t \\ 3 & l & 3 \end{Bmatrix} \langle 3 | r | l \rangle \langle 3 | r^t | l \rangle \left\langle 3 \| C^{(1)} \| l \right\rangle \\ & \times \left\langle l \| C^{(t)} \| 3 \right\rangle \langle f^n J_i M_i | U_{p+q}^{(k)} | f^n J_f M_f \rangle \end{aligned} \tag{5.36}$$

In order to calculate the above matrix element, three approximations have to be introduced. First, it is assumed that the states of $\varphi(nl)$ are degenerate with respect to all the quantum numbers except nl. This is a rather rough assumption because, for example, if the intermediate states are those of $4f^{\,n-1}5d$ configuration, the energy range of this configuration is from 50,000 cm^{-1} to at least 100,000 cm^{-1}. Hence this approximation will introduce considerable errors in the final result. The second approximation is

$$E(f^n J_f M_f) - E(\varphi(nl)) = E(f^n J_i M_i) - E(\varphi(nl)) \tag{5.37}$$

This is also a very rough approximation, because the errors introduced by taking, for example, a transition from a initial state at about 20,000 cm^{-1} higher than the ground state to the final ground state, the center of the intermediate $4f^{\,n-1}5d$

configuration state assumed to be at 75,000 cm^{-1}, the difference of $E(f^{\,n}J_fM_f)$ − $E(\varphi(nl))$ and $E(f^{\,n}J_iM_i)$ − $E(\varphi(nl))$ can be larger than 20%. However, it makes possible to combine the two summations in (5.35) into one term and neglect all details of the intermediate states, by using $\sum_{\varphi(nl)} |\varphi(nl)\rangle\langle\varphi(nl)| = 1$. The third approximation is that all the crystal field energy levels of the initial and final multiplets have the same population. However, the different crystal field energy level is in fact populated according to the Boltzmann distribution. On the other hand, from the property of 3-*j* symbol explained in Chap. 2, one has

$$\begin{pmatrix} 1 & k & t \\ q & -p-q & p \end{pmatrix} = (-1)^{1+k+t} \begin{pmatrix} t & k & 1 \\ p & -p-q & q \end{pmatrix} \tag{5.38}$$

Therefore, only when k is even, the two parts of (5.35) do not counteract each other, because t is an odd number and $1 + t + k$ should be even. From the tangle condition of 6-*j* symbol in (5.36) it can be seen that k should satisfy $0 \le k \le 6$, so that $k = 2$, 4, and 6.

Further calculation of (5.35) can be found in the papers of Judd and Ofelt [2, 6] as well as the books on the angular momentum theory and irreducible tensor method [8, 9]. The final result is simply introduced in the following. Owing to the fact that the J–O three-parameter method is used to calculate the electric-dipole transition probability of multiplet to multiplet, the probability should be summed over different crystal field levels within the same multiplet. Assuming the same population in all the crystal field levels of the same multiplet, it can be shown that the square of transition matrix element in (5.33) can be expressed as

$$S^{(q)}(J_i \to J_f) = \left|\left\langle J_f \middle| D_q^{(1)} \middle| J_i \right\rangle\right|^2 = \sum_{k=2,4,6} \Omega_k^{(q)} \left|\left\langle [\alpha SLJ]_f \middle\| U^{(k)} \middle\| [\alpha SLJ]_i \right\rangle\right|^2 \tag{5.39}$$

The square of reduced matrix element $\left|\left\langle [\alpha SLJ]_f \middle\| U^{(k)} \middle\| [\alpha SLJ]_i \right\rangle\right|^2$ can be calculated by intermediate approximation wave functions or found from the tables in Kaminskii's book (Ref. 7 of Chap. 2). The J–O parameters $\Omega_k^{(q)}$ are determined by the spectroscopic experimental method, because the odd crystal field potential and odd lattice vibrational mode contributions are difficult to be calculated theoretically.

Combining (5.33) with (5.39) and remembering that the degeneracy of the initial state is $2J_i + 1$, it gives:

$$\begin{aligned}\Gamma^{(q)} &= \int \kappa^{(q)}(\lambda)\mathrm{d}\lambda = \frac{N_1}{2J_\mathrm{i}+1}\frac{8\pi^3\bar{\lambda}e^2}{3hc}\frac{\left(n_q^2+2\right)^2}{9n_q}\sum_{k=2,4,6}\Omega_k^{(q)}\left|\left\langle[\alpha SLJ]_\mathrm{f}\|U^{(k)}\|[\alpha SLJ]_\mathrm{i}\right\rangle\right|^2 \\ &= \frac{N_1}{2J_\mathrm{i}+1}\frac{8\pi^3\bar{\lambda}e^2}{3hc}\frac{\left(n_q^2+2\right)^2}{9n_q}S^{(q)}(J_\mathrm{i}\to J_\mathrm{f})\end{aligned} \tag{5.40}$$

where $\Gamma^{(q)}$ is referred to as absorbance of polarization q, then the experimental line strength S becomes

$$S^{(q)}(J_i \to J_f) = \frac{2J_i+1}{N_1}\frac{3hc}{8\pi^3\bar{\lambda}e^2}\frac{9n}{(n^2+2)^2}\Gamma^{(q)} \tag{5.41}$$

By using the determined experimental line strength $\Gamma^{(q)}$, $S_q(J_i \to J_f)$ can be calculated by (5.41) and then $\Omega_k^{(q)}$ can be obtained by the fitting method according to (5.39).

It should be reminded that the J–O parameters are used for the electric-dipole transition of rare earth ions. The magnetic-dipole transition probability between some pairs of energy levels of some rare earth ions can reach an order of magnitude comparable with that of electric-dipole transition. For example, the $^4I_{15/2} \to {}^4I_{13/2}$ transition of Er^{3+} ions and the $^4I_8 \to {}^4I_7$ transition of Ho^{3+} ions. The experimental spectra include the contribution of the magnetic-dipole transition, and how to correctly calculate J–O parameters in this case, which will be explained in the next section.

Now it is possible to demonstrate the selection rule of electric-dipole transition about quantum number JLS for rare earth ion in materials, which has been mentioned in Sect. 4.3. As has been demonstrated above, the k should be 2, 4, and 6. On the other hand, the reduced matrix element of unit operator in (5.39) includes a 6-j symbol $\begin{Bmatrix} J & J' & k \\ L' & L & S \end{Bmatrix}$ (see (2.52)). From its triangle condition, it can be seen that the transition allowed for $\Delta J \le 6$, $\Delta L \le 6$ and in case of J or $J' = 0$, only $\Delta J =$ 2, 4, 6 is allowed. $J = 0 \leftrightarrow J' = 0$ is forbidden except in the case of the matrix element of zero-order unit tensor operator $U^{(0)}$ exist, which will be discussed later, while by the δ_{SS}' factor in (2.52) $\Delta S = 0$ should be satisfied.

The interaction between different multiplets with different J values resulted from even parity crystal field, that is J mixing effect, has not been taken into account in the above calculation. Include this J mixing effect, and then the square of reduced matrix element of unit tensor operator should be replaced by the following expression [10, 13]

$$
\begin{aligned}
\left|\left\langle [\alpha SLJ]_f \middle\| U_{eff}^{(k)} \middle\| [\alpha SLJ]_i \right\rangle\right|^2 &\equiv \left|\left\langle \Psi_f J_f \middle\| U_{eff}^{(k)} \middle\| \Psi_i J_i \right\rangle\right|^2 \\
&= \sum_{\Psi'_f J'_f} \sum_{\Psi'_i J'_i} \frac{\left|a\left(\Psi_f J_f, \Psi'_f J'_f\right)\right|^2}{2J'_f + 1} \frac{\left|a\left(\Psi_i J_i, \Psi'_i J'_i\right)\right|^2}{2J'_i + 1} \\
&\quad \times \left|\left\langle \Psi'_f J'_f \middle\| U^{(k)} \middle\| \Psi'_i J'_i \right\rangle\right|^2
\end{aligned}
\tag{5.42}
$$

The coefficients $|a(\Psi J, \Psi' J')|^2/(2\,J'+1)$ is the J mixing coefficients that take into account the mixing of multiplets $\Psi' J'$ and ΨJ. The summation of above equation involves all the multiplet $\Psi' J'$, which is near to the multiplet ΨJ, and the multiplet ΨJ itself. When $\Psi' J' \neq \Psi J$, one has

$$
|a(\Psi J, \Psi' J')|^2 = \sum_{k=2,4,6} S_k^2 \frac{|\langle \Psi J \| U^{(k)} \| \Psi' J' \rangle|^2}{[E(\Psi J) - E(\Psi' J')]^2}
$$

In case of $\Psi' J' = \Psi J$, the J mixing coefficients should be calculated by

$$
|a(\Psi J, \Psi' J')|^2 = (2J + 1)\left[1 - \sum_{\psi' J' \neq \psi J} \frac{|a(\Psi J, \Psi' J')|^2}{2J' + 1}\right]
$$

The key for the calculation of J mixing coefficients is the calculation of crystal field strength coefficient S_k^2, which should be calculated by means of the data of crystal field energy level, because S_k^2 is obtained by the following expression

$$
\Delta^2(\Psi J) = \frac{1}{2J+1} \sum_n [E(\Psi J n) - \bar{E}(\Psi J)]^2 = \frac{1}{2J+1} \sum_{k=2,4,6} S_k^2 |\langle \Psi J \| U^{(k)} \| \Psi J \rangle|^2
$$

where $E(\Psi J n)$ is nth crystal field energy level of multiplet ΨJ and $\bar{E}(\Psi J)$ is the center of all these energy levels.

The practical work of laser material research is not always necessary to determine precisely the crystal field energy level position of the multiplets obtained by low temperature spectral measurement. Therefore, it is impossible to use the above equations to calculate line strength S including J mixing effect. However, this effect works only in the case that the strength of even crystal field is strong enough and the energy interval between different multiplets is small. For the Nd^{3+}:YVO_4 crystal, the RMF error of the absorption intensity parameter S of the $^4I_{9/2}$ multiplet measure at 300 K, taking into account the J mixing effect only to reduce its value from 0.152 to 0.147 [13]. Certainly, with J mixing effect some of the forbidden magnetic-dipole transition can have certain intensity.

Although Judd [2] and Ofelt [6] neither take into account the effect of wave function mixing by ligand wave functions with different parity nor the mixing contributions of electron-electron Coulomb interaction with spin-orbital interaction in higher approximation, (5.39) is a generally applicable equation. During the past 40 years, after considering all the above mixing effects as well as the dynamic interaction, electron correlation interaction, and relativistic effect, and calculating the second and third-order approximation terms, the (5.39) is still correct. The odd k term introduced by mass polarization effect is small and it has its influence only in the case that the transition is forbidden by standard selection rule; for example, the transition from the multiplet of $J = 0$ to those odd number J in Eu^{3+} ions [14]. Other odd k term is very small.

After the spectral measurement, the data of absorbance can be calculated from absorption bands. Usually the number of absorbance data obtained is larger than that of Ω_k ($k = 1, 2, 3$). The reduced matrix elements of unit operator can be found directly from the literature or calculated by intermediate-coupling wave function and the data are given in the table of literature [15]. (Some of these usually used in laser materials can be found in Appendix D.) By means of (5.41), several equations can be listed and then the fixing method is used to calculate the J–O parameters.

As pointed out in Sect. 5.2, for the anisotropic crystals, the absorbance in different direction is usually different, which corresponds to different strength parameter. For a biaxial crystal, there are three kinds of absorption spectra with polarization direction parallel to X, Y, and Z crystallophysics axes and for a uniaxial crystal there are π spectrum with polarization direction parallel to the optics axis and σ spectrum with polarization direction perpendicular to the optics axis. Using the data of these spectra—$\Omega_k^{(X)}$, $\Omega_k^{(Y)}$, and $\Omega_k^{(Z)}$—$\Omega_k^{(\pi)}$ and $\Omega_k^{(\sigma)}$ can be calculated and then the emission probability in different polarization directions can be obtained. However, it should be noted that in the evaluation of radiative lifetime, the theorem on the total probability should be used, otherwise the calculated transition probability will have a deviation of three times to the actual value [16–18]. If the transition probability for q polarization is P_q, the total transition probabilities P should be

$$P = \frac{1}{3}\sum_q P^{(q)} \tag{5.43}$$

and so in the general calculation, Ω_k should be

$$\Omega_k = \frac{1}{3}\sum_q \Omega_k^{(q)} \tag{5.44}$$

When one uses (4.23) to calculate spontaneous transition probability in any polarization direction, the integral of the absorbing line strength is related to the integral of the absorbance over all the solid angle. Thus by (5.40) we have,

$$\frac{1}{4\pi}\int S(J \to J')\mathrm{d}\Omega_p = \frac{2J+1}{N}\frac{3hc}{8\pi^3\bar{\lambda}}\frac{9n}{(n^2+2)^2}\frac{1}{4\pi}\int \Gamma \mathrm{d}\Omega \tag{5.45}$$

where the choice of coordinates is arbitrary, the variables of the integral in the left-hand side can be in the polarization direction θ_p, φ_p and the variables of the integral in the right-hand side can be in the light beam direction θ, φ. Both integrals are integrated over solid angle 4π.

Kaminskii [17] discussed the relation between fluorescence branching ratio and the ratio of $X = \Omega_4/\Omega_6$ for Nd^{3+} ions. The ratios of fluorescence branching for the transitions $^4F_{3/2} \to {}^4I_{9/2}$, $^4I_{11/2}$, $^4I_{13/2}$, and $^4I_{15/2}$ can be obtained by the curves given. They can also be obtained by the following relations given by Lomheim et al. [11]

$$\begin{aligned}\beta_{0.9\,\mu\mathrm{m}} &= (0.319X + 0.077)/, \beta_{1.06\,\mu\mathrm{m}} = (0.105X + 0.287)/A,\\ \beta_{1.3\,\mu\mathrm{m}} &= 0.09/A, \beta_{1.8\,\mu\mathrm{m}} = 0.002/A, A = 0.425X + 0.457\end{aligned}$$

The following relationship between the X value and the cell parameter a_0 of garnet crystal can be obtained

$$X = La_0(1 + C\Delta a_0) + 0.128$$

where Δa_0 is the difference of lattice parameter a_0 between the discussed crystal and the lutetium aluminum garnet crystal. For the neodymium-doped rare earth garnet, one has $L = 8.04 \times 10^{-3}\ \mathrm{nm}^{-4}$, $C = 10.1\ \mathrm{nm}^{-1}$. The calculated results of five neodymium-doped garnets are as follows:

Crystal	$Lu_3Al_5O_{12}$	$Y_3Al_5O_{12}$	$Lu_3Ga_5O_{12}$	$Y_3Ga_5O_{12}$	$Gd_3Ga_5O_{12}$
a_0 (nm)	1.1915	1.201	1.2183	1.2777	1.2376
Experimental value of X	0.29	0.30	0.35	0.38	0.41
Calculated value of X	0.29	0.31	0.35	0.38	0.40

In many cases, (5.39) gives us a satisfactory analysis of experimental results. However, negative values of Ω_2 have been found for some ions, for example Pr^{3+}, and it is contradictory to the theory of Judd and Ofelt. Occasionally, two sets of parameters are needed; one used to analyze the transitions between low energy states, and the other used to analyze the transitions between high energy states. This problem comes from the approximate hypotheses mentioned above in the derivation of (5.39), of which the influence of the second hypotheses may be the largest.

To deal with this problem, (5.39) should be modified to [12].

$$S(J_i \to J_f) = \sum_{k=1,2,3,4,5,6} \left[\frac{1}{E(f^n J_i) - E(\varphi(nl))} + (-1)^k \frac{1}{E(f^n J_f) - E(\varphi(nl))} \right]^2 \times \frac{[E(f^n) - E(\varphi(nl))]^2}{4} \Omega_k \left| \left\langle [\alpha SLJ]_f \| U^{(k)} \| [\alpha SLJ]_i \right\rangle \right|^2 \tag{5.46a}$$

where the strength parameters with $k = 1, 3, 5$ have been taken into account. However, this modification is still in the theoretical scheme of Judd and Ofelt three-parameter method. By using this expression of $S(J_i \to J_f)$, the problem of negative Ω_2 values can be solved but the difference between experiment and calculated results is still large. In order to reduce the error, the authors of paper [12] adopted a method called normalized method. By using their method, in the least-square fitting the root-mean square deviation is calculated not by

$$\mathrm{RMS}\Delta S = \left(\frac{\sum_{i=1}^{N} \left(S_i^{exp} - S_i^{cal} \right)^2}{N-3} \right)^{1/2} \tag{5.46b}$$

but by

$$\mathrm{RMS}\Delta S = \left(\frac{\sum_{i=1}^{N} \left(\frac{S_i^{exp} - S_i^{cal}}{S_i^{exp}} \right)^2}{N-3} \right)^{1/2} \tag{5.46c}$$

Dunina [20] took into account the non-orthogonal wave function and obtained another expression of line strength S

$$S(J_i \to J_f) = \sum_{k=2,4,6} \Omega_k \left| \left\langle [\alpha SLJ]_f \| U^{(k)} \| [\alpha SLJ]_i \right\rangle \right|^2 \times \left[1 + \frac{E_{J_i} + E_{J_f} - 2\bar{E}_{4f^2}}{E_{4f4d} - \bar{E}_{4f^2}} \right] \tag{5.46d}$$

where E_{4f4d} is the lowest $4f5d$ configuration energy level position of Pr^{3+} ion and $\bar{E}_{4f^2}$ is the average energy level position of $4f^2$ configuration. However, by (5.46b) and (5.46d) we can solve the problem of negative Ω_2 but the relative error is still large. If one uses (5.46c) and (5.46d) to evaluate J–O parameters, not only the problem of negative Ω_2 can be solved but also the relative error can be reduced. This situation can be seen in the crystal Pr^{3+}:GdCOB [21] and Pr^{3+}:LaB_3O_6 [22].

A problem in the calculation of spectral intensity parameter must be mentioned. It is the existing parameter Ω_0. Although the matrix element of zero-order unit tensor operator $U^{(0)}$ does not appear in the table published by Nielson and Koster [15], Piepho and Schatz [23] have given its expression

$$\left\langle l^n SL \| U^{(0)} \| l^n S'L' \right\rangle = \delta_{LL'}\delta_{SS'} n(2L+1)^{1/2}(2l+1)^{-1/2} \tag{5.47}$$

It can be seen that this matrix element exists only in the case that the quantum numbers L and S of initial and final states are equal to each other. On the other hand, it can be seen by (5.36), when $k = 0$, the triangle condition of 3-j symbol requires $t = 1$, so that only in case $B_{1p} \neq 0$ the matrix element $U^{(0)}$ is non-zero. When the ions are at a position symmetry of point group G, the condition of $B_{1p} \neq 0$ is that the irreducible representation 1^- of group O_3 resolves to the irreducible representations of group G that includes the identity representation Γ_1. From the correlation table in Appendix B, it can be seen that one-order odd crystal field component appeared only in the following ten point groups: C_s, C_1, C_2, C_3, C_4, C_6, C_{2v}, C_{3v}, C_{4v}, and C_{6v}. Moreover, because of (5.47), the effect of Ω_0 should be considered only in the case that the intermediate-coupling interaction is strong and both the initial and final multiplets have a larger proportion of the same LS intermediate-coupling wave function. One example is Eu^{3+}:LaF_3 crystal, in which Eu^{3+} ions occupy C_2 point group position. The transitions of $^5D_0 \rightarrow {}^7F_0$, $^5D_1 \rightarrow {}^7F_1$, and $^5D_2 \rightarrow {}^7F_2$ all have the contribution of Ω_0. Among these, the transition of $^5D_0 \rightarrow {}^7F_0$ only results from the effect of Ω_0. By the fitting of spectral data, it can be found that in this sample Ω_0 is smaller than Ω_2 and Ω_4 by two orders of magnitude, while smaller than Ω_6 by one order of magnitude [24].

The three-parameter method of Judd-Ofelt can only be used in the calculation of spectroscopic parameters for multiplet-to-multiplet transitions. However, nowadays, the rare earth spectroscopic measurement, especially those performed at low temperature, can give us well-resolved spectral lines of the transitions between crystal field levels. In order to evaluate this kind of fine structure spectral lines, an all-parameter method should be used. The primary all-parameter method was introduced by Axe in 1963 [25]. It adopts $B^k_{tp} \equiv A_{tp}Y_{tk}$ as parameters, where $k = 2$, 4, 6 and t is limited to be $k \pm 1$, p can be 0, ±1, ±2, …, $\pm t$ (the number depends on the symmetry of the crystal field.). Newman [26] pointed out that Axe's method is not generally applicable, because it is assumed that the single-electron approximation and the overlapping approximation should be satisfied by the interactions between rare earth ions and their ligands. It does assume that all of the RE-L interactions are cylindrical and independent of each other. If these conditions are not satisfied, then new parameters should be added. In fact, the restriction on t will be lost, and let it can be equal to k. Reid and Richardson developed and applied the all-parameter method in a series of papers [27–31]. According to their method, electric-dipole line strength for a transition from ψ_i to ψ_f can be written as

$$S_{i\rightarrow f} = \left| \sum_{ktp} B^k_{tp} \sum_{lq} \begin{pmatrix} k & 1 & t \\ l & -q & -p \end{pmatrix} (-1)^{k-1+p+q}(2t+1)^{1/2} \left\langle \psi_i \middle| U_l^{(k)} \middle| \psi_f \right\rangle \right|^2 \tag{5.48}$$

where $\begin{pmatrix} k & 1 & t \\ l & -q & -p \end{pmatrix}$ is the 3-j symbol and the unit tensor matrix element $\langle \psi_i | U_l^{(k)} | \psi_f \rangle$ can be expressed as the product of 3-j symbol and reduced tensor matrix elements $\langle \psi_i \| U^{(k)} \| \psi_f \rangle$ by using Wigner–Eckart theorem. The numbers of the parameters B_{tp}^k can reach 81 in general cases, where $k = 2, 4, 6$, and t is limited to be k and $k \pm 1$, while p can be $0, \pm 1, \pm 2, \ldots, \pm t$. Considering the crystal field symmetry restrictions, these numbers can be reduced differently in different symmetry situations. If the rare earth ions occupy point group O positions, there are only two parameters while for the point group $D_{2\mathrm{d}}$ it should be 10. In case of a lower symmetry, the parameters increase rapidly. For the point groups D_3 and D_2, there are 12 and 18 parameters, respectively. In order to evaluate the line strength with so many parameters, one should have low temperature spectral data with high-enough resolution. Burdick et al. [32] used the data of 97 well-resolved absorption spectral lines of Nd^{3+}:YAG crystal at 10 K to calculate 18 parameters B_{tp}^k of D_2 symmetry. Stefan et al. [33] used this method for the line strength evaluation of $Er^{3+}:Cs_3Lu_2Cl_9$ crystal. They measured the polarization absorption spectra at 10 and 16 K as well as the up-conversion fluorescence and excitation spectra at 4.2 K and used data of 95 spectral lines to calculate 12 parameters B_{tp}^k in $C_{3\mathrm{v}}$ symmetry. The obtained parameters were used to evaluate the line strengths and to simulate the absorption and emission lines not observed previously.

It should be mentioned that to calculate the transitions between crystal field levels, there is a method introduced by the authors with parameters much less than those of the all-parameter method [34–36]. The relative line strengths and line to line transition possibilities were calculated for $NdAl_3(BO_3)_4$, $Nd^{3+}:YVO_4$, and $Nd^{3+}:YAl_3(BO_3)_4$ crystals as well as $Er^{3+}:LiYF_4$ crystal [37]. In the latter case, the excited-state absorption (ESA) spectra have also been calculated with rather good results by using data from ground state absorption measurements.

5.4 Spectroscopic Parameter Calculation of Rare Earth Ion in Crystal

Let us discuss a two energy level system in a closed cavity. Its low level has N_1 population, g_1 degeneracy with energy E_1 and its high level has N_2 population, g_2 degeneracy with energy E_2. According to Boltzmann distribution law, we have

$$\frac{N_2}{N_1} = \frac{g_2}{g_1} \exp\left(-\frac{E_2 - E_1}{k_{\mathrm{B}} T}\right) \tag{5.49}$$

When this system interacts with a radiation field having frequency satisfying $\omega = (E_2 - E_1)/\hbar$, the electrons in the high level have spontaneous and stimulated emission and those in the low level have stimulated absorption. The up and down transitions will be finally in equilibrium. From the earlier discussion, it can be seen that the probabilities of stimulated emission and absorption are proportional to the energy density of the radiation field (it can be written as $B_{ij}\mathrm{d}(\omega)$), but the spontaneous emission probability is independent of the radiation field. Hence one has the following relation

$$\mathrm{d}(\omega)(N_1B_{12} - N_2B_{21}) = N_2A_{21} \tag{5.50}$$

By using (5.49) and (5.50), $\mathrm{d}(\omega)$ becomes

$$\mathrm{d}(\omega) = \frac{N_2A_{21}}{(N_1B_{12} - N_2B_{21})} = \frac{A_{21}}{B_{12}\frac{g_1}{g_2}\exp(\hbar\omega/k_BT) - B_{21}} \tag{5.51}$$

According to Plank formula, the energy density of the radiation field in a closed cavity is

$$\mathrm{d}(\omega) = \frac{\hbar\omega^3n^3}{\pi^2c^3}\frac{1}{\exp(\hbar\omega/k_\mathrm{B}T) - 1} \tag{5.52}$$

By comparing (5.51) with (5.52) and by using relation $g_1B_{12} = g_2B_{21}$ similar to (5.8), one obtains a relation between the Einstein coefficients

$$B_{21} = \frac{\pi^2c^3}{\hbar\omega^3n^3}A_{21} = \frac{4\pi^2}{3\hbar^2n^2g_2}\left(\frac{n^2+2}{3}\right)^2\sum_{n,m}|\mu_{2_n,1_m}|^2 \tag{5.53a}$$

Obviously, the expression of B_{12} is

$$B_{12} = \frac{4\pi^2}{3\hbar^2n^2g_1}\left(\frac{n^2+2}{3}\right)^2\sum_{n,m}|\mu_{1_m,2_n}|^2 \tag{5.53b}$$

By using (5.2), (5.30), (5.39) and the formula taking into account the average over initial states and summation over final states as expressed by (5.7), at the same time referring to formula (5.53a, 5.53b), Einstein coefficient B can be written as

$$B^{\mathrm{ed}}(J_i \to J_f) = \frac{4\pi^2e^2}{3\hbar^2n^2(2J_i+1)}\left(\frac{n^2+2}{3}\right)^2S^{\mathrm{ed}}(J_i \to J_f) \quad . \tag{5.54a}$$

Similarly, the corresponding relation for magnetic-dipole transition can be obtained as follows

$$B^{\mathrm{md}}(J_i \to J_f) = \frac{4\pi^2e^2}{3\hbar^2(2J_i+1)}S^{\mathrm{md}}(J_i \to J_f) \tag{5.54b}$$

It can also be shown that the spontaneous transition probabilities of electric-dipole transition can be expressed as

$$\begin{aligned} A^{ed(q)}(J_i \to J_f) &= \frac{64\pi^4 e^2}{3h(2J_i+1)\bar{\lambda}^3} \frac{n_q\left(n_q^2+2\right)^2}{9} S^{ed(q)}(J_i \to J_f) \\ &= 8.047 \times 10^9 (\text{cm/s}) \\ &\quad \times \frac{n_q\left(n_q^2+2\right)^2}{(2J+1)\bar{\lambda}^3} \sum_{k=2,4,6} \Omega_k^{(q)} \left| \left\langle [\alpha SLJ]_f \| U^{(k)} \| [\alpha SLJ]_i \right\rangle \right|^2 \end{aligned} \tag{5.55}$$

The magnetic-dipole transition between multiplets in the same configuration is allowed and different from the electric-dipole transition aroused by odd parity perturbation which is independent of the structure of the host. Its difference in different hosts resulted from the variation of free ion parameters produced by the nephelauxetic effect, or the influence of J mixing effect can be neglected [38]. Therefore, the magnetic-dipole transition probability in different hosts can be considered as the same. The oscillator strength in vacuum for magnetic-dipole transition f_{md} can be found in the literature [38], then S_{md} can be calculated by the following expression

$$S_{md} = \frac{3h(2J_i+1)\bar{\lambda}}{8\pi^2 mc} f_{md} = 9.2189 \times 10^{-12} (\text{cm})(2J_i+1)\bar{\lambda} f_{md} \tag{5.56}$$

where S_{md} can also be expressed as

$$S_{\text{md}} = \left(\frac{h}{4\pi mc}\right)^2 \left| \left\langle [\alpha SLJ]_i \| \boldsymbol{L} + 2\boldsymbol{S} \| [\alpha SLJ]_f \right\rangle \right|^2 \tag{5.57}$$

S_{md} in (5.57) is the result of summation over crystal field energy levels of the multiplet and so is independent of the wave functions of crystal field energy levels. If one discusses the transition between two crystal field energy levels, then the line strength will be dependent on the wave functions of crystal field energy levels and so is dependent on the point group symmetry of the ions in the host. Its transition selection rule is described in Chap. 4. The transition rate of spontaneous emission can be expressed as

$$\begin{aligned} A^{\text{md}}(J_i \to J_f) &= \frac{8\pi^2 e^2 n^2}{mc\bar{\lambda}^2} f^{\text{md}}(J_i \to J_f) = \frac{64\pi^4 e^2}{3h(2J_i+1)\bar{\lambda}^3} n^3 S_{\text{md}} \\ &= 7.242 \times 10^{10} (\text{cm/s}) \\ &\quad \times \frac{n^2}{(2J+1)\bar{\lambda}^3} \sum_{[\alpha SL]_i [\alpha SL]_j} C([\alpha SL]_i) C\left([\alpha SL]_j\right) \left| \left\langle [\alpha SLJ]_f \| \mathbf{L} + 2\mathbf{S} \| [\alpha SLJ]_f \right\rangle \right|^2 \end{aligned} \tag{5.58}$$

The square of matrix elements in the above equation can be calculated as follows:
When $J_f = J_i - 1$

$$\left|\left\langle[\alpha SLJ]_i \| \boldsymbol{L}+2\boldsymbol{S} \| [\alpha SLJ]_f\right\rangle\right|^2$$
$$= \delta(\alpha_i\alpha_f)\delta(S_iS_f)\delta(L_iL_f)\left\{\frac{\left[(S_i+L_i+1)^2-J_i^2\right]\left[J_i^2-(L_i-S_i)^2\right]}{4J_i}\right\}^{1/2}$$

when $J_f = J_i = J$

$$|\langle[\alpha SL]J \| \boldsymbol{L}+\boldsymbol{2S} \| [\alpha' S' L']J'\rangle| = \delta(\alpha,\alpha')\delta(S,S')\delta(L,L')$$
$$\times\left[\frac{(2J+1)}{4J(J+1)}\right]^{1/2}\times[3J(J+1)+S(S+1)-L(L+1)]$$

$$\left|\left\langle[\alpha SLJ]_i \| \boldsymbol{L}+2\boldsymbol{S} \| [\alpha SLJ]_f\right\rangle\right|^2 = \delta(\alpha_i\alpha_f)\delta(S_iS_f)\delta(L_iL_f)$$
$$\times\left[\frac{(2J_i+1)}{4J_i(J_i+1)}\right]\times[3J_i(J_i+1)+S_i(S_i+1)-L_i(L_i+1)]$$

when $J_f = J_i + 1$

$$\left|\left\langle[\alpha SLJ]_i \| \boldsymbol{L}+2\boldsymbol{S} \| [\alpha SLJ]_f\right\rangle\right|^2$$
$$= \delta(\alpha_i\alpha_f)\delta(S_iS_f)\delta(L_iL_f)\left\{\frac{\left[(S_i+L_i+1)^2-(J_i+1)^2\right]\left[(J_i+1)^2-(L_i-S_i)^2\right]}{4(J_i+1)}\right\}$$

where $C([\alpha SL]_i)$ and $C([\alpha SL]_f)$ are the coefficients of the intermediate-coupling wave functions for initial and final multiplets, respectively.

The total transition rates in case of both the electric-dipole and the magnetic-dipole transitions are allowed will be

$$A(J_i \to J_f) = A^{\mathrm{ed}}(J_i \to J_f) + A^{\mathrm{md}}(J_i \to J_f)$$

For uniaxial crystals

$$A(J_i \to J_f) = \frac{1}{3}A^{(\pi)}(J_i \to J_f) + \frac{2}{3}A^{(\sigma)}(J_i \to J_f)$$

For biaxial crystals

$$A(J_i \rightarrow J_f) = \frac{1}{3}A^{(x)}(J_i \rightarrow J_f) + \frac{1}{3}A^{(y)}(J_i \rightarrow J_f) + \frac{1}{3}A^{(z)}(J_i \rightarrow J_f)$$

The oscillator strength of electric-dipole transition will be expressed as

$$\begin{aligned} f^{\mathrm{ed}(q)}(J_i \rightarrow J_f) &= \frac{8\pi^2 mc}{3h(2J_i+1)\bar{\lambda}} \frac{(n^2+2)^2}{9n} \sum_{k=2,4,6} \Omega_k^{(q)} \left| \left\langle [\alpha SLJ]_f \right| \left| U^{(k)} \right| \left| [\alpha SLJ]_i \right\rangle \right|^2 \\ &= 1.205 \times 10^{10}(\mathrm{cm}^{-1}) \times \frac{(n^2+2)^2}{n(2J_i+1)\bar{\lambda}} \sum_{k=2,4,6} \Omega_k^{(q)} \left| \left\langle [\alpha SLJ]_f \right| \left| U^{(k)} \right| \left| [\alpha SLJ]_i \right\rangle \right|^2 \end{aligned} \tag{5.59a}$$

The oscillator strength of magnetic-dipole transition will be expressed as

$$\begin{aligned} & f^{\mathrm{md}}(J_i \rightarrow J_f) \\ &= \frac{nh}{6mc(2J_i+1)\bar{\lambda}} \\ &\quad \times \sum_{[\alpha SL]_i [\alpha SL]_j} C\left([\alpha SL]_i\right) C\left([\alpha SL]_j\right) \left| \langle [\alpha SLJ]_i \| \mathbf{L} + 2\mathbf{S} \| [\alpha SLJ]_f \rangle \right|^2 \\ &= n f_{\mathrm{md}'}(J_i \rightarrow J_f) \end{aligned} \tag{5.59b}$$

The total oscillator strength is

$$f(J \rightarrow J') = f^{\mathrm{ed}}(J \rightarrow J') + f^{\mathrm{md}}(J \rightarrow J')$$

For uniaxial crystals

$$f(J \rightarrow J') = \frac{1}{3} f^{(\pi)}(J \rightarrow J') + \frac{2}{3} f^{(\sigma)}(J \rightarrow J')$$

For biaxial crystals

$$f(J \rightarrow J') = \frac{1}{3} f^{(x)}(J \rightarrow J') + \frac{1}{3} f^{(y)}(J \rightarrow J') + \frac{1}{3} f^{(z)}(J \rightarrow J')$$

Note that the above equations can be used to calculate absorption and emission oscillator strengths but the degeneracy degree of initial states and the average wave lengths should correspond to the absorption spectrum and emission spectrum, respectively.

In uniaxial or biaxial crystals, the magnetic-dipole transition is also polarized. In the J–O parameter calculation of different polarization direction, the magnetic-dipole oscillator strength that should be subtracted is not the total oscillator strength of magnetic-dipole transition with all the polarization directions as expressed in (5.59b). Taking uniaxial crystals as an example, it is necessary to measure three different polarization spectra, the α, σ, and π spectra, as mentioned in Chap. 4. If f_{md} is the total oscillator strength of the magnetic-dipole transition in vacuum, f^{md} expresses the total oscillator strength of magnetic-dipole transition between two multiplets in the crystals, then $f^{\mathrm{md}} = nf_{\mathrm{md}}$. One has the following relation

$$f^{\mathrm{md}} = \frac{2}{3} f^{\mathrm{md}}_{H\perp c} + \frac{1}{3} f^{\mathrm{md}}_{H//c}$$

where $f^{\mathrm{md}}_{H\perp c}$ and $f^{\mathrm{md}}_{H//c}$ are the oscillator strengths of magnetic-dipole transition when the magnetic field perpendicular and parallel to the c axis, respectively. From the σ spectrum measurement, the oscillator strength $f^{(\sigma)}_{\mathrm{exp}}$ can be expressed as

$$f^{(\sigma)}_{\mathrm{exp}} = f^{\mathrm{ed}(\sigma)} + f^{\mathrm{md}}_{H//c}$$

From the π spectrum measurement, the oscillator strength $f^{(\pi)}_{\mathrm{exp}}$ can be expressed as

$$f^{(\pi)}_{\mathrm{exp}} = f^{\mathrm{ed}(\pi)} + f^{\mathrm{md}}_{H\perp c}$$

Similarly, in addition to the π polarized oscillator strength of electric-dipole transition, it includes the oscillator strength oscillator strength of magnetic-dipole transition with magnetic field perpendicular to the optical axis. The physical quantities of $f^{\mathrm{md}}_{H\perp c}$ and $f^{\mathrm{md}}_{H//c}$ cannot be obtained by calculation or direct measurement. In order to obtain data of $f^{\mathrm{ed}(\sigma)}$ and $f^{\mathrm{ed}(\pi)}$, the α spectrum should be measured, because

$$f^{(\alpha)}_{\mathrm{exp}} = f^{\mathrm{ed}(\sigma)} + f^{\mathrm{md}}_{H\perp c}$$

Therefore

$$f^{\mathrm{ed}(\sigma)} = \frac{2}{3} f^{(\alpha)}_{\mathrm{exp}} + \frac{1}{3} f^{(\sigma)}_{\mathrm{exp}} - f^{\mathrm{md}} \tag{5.60a}$$

$$f^{\mathrm{ed}(\pi)} = f^{(\pi)}_{\mathrm{exp}} + \frac{1}{3}\left[f^{(\sigma)}_{\mathrm{exp}} - f^{(\alpha)}_{\mathrm{exp}}\right] - f^{md} \tag{5.60b}$$

$$f^{\mathrm{ed}} = \frac{2}{3} f^{\mathrm{ed}(\sigma)} + \frac{1}{3} f^{\mathrm{ed}(\pi)} = \frac{1}{3} f^{(\alpha)}_{\mathrm{exp}} + \frac{1}{3} f^{(\sigma)}_{\mathrm{exp}} + \frac{1}{3} f^{(\pi)}_{\mathrm{exp}} - f^{\mathrm{md}} \tag{5.60c}$$

The oscillator strength expressed in the above equations should be used in the calculation of J–O parameters. The absorption oscillator strength $f^{(q)}_{\mathrm{exp(abs)}}$ can be calculated by using the following expression

$$f_{\exp(\mathrm{abs})}^{(q)} = \frac{mc^2}{\pi e^2 \bar{\lambda}_{\mathrm{abs}}^2 N_1} \int \kappa^{(q)}(\lambda)\mathrm{d}\lambda = \frac{1.106 \times 10^{12}(\mathrm{cm}^{-1})}{\bar{\lambda}_{\mathrm{abs}}^2 N_1} \int \kappa^{(q)}(\lambda)\mathrm{d}\lambda \quad (5.61)$$

As shown by (5.43), the effective J–O parameters for biaxial and uniaxial crystals should be written as

$$\Omega_k = \frac{1}{3}\left(\Omega_k^{(X)} + \Omega_k^{(Y)} + \Omega_k^{(Z)}\right) \quad (5.62\mathrm{a})$$

$$\Omega_k = \frac{1}{3}\Omega_k^{(\pi)} + \frac{2}{3}\Omega_k^{(\sigma)} \quad (5.62\mathrm{b})$$

The emission cross-section can be introduced as follows: the emission cross-section σ_{21} and the absorption cross-section σ_{12} are related in a familiar way as that of (5.8), and so the emission cross-section σ_{21} has the following expression

$$\begin{aligned}\sigma_{21}^{\mathrm{ed}(q)}(\omega_k) &= \sigma_{12}^{\mathrm{ed}(q)}(\omega_k)\frac{g_1}{g_2} \\ &= \frac{4\pi^2\omega_k e^2}{3\hbar c(2J_2+1)} \frac{(n^2+2)^2}{9n_q} \sum_{k=2,4,6} \Omega_k^{(q)} \left|\left\langle [\alpha SLJ]_1 \middle\| U^{(k)} \middle\| [\alpha SLJ]_2 \right\rangle\right| g(\omega_k)\end{aligned} \quad (5.63\mathrm{a})$$

$$\sigma_{21}^{\mathrm{md}}(\omega_k) = \sigma_{12}^{\mathrm{md}}(\omega_k)\frac{g_1}{g_2} = \frac{4\pi^2\omega_k e^2 n}{3\hbar c(2J_2+1)} S^{\mathrm{md}}(J_2 \to J_1) g(\omega_k) \quad (5.63\mathrm{b})$$

It should be noted that (5.63a, 5.63b) can only be used in absorption and emission between two very sharp energy levels. In the solid laser materials, transitions related to absorption and emission bands usually occur between two multiplets constituted by sublevels E_i with degeneracy g_i and sublevels E_j with degeneracy g_j, respectively. In the calculation of both absorption and emission cross-sections the fact that the ions are distributed according to Boltzmann law in the related sublevels of multiplets should be taken into account. In other words, the absorption and emission cross-sections should be averaged over different levels with average weight of $g_j e^{-\varepsilon_j/k_\mathrm{B}T}$ and $g_i e^{-\varepsilon_i/k_\mathrm{B}T}$ for the high and low multiplet, respectively, and then the cross-sections can be expressed as

$$\sigma_{\mathrm{em}}(\omega) = \frac{\sum_{ji} g_j e^{-\varepsilon_j/k_B T} \sigma_{ji}}{\sum_j g_j e^{-\varepsilon_j/k_\mathrm{B}T}} \quad (5.64\mathrm{a})$$

$$\sigma_{\mathrm{abs}}(\omega) = \frac{\sum_{ji} g_i e^{-\varepsilon_i/k_B T} \sigma_{ij}}{\sum_i g_i e^{-\varepsilon_i/k_\mathrm{B}T}} \quad (5.64\mathrm{b})$$

The energies of the high and low multiplets, calculated from the bottom of each multiplet, are denoted by ε_j and ε_i, respectively. If we change the zero point of energy level to calculate from the ground state, then the following relation should be used

$$\varepsilon_j = E_j - E_{ZL}, \quad \varepsilon_i = E_i, \quad E_j - E_i = \hbar\omega, \quad \varepsilon_j - \varepsilon_i = \hbar\omega - E_{ZL}$$

and then

$$\begin{aligned}\sum_{ji} g_j e^{-\varepsilon_j/k_B T}\sigma_{ji} &= \sum_{ji} g_j e^{-(\varepsilon_j-\varepsilon_i)/k_B T} e^{-\varepsilon_i}\sigma_{ji} = \exp[(E_{ZL} - \hbar\omega)/k_B T]\sum_{ji} g_j e^{-\varepsilon_i/k_B T}\sigma_{ji} \\ &= \exp[(E_{ZL} - \hbar\omega)/k_B T]\sum_{ji} g_i e^{-\varepsilon_i/k_B T}\sigma_{ij}\end{aligned}$$

The relation $g_i\sigma_{ij} = g_j\sigma_{ji}$ has been used in the last step of the above equation. By means of (5.64a) and (5.64b), the following expressions can be obtained.

$$\sigma_{\mathrm{em}}(\omega) = \sigma_{\mathrm{abs}}(\omega)\frac{Z_l}{Z_u}\exp[(E_{ZL} - \hbar\omega)/k_B T] \tag{5.65a}$$

$$\sigma_{\mathrm{em}}(\lambda) = \sigma_{\mathrm{abs}}(\lambda)\frac{Z_l}{Z_u}\exp[(E_{ZL} - hc/\lambda)/k_B T] \tag{5.65b}$$

For anisotropic crystals, the absorption and emission cross-sections in q polarization direction have the following relations

$$\sigma_{\mathrm{em}}^{(q)}(\omega) = \sigma_{\mathrm{abs}}^{(q)}(\omega)\frac{Z_l}{Z_u}\exp[(E_{ZL} - \hbar\omega)/k_B T] \tag{5.65c}$$

$$\sigma_{\mathrm{em}}^{(q)}(\lambda) = \sigma_{\mathrm{abs}}^{(q)}(\lambda)\frac{Z_l}{Z_u}\exp[(E_{ZL} - hc/\lambda)/k_B T] \tag{5.65d}$$

The above absorption and emission cross-sections include the contributions from both of the electric and magnetic transitions. Where $Z = \sum_t g_t \exp(-\varepsilon_t/k_B T)$, Z_u denotes the function for high multiplet with $t = j$ and Z_l denotes the function for low one with $t = i$. The noreciprocity introduced by electron–phonon interaction [40] has not been taken into account in the discussion.

In order to use (5.65a, 5.65b, 5.65c, 5.65d) to calculate emission cross-section, it is necessary to have the data of crystal field energy levels of the high and low multiplets, which can only be obtained from the low temperature spectra. However, the Z_l/Z_u in (5.65a, 5.65b, 5.65c, 5.65d) and then emission cross-section can also be calculated from the absorption spectrum and the radiative lifetime by using the formula introduced as follows.

By (5.5a, 5.5b) and (5.14a, 5.14b), it can be shown that the electric and magnetic transitions have the same relation as the probability of spontaneous dipole transition with the emission cross-section. The related formulas for anisotropic crystal are as follows

$$\sigma_{\text{em}}^{(q)}(\omega) = \frac{\pi^2 c^2}{\omega^2 n^2} A_{21}^{(q)} g(\omega) \quad \sigma_{\text{em}}^{(q)}(v) = \frac{c^2}{8\pi v^2 n^2} A_{21}^{(q)} g(v) \quad \sigma_{\text{em}}^{(q)}(\lambda) = \frac{\lambda^4}{8\pi n^2 c} A_{21}^{(q)} g(\lambda) \tag{5.66}$$

Integrated over ω then

$$\int \sigma_{\text{em}}^{(q)}(\omega)\omega^2 d\omega = \frac{\pi^2 c^2}{n^2} A_{21}^{(q)}$$

Two sides of (5.65c) multiplied by ω^2 and integrated over ω, by using above equation, it can be shown that

$$Z_l/Z_u = \frac{\pi^2 c^2 \exp(-E_{\text{ZL}}/k_B T) A_{21}^{(q)}}{n^2 \int \sigma_{\text{em}}^{(q\cdot)}(\omega)\omega^2 \exp(-\hbar\omega/k_B T) d\omega}$$

therefore

$$\sigma_{\text{em}}^{(q)} = \frac{\pi^2 c^2 \sigma_{\text{abs}}^{(q)}(\omega) A_{21}^{(q)}}{n^2 \int \sigma_{\text{abs}}^{(q\cdot)}(\omega)\omega^2 \exp(-\hbar\omega/k_B T) d\omega} \tag{5.67a}$$

Change the variable from ω to λ, then

$$\sigma_{\text{em}}^{(q)}(\lambda) = \frac{1}{8\pi n^2 c} \frac{\exp(-hc/(\lambda k_{\text{B}} T)) A_{21}^{(q)}}{\int \sigma_{\text{abs}}^{(q)}(\lambda)\lambda^{-4} \exp(-hc/(\lambda k_{\text{B}} T)) d\lambda} \sigma_{\text{abs}}^{(q)}(\lambda) \tag{5.67b}$$

In the calculation of the emission cross-section, (5.65a, 5.65b, 5.65c, 5.65d) and (5.67a, 5.67b) are usually referred to as reciprocity method. The emission cross-section can also be calculated directly by the data of fluorescence spectra, it is called Fuchtbauer-Ladenburg equation, which will be introduced in the following.

The spontaneous transition probability from a high multiplet 2 to a low multiplet 1, with wavelength λ, can be expressed as

$$A_{21}^{(q)}(\lambda) = \sum_{ji} f_j A_{ji}^{(q)} \delta\left(E_j - E_i - \frac{hc}{\lambda}\right) \tag{5.68}$$

In the above equation, the indexes of j and i are used to denote the crystal field energy levels of multiplets 2 and 1, respectively. f_j is the Boltzmann distribution factor of the ions in high multiplet 2. Substituting (5.68) into the third equation of (5.66), one obtains

$$\sigma_{\text{em}\cdot}^{(q)}(\lambda) = \frac{\lambda^4}{8\pi n_q^2 c} \sum_{ji} f_j A_{ji}^{(q)} \delta\left(E_j - E_i - \frac{hc}{\lambda}\right) g(\lambda) \tag{5.69}$$

If the sensitivity of the fluorescence detective system is G and the ion number of the high multiplet is N_u, then the total detected fluorescence intensity polarized in q direction will be

$$I^{(q)}(\lambda) = GN_u \sum_{ji} f_j A_{ji}^{(q)} \delta\left(E_j - E_i - \frac{hc}{\lambda}\right) g(\lambda) \times \frac{hc}{\lambda} \tag{5.70}$$

The total photon number in the spectrum polarized in q direction can be expressed as

$$\frac{1}{hc} \times \int_{\substack{\text{spectrum} \\ 2 \to 1}} \lambda I^{(q)}(\lambda) d\lambda = GN_u A_{21}^{(q)}$$

so that

$$GN_u = \int_{\substack{\text{spectrum} \\ 2 \to 1}} \lambda I(\lambda) d\lambda / hc A_{21}^{(q)} \tag{5.71}$$

By substituting (5.71) into (5.70) and comparing with (5.69), one can obtain the following expression referred to as Fuchtbauer-Ladenburg (F-L) equation finally.

$$\sigma_{\text{em}}^{(q)}(\lambda) = \frac{\lambda^5 A_{21}^{(q)} I^{(q\cdot)}(\lambda)}{8\pi n_q^2 c \int_{\substack{\text{spectrum} \\ 2 \to 1}} \lambda I^{(q\cdot)}(\lambda) d\lambda} \tag{5.72}$$

If the variable is chosen to be the frequency, then the corresponding expression is

$$\sigma_{\text{em}}^{(q)}(v) = \frac{c^2 A_{21}^{(q)} I_q(v)}{8\pi n_q^2 v^3 \int_{\substack{\text{spectrum} \\ 2 \to 1}} \frac{I_q(v)}{v} dv} \tag{5.73}$$

For the cubic crystals and the glasses, the emission cross-section of active ions can be calculated by the following formulas using fluorescence lifetime τ_f, fluorescence branching ratio β, and fluorescence quantum efficiency η

$$\sigma_{\mathrm{em}}(\lambda) = \frac{\beta\eta\lambda^5 I(\lambda)}{8\pi n^2 c\tau_f \int\limits_{\substack{\text{spectrum}\\ 2\to 1}} \lambda I(\lambda)\mathrm{d}\lambda}, \quad \sigma_{\mathrm{em}}(v) = \frac{\beta\eta c^2 I(v)}{8\pi n^2 \tau_f v^3 \int\limits_{\substack{\text{spectrum}\\ 2\to 1}} \frac{I(v)}{v}\mathrm{d}v}$$

These expressions are the same as those introduced by Aull [40] but without the Boltzmann distribution factor f_j appearing in the denominators. It should be noted that these expressions are based on the supposition that the sensitivity of the detection is the same for the whole fluorescence spectrum of the transition between the two multiplets. It is also useful to mention the fact that in order to measure the fluorescence lifetime accurately, it is necessary to eliminate the effect of radiation trapping, especially for the three-level system or quasi-three-level system like that of Yb^{3+} ions, in which this effect is more significant [41]. In these cases in order to reduce the error of fluorescence lifetime measurement, it is usual to use powder sample with the grain diameter about 100 μm immersing in a liquid having refractive index near that of the sample. In this way, the internal reflection of the fluorescent light by the powder grains can be reduced so that the effect of radiation trapping is partially eliminated.

The calculation of radiative lifetime for Yb^{3+} ions should be mentioned. For this ion, it is impossible to use absorption spectral data to calculate J–O parameter and then calculate transition probability. In this case, radiative lifetime is usually calculated by emission cross-section obtained by means of F-L formula or reciprocity method. By integrating both sides of the second equation of (5.66) over wavelength, then for the transition from multiplet J_i to multiplet J_j, one has

$$A^{(q)}(J_i \to J_j) = \frac{8\pi n^2 c}{\bar{\lambda}^4} \int \sigma^{(q)}_{J_i \to J_j}(\lambda) d\lambda \tag{5.74}$$

The transition probabilities for uniaxial and biaxial crystals should be calculated using the following formulas, respectively

$$A(J_i \to J_f) = \frac{1}{3}A^{(\pi)}(J_i \to J_f) + \frac{2}{3}A^{(\sigma)}(J_i \to J_f)$$
$$A(J_i \to J_f) = \frac{1}{3}A^{(x)}(J_i \to J_f) + \frac{1}{3}A^{(y)}(J_i \to J_f) + \frac{1}{3}A^{(z)}(J_i \to J_f)$$

Finally, the radiative lifetime τ_r and fluorescence branching ratio β_{ij} can be calculated by the following expressions, respectively

$$\tau_r(J_i) = \frac{1}{\sum_{J_j} A(J_i \to J_j)}, \quad \beta_{ij} = \frac{A(J_i \to J_j)}{\sum_{J_j} A(J_i \to J_j)} \tag{5.75}$$

5.5 Hypersensitive Transitions

Just as pointed out earlier, the 4*f* electrons of rare earth ions are shielded by its outside filled shell $5s^2 5p^6$, so that the transition intensities between different energy levels for trivalent rare earth ion are not much affected by its host. The dipole transition intensity of a particular transition channel for a rare earth ion in different hosts does not vary more than a factor of 2–3. However, a few transition channels are very sensitive to the environment of rare earth ion; these are so-called hypersensitive transition. All known hypersensitive transitions obey the selection rules of $\Delta S = 0$, $|\Delta L| \leq 2$, $|\Delta J| \leq 2$. These selection rules are the same as those of electric-quadrupole transition, but their intensity is several order magnitude larger than that of electric-quadrupole transition; therefore, it is also called pseudo-quadrupole transition. Hypersensitive transition channels of six rare earth ions commonly used in laser materials are shown in Table 5.1. Among these, some transitions do not obey the selection rule $\Delta S = 0$ seemingly, however, their intermediate-coupling wave function does include the component satisfying this selection rule. For example, the intermediate-coupling wave function of $^1G_4$ of the Tm^{3+} ion includes 31.8% $^3F_4$ wave function and 56.7% $^3H_4$ wave function which have the spin quantum number the same as that of the energy level $^3H_6$.

Among these hypersensitive transition channels, the $^3F_4$–$^3H_6$ transition of Tm^{3+} ion is the channel for 2 μm laser which has important applications; the study on the relation between its transition probability and the composition as well as the structure of the hosts has important significance.

Regarding the hypersensitive transition mechanism, different authors have different opinions. Judd [42] considered that this transition is associated with large value of $U^{(2)}$ matrix element. The line strength S of electric-dipole transition

Table 5.1 Hypersensitive transition channels of six rare earth ions commonly used in laser materials

Ion	Transition channel	Wave number (cm^{-1})	Ion	Transition channel	Wave number (cm^{-1})
Pr^{3+}	$^3F_2$–$^3H_4$	4950	Er^{3+}	$^4G_{11/2}$–$^4I_{15/2}$	26400
Nd^{3+}	$^4G_{5/2}$–$^4I_{9/2}$	17200		$^2H_{11/2}$–$^4I_{15/2}$	19200
Ho^{3+}	$^5G_6$–$^5I_8$	22150	Tm^{3+}	$^3F_4$–$^3H_6$	5600
	$^3H_6$–$^5I_8$	27700		$^3H_4$–$^3H_6$	12550
Dy^{3+}	$^6F_{11/2}$–$^6H_{15/2}$	7550		$^1G_4$–$^3H_6$	21050

includes the reduced matrix element of unit operator $U^{(k)}$ as shown in (5.39) and the expression of this matrix element can be seen in (2.52) as

$$\langle f^n SLJ||U^{(k)}||f^n S'L'J'\rangle = (-1)^{S+L'+J+k}[(2J+1)(2J'+1)]\begin{Bmatrix} J & J' & k \\ L' & L & S \end{Bmatrix}\langle f^n SL||U^{(k)}||f^n S'L'\rangle\delta_{SS'}$$

where the δ function and the triangle condition of 6-j symbol for $k = 2$ require that $\Delta S = 0$, $|\Delta L| \leq 2$, and $|\Delta J| \leq 2$ should be satisfied. This is just the selection rules of hypersensitive transition. Therefore, the hypersensitive transition not only demands a large reduced matrix element of unit operator $U^{(2)}$ but also a large J–O parameter Ω_2. The contributions of odd crystal field and odd lattice vibration to the J–O parameter Ω_k and Ω_k' are expressed as [2]

$$\begin{aligned}\Omega_k &= (2k+1)\sum_{tp}|B_{tp}|^2\Xi^2(t,k)(2t+1)^{-1}, \Omega'_k \\ &= (2k+1)\sum_{tp}|\langle n|Q|n'\rangle|^2\rho(n)\left|\frac{\partial B_{tp}}{\partial Q}\right|^2\Xi^2(t,k)(2t+1)^{-1}\end{aligned}$$

where $\Xi(t,k)$ is a factor that depends only on the atomic parameters, $B_{tp}(t = 1, 3, 5)$ is the odd parity crystal parameters and n and n' are lattice vibration quantum numbers of initial and final states, respectively, and $\rho(n)$ is lattice vibration state density. It can be shown that the expression of J–O parameter Ω_k includes the following 3-j symbol as a factor

$$\begin{pmatrix} 1 & k & t \\ q & -p-q & p \end{pmatrix}$$

Then by the triangle condition of 3-j symbol it can be shown that for $\Omega_2(k = 2)\neq 0$, $3 \geq t \geq 1$ should be satisfied, that is the quantity of Ω_2 is determined by the odd crystal field parameters B_{1p} and B_{3p}. From the data of $\Xi(t, k)$ obtained by Krupke [1] calculated using free ion wave function, it can be seen that in order to have Ω_2 larger than Ω_4 and Ω_6, a large one-order crystal field component B_{1p} is necessary. The study of Krupke [1] has shown that Ω_2 is mainly produced by the odd crystal field potential, while for the Ω_4 and Ω_6, the contribution of odd vibrational modes accounts for a large proportion. Therefore, it is reasonable to consider that the one-order crystal field potential is the important micro-mechanism of hypersensitive transition. The point group of crystal having crystal field component B_{1p} is the same as the situation of reduced matrix element $U^{(0)}$; that is the following ten point groups: C_s, C_1, C_2, C_3, C_4, C_6, C_{2v}, C_{3v}, C_{4v}, and C_{6v}. Thereafter many authors proposed other mechanisms, among those like inhomogeneous dielectric model, ligand-polarization model and covalence model, and so on. However, these models cannot be unified to interpret hypersensitive transition phenomena in different dielectrics. For example, according to ligand polarization model, [43] the hypersensitive transition can occur when active ions occupy D_{3h} point group position, but

Krupke [1] showed that the rare earth ion in LaF_3 crystal occupy D_{3h} point group position; its Ω_2 value is much smaller than those of Ω_4 and Ω_6, while Nd^{3+} in Y_2O_3 crystal occupy C_2 position—one of symmetry positions having crystal field component B_{1p}; its $^4G_{5/2} \rightarrow {}^4I_{9/2}$ transition intensity is 12 times that in LaF_3 crystal. Meanwhile for other non-hypersensitive transition channel, the transition intensity of Nd^{3+}:Y_2O_3 is only two times that of Nd^{3+}:LaF_3. It suggests that the hypersensitive transition of Nd^{3+} ions does not occur in D_{3h} point group position. Recently, Russian authors [44–46] studied the Nd^{3+}, Tm^{3+}, Ho^{3+}, and Dy^{3+} ions in YAG and GGG garnet crystals in which the ions can only occupy D_2 position and the $Ca_3(NbGa)_5O_{12}$ garnet crystal in which the ions can occupy C_2, C_{2v}, and C_1 positions as well as the Y_2O_3 and KGd $(WO_4)_2$ crystals in which the ions occupy C_2 position. They measured the oscillator strength and calculated the J–O parameters of related hypersensitive transition channels. Their experimental results support the B_{1p} crystal field component model.

The intensity of hypersensitive transition has obvious relationship with the bond length and coordination number between the rare earth ion and their ligands [47]. The shorter the bond length and the higher the coordination number, the stronger is the hypersensitive transition. For example, the intensities of hypersensitive transition for Nd^3 in three oxide crystals vary according to the sequence of YAG < YAP < Y_2O_3.

References

1. W.F. Krupke, Phys. Rev. **145**, 325 (1966)
2. B.R. Judd, Phys. Rev. **127**, 750 (1962)
3. L.J.F. Broer, G.J. Gortter, J. Hoogschagen, Physica **11**, 231 (1945)
4. M. Born, E. Wolf, *Principle of Optics*, 5th edn. (Pergamon Press, Oxford, 1975)
5. Z.D. Luo, X.Y. Chen, T.J. Zhao, Opt. Commun. **134**, 415 (1997)
6. G.S. Ofelt, J. Chem. Phys. **37**, 511 (1962)
7. X.Y. Chen, Z.D. Luo, J. Phys. Condens. Matter **13**, 1171 (2001)
8. B.R. Judd, *Operator Techniques in Atomic Spectroscopy* (McGraw-Hill Inc., New York, 1963)
9. A.R. Edmonds, *Angular Momentum in Quantum Mechanics* (Princeton University Press, Princeton, 1957)
10. S.D. Xia, Y.M. Chen, Sci. Sin. (Series A) **28**, 950–974 (1985)
11. T.S. Lomheim, L.G. Deshazer, Opt. Commun. **24**, 89 (1978)
12. P. Goldner, F. Auzel, J. Appl. Phys. **79**, 7972 (1996)
13. X.Y. Chen, Z.D. Luo, Chinese J. Struct. Chem. **16**, 358 (1997)
14. L. Smentek, B.G. Wybourne, B.A. Hess Jr., J. Alloys Comp. **323–324**, 645 (2001)
15. C.W. Nielson, G.F. Koster, *Spectroscopic coefficients for the p^n, d^n and f^n configurations* (M. I. T. Press, Cambridge, Massachusetts, 1963)
16. H. Dai, O.M. Stafsudd, J. Phys. Chem. Solids **52**, 367 (1991)
17. H. Dai, O.M. Stafsudd, B. Dunn, Appl. Opt. **30**, 4330 (1991)
18. A. Brinier, W. Ych, F. Pz et al., J. Appl. Phys. **98**, 123528 (2005)
19. A.A. Kaminskii, L. Li, Phys. Stat. Sol (A), **26**, 593 (196)
20. W.T. Dunina, A.A. Kaminskii, A.A. Kornienko et al., Sov. Phys. Solid State **32**, 289 (1990)
21. A. Brinier, I.V. Kityk, J. Appl. Phys. **90**, 232 (2001)

22. F.B. Xiong, X.Q. Lin, Z.D. Luo et al., J. Appl. Phys. **99**, 064905 (2006)
23. S.B. Piepho, P.N. Schatz, *Group Theory in Spectroscopy with Applications to Magnetic Circular Dichroism* (Wiley, New York, 1983)
24. M.J. Weber, in *Optical Properties of Ions in Crystals*, ed. by H.M. Crosswhite, H.W. Moos. (Wiley, New York, 1967)
25. J.D. Axe, J. Chem. Phys. **39**, 1154 (1963)
26. D.J. Newman, G. Balasubramanian, J. Phys. C **8**, 37 (1975)
27. M.F. Reid, F.S. Richardson, J. Chem. Phys. **71**, 5735 (1983)
28. M.F. Reid, F.S. Richardson, J. Phys. Chem. **88**, 3579 (1984)
29. M.F. Reid, J.J. Dallara, F.S. Richardson, J. Chem. Phys. **79**, 5743 (1983)
30. M.F. Reid, J. Chem. Phys. **87**, 6388 (1987)
31. M.F. Reid, J. All. Com. **180**, 93 (1992)
32. G.W. Burdick, C.K. Jayasankar, F.S. Richardson, M.F. Reid, Phys. Rev. B **50**, 16309 (1994)
33. R.L. Stefan, H.U. Gudel, M.P. Hehlen, J.R. Quagliano, Phys. Rev. B **57**, 15229 (1998)
34. Y.D. Huang, Z.D. Luo, J. Phys. Condens. Matter **5**, 1581 (1993)
35. Z.D. Luo, Y.D. Huang, J. Phys. Condens. Matter **6**, 3737 (1994)
36. X.Y. Chen, Z.D. Luo, J. Phys. Condens. Matter **10**, 5147 (1998)
37. Y.D. Huang, Ph.D. Dissertation, University of Strathclyde, U.K., 1997
38. C.L. Görller-Warlrand, A. Fluyt, Ceulemans et al., J. Chem. Phys. **95**, 3099 (1991)
39. W.T. Carmall, P.R. Field, K. Rajnak, J. Chem. Phys. **49**, 4412 (1968)
40. B.F. Aull, H.P. Jenssen, IEEE J Quant. Elect. QE **18**, 925 (1982)
41. D.S. Sumida, T.Y. Fan, Opt. Lett. **17**, 1343 (1994)
42. B.R. Judd, J. Chem. Phys. **44**, 839 (1966)
43. S.F. Mason, R.D. Peacock, B. Stewart, Chem. Phys. Lett. **29**, 149 (1974)
44. P.A. Ryabochkina, E.V. Bolshakova, S.N. Ushakov et al., J. Lum. **132**, 240 (2012)
45. P.A. Ryabochkina, S.A. Antoshkina, E.V. Bolshakova et al., J. Lum. **132**, 1900 (2012)
46. P.A. Ryabochkina, S.A. Antoshkina, S.A. Klimin et al., J. Lum. **138**, 32 (2013)
47. D.E. Henrie, R.L. Fellows, G.R. Choppin, Coord. Chem. Rev. **18**, 199 (1976)

Chapter 6
Phonon and Spectral Line

The lattice vibration has many effects on spectroscopic properties of active ions in solids, including inducing of electric-dipole transition and vibronic-electronic transition, broadening of spectral line, shifting of spectral line position, introducing of non-radiative transition, and transforming of excited energy between ions. Therefore, studying the interaction of electron with lattice vibration is one of the most important problems in the physics of solid-state laser material. During the discussion of this chapter, the conception of phonon by quantization of lattice vibration will be introduced; the effects of phonon emission and absorption on the spectral line shape, spectral linewidth, and spectral line shifting will be discussed. In these discussions, the concentration of active ions is very low and the interaction between different active ions can be neglected, and the interaction of electronic states of active ions with the electronic band of the crystals is neglected too. The non-radiative transition inside rare earth ions introduced by electron-phonon interaction will be discussed in Chap. 8, while the phonon-assisted energy transfer will be discussed in Chap. 9.

6.1 Quantization of Lattice Vibration—Phonon

Theoretically, the starting point for studying lattice vibration is to solve the Schrödinger equation in quantum mechanics. Before the discussion, one important approximation should be assumed. It aims at the separation of the movement of electrons and the lattice vibration. As well known, the mass of the electron is much less than that of the nucleus. Even for hydrogen atom, the mass of the electron is only one part of 1840 of that of the nucleus. From classical point of view, the movement of atoms (ions) is much slower than that of the electrons. Therefore, in the same period of time, the ions have just moved for a very short distance, while the electrons have moved around the nucleus for many circles. It is reasonable to solve the problem of the electronic movement in fixed positions of the ions and then

Z. Luo and Y. Huang, *Physics of Solid-State Laser Materials*, Springer Series in Materials Science 289, https://doi.org/10.1007/978-981-32-9668-8_6

solve the problem of the lattice vibration in some kinds of average electronic field. This is referred to as the adiabatic approximation. By the formulas of classical and quantum mechanics, before taking into account the electron–lattice interaction, the total energy of a system of n electrons with mass m_e and N lattice ions with mass m_α is [1]

$$E = \frac{m}{2}\sum_{i=1}^{n}\dot{\boldsymbol{r}}_i^2 + \frac{1}{2}\sum_{\alpha=1}^{N} m_\alpha \dot{\boldsymbol{R}}_\alpha^2 + \frac{1}{2}\sum_{i\neq j}\sum_{j}\frac{e^2}{|\boldsymbol{r}_i - \boldsymbol{r}_j|} + \frac{1}{2}\sum_{\alpha\neq\beta}\sum_{\beta}\frac{e^2 Z_\alpha Z_\beta}{|\boldsymbol{R}_\alpha - \boldsymbol{R}_\beta|} \tag{6.1}$$

Changing the classical variables into quantum mechanics operators, the partial differential operators Δ_i and Δ_α are expressed as

$$\Delta_i = \frac{\partial^2}{\partial x_i} + \frac{\partial^2}{\partial y_i} + \frac{\partial^2}{\partial z_i}, \Delta_\alpha = \frac{\partial^2}{\partial X_\alpha} + \frac{\partial^2}{\partial Y_\alpha} + \frac{\partial^2}{\partial Z_\alpha} \tag{6.2}$$

The last two terms in (6.1) are interaction potential usually denoted by $V(r, R)$

$$V(r,R) = \frac{1}{2}\sum_{i\neq j}\sum_{j}\frac{e^2}{|\boldsymbol{r}_i - \boldsymbol{r}_j|} + \frac{1}{2}\sum_{\alpha\neq\beta}\sum_{\beta}\frac{e^2 Z_\alpha Z_\beta}{|\boldsymbol{R}_\alpha - \boldsymbol{R}_\beta|} \tag{6.3}$$

By using operator expression of (6.2) and potential expression of (6.3), the Hamiltonian of the system will be

$$H = -\frac{\hbar^2}{2m_e}\sum_i \Delta_i - \frac{\hbar^2}{2}\sum_\alpha \frac{\Delta_\alpha}{m_\alpha} + V(r,R)$$

and then the Schrödinger equation is expressed as

$$-\frac{\hbar^2}{2m_e}\sum_i \Delta_i \Psi(r,R) - \frac{\hbar^2}{2}\sum_\alpha \frac{\Delta_\alpha \Psi(r,R)}{m_\alpha} + V(r,R)\Psi(r,R) = E\Psi(r,R) \tag{6.4}$$

In adiabatic approximation, the movement of electron and the lattice system can be dealt with as two independent systems (their interaction will be introduced later). The zero-order wave function can be written as

$$\Psi(r,R) = \Phi(r,R)\varphi(R) \tag{6.5}$$

where $\Phi(r, R)$ is an electronic wave function depending on R (the coordinate of the lattice ions) as a parameter and $\varphi(R)$ is lattice wave function. Substituting (6.5) into (6.4) and neglecting the dependency of the electronic wave function on the lattice coordinates, that is, the following so-called non-adiabatic term

$$-\frac{\hbar^2}{2}\sum_{\alpha}\frac{1}{m_\alpha}[\varphi(R)\Delta_\alpha\Phi(r,R)+2\nabla_\alpha\varphi(R)\nabla_\alpha\Phi(r,R)] \tag{6.6}$$

then the equation becomes

$$-\frac{\hbar^2}{2m_e}\varphi(R)\sum_i\Delta_i\Phi(r,R)-\frac{\hbar^2}{2}\sum_\alpha\frac{\Phi(r,R)\Delta_\alpha\varphi(R)}{m_\alpha}+V(r,R)\Phi(r,R)\varphi(R)$$
$$=E\Phi(r,R)\varphi(R)$$

Let

$$-\frac{\hbar^2}{2m_e}\varphi(R)\sum_i\Delta_i\Phi(r,R)+V(r,R)\Phi(r,R)\varphi(R)=W(R)\Phi(r,R)\varphi(R)$$

That is

$$-\frac{\hbar^2}{2m_e}\sum_i\Delta_i\Phi(r,R)+V(r,R)\Phi(r,R)=W(R)\Phi(r,R) \tag{6.7}$$

Equation (6.7) is in fact an electronic equation. In this way, the separation between electronic coordinates and those of the lattice ions is done. The related equation for the lattice ions is

$$-\frac{\hbar^2}{2}\sum_\alpha\frac{\Delta_\alpha\varphi(R)}{m_\alpha}+W(R)\varphi(R)=E\varphi(R) \tag{6.8}$$

In the following, (6.8) will be solved when the lattice ions have only a small vibration around their equilibrium positions. In this discussion, the coordinates of the lattice ions should be labeled clearly. For example, the unit cell of the lattice ion is numbered by n, and the class of the ion in the cell is numbered by α (note that the meaning of n and α are different from those in (6.1)). The lattice potential can be expanded as a Taylor series near the equilibrium position. Except for the zero-order term, which is independent of the coordinates of the displacement, the first-order term representing the force subject by the lattice points is equal to zero in the equilibrium. If the terms higher than the second order are temporarily neglected, then the potential can be expressed as

$$\begin{aligned}W(R)&=\frac{1}{2}\sum_{\substack{n\alpha i\\ n'\alpha' i'}}\frac{\partial^2 W}{\partial R_{n\alpha i}\partial R_{n'\alpha' i'}}\bigg|_{R(0)}[R_{n\alpha i}-R_{n\alpha i}(0)][R_{n'\alpha' i'}-R_{n'\alpha' i'}(0)]\\&=\tfrac{1}{2}\sum_{\substack{n\alpha i\\ n'\alpha' i'}}f_{ii'}(n\alpha,n'\alpha')u_i(n\alpha)u_{i'}(n'\alpha')\end{aligned} \tag{6.9}$$

where the second-order derivative should adopt the value at the equilibrium position. $f_{ii'}(n\alpha, n'\alpha')$ refers to the force constant and is a real number. Their symmetric properties can be seen from the definition.

$$f_{ii'}(n\alpha, n'\alpha') = f_{i'i}(n'\alpha', n\alpha)$$

where i denotes the coordinate components i = 1, 2, 3. The Hamiltonian of the lattice system can be written as a sum of kinetic energy and potential energy

$$H = \frac{1}{2}\sum_{n\alpha i} m_\alpha \dot{u}_i^2(n\alpha) + \frac{1}{2}\sum_{\substack{n\alpha i \\ n'\alpha' i'}} f_{ii'}(n\alpha, n'\alpha')u_i(n\alpha)u_{i'}(n'\alpha') \tag{6.10}$$

The above formula has the same form as that of the Hamiltonian of the harmonic oscillator $\left(\frac{1}{2}mv^2 + \frac{1}{2}m\omega^2u^2\right)$. It means that the lattice vibration consists of a series of harmonic oscillators. In the following, the method of second quantization is used to describe the lattice vibration.

In the Hamiltonian (6.10), the second term on the right-hand side is the potential of lattice system obviously. Then the motion equation of the lattice can be written as

$$M_\alpha \ddot{u}_i(n\alpha) = -\frac{\partial W}{\partial u_i(n\alpha)} = -\sum_{n'\alpha' j'} f_{ii'}(n\alpha, n'\alpha')u_{i'}(n'\alpha') \tag{6.11}$$

The solution of this equation can be written as follows, because the lattice vibration is periodicity in time. Its displacement can be expressed as

$$u_i(n\alpha) = \frac{1}{\sqrt{m_\alpha}} q_i(n\alpha)\exp(-i\omega t) \tag{6.12}$$

Substituting the above formula into (6.11), then

$$\omega^2 q_i(n\alpha) = \sum_{n'\alpha' i'} D_{ii'}(n\alpha, n'\alpha')q_{i'}(n'\alpha') \tag{6.13}$$

where

$$D_{ii'}(n\alpha, n'\alpha') = \frac{f_{ii'}(n\alpha, n'\alpha')}{\sqrt{m_\alpha m_{\alpha'}}}$$

If the crystal has N unit cells and each cell has r atoms, then (6.13) is a simultaneous equation with $3rN$ dimensions and there are $3rN$ solutions of ω^2

(therefore have $3rN$ normal vibrational modes). By using the translation symmetry of the lattice, the function of $q(n\alpha)$ can be written as

$$q_i(n\alpha) = s_i(\alpha)\exp(i\boldsymbol{k}\cdot\boldsymbol{R}_n) \tag{6.14}$$

where $\boldsymbol{k}$ is a reciprocal vector. Equation (6.13) then becomes

$$\omega^2 s_i(\alpha) = \sum_{\alpha' i'}\left\{\sum_{n'-n}\frac{f_{ii'}(n'-n,\alpha\alpha')}{\sqrt{m_\alpha m_{\alpha'}}}\exp[i\boldsymbol{k}\cdot(\boldsymbol{R}_n-\boldsymbol{R}_n)]\right\}s_{i'}(\alpha') \tag{6.15}$$

The translation symmetry has been used to write down (6.15). In this case, $f_{ii'}(n\alpha, n'\alpha')$ depends only on the relative position of the cell, that is $f_{ii'}(n\alpha, n'\alpha') = f_{ii'}(n'-n,\alpha\alpha')$. If $D_{ii'}(\boldsymbol{k},\alpha\alpha')$ is used to represent the quantities inside the curly bracket in (6.15), then

$$\omega^2 s_i(\alpha) = \sum_{\alpha' i'} D_{ii'}(\boldsymbol{k},\alpha\alpha')s_{i'}(\alpha') \tag{6.16}$$

Equation (6.16) is a $3r$-dimensional simultaneous equation. It has $3r$ solutions for each $\boldsymbol{k}$ vector

$$\omega \equiv \omega_i(\boldsymbol{k}) \quad (i = 1, 2, \ldots, 3r)$$

The number of $\boldsymbol{k}$ vectors is the cell number N of the crystal and so the number of total vibration mode will be $3Nr$. It can be shown that because of the translation symmetry of the crystal, these $3Nr$ modes are N-fold degeneracy. Every frequency ω_i has a corresponding s_i which can be normalized to become $e_i(\boldsymbol{k},\alpha)$ and simply expressed as

$$s_i(\alpha) = e_i(\boldsymbol{k},\alpha) \tag{6.17}$$

The basic characteristics of the lattice vibration are obtained by using a model with two kinds of ions in a cell, which can be found in many textbooks in solid-state physics (e.g. [2]). For the sake of brevity, the details will not be introduced while the main results will be mentioned. When there are two kinds of ions in the cell ($r = 2$), the lattice system has three acoustic modes and three optical modes. The character of the acoustic mode is that two kinds of ions vibrate in the same direction, while for the optical mode, the vibration directions of two ions with opposite sign of electric charge are opposite. The later vibrational mode will produce a net electric-dipole moment and so can interact with radiation field. In general cases, there are three acoustic modes and $3r$–3 optical modes.

The displacement of the lattice ions can be expressed by (6.12), (6.14), and (6.17) as a linear combination of the displacements of all the harmonic vibrational modes [3]

$$u_{n\alpha i}(t) = \frac{1}{\sqrt{Nm_\alpha}} \sum_{jk} Q_j(\boldsymbol{k}, t) e_i^{(j)}(\boldsymbol{k}, \alpha) \exp(i\boldsymbol{k} \cdot \boldsymbol{R}_n) \tag{6.18}$$

where the time dependency of the displacement has been put into the combination coefficients $Q_j(\boldsymbol{k},t)$, which are the so-called normal coordinates. It can be seen by the free electron theory that $\frac{1}{\sqrt{N}} exp(i\boldsymbol{k} \cdot \boldsymbol{R}_n)$ is the wave function of free electron in the lattice and satisfies the following orthogonal relation [1]:

$$\sum_n exp[i(\boldsymbol{k} - \boldsymbol{k}') \cdot \boldsymbol{R}_n] = N\delta_{\boldsymbol{k}-\boldsymbol{k}', \boldsymbol{K}_m} \tag{6.19}$$

where $\boldsymbol{K}_m$ is the period vector in $\boldsymbol{k}$ space. The basis vectors can be chosen to make the solution $e_i(\boldsymbol{k},\alpha)$ of (6.16) to satisfy the following relation

$$\sum_{\alpha i} e_i^{(j)}(\boldsymbol{k}, \alpha) e_i^{(j')*}(\boldsymbol{k}, \alpha) = \delta_{jj'} \tag{6.20}$$

Owing to the fact that the displacement $u_{n\alpha i}(t)$ of the lattice ions must be a real number, then by (6.18), one has

$$e_i^{(j)*}(\boldsymbol{k}, \alpha) = e_i^{(j)}(-\boldsymbol{k}, \alpha), Q_j^*(\boldsymbol{k}, t) = Q_j(-\boldsymbol{k}, t)$$

Substituting (6.18) into (6.10) and using (6.15) and (6.19), the Hamiltonian of the system can be expressed as

$$H = \frac{1}{2} \sum_{jk} \left[\dot{Q}_j^*(\boldsymbol{k}, t) \dot{Q}_j(\boldsymbol{k}, t) + \omega_j^2(\boldsymbol{k}) Q_j^*(\boldsymbol{k}, t) Q_j(\boldsymbol{k}, t) \right] \tag{6.21}$$

Then the Lagrange function becomes

$$L = \frac{1}{2} \sum_{jk} \left[\dot{Q}_j^*(\boldsymbol{k}, t) \dot{Q}_j(\boldsymbol{k}, t) - \omega_j^2(\boldsymbol{k}) Q_j^*(\boldsymbol{k}, t) Q_j(\boldsymbol{k}, t) \right]$$

Obviously, the normal momentum $P_j(\boldsymbol{k}, t)$ which is conjugated to normal coordinate $Q_j(\boldsymbol{k}$, t) will be

$$P_j(\boldsymbol{k}, t) = \frac{\partial L}{\partial \dot{Q}_j^*(\boldsymbol{k}, t)} = \dot{Q}_j(\boldsymbol{k}, t), P_j^*(\boldsymbol{k}, t) = \frac{\partial L}{\partial \dot{Q}_j(\boldsymbol{k}, t)} = \dot{Q}_j^*(\boldsymbol{k}, t)$$

then (6.21) can be written as

$$H = \frac{1}{2}\sum_{jk}\left[P_j^*(\boldsymbol{k},t)P_j(\boldsymbol{k},t) + \omega_j^2(\boldsymbol{k})Q_j^*(\boldsymbol{k},t)Q_j(\boldsymbol{k},t)\right] \tag{6.22}$$

Equation (6.22) is exactly the Hamiltonian of a set of harmonic oscillators. The normal coordinates and normal momentum satisfy the commutative relation in quantum mechanics

$$\left[Q_j(\boldsymbol{k},t), P_{j'}(\boldsymbol{k}',t)\right] = i\hbar\delta_{\boldsymbol{k}\boldsymbol{k}'}\delta_{jj'} \quad \left[Q_j(\boldsymbol{k},t), Q_{j'}(\boldsymbol{k},t)\right] = 0 \quad \left[P_j(\boldsymbol{k},t), P_{j'}(\boldsymbol{k},t)\right] = 0 \tag{6.23}$$

If one introduces the following transformation

$$Q_j(\boldsymbol{k},t) = \left(\frac{\hbar}{2\omega_j(\boldsymbol{k})}\right)^{1/2}\left[b_j(\boldsymbol{k}) + b_j^+(-\boldsymbol{k})\right],$$
$$P_j(\boldsymbol{k},t) = \left(\frac{\hbar\omega_j(\boldsymbol{k})}{2}\right)^{1/2}\frac{1}{i}\left[b_j(-\boldsymbol{k}) - b_j^+(\boldsymbol{k})\right]$$

that is

$$b_j^+(\boldsymbol{k}) = \left[2\hbar\omega_j(\boldsymbol{k})\right]^{-1/2}\left[\omega_j(\boldsymbol{k})Q_j^*(k,t) - iP_j(\boldsymbol{k},t)\right],$$
$$b_j(\boldsymbol{k}) = \left[2\hbar\omega_j(\boldsymbol{k})\right]^{-1/2}\left[\omega_j(\boldsymbol{k})Q_j(\boldsymbol{k},t) + iP_j^*(\boldsymbol{k},t)\right]$$

then the following commutative relations can be proofed by (6.23)

$$\left[b_j(\boldsymbol{k}), b_j(\boldsymbol{k}')\right] = \left[b_j^+(\boldsymbol{k}), b_j^+(\boldsymbol{k})\right] = 0, \left[b_j(\boldsymbol{k}), b_j^+(\boldsymbol{k})\right] = \delta_{\boldsymbol{k}\boldsymbol{k}'}\delta_{jj'} \tag{6.24}$$

For example, the last commutative relation can be demonstrated as follows

$$\begin{aligned}
\left[b_j(\boldsymbol{k}), b_j^+(\boldsymbol{k}')\right] &= b_j(\boldsymbol{k})b_j^+(\boldsymbol{k}') - b_j^+(\boldsymbol{k}')b_j(\boldsymbol{k}) \\
&= \left[2\hbar\omega_j(\boldsymbol{k})\right]^{-1}\Big(\Big\{\left[\omega_j(\boldsymbol{k})Q_j(\boldsymbol{k},t) + iP_j^*(\boldsymbol{k},t)\right]\left[\omega_{j'}(\boldsymbol{k}')Q_j^*(\boldsymbol{k}',t) - iP_j(\boldsymbol{k}',t)\right]\Big\} \\
&\quad -\Big\{\left[\omega_{j'}(\boldsymbol{k}')Q_j^*(\boldsymbol{k}',t) - iP(\boldsymbol{k}',t)\right]\left[\omega_j(\boldsymbol{k})Q_j(\boldsymbol{k},t) + iP_j^*(\boldsymbol{k},t)\right]\Big\}\Big) \\
&= \left[2\hbar\omega_j(\boldsymbol{k})\right]^{-1}\Big\{-i\omega_j(\boldsymbol{k})\left[Q_j(\boldsymbol{k},t), P_j(\boldsymbol{k}',t)\right] - i\omega_{j'}(\boldsymbol{k}')\left[Q_j^*(\boldsymbol{k},t), P_j^*(\boldsymbol{k}',t)\right]\Big\} \\
&= \frac{i}{2\hbar\omega_j(\boldsymbol{k})}\left[-i\omega_j(\boldsymbol{k}) \times i\hbar\delta_{\boldsymbol{k}\boldsymbol{k}'}\delta_{jj'} - i\omega_{j'}(\boldsymbol{k}') \times i\hbar\delta_{\boldsymbol{k}\boldsymbol{k}'}\delta_{jj'}\right] = \delta_{\boldsymbol{k}\boldsymbol{k}'}\delta_{jj'}
\end{aligned}$$

Equation (6.23) has been used in the above calculation. The Hamiltonian expressed by (6.22) becomes

$$H = \sum_{jk} \hbar\omega_j(\boldsymbol{k})\left[b_j^+(\boldsymbol{k})b_j(\boldsymbol{k}) + \frac{1}{2}\right] + \sum_{jk} i\omega_j^2(\boldsymbol{k})\left[Q_j(\boldsymbol{k},t)P_j(\boldsymbol{k},t) - Q_j(-\boldsymbol{k},t)P_j(-\boldsymbol{k},t)\right]$$

The summation is over all the $\boldsymbol{k}$ vectors. However, every $\boldsymbol{k}$ vector has a corresponding vector of $-\boldsymbol{k}$, and so the second term in the above equation is equal to zero. Therefore

$$H = \sum_{jk} \hbar\omega_j(\boldsymbol{k})\left[b_j^+(\boldsymbol{k})b_j(\boldsymbol{k}) + \frac{1}{2}\right] \tag{6.25}$$

By using quantum number (phonon number in this situation) representation introduced first by Dirac [4], the ground state is denoted by $|0>$ and the state having n phonons is denoted by $|n>$, then it can be shown by (6.24) and (6.25) that

$$\begin{aligned} b_j(\boldsymbol{k})\left|n_j(\boldsymbol{k})\right\rangle &= \sqrt{n_j(\boldsymbol{k})}\left|n_j(\boldsymbol{k}) - 1\right\rangle, b_j^+(\boldsymbol{k})n_j(\boldsymbol{k}) \\ &= \sqrt{n_j(\boldsymbol{k}) + 1}\left|n_j(\boldsymbol{k}) + 1\right\rangle, b_j^+(\boldsymbol{k})b_j(\boldsymbol{k})\left|n_j(\boldsymbol{k})\right\rangle = n_j(\boldsymbol{k})\left|n_j(\boldsymbol{k})\right\rangle \end{aligned} \tag{6.26}$$

Equation (6.26) shows that the effect of operator $b_j(\boldsymbol{k})$ is to eliminate one phonon from the state and so it is called annihilation operator. The effect of $b_j^+(\boldsymbol{k})$ is to create one phonon and so it is called creation operator. $|n_j(\boldsymbol{k})>$ is the eigenstate of operator $b_j^+(\boldsymbol{k})b_j(\boldsymbol{k})$ with eigenvalue $n_j(\boldsymbol{k})$ and $b_j^+(\boldsymbol{k})b_j(\boldsymbol{k})$ is a phonon number operator. The commutative relation (6.24) is satisfied by the creation operator and the annihilation operator. Therefore, there is no any restriction on the phonon number of any phonon mode. Hence phonon is a Boson in statistical physics. In this way, (6.25) means that besides the zero energy, the lattice system consisted of $3Nr$ harmonic oscillators and each has energy $\hbar\omega_i(\boldsymbol{k})$. The total energy in a certain state will be

$$E(n_1, n_2, n_3, \ldots n_{3Nr}) = \sum_{i=1}^{3Nr}\left(n_i + \frac{1}{2}\right)\hbar\omega_i \tag{6.27}$$

where the harmonic oscillator is labeled by the index i and the summation is over all the $3Nr$ harmonic oscillators.

It should be pointed out that in the above discussion the terms higher than the second order in the expansion of (6.9) have been neglected. This is the so-called harmonic approximation. By this approximation, there is no interaction between different normal modes. The terms higher than second order will introduce the interaction between different normal modes and so are the mechanisms to reach thermal equilibrium.

The probability to find the system in a state with energy E is proportional to $\exp(-E/k_BT)$ in thermal equilibrium (k_B is Boltzmann constant). Naturally, the average phonon number of certain phonon mode with frequency ω_i, can be expressed as

$$\bar{n}_i = \frac{\sum_{n_i} n_i \exp(-E_i/k_BT)}{\sum_{n_i} \exp(-E_i/k_BT)} = \frac{\sum_{n_i} n_i \exp(-n_i\hbar\omega_i/k_BT)}{\sum_{n_i} \exp(-\hbar\omega_i/k_BT)}$$

where $E_i = \left(n_i + \frac{1}{2}\right)\hbar\omega_i$, and the common factor $\exp(-\hbar\omega_i/2k_BT)$ in the numerator and denominator has been reduced. The above summation is over the phonon number of a certain phonon mode. Obviously,

$$\sum_{n_i} \exp(-n_i\hbar\omega_i/k_BT) = \frac{1}{[1 - \exp(-\hbar\omega_i/k_BT)]}$$

On the other hand

$$\begin{aligned}\sum_n n_i \exp(-\hbar\omega_i n_i/k_BT) &= \exp(-\hbar\omega_i/k_BT)\frac{d\sum_{n_i}\exp(-\hbar\omega_i n_i/k_BT)}{d[\exp(-\hbar\omega_i/k_BT)]} \\ &= \exp(-\hbar\omega_i/kT)\frac{d\left\{\frac{1}{[1-\exp(-\hbar\omega_i/k_BT)]}\right\}}{d[\exp(-\hbar\omega_i/k_BT)]} \\ &= \frac{\exp(-\hbar\omega_i/k_BT)}{[1 - \exp(-\hbar\omega_i/k_BT)]^2}\end{aligned}$$

Therefore

$$\bar{n}_i = \frac{1}{\exp(\hbar\omega_i/k_BT) - 1} \tag{6.28}$$

The above formula is often used to discuss the spectroscopic problem related to lattice vibration. For every phonon mode, one can obtain a formula for average phonon number like (6.28).

Now turn to the problem of phonon state density. Similar to the photon state density, they all involve the state number in $\boldsymbol{k}$ space. The periodicity of the lattice requires that k_x, k_y, and k_z should be integer multiples of $2\pi/L$. An allowed $\boldsymbol{k}$ value in the volume element $(2\pi/L)^3$ of $\boldsymbol{k}$ space corresponds to a phonon state. The total volume of the $\boldsymbol{k}$ space is $4\pi k^3/3$ and so the total state number is $(L/2\pi)^3 \times 4\pi k^3/3$. For the acoustic phonon $k = \omega/v$, where v is the velocity of sound, the total state number will be $N_m = V\omega^3/6\pi^2v^3$, and so the state density can be expressed as

$$D(\omega) = \frac{dN_m}{d\omega} = \frac{V\omega^2}{2\pi^2v^3} \tag{6.29}$$

It has a similar dependent relation on the frequency and wave velocity as that for the photon state density. However, they have a big difference in the orders of magnitude, because the velocity of the light is 3×10^{10} cm/s but that of the sound in the solid is about 5×10^5 cm/s and the frequency of the light is about 100–1000 times that of the sound. On the other hand, the energy of photon is many orders of magnitude higher than that of the phonon. Therefore, at a comparable power density level, the phonon number can be a dozen order of magnitude larger than the photon number. For example, in a ruby crystal, when the power density of laser light reaches 1 W/cm^2, the photon number per cm^3 is only 10^8 order of the magnitude. However, for a solid with high hardness (Debye temperature about 1000 K), the phonon number per cm^3 can be 10^{23} orders of magnitude. This is the reason why the lattice vibration has a strong effect on the spectroscopic properties of the solid materials.

6.2 Phonon Emission and Absorption in the Optical Transition

It can be seen from (6.7) that the potential energy of the lattice vibration is an important potential energy term in the electron Schrödinger equation. On the other hand, the phonon density is very high. Consequently, the local electronic states are strongly affected by the lattice vibration. The lifetime of the electrons in their initial state is much longer than the time to reach thermal equilibrium by their interaction with phonons, so that the electron and phonon systems are in a thermal equilibrium before the electron transition. At the moment of the transition to the final state, the electron distribution of the initial and the final states has a considerable difference but the configuration of the lattice has no substantial change, because compared to the lattice vibration, the process of electron transition is a very rapid process. Therefore, at the moment the electron transits to its final state, the electron system and phonon system are not in the thermal equilibrium. In this circumstance, the electron will release or absorb phonon to reach thermal equilibrium of electron-phonon system; this is so called lattice relaxation. By the language of quantum mechanics, the vibrational wave functions with different vibrational quantum number of initial state and those of the final state are generally not orthogonal, because the potential functions introduced to the vibrational Schrödinger equation are different for these two states. In this case, the phonon number will be changed in the transition process (phonon emission or absorption). In the following discussion, we will use a single-frequency model supposing that the phonon modes interacting with the electronic system all have the same frequency and demonstrate by the method of Huang and Rhys [5].

Referring to the transition probability expressions in Chap. 4, taking into account the summation over final states and average over initial states, if we are not concerned about the dependence of the frequency and reflective index, then the following line shape function can be introduced

$$F(E) = \underset{i}{Av} \sum_{f} |\langle \Psi_f | \boldsymbol{M} | \Psi_i \rangle|^2 \delta[E - (E_f - E_i)] \tag{6.30}$$

where for electric-dipole transition, $\boldsymbol{M}$ is the electric-dipole moment. In the case studied, the electrons interact with the phonon system and act as one part of a composite system, and so their wave functions are not as simple as those in Chap. 3, but should be (6.5). The phonon wave functions $\varphi(R)$ are designated by phonon number n. Therefore, the wave functions of the initial and the final states of the composite system can be simply denoted by $|in'>$ and $|fn>$, respectively. Summation over the final states and average over the initial states of (6.30) can then be explicitly expressed as

$$F(E) = \underset{n'}{Av} \sum_{n} |\langle fn | \boldsymbol{M} | in' \rangle|^2 \delta[E - (E_{fn} - E_{in'})] \tag{6.31}$$

where the energy eigenvalue E_{km} ($k = i$ or f, $m = n$ or n') is the sum of electron energy and those of the phonons. In the single-frequency model

$$E_{km} = E_k^e + \sum_s \left(m_s + \frac{1}{2} \right) \hbar\omega \tag{6.32}$$

where s is the label of different phonon modes having the same frequency ω which is used to replace the phonon labels of j and $\boldsymbol{k}$. The electric-dipole moment $\boldsymbol{M}$ is a function of electronic coordinates only, but the electronic wave functions $\Phi(r, R)$ is not only a function of the electronic coordinates but also parameterized dependence on the lattice coordinates. Condon approximation will be adopted to deal with this problem, which assumes that the matrix element $\langle \Phi_f(r, R) | \boldsymbol{M} | \Phi_i(r, R) \rangle = M_{fi}$ is independent of the lattice coordinates. In the harmonic approximation the interaction between different phonons can be neglected and different phonon modes are independent of each other. Therefore, the wave function for the phonon system is a product of phonon wave functions of all the phonon modes. As a result, the line shape function representing the intensity of a spectral line is

$$F(E) = |M_{fi}|^2 \underset{n'}{Av} \sum_{n} \prod_{s} [\langle \varphi_{fm} | \varphi_{in'} \rangle]^2 \delta[E - (E_{fn} - E_{in})] \tag{6.33}$$

In order to obtain the line shape function, the important point is to calculate the overlap integral $<\varphi_{fn}|\varphi_{in'}>$. For this purpose, the effect of the electronic movement on the coordinates of lattice vibration should be taken into account. It is necessary to introduce the interaction Hamiltonian of the electrons with phonons as well as the first-order approximation expression of the potential function $W(R)$. The normal coordinate Q is generally used to replace the coordinate R and a linear term of the interaction potential of the electrons with phonons is assumed. If $N_m = 3Nr$ represents the total numbers of the phonon mode, then the interaction Hamiltonian expanded to the linear term of normal coordinates Q_s (i.e. the aforementioned $Q_j(\boldsymbol{k},t)$) will be

$$H_{\text{ep}} = \frac{1}{\sqrt{N_m}} \sum_s v_s(r) Q_s \tag{6.34}$$

where $v_s(r)$ is the first derivative of the potential with respect to the phonon coordinate. The effect of this interaction is to add the following potential to the system

$$\langle \Phi_i(r,Q)|H_{\text{ep}}|\Phi_i(r,Q)\rangle = \frac{1}{\sqrt{N_m}} \sum_s \langle \Phi_i(r,Q)|v_s|\Phi_i(r,Q)\rangle Q_s$$

Let

$$\langle \Phi_i(r,Q)|v_s|\Phi_i(r,Q)\rangle/\omega^2 \equiv \Delta_{is}$$

It should be noted that Δ_{is} is not the differentiate operator Δ_i used previously and its meaning can be clearly seen in (6.38). Since the interaction potential $\omega^2 Q_s^2/2$ not being calculated, the lattice vibrational potential produced by electronic movement can be written as

$$W_i(Q) = W_i(0) + \frac{1}{2} \sum_s \omega^2 Q_s^2 + \frac{1}{\sqrt{N_m}} \sum_s \omega^2 \Delta_{is} Q_s \tag{6.35}$$

Substituting it into (6.8), the wave equation of the lattice vibration can be written as

$$\left\{ W_i(0) + \sum_s \left[\frac{1}{2} \left(-\hbar \frac{\partial^2}{\partial Q_s^2} + \omega^2 Q_s^2 \right) + \frac{1}{\sqrt{N_m}} \sum_s \omega^2 \Delta_{is} Q_s \right] \right\} \varphi_{in}(Q) = E_{in} \varphi_{in}(Q) \tag{6.36}$$

After the following coordinates transformation

$$Q_{is} = Q_s + \frac{1}{\sqrt{N_m}} \Delta_{is}$$

a typical simple harmonic oscillator equation is obtained

$$\left\{ W_i + \sum_s \left[\frac{1}{2} \left(-\hbar \frac{\partial^2}{\partial Q_{is}^2} + \omega^2 Q_{is}^2 \right)_s \right] \right\} \varphi_{in}(Q) = E_{in} \varphi_{in}(Q) \tag{6.37}$$

where the zero-point energy is

$$W_i = W_i(0) + \frac{1}{N_m}\sum_s \frac{1}{2}\omega^2 \varDelta_{is}^2$$

and the lattice vibrational wave functions can be written as

$$\varphi_{in}(Q) = \prod_s \varphi_{n_s}\left(Q_s + \frac{1}{\sqrt{N_m}}\varDelta_{is}\right) \tag{6.38}$$

Equation (6.38) shows that due to the lattice relaxation, the coordinates of the lattice vibrational wave function have different shifts for different electronic energy levels, and the overlap integrals $<\varphi_{fn}|\varphi_{in'}>$ are generally not equal to zero if $n \neq n'$. For the sake of convenience, a factor of the product can be calculated at first.

$$\int \varphi_{n'_s}\left(Q_s + \frac{1}{\sqrt{N_m}}\Delta_{is}\right)\varphi_{n_s}\left(Q_s + \frac{1}{\sqrt{N_m}}\varDelta_{fs}\right)dQ_s \tag{6.39}$$

In order to do this calculation; it is better to shift the origin of the coordinates by introducing

$$Q_{is} = Q_s + \frac{1}{\sqrt{N_m}}\varDelta_{is}$$

Then

$$Q_s + \frac{1}{\sqrt{N_m}}\varDelta_{fs} = Q_s + \frac{1}{\sqrt{N_m}}\varDelta_{is} + \frac{1}{\sqrt{N_m}}\varDelta_{fs} - \frac{1}{\sqrt{N_m}}\varDelta_{is} = Q_{is} + \frac{1}{\sqrt{N_m}}\varDelta_{sfi}$$

where the symbol $\Delta_{sfi} = \Delta_{fs} - \Delta_{is}$ is introduced. To simplify the mathematics expression, symbol η is used to denote $\frac{1}{\sqrt{N_m}}\varDelta_{sfi}$, at the same time, the index s is omitted, then (6.39) becomes

$$\int \varphi_{n'}(Q_i)\varphi_n(Q_i + \eta)dQ_i \tag{6.40}$$

where $\varphi_n(Q_i + \eta)$ should be expanded into a Taylor series of η, then one has

$$\begin{aligned}&\int \varphi_{n'}(Q_i)\varphi_n(Q_i)dQ_i + \eta\int \varphi_{n'}(Q_i)\frac{\partial}{\partial Q_i}\varphi_n(Q_i)dQ_i \\ &+ \frac{1}{2!}\eta^2\int \varphi_{n'}(Q_i)\frac{\partial^2}{\partial Q_i^2}\varphi_n(Q_i)dQ_i + \cdots \\ &+ \frac{1}{k!}\eta^k\int \varphi_{n'}(Q_i)\frac{\partial^k}{\partial Q_i^k}\varphi_n(Q_i)dQ_i + \cdots\end{aligned} \tag{6.41}$$

The next step in the calculation of (6.41) is to study the detailed expression of the wave function of harmonic oscillators. By the formula of lattice dynamics [6], we have

$$\varphi_n(Q_i) = \left(\frac{\omega}{\pi\hbar}\right)^{1/4} \frac{1}{\sqrt{2^n n!}} e^{-\frac{1}{2}\frac{\omega Q_i^2}{\hbar}} H_n\left(\sqrt{\frac{\omega}{\hbar}} Q_i\right) \tag{6.42}$$

From the textbook of mathematical physics [7], the Hermitian polynomial $H_n\left(\xi \equiv \sqrt{\frac{\omega}{\hbar}} Q_i\right)$ can be obtained by the expansion of the generating function $\Xi(t, \xi)$ as follows

$$\Xi(t, \xi) = \exp\left(2t\xi - t^2\right) = \sum_{n=0}^{\infty} H_n(\xi) \frac{t^n}{n!}$$

It is obvious that from the above formula, one can obtain

$$\frac{\partial \Xi}{\partial \xi} = 2t\Xi, \frac{\partial \Xi}{\partial t} = -2(t - \xi)\Xi$$

Substituting the expanded formula of Ξ (t, ξ) into above equations and comparing the coefficients of the same $t^n/n!$ terms, the following useful recursion formulas are obtained

$$H_n'(\xi) = 2nH_{n-1}(\xi), \xi H_n(\xi) = \frac{1}{2} H_{n+1}(\xi) + nH_{n-1}(\xi) \tag{6.43}$$

The expression of the first derivative of phonon wave function is easily obtained by (6.42) and (6.43)

$$\frac{\partial}{\partial \bar{Q}_i} \varphi_n(Q_i) = \left(\frac{\omega}{2\hbar}\right)^{1/2} \left[(n)^{1/2} \varphi_{n-1}(Q_i) - (n+1)^{1/2} \varphi_{n+1}(Q_i)\right] \tag{6.44}$$

Differentiating both sides of the above equation with respect to Q_i once again, it can be seen that the second derivative of the phonon wave function includes the terms of $\varphi_{n-2}(Q_i)$ and $\varphi_n(Q_i)$. Reasoning by analogy, $\partial^k \varphi_n(Q_i)/\partial Q_i^k$ includes the term of $\varphi_{n-k}(Q_i)$, and its coefficient is

$$\left(\frac{\omega}{2\hbar}\right)^{k/2} \left[\frac{n!}{(n-k)!}\right]^{1/2} \tag{6.45}$$

If k is an even number, it includes also the term of $\varphi_n(Q_i)$. The general expression of its coefficient is rather complicated. In the case of $\varphi_n(Q_i) = \varphi_0(Q_i)$, it can be shown that the coefficient of the term of η^k in (6.41) is

$$\left(\frac{-1}{2}\right)^{k/2}\left(\frac{\omega}{2\hbar}\right)^{k/2}\left(\frac{1}{(k/2)!}\right) \tag{6.46}$$

Let us consider the line shape factor at very low temperature. Obviously, in the initial electronic state, the lattice should be in zero-point vibrational state, that is, $n'_s = 0$. When electrons transit from a high energy state to a low energy state (emit the photons), the phonon number of all the modes can be increased. If the total number of phonon increased is p, ω_r denotes the photon frequency and ω expresses the phonon frequency. The conservation of the energy requires that the variation of the electronic energy should be equal to the photon energy plus the total energy of the phonons, that is, $\Delta E_{fi} = \hbar\omega_r + p\hbar\omega$ or $\hbar\omega_r = \Delta E_{fi} - p\hbar\omega$. Similarly, in the process of radiation absorption, $\hbar\omega_r = \Delta E_{fi} + p\hbar\omega$.

Suppose p phonons generated in the transition process belong to any different g modes which have phonon number increase of $n_{s1}, n_{s2}, \ldots, n_{sg}$, respectively, while the phonon numbers of other modes have no any change. The relation $n_{s1} + n_{s2} + \cdots + n_{sg} = p$ should be satisfied (the maximum g is p, corresponding to the special situation of each mode has only one phonon increase). In the case studied, the initial state is $n'_s = 0$, then only terms in (6.41) with k equal to or larger than n are different from zero, because the derivative of wave function $\varphi_n(Q_i)$ with respect to Q_i only consists of the terms with $\varphi_l(Q_i)$ $(l > 0)$, when $k < n$. In this case, $\varphi_l(Q_i)$ $(l > 0)$ orthogonal to $\varphi_0(Q_i)$ leads the integral of related terms in (6.41) equal to zero. Note in the expansion of (6.41), the high-order derivative terms include also the wave functions of lower vibrational quantum number; therefore, for a certain phonon number variation, the result of (6.41) consists of a series of terms with power of η from lower to higher. Owing to the fact that N_m is a very large number and so η is a very small quantity, therefore it is only necessary to consider the term having lowest power of η, that is, the term with the power k equal to the phonon number variation. Consequently,

$$\begin{aligned}
&\left|\prod_s \langle \varphi_{n_s}(Q_i)|\varphi_{n_s}(Q_i+\eta_s)\rangle\right|^2 \\
&= \frac{1}{n_{s1}!}\left(\frac{\omega\eta_1^2}{2\hbar}\right)^{n_{s1}} \times \frac{1}{n_{s2}!}\left(\frac{\omega\eta_2^2}{2\hbar}\right)^{n_{s2}} \times \cdots \times \frac{1}{n_{sg}!}\left(\frac{\omega\eta_g^2}{2\hbar}\right)^{n_{sg}} \\
&\quad \times \left\{\prod_s\left[1 - \frac{1}{2}\frac{\omega\eta_s^2}{2\hbar} + \frac{1}{2!}\left(-\frac{1}{2}\frac{\omega\eta_s^2}{2\hbar}\right)^2 + \cdots\right]\right\}^2
\end{aligned} \tag{6.47}$$

where the first g factors are coming from terms in expansion expression (6.41) in which the phonon number that increased for the first mode is n_{s1}, second mode is n_{s2}, and so on. Of course, the total phonon number increasing is p. The expressions of these g factors corresponding to $n'_s = 0$ and k equal to $n_{s1}, n_{s2}, \ldots, n_{sg}$, respectively, are calculated by using (6.41) and (6.44). Following these factors, there is the square of a continued product, which comes from the contributions of the modes without phonon number variation in the transition. Each factor in this continued product corresponds to the contribution from one mode. In (6.41), all the terms with even powers of η (even-order derivative $\partial^k \varphi_n(Q_i)/\partial Q_i^k$) have a contribution to this factor. Therefore, each factor is a summation of a series terms and their coefficients are expressed by (6.46). The square of this continued product becomes

$$\left\{\prod_s\left[1-\frac{1}{2}\frac{\omega\eta_s^2}{2\hbar}+\frac{1}{2!}\left(-\frac{1}{2}\frac{\omega\eta_s^2}{2\hbar}\right)^2+\cdots\right]\right\}^2=\prod_s\left[exp\left(-\frac{1}{2}\frac{\omega\eta_s^2}{2\hbar}\right)\right]^2$$
$$=\exp\left(-\sum_s\frac{\omega\eta_s^2}{2\hbar}\right)$$

Because the total mode number N_m is very large, to exclude g modes having phonon number variation will not actually alter the summation of the $\omega\eta_s^2/2\hbar$ over all the vibrational modes, therefore the above summation should be over all the vibrational modes s. On the other hand, the line shape factor, as shown in (6.33), should be summed over all possible final states. To include all kinds of final state, n_{s1}, n_{s2}, …, n_{sg} must be selected to be any of the N_m modes. Hence, each of the g factors in (6.47) should be summed over all the N_m modes. The order of these factors is of no significance, and so the result should be divided by permutation number of $p!/n_{s1}!n_{s2}!\ldots n_{sg}!$ of the p units with the same n_{s1}, n_{s2}, …, n_{sg} units among them. It is obviously

$$\left|\prod_s\langle\varphi_0(Q_{is})|\varphi_{n_s}(Q_{is}+\eta_s)\rangle\right|^2=\frac{1}{n_{s1}!}\left(\sum_s\omega\eta_s^2/2\hbar\right)^{n_{s1}}$$
$$\times\frac{1}{n_{s2}!}\left(\sum_s\omega\eta_s^2/2\hbar\right)^{n_{s2}}\times\cdots\times\frac{1}{n_{sg}!}\left(\sum_s\omega\eta_s^2/2\hbar\right)^{n_{sg}}$$
$$\times\exp\left(-\sum_s\frac{\omega\eta_s^2}{2\hbar}\right)\times\left(p!/n_{s1}!n_{s2}!\ldots n_{sg}!\right)^{-1}$$
$$=\exp\left(-\sum_s\frac{\omega\eta_s^2}{2\hbar}\right)\left(\sum_s\omega\eta_s^2/2\hbar\right)^p/p!$$

Therefore, the line shape function takes the following form

$$F(E)=\left|M_{fi}\right|^2e^{-S}\left(\frac{S^p}{p!}\right) \tag{6.48}$$

where the Huang–Rhys factor S is defined as

$$S = \sum_s \frac{\omega \eta_s^2}{2\hbar} = \frac{1}{N_m} \sum_s \frac{\omega \Delta_{sfi}^2}{2\hbar} \tag{6.49}$$

Another method to obtain (6.48) is from the differential expression of Hermitian polynomial

$$\mathrm{H}_n(\xi) = (-1)^n e^{\xi^2} \frac{d^n}{d\xi^n} e^{-\xi^2}$$

A differential expression of the harmonic oscillator wave function can be obtained as

$$\varphi_n(\bar{Q}_i) = \left(\frac{\omega}{\pi\hbar}\right)^{1/4} \frac{(-1)^n}{\sqrt{2^n n!}} e^{\frac{\xi^2}{2}} \frac{d^n}{d\xi^n} e^{-\xi^2}$$

where $\xi \equiv \sqrt{\frac{\omega}{\hbar}} \bar{Q}_i$. Calculating the overlap integral using the above formula, the (6.48) can be obtained by the way adopted by Di Bartolo [8]. It should be pointed out that this single-mode and single-frequency approximation can be generalized to multi-mode and single-frequency situation according to the method described above.

From (6.49), one has

$$S\hbar\omega = \sum_s \frac{1}{2} \omega^2 \eta_s^2 = \sum_s \frac{1}{2} \omega^2 \Delta Q_s^2 \tag{6.50}$$

It means that $S\hbar\omega$ corresponds to the total lattice relaxation energy generated by the shift of the origin of vibrational coordinates during the electronic transition. It can be seen by the curve of the line shape function versus the S factor that the peak of the spectrum is at the point $S = 1$ when S factor of the system is equal to 1, and the peak of the spectrum will be at the point $S = k$ when S factor of the system is equal to k. It shows that the electrons have the maximum transition probability when the lattice relaxation energy is equal to the phonon energy absorbed or emitted in the transition.

If it is not at very low temperature, there are also phonons excited in the initial electronic states, then $n'_s \neq 0$. In the electronic transition, the phonon number of some vibrational modes is increased while those of the other modes can be decreased. The line shape function in this case can be calculated by assuming that there is only one phonon increased or decreased in each of the modes having phonon number variation and those modes with several phonons variation can be dealt with as each mode has one phonon variation. Obviously, when the initial phonon number is n'_s, by using (6.44) to calculate expansion (6.41), for those modes with one phonon increased, by using (6.45) and let $n = n'_s + 1$ and $k = n'_s$, it can be shown that each g factor $\omega\eta_i^2/2\hbar (i = 1, 2, \ldots, g)$ in (6.47) should be multiplied by $n'_s + 1$; for those modes with one phonon decreased, $n = n'_s$ and

$k = n'_s - 1$, the corresponding factor $\omega\eta^2/2\hbar$ in (6.47) should be multiplied by n'_s; and for those modes without phonon number variation, that is $n = n'_s$, calculated by (6.41) and (6.44), it can be shown that each factor of $\omega\eta_s^2/2\hbar$ in the continued product of (6.47) should be multiplied by $2n'_s + 1$. If the total phonon number increase is p, then the number of modes with one phonon increased is $p + k$ and those with one phonon decreased is k, while k can be any integer corresponding to different distribution of the phonon number variation. The result, obviously, should be summed over all possible k. Therefore, referring to (6.48), the line shape function before statistical average takes the form

$$\left|M_{ij}\right|^2 e^{-S(2n'_s+1)} \sum_k \frac{\left[S(n'_s+1)\right]^{p+k} (Sn'_s)^k}{(p+k)!k!}$$

The initial phonon number should be statistical average according to Boltzmann distribution law; that is the phonon number n'_s should be multiplied by the factor $\frac{\exp(-n_s\hbar\omega/k_BT)}{\sum_s \exp(-n_s\hbar\omega/k_BT)}$. As shown earlier, it corresponds to replace n'_s by $\bar{n} = 1/[\exp(\hbar\omega/k_BT) - 1]$, and so the emission line shape function at temperature T is

$$F(E - \Delta E_{fi} + p\hbar\omega) = \left|M_{fi}\right|^2 e^{-S(2\bar{n}+1)} \sum_k \frac{[S(\bar{n}+1)]^{p+k} (S\bar{n})^k}{(p+k)!k!} \tag{6.51}$$

The shape of this function obviously depends on temperature. Figure 6.1 is the relation of the relative strength of the spectral line versus the number of phonon emitted at $T = 0$ K. When the electron–phonon coupling is very weak (e.g. in the case of trivalent rare earth ions in solid), the value of S is assumed to be nearly zero. The spectral line is a single peak structure, like that shown in Fig. 6.1a. For stronger electron–phonon coupling (for example in the case of trivalent iron group ions in solid), the spectral line has a multi-peak structure, like that shown in Fig. 6.1b and it

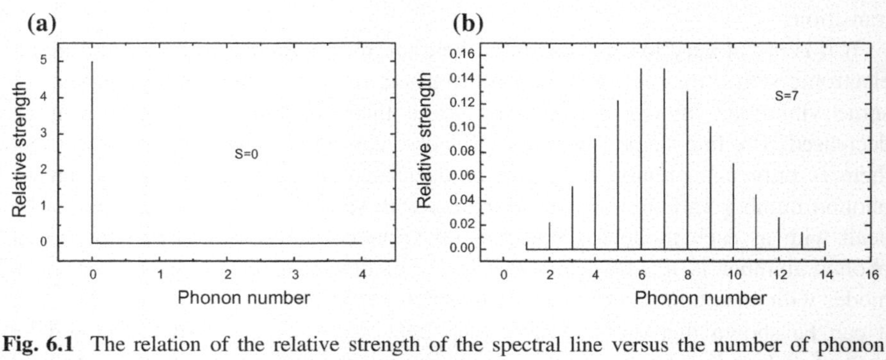

Fig. 6.1 The relation of the relative strength of the spectral line versus the number of phonon emitted at low temperature limit: **a** $S = 0$; **b** $S = 7$

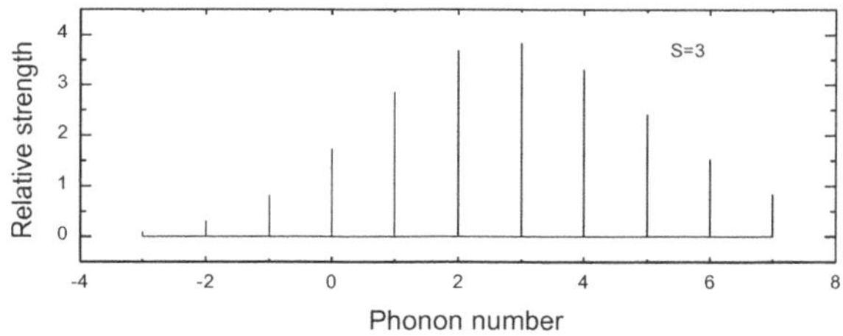

Fig. 6.2 The relation between relative strength of the spectral line and the number of phonon emitted and absorbed at room temperature for the system with Huang–Rhys factor $S = 3$

is broadened by the electron–phonon coupling. Therefore, it is generally impossible to observe the single spectral peaks corresponding to different phonon emitted or absorbed, which can be seen at very low temperature, because the broadening effect of the spectral line is enhanced rapidly as temperature rise. This will be discussed in detail later.

At a higher temperature, the phonon number in initial electronic states is not equal to zero and there are phonons emitted and absorbed in the electronic transition. The relation between the relative strength of the spectral line and the number of phonon emitted and absorbed is shown in Fig. 6.2, for the case of $S = 3$ and phonon energy $\hbar\omega = 250\,\text{cm}^{-1}$ at room temperature $T = 300$ K. Negative phonon number corresponds to phonon absorption.

Note that in Figs. 6.1 and 6.2, the curves are the relative line strength without taking into account the effect of line broadening. Actually, the phonon always has a frequency distribution over a certain frequency range. As that discussed by K. Huang [9], if the contribution of the single phonon transition to line shape function in frequency range $d\bar{\omega}$ is $e^{-S}\sigma_1(\bar{\omega})d\bar{\omega}$, then that for the two-phonon transition will be

$$\frac{1}{2!}e^{-S}\sigma_2(\bar{\omega})d\bar{\omega}$$

where $\sigma_2(\bar{\omega})$ is the following convolution integral

$$\sigma_2(\bar{\omega}) = \int \sigma_1(\bar{\omega} - \omega_{s1})\sigma_1(\omega_{s1})d\omega_{s1}$$

These relations can be generalized to multi-phonon situations. It can be shown that the contribution of the multi-phonon transition to line shape function in frequency range $d\bar{\omega}$ is

$$\frac{1}{n!}e^{-S}\sigma_n(\bar{\omega})d\bar{\omega}$$

Therefore, the phonon side band corresponding to high-order phonon transition has a wider bandwidth, while the line corresponding to $p = 0$ has a higher relative intensity than that shown in Figs. 6.1 and 6.2. Generally, the part of the spectrum corresponding to smaller p has a higher relative intensity. However, for the system with smaller S factor, all the peaks of the phonon side band still can be seen. On the other hand, for many strong electron–phonon coupling systems, the spectral peaks of the phonon side band corresponding to $p > S$ cannot be seen even at a very low temperature.

In Condon approximation, it can be shown that the integral area of spectral band, consisting of the zero phonon line and all the phonon side bands, is independent of the temperature [10]. The intensity of zero phonon line decreases, while the area of the other part of the phonon side band increases with the rising of temperature. The relation between the zero phonon line intensity and the temperature can be expressed as $I(T) = I_0 \exp[-2Sk_BT/\hbar\bar{\omega}]$ at high temperature. The larger the Huang–Rhys factor S and the lower the average phonon energy $\hbar\bar{\omega}$, the more rapidly weakened is the intensity of zero phonon line. At intermediate temperature region, this relation is a complex function but the tendency is still: the intensity of the zero phonon line decreases with the increase of temperature and this decrease is more significant for the situation of larger S factor. In the case of laser material activated by trivalent rare earth ions, the electron–phonon interaction is very weak (S factor is very small) and so the intensity variation of the zero phonon line with the temperature is not obvious. However, the transition metal ion in the laser material have a much larger S factor and so the above variation is very significant that one often fails to see the zero phonon line at room temperature.

The "barycenter" of a multi-phonon emission spectrum was discussed by K. Huang [9] using the method of Fourier transformation and its result was

$$E_e = W_{fi} - S\hbar\bar{\omega}_s \tag{6.52}$$

For absorption spectrum, it can be shown that

$$E_a = W_{fi} + S\hbar\bar{\omega}_s$$

where the average phonon energy $\hbar\bar{\omega}$ can be expressed as

$$\hbar\bar{\omega}_s = \frac{1}{S}\sum_s \left[\left(\frac{\omega_s}{2\hbar}\right)\bar{\Delta}_{sfi}^2\right](\hbar\omega_s) \tag{6.53}$$

and

$$\bar{\Delta}_{sfi} = \frac{\Delta_{sfi}}{\sqrt{N_m}}$$

Obviously, (6.52) shows that the separation of the "barycenter" of the emission spectrum from the difference of electronic energy levels is equal to the phonon energy multiplied by S, the same as the situations in single-frequency mode (Figs. 6.1 and 6.2). By using (6.52), the Stokes shift can be defined as

$$E_S = 2S\hbar\bar{\omega}_s \tag{6.54}$$

An approximation method of effective phonon model with phonon frequency $\omega_{\rm eff}$ (belonging to single-frequency approximation) is often used in the study of spectroscopic properties of solid laser materials. By this approximation, it can be shown that the square of full width at half maximum of the spectrum is [9]

$$W^2 = 8(\ln 2)S\left(\hbar\omega_{eff}\right)^2\coth\left(\hbar\omega_{eff}/2k_BT\right) \tag{6.55}$$

The above formulas are very useful for the calculation of spectroscopic parameters.

6.3 Main Mechanisms of the Thermal Spectral Line Broadening and Shifting

The spectral linewidth mentioned in the previous section is generated by the optical transition accompanied by multi-phonon emission or absorption. This section will discuss the thermal broadening and shifting of the zero phonon line. This broadening and shifting will have a direct effect on the thermal stability and tuning range of laser materials. The frequency of the spectral line is determined by the relative position of the initial state and the final state. The thermal shifting and broadening of the spectral line are the combined results of the thermal shifting and broadening of the initial and final energy levels involved in the electronic transition. As temperature increases, the lattice expansion, the spin orbit coupling constant, the Slater parameter, and the static crystal field parameter also change, which are one of the reasons for thermal broadening and shifting of the spectral line. On the other hand, the interaction between electron and lattice vibration will also change the position and width of the initial and final energy levels. Generally speaking, the broadening and shifting caused by the previous action are relatively small. The peak position of the spectral line shifts with the temperature, and the increase of its width with the temperature are mainly caused by the interaction of the electron with phonon. Defects and stresses in crystals also produce broadening and shifting of spectral lines, but this mechanism is generally independent of temperature. Therefore, the effect of lattice vibration on the thermal broadening and shifting of the spectral lines should be studied in detail.

The effect of the electron–phonon interaction mechanism on the spectral linewidth can be summarized into two aspects: first, the electron transits to other energy levels, thus shortens the lifetime of the electronic energy level, and broadens the width of the electronic energy level; second, the phonon absorption and emission in different crystal field energy levels in the same multiplet as well as the Raman scattering in the same crystal field energy level will extend the width of the electronic energy level in the time period shorter than its lifetime. For the Raman scattering of phonon, the frequency of emitted phonon can be lower or higher than the frequency of absorbed phonon.

The first kind of contribution to the linewidth can be calculated by

$$\Delta E\left(\mathrm{cm}^{-1}\right) = \frac{5.3 \times 10^{-12}}{\tau(\mathrm{sec})} \tag{6.56}$$

where τ is the lifetime of the electronic energy level. Equation (6.56) is obtained by uncertainty relation $\Delta E \Delta t = \hbar$ and the energy is expressed as the unit of wave number according to 1 erg = 5.035×10^{15} cm^{-1}. This linewidth is called the natural linewidth in spectroscopy. Actually, this kind of effect is not the main contribution of the linewidth for zero phonon lines. One can see it clearly from the example of the pure electronic transition line R_1 of the trivalent chromium in ruby. The lifetime of the excited state in this case is 3 ms at room temperature. Calculating by (6.56), the linewidth will be 2×10^{-9} cm^{-1}. However, the linewidth observed was 10 cm^{-1} in fact.

In strong electron–phonon coupling system such as transition-metal ion system, the normal vibrational frequencies for ground and excited states can be different due to different force constants. This will result in spectral thermal line broadening. The non-harmonic lattice vibrational also brings about the thermal broadening and sift of the phonon side band. This information can refer to the literature [10].

This book mainly concerns the weak electron–phonon coupling system such as trivalent rare earth ions in solid-state laser materials. In these materials only zero phonon lines have observable intensity. From the above analysis it can be seen that in these materials the main mechanism of energy level broadening is phonon absorption and emission as well as the Raman scattering, which does not change the lifetime of the electronic energy level. It is equivalent to that the electrons of the initial and final states are oscillating with the frequency of single-phonon emission, absorption, or Raman scattering. This oscillation corresponds to an additional energy to the electronic system. The additional energy is equal to the probability of above processes multiplied by the Planck constant h. As to the spectral line thermal shifting, besides the crystal field weakened by lattice thermal expansion, the phonon transition effect on the energy of electronic system, and the modulation of the Coulomb interaction between electrons and the spin–orbital coupling all have their contributions. Many theoretical models have been introduced to explain thermal broadening and shifting of spectral line. However, the model proposed by McCumber and Sturge [11] is highly regarded as an enormously influential work, which is concise and clear, and its main conclusions can be applied quite generally [12]. Therefore we will use this model in the discussion of line broadening and shifting, but introduce the modification of mass difference of the host ions. It should be pointed out that all the different models published so far include a series of significant simplification of the real physical reality (e.g. to use long wavelength approximation of the lattice vibration and the Debye distribution of phonon frequency), and so can only be used to describe approximately the related phenomena.

The method other than the perturbation theoretical can also be used to study these problems, but the mathematical tool of many-body theory should be used [13].

It should be pointed out that the broadening and shifting of the spectral line are caused by the transitions between the neighboring crystal field levels with separation from 10^0 to 10^2 cm^{-1}. Generally, the energy of optical phonon has a minimum order of magnitude of about 10^3 cm^{-1}, while the energy of acoustic phonon can be from 0 cm^{-1} to cut-off energy. Therefore, the same as the other authors, the phonon used in the following discussion is the acoustic phonon.

6.4 The Contribution of Single-Phonon Absorption (Emission) to the Spectral Linewidth

To calculate the transition probabilities of the single-phonon absorption and emission, one should, at first, expand the energy of the electron–phonon interaction as a series of the strain. The strain can be expressed as a linear combination of phonon annihilation operator and creation operator as will be shown in the following. In an anisotropy crystal, the strain is a tensor obviously. However, in the study of thermal vibrational effect on the spectral linewidth, the direction problem can be neglected and so the displacement of the lattice can be expressed by a scalar quantity u. It is the quantity in (6.18), which does not include the basic vector $e_i^{(j)}(\boldsymbol{k}, \alpha)$ (Many authors adopted this expression; for example, Di Bartolo [14].)

$$u(R_\alpha) = \sum_k \left(\frac{\hbar}{2M\omega_k}\right)^{1/2} \left(e^{i\boldsymbol{k}\boldsymbol{R}_\alpha} b(\boldsymbol{k}) + e^{-i\boldsymbol{k}\boldsymbol{R}_\alpha} b^+(\boldsymbol{k})\right)$$

where $M = Nm$ is used to denote the mass of the crystal and for the acoustic mode $k = \omega_k/v$. Therefore, one has

$$\varepsilon \approx \left.\frac{\partial u}{\partial R_\alpha}\right|_{R_\alpha=0} = i\sum_{\boldsymbol{k}} \left(\frac{\hbar\omega_k}{2Mv^2}\right)^{1/2} [b(\boldsymbol{k}) - b^+(\boldsymbol{k})]$$

However, the assumption that all the ions in the host crystal have the same mass is too simple and cannot account for the actual situation of the host crystal. In fact, the laser materials are made up of more than two kinds of ions having different mass. The mass difference in the host crystal should be taken into account as pointed out by the authors [15]. Considering a crystal consisting of N molecules, each is constituted of not only n_α ions of mass m_α but also n_β ions of mass m_β. The number of ions in a molecule is $n = n_\alpha + n_\beta$. The mass of each molecule is $m = (n_\alpha m_\alpha + n_\beta m_\beta)$. The total number of ions in the crystal is nN and the mass of the crystal is $M = N(n_\alpha m_\alpha + n_\beta m_\beta)$. By using the above expression of the strain, the expression of the strain for the case of all the host ions have the same mass of m_α and that for the case of all the host ions have the same mass of m_β can be written as follows

$$\varepsilon_\alpha = i\sum_k \left(\frac{\hbar\omega_k}{2v^2}\right)^{1/2}\left(\frac{1}{\sqrt{nNm_\alpha}}\right)[b(\boldsymbol{k}) - b^+(\boldsymbol{k})],\ \varepsilon_\beta = i\sum_k \left(\frac{\hbar\omega_k}{2v^2}\right)^{1/2}\left(\frac{1}{\sqrt{nNm_\beta}}\right)[b(\boldsymbol{k}) - b^+(\boldsymbol{k})]$$

The average strain becomes

$$\bar{\varepsilon} = \frac{1}{n}(n_\alpha\varepsilon_\alpha + n_\beta\varepsilon_\beta) = iD\sum_k \left(\frac{\hbar\omega_k}{2Mv^2}\right)^{1/2}[b(\boldsymbol{k}) - b^+(\boldsymbol{k})]$$

where

$$D = \left[\frac{m\left(n_\alpha\sqrt{m_\beta} + n_\beta\sqrt{m_\alpha}\right)^2}{n^3 m_\alpha m_\beta}\right]^{1/2}$$

If there are three kinds of host ion with different masses in one crystal, the factor D can be derived by the same process as

$$D = \left[\frac{m\left(n_\alpha\sqrt{m_\beta m_\gamma} + n_\beta\sqrt{m_\alpha m_\gamma} + n_\gamma\sqrt{m_\alpha m_\beta}\right)^2}{n^3 m_\alpha m_\beta m_\gamma}\right]^{1/2}$$

where $n = n_\alpha + n_\beta + n_\gamma$, $m = n_\alpha m_\alpha + n_\beta m_\beta + n_\gamma m_\gamma$. Generalize the above formula, and it can be obtained that for the situation of four kinds of host ion, one has

$$D = \left[\frac{m\left(n_\alpha\sqrt{m_\beta m_\gamma m_\delta} + n_\beta\sqrt{m_\alpha m_\gamma m_\delta} + n_\gamma\sqrt{m_\alpha m_\beta m_\delta} + n_\delta\sqrt{m_\alpha m_\beta m_\gamma}\right)^2}{n^2 m_\alpha m_\beta m_\gamma m_\delta}\right]^{1/2}$$

Expanding the energy of the electron–phonon interaction in terms of the local strain, then

$$V_{e-p} = V_1\bar{\varepsilon} + V_2\bar{\varepsilon}^2 + \cdots \tag{6.57}$$

The Hamiltonian of single-phonon transition expressed by linear term of ε in (6.57) will be

$$H_{e-sp} = iV_1 D \sum_{\boldsymbol{k}} \left(\frac{\hbar\omega_k}{2Mv^2}\right)^{1/2} [b(\boldsymbol{k}) - b^+(\boldsymbol{k})] \tag{6.58}$$

According to Fermi golden rule, the transition probability of single-phonon absorption or emission will be

$$W_{\rm sp} = \int \frac{2\pi}{\hbar} \left|\langle f|H_{\rm e-sp}|i\rangle\right|^2 \rho(E_f = E_i) dE_f \tag{6.59}$$

For the phonon absorption, the initial state is $|i,n_{\boldsymbol{k}}>$ while the final state is $|f,n_{\boldsymbol{k}}-1>$. The matrix element for the transition from energy level i to energy level f will be

$$\langle f|H_{e-sp}|i\rangle = iD\langle f|V_1|i\rangle \sum_k \left(\frac{\hbar\omega_k}{2Mv^2}\right)^{1/2} \sqrt{n_k} \tag{6.60}$$

The total transition probability of single-phonon absorption is

$$W_{\rm sp} = \frac{2\pi}{\hbar} \int \sum_{E_f^e > E_i^e} \left|\langle f|H_{\rm e-sp}|i\rangle\right|^2 \rho(E_f = E_i) dE_f \tag{6.61}$$

The problem now is to determine the final state density $\rho(E_f = E_i)$; it can be written as

$$\rho(E_f = E_i) = \rho(E_f)\delta(E_f - E_i)$$

where E_f and E_i are the final and initial state energies of the electron–phonon system. The state density of the final state is a product of the electronic state density and the phonon state density, therefore one has

$$\rho(E_f = E_i)dE_f = \rho\left(E_f^{\rm e}\right)\rho\left(E_f^{\rm p}\right)\delta\left[\left(E_f^{\rm e} + E_f^{\rm p}\right) - \left(E_i^{\rm e} + E_i^{\rm p}\right)\right] dE_f^{\rm e} dE_f^{\rm p}$$

The system absorbs a phonon transition to the final state f, $E_f^{\rm e} \rangle E_i^{\rm e}$, meanwhile $E_f^{\rm p} \langle E_i^{\rm p}$, therefore

$$\delta\left[\left(E_f^{\rm e} + E_f^{\rm p}\right) - \left(E_i^{\rm e} + E_i^{\rm p}\right)\right] = \frac{1}{\hbar}\delta\left(\omega_{fi}^{\rm e} - \omega_k\right)$$

$$\rho\left(E_f^{\rm p}\right) dE_f^{\rm p} = \rho(\omega_k) d(\omega_k)$$

Because both E_f^e and E_i^e are discrete electronic levels and the transition is the emission or absorption of a phonon having particular frequency ω_0, the energy conservation law must be observed, so

$$\rho\left(E_f^e\right)dE_f^e = \delta\left(\omega_{fi}^e - \omega_0\right)d\omega_{fi}^e$$

where $\omega_{fi}^e = \left(E_f^e - E_i^e\right)/\hbar$.

By using the above formula of phonon state density and the Debye approximation considering that the acoustic mode consisting of two transverse vibrational mode of phonon velocity v_t and one longitudinal vibrational mode of phonon velocity v_l, it can be obtained that

$$\rho(\omega_k) = \begin{cases} \frac{V\omega_k^2}{2\pi^2}\left(\frac{1}{v_l^3} + \frac{2}{v_t^3}\right) & \omega_k \le \omega_D \\ 0 & \omega_k \rangle \omega_D \end{cases} \tag{6.62}$$

where $\hbar\omega_D = k_B T_D$, T_D is referred to as Debye temperature. Substituting these results to (6.61) and integral over ω_{fi}^e and ω_k, considering that in the thermal equilibrium, the phonon number is

$$\bar{n} = \frac{1}{e^{\hbar\omega_k/k_B T} - 1} = \frac{1}{e^{\hbar\omega_0/k_B T} - 1}$$

then the transition probability of one phonon absorption takes the form

$$W_{sp}^a = \sum_{f>i} \frac{(\omega_0)^3 D^2}{2\pi\rho\hbar}\left(\frac{1}{v_l^5} + \frac{2}{v_t^5}\right)|\langle f|V_1|i\rangle|^2 \frac{1}{e^{\hbar\omega_0/k_B T} - 1} \tag{6.63}$$

where $\rho = M/V$. Similarly, it can be shown that for phonon emission

$$W_{sp}^e = \sum_{f\langle i} \frac{(\omega_0)^3 D^2}{2\pi\rho\hbar}\left(\frac{1}{v_l^5} + \frac{2}{v_t^5}\right)|\langle f|V_1|i\rangle|^2 \left(1 + \frac{1}{e^{\hbar\omega_0/kT} - 1}\right) \tag{6.64}$$

These kinds of phonon absorption and emission are reversible and it does not change the lifetime of electronic energy levels. According to the analysis given in the last section, a broadening of the initial state i resulting from the phonon absorption and emission will be $\Delta E_{sp}^a = W_{sp}^a h$ and $\Delta E_{sp}^e = W_{sp}^e h$, respectively. In the energy unit of wave number cm^{-1}, 1erg = 5.035 × 10^{15} cm^{-1}, then $h = 6.62620 \times 10^{-27}\text{erg} \cdot \text{sec} = 3.3362917 \times 10^{-11}\text{cm}^{-1} \cdot \text{sec} = \frac{1}{c}$, therefore

$$\Delta E_{\text{sp}}^a\left(\text{cm}^{-1}\right) = \sum_{f>i} \frac{1}{c}\frac{(\omega_0)^3 D^2}{2\pi\rho\hbar}\left(\frac{1}{v_l^5} + \frac{2}{v_t^5}\right)|\langle f|V_1|i\rangle|^2 \frac{1}{e^{\hbar\omega_0/k_B T} - 1} \tag{6.65}$$

$$\Delta E_{\mathrm{sp}}^{e}\left(\mathrm{cm}^{-1}\right) = \sum_{f<i} \frac{1}{c} \frac{(\omega_0)^3 D^2}{2\pi\rho\hbar} \left(\frac{1}{v_l^5} + \frac{2}{v_t^5}\right) |\langle f|V_1|i\rangle|^2 \left(1 + \frac{1}{e^{\hbar\omega_0/k_B T} - 1}\right) \tag{6.66}$$

where β_{fi} is usually used in the literature to denote the product of the physical quantities in the above formulas

$$\beta_{fi} = \frac{1}{c} \frac{(\omega_0)^3 D^2}{2\pi\rho\hbar} \left(\frac{1}{v_l^5} + \frac{2}{v_t^5}\right) |\langle f|V_1|i\rangle|^2$$

Neglecting the difference between the velocity of transverse and longitude acoustic waves, the above formula can be written as

$$\beta_{fi} = \frac{3}{c} \frac{(\omega_0)^3 D^2}{2\pi\rho\hbar v^5} |\langle f|V_1|i\rangle|^2 \tag{6.67}$$

Hence

$$\Delta E_{\mathrm{sp}}\left(\mathrm{cm}^{-1}\right) = \sum_{f>i} \beta_{fi} \frac{1}{e^{\hbar\omega_0/k_B T} - 1} + \sum_{f<i} \beta_{if} \frac{1}{e^{\hbar\omega_0/k_B T} - 1} + \sum_{f<i} \beta_{if} \tag{6.68}$$

where the first term is the contribution of the phonon absorption and the following two terms are those of the phonon emission. At low temperature, two $1/\left(e^{\hbar\omega_0/k_B T} - 1\right)$ factors in the summations of expression (6.68) are much smaller than 1 and so the main contribution comes from the third term. In this case, when the final state of emission or absorption is a higher crystal field level of the final multiplet, there are more channels to carry out the phonon spontaneous emission, that is the third term of (6.68) has a larger value. Therefore the width of low temperature emission spectral line at longer wavelength is wider than that at shorter wavelength, while the width of low temperature absorption spectral line at longer wavelength is narrower than that at shorter wavelength. The estimation of the magnitude of the thermal linewidth produced by this mechanism as well as the comparison with those by other mechanisms will be given later.

6.5 The Contribution of Phonon Raman Scattering to the Spectral Linewidth

The perturbation theory will be used to introduce the related formulas. Readers can refer to the paper published by Skinner and Hsu to know the method of non-perturbation theory [13]. Raman scattering is a second perturbation process. It involves second-order perturbation of the first-order term of electron–phonon interaction Hamiltonian (see (6.58)) and first-order perturbation of the following second-order term

$$H_{\mathrm{e-dp}} = -V_2 \frac{\hbar D^2}{2Mv^2} \sum_{kl} \sqrt{\omega_k \omega_l} (b(\boldsymbol{k}) - b^+(\boldsymbol{k}))(b(\boldsymbol{l}) - b^+(\boldsymbol{l})) \tag{6.69}$$

Note that the mass difference factor D has been introduced in the above expression. Similar to (6.59), the transition probability of the Raman scattering can be calculated by

$$W_{\mathrm{R}} = \frac{2\pi}{\hbar^2} \int |\langle f|H_{\mathrm{R}}|i\rangle|^2 \rho(\omega_f) d\omega_f \tag{6.70}$$

In the following discussion, the electronic wave functions are labeled by Greek letter. Because the Raman scattering process of an electronic level is discussed, the initial and final states are the different phonon states of the same electronic state denoted by α.

According to second-order perturbation theory, the transition matrix element appeared in (6.70) can be expressed as

$$\begin{aligned}
\langle f|H_{\mathrm{R}}|i\rangle &= \langle \alpha; n_k - 1, \ldots n_l + 1|H_{\mathrm{R}}|\alpha; n_k \ldots n_l\rangle \\
&= \sum_{\gamma \neq \alpha} \frac{\langle \alpha; n_k - 1, \ldots n_l + 1|H_{\mathrm{e-sp}}|\gamma; n_k - 1 \ldots n_l\rangle\langle \gamma; n_k - 1 \ldots n_l|H_{\mathrm{e-sp}}|\alpha; n_k \ldots n_l\rangle}{E_\alpha - (E_\gamma - \hbar\omega_k)} \\
&+ \sum_{\gamma \neq \alpha} \frac{\langle \alpha; n_k - 1, \ldots n_l + 1|H_{\mathrm{e-sp}}|\gamma; n_k, \ldots n_l + 1\rangle\langle \gamma; n_k, \ldots n_l + 1|H_{\mathrm{e-sp}}|\alpha; n_k \ldots n_l\rangle}{E_\alpha - (E_\gamma + \hbar\omega_l)} \\
&+ \langle \alpha; n_k - 1, \ldots n_l + 1|H_{\mathrm{e-dp}}|\alpha; n_k \ldots n_l\rangle
\end{aligned} \tag{6.71}$$

Substituting the first-order term of the electron–phonon interaction Hamiltonian (6.58) and that of the second-order term (6.70) into above equation, the following can be obtained

$$\begin{aligned}
\langle f|H_{\mathrm{R}}|i\rangle = \frac{\hbar\sqrt{\omega_k\omega_l}D^2}{2Mv^2} \Bigg\{ \sum_{\gamma \neq \alpha} \Bigg[&\frac{|\langle \alpha|V_1|\gamma\rangle|^2}{E_\alpha - (E_\gamma - \hbar\omega_k)} \langle n_k - 1, n_l + 1|b^+(\boldsymbol{l})b(\boldsymbol{k})|n_k, n_l\rangle \\
+ &\frac{|\langle \alpha|V_1|\gamma\rangle|^2}{E_\alpha - (E_\gamma + \hbar\omega_l)} \langle n_k - 1, n_l + 1|b^+(\boldsymbol{l})b(\boldsymbol{k})|n_k, n_l\rangle \Bigg] \\
+ &\langle \alpha|V_2|\alpha\rangle\langle n_k - 1, n_l + 1|(b(\boldsymbol{k})b^+(\boldsymbol{l}) + b^+(\boldsymbol{l})b(\boldsymbol{k}))|n_k, n_l\rangle \Bigg\}
\end{aligned} \tag{6.72}$$

The next step is to substitute the matrix elements of phonon creation and annihilation operators obtained previously. $E_\alpha - (E_\gamma - \hbar\omega_k)$ and $E_\alpha - (E_\gamma + \hbar\omega_l)$

in the denominators of the terms in the above equation can be approximately replaced by $E_\alpha - E_\gamma$ because $\hbar\omega_l \approx \hbar\omega_k \ll |E_\alpha - E_\gamma|$ (in this way, the errors introduced by those terms with $E_\alpha - E_\gamma > 0$ will partly be cancelled by those with $E_\alpha - E_\gamma < 0$). Hence the following formula resulted

$$\begin{aligned} \langle f|H_R|i\rangle &\approx \frac{\hbar\sqrt{\omega_k\omega_l n_k(n_l+1)}D^2}{Mv^2}\left(\sum_{\gamma\neq\alpha}\frac{|\langle\alpha|V_1|\gamma\rangle|^2}{E_\alpha - E_\gamma} + \langle\alpha|V_2|\alpha\rangle\right) \\ &= \chi_m\sqrt{\omega_k\omega_l n_k(n_l+1)} \end{aligned} \tag{6.73}$$

where

$$\chi_m = \frac{\hbar D^2}{Mv^2}\left(\sum_{\gamma\neq\alpha}\frac{|\langle\alpha|V_1|\gamma\rangle|^2}{E_\alpha - E_\gamma} + \langle\alpha|V_2|\alpha\rangle\right) \tag{6.74}$$

For the Raman scattering involving emit and absorb phonons with the same frequency, the final state density can be expressed as

$$\rho(\omega_f)d\omega_f = \rho(\omega_k)\rho(\omega_l)\delta(\omega_l - \omega_k)d\omega_l d\omega_k$$

Unlike the previous section, where only certain phonon energy can match the interval of two electronic energy levels, the initial and final states of the Raman scattering process are the same electronic energy level, and there is no limitation on the energy of the scattering phonon. By using Debye approximation, the integral of (6.70) should be from the phonon frequency 0 to ω_D. By means of the formula of phonon state density (6.62), supposing the transverse acoustic mode and the longitude mode have the same velocity and using the thermal equilibrium phonon number expression, (6.70) becomes

$$W_R = \frac{2\pi}{\hbar^2}\chi_m^2\frac{9V^2}{4\pi^4 v^6}\int_0^{\omega_D}\frac{\omega_k^6 e^{\hbar\omega_k/k_BT}}{\left(e^{\hbar\omega_k/k_BT}-1\right)^2}d\omega_k \tag{6.75}$$

After the variable transformation $x = \hbar\omega_k/k_BT$, it can also be expressed as

$$W_R = \frac{9V^2}{2\pi^3\hbar^2 v^6}\chi_m^2\left(\frac{k_BT_D}{\hbar}\right)^7\left(\frac{T}{T_D}\right)^7\int_0^{T_D/T}\frac{x^6e^x}{(e^x-1)^2}dx \tag{6.76}$$

Similar to the introduction of (6.65) and (6.66) from (6.63) and (6.64), the formula of linewidth as a function of temperature can be deduced

$$\Delta E(T)(\mathrm{cm}^{-1}) = \bar{\alpha}\left(\frac{T}{T_D}\right)^7 \int_0^{T_D/T} \frac{x^6 e^x}{(e^x - 1)^2} dx \equiv \bar{\alpha} F(T/T_D) \tag{6.77}$$

where

$$\bar{\alpha} = \frac{1}{c} \frac{9D^4}{2\pi^3 \rho^2 v^{10}} \left(\frac{k_B T_D}{\hbar}\right)^7 \left(\sum_{\gamma \neq \alpha} \frac{|\langle \alpha | V_1 | \gamma \rangle|^2}{E_\alpha - E_\gamma} + \langle \alpha | V_2 | \alpha \rangle\right)^2 \tag{6.78}$$

where $\bar{\alpha}/T_D^7$ is an adjustable parameter depending on the density of the crystal, the phonon velocity, and the electron–phonon coupling strength. The value of this parameter is proportional to D^4; therefore the mass difference of the host ions has a much obvious effect on the thermal line broadening than that in the case of phonon emission and absorption mechanisms. The value of $\bar{\alpha}$ is generally less than 300 cm^{-1} for rare earth ions and is different for different ones. The value reaches the maximum for Ce^{3+} ion and then decreases gradually down to Gd^{3+} ion reaching the minimum, but it will increase again gradually up to Yb^{3+} ion reaching the maximum. This variation tendency reflects the U shape variation tendency of electron-phonon interaction strength. Ellens [16–18] measured the value of $\bar{\alpha}$ for nine rare earth ions Ce^{3+}, Pr^{3+}, Nd^{3+}, Eu^{3+}, Gd^{3+}, Tb^{3+}, Er^{3+}, Tm^{3+}, and Yb^{3+} in $LiYF_4$ crystal. Their experimental results are in agreement with the above variation tendency. Hellwege [19] thought that it is due to the fact that the total spin quantum number S is the same for the rare earth ions in the corresponding positions of two sides of Gd^{3+} ion but he did not explain how the symmetry tendency of S results in the variation tendency of electron-phonon interaction strength. According to the phenomenon of J–O parameters, Ω_4 and Ω_6 in Y_2O_3 crystal for the ions in the lanthanum series are larger in the front and the end, while smaller in the middle. Krupke [20] considered that the variation tendency of electron-phonon interaction strength is the reflection of variation tendency of J–O parameters. However, in crystals $LiYF_4$, LaF_3, and $YAlO_3$ the variation tendency of J–O parameters is different from that of Y_2O_3 crystal. Ellens and his coworkers [18] suggested the following two mechanisms: the first is the lanthanide contraction. With the increase of atomic number, the electronic radius of the $4f$ electron decreases, which weakens the interaction with the surrounding coordination ions; and the second is the parameter σ related to the $4f$ electrons shielded or screened by the 5 s and $5p$ electrons, which decreases gradually from Ce^{3+} ion (σ = 1.1) to Yb^{3+} ion (σ = 0.6). For the ions on the left of the Gd^{3+} ion, the reduction of the electronic radius plays a major role on the variation tendency of electron-phonon interaction strength, while the ions on the right of the Gd^{3+} ion, the decrease of the shielding parameters plays a major role on the variation tendency. As a result, the U shape variation tendency of electron-phonon interaction strength is created. Actually, one can explain this variation tendency from another point of view. The average energy level separation between the f^n configuration and the $4f^{n-1}5d$ configuration is minimum for Ce^{3+} ion, then increases gradually and reaches its maximum at Gd^{3+} ion. For the right side of Gd^{3+} ions, this separation decreases gradually and reaches its minimum at Yb^{3+} ion (it can be seen in Fig. 1.5). The variation of $4f^n$–$4f^{n-1}5d$ energy level separation results in the variation of the proportion of the component of $5d$ wave function, which has a stronger interaction with its ligands, in the $4f$ wave function. The proportion of the component of $5d$ wave function can certainly measure the strength of electron–phonon interaction. Therefore, the U shape variation tendency of the strength of electron–phonon interaction is a reflection of the U shape variation of $4f^n$–$4f^{n-1}5d$ separation. For transition metal ions, the electron–phonon interaction is much stronger and $\bar{\alpha} \geq 300\,\text{cm}^{-1}$. Remember that its detailed values are different for different crystals (see the examples in Sect. 6.7). Different values of $(T/T_D)^7$ multiplied by the integral value of (6.76) are shown in Appendix G. As an example, at very low temperature when $T_D/T = 20$ ($T_D = 500$ K, $T = 25$ K), $F(T/T_D)$ takes 5.72×10^{-7}, even if $\bar{\alpha} = 300\,\text{cm}^{-1}$, ΔE only has $1.8 \times 10^{-4}\,\text{cm}^{-1}$ while when $T_D/T = 1.7$ ($T_D = 500$ K, $T = 300$ K) $F(T/T_D)$ has a value of 5.8×10^{-2} and ΔE will reach 17 cm^{-1} with the same $\bar{\alpha}$ value.

The contributions of the above-mentioned mechanisms change with temperature. At very low temperature, the single-phonon emission and absorption have the major contribution and that of Raman scattering can be neglected. At higher temperature, the main contribution comes from Raman scattering, which covers the entire phonon spectrum and is independent of the energy level separation of the light transition.

It should be pointed out that the thermal line broadening produced by electron–phonon interaction belongs to the so-called homogenous broadening. In a crystal with considerable defects, the line broadening of the active ions is often dominated by inhomogeneous broadening. In this kind of crystal, different ions can be in different positions with different strains, different lattice distortions, and so different crystal field potentials. The spectrum observed is actually a superposition of the different spectra produced by the ions in different positions. Owing to the random distributions of the strains, defects, and impurity ions, the inhomogeneous broadening spectral line has Gaussian line shape with the following line shape function

$$g(\omega) = \frac{2}{\Delta\omega}\left(\frac{\ln 2}{\pi}\right)^{1/2} \exp\left\{-\left[\frac{2(\omega - \omega_0)}{\Delta\omega}\sqrt{\ln 2}\right]^2\right\} \tag{6.79}$$

However, each homogenous broadening spectral line has Lorentz line shape and its line shape function is

$$g(\omega) = \frac{2}{\pi\Delta\omega}\frac{1}{1 + \left(\frac{\omega-\omega_0}{\Delta\omega/2}\right)^2} \tag{6.80}$$

In the general case, the spectral line has a Voigt line shape, which is a convolution integral of the Lorentz lines with a Gaussian frequency distribution. When the width of the Gaussian distribution is zero (in case of an ideal crystal without inhomogeneous), the line shape becomes Lorentz. On the other hand, when the width of Lorentz line tends to zero, the spectral line takes the shape of Gaussian. This is the line shape at very low temperature, in which the effect of lattice vibration is very weak. At high temperature, the linewidth of the Lorentz components can be much larger than that of the Gaussian in the composite Voigt line and so the integral line shape is practically Lorentz. This is the reason why the spectral lines of the active ions at room temperature in a reasonable perfect crystal that can be described by Lorentz line shape. For the crystals including defects and impurity ions, based on a theoretical model, Orth [21] obtained the result: when the density of defect and/or impurity ion is low, the line shape of the spectral line is still Lorentz; while the density of defect and/or impurity ion is high, the line shape is Gaussian.

6.6 Calculation of the Thermal Shifting of Spectral Lines

To discuss thermal shifting of the spectral lines, it is natural to think that the thermal expansion is the main mechanism, because thermal expansion will weaken the crystal field and shorten the energy level separations. Certainly, for the so-called "soft crystal" having large thermal expansion coefficient, this mechanism has the major contribution. However, it is not true for the "hard crystal" like YAG. Actually, the calculation results according to the mechanism of thermal expansion cannot account for the experimental data of thermal shifting, and even obtain a negative result in some situations. Therefore, the electron–phonon interaction can be considered as the major mechanism of the thermal shifting of spectral lines.

Many authors proposed a variety of different theoretical models to deal with this problem, but as pointed out before, most of these models have not been universally accepted. So we still use the McCumber and Sturg model [11] to calculate thermal shifting of the spectral lines. In order to realize the relation between thermal shifting of the spectral line and the temperature, it is enough to calculate the diagonal matrix elements of electron-phonon interaction Hamiltonian by the general method, that is, to calculate the increase of the electronic level energy generated by the electron-phonon interaction. The electron–phonon interaction energy is expanded in terms of the normal coordinates of lattice vibration to second-order term and expressed as the sum of a single-phonon Hamiltonian $H_{\mathrm{e-sp}}$ and a double-phonon Hamiltonian $H_{\mathrm{e-dp}}$

$$H_{\mathrm{e-p}} = H_{\mathrm{e-sp}} + H_{\mathrm{e-dp}}$$

where $H_{\mathrm{e-sp}}$ and $H_{\mathrm{e-dp}}$ have been introduced in (6.58) and (6.69).

To calculate the diagonal matrix elements in second-order perturbation approximation, the double-phonon Hamiltonian (it is already a second-order quantity) and the single-phonon Hamiltonian should be calculated to zero-order and first-order terms, respectively, that is

$$\delta E_i = \sum_j \frac{\langle i|H_{\mathrm{e-sp}}|j\rangle\langle j|H_{\mathrm{e-sp}}|i\rangle}{E_i - E_j} + \langle i|H_{\mathrm{e-dp}}|i\rangle \tag{6.81}$$

In the initial, intermediate, and final states, the electron–phonon wave functions of the system are expressed as

$$\langle i| = \langle \alpha|\langle n_k|,\ \langle j| = \langle \beta|\langle n_k + 1| \text{ and } \langle i| = \langle \beta|\langle n_k - 1|$$

where α and β are used to denote the electronic wave functions of the initial and final states, respectively. By substituting (6.58) and (6.69) into (6.81), the following expression can be obtained

$$\delta E_i = \frac{\hbar D^2}{2Mv^2}\left\{\sum_{\beta,k}\omega_k\left[\frac{\langle\alpha|V_1|\beta\rangle\langle\beta|V_1|\alpha\rangle}{E_\alpha-(E_\beta+\hbar\omega_k)}\langle n_k|b_k b_k^+|n_k\rangle + \frac{\langle\alpha|V_1|\beta\rangle\langle\beta|V_1|\alpha\rangle}{E_\alpha-(E_\beta-\hbar\omega_k)}\langle n_k|b_k^+ b_k|n_k\rangle\right]\right.$$
$$\left. + \sum_k \omega_k|\langle\alpha|V_2||\alpha\rangle\langle n_k|b_k b_k^+ + b_k^+ b_k|n_k\rangle\right\} \tag{6.82}$$

By using matrix elements of creation and annihilation operators, it becomes

$$\delta E_i = \frac{\hbar D^2}{2Mv^2}\left\{\sum_{\beta,k}\omega_k|\langle\alpha|V_1|\beta\rangle|^2\left[\frac{n_k+1}{E_\alpha-(E_\beta+\hbar\omega_k)} + \frac{n_k}{E_\alpha-(E_\beta-\hbar\omega_k)}\right]\right.$$
$$\left. + \sum_k \omega_k\langle\alpha|V_2|\alpha\rangle(1+2n_k)\right\} \tag{6.83}$$

The terms independent of the temperature should be subtracted, because what we are concerned is the thermal shifting of the spectral line. In this way, the shifting of the energy level i should be

$$\delta E_i = \frac{\hbar D^2}{2Mv^2}\sum_k \omega_k n_k\left\{\sum_{\beta\neq\alpha}|\langle\alpha|V_1|\beta\rangle|^2\left[\frac{1}{E_\alpha-(E_\beta+\hbar\omega_k)} + \frac{1}{E_\alpha-(E_\beta-\hbar\omega_k)}\right] + 2\langle\alpha|V_2|\alpha\rangle\right\} \tag{6.84}$$

Two cases of $|E_i - E_j| \gg \hbar\omega_D$ and $|E_i - E_j| \le \hbar\omega_D$ will be analyzed separately in the following.

For the case of $|E_i - E_j| \gg \hbar\omega_D$, (6.84) can be reduced to

$$\delta E_i' = a\sum_k \omega_k n_k \tag{6.85}$$

where

$$a = \frac{\hbar D^2}{Mv^2}\left[\sum_{\beta\neq\alpha}\frac{|\langle\alpha|V_1|\beta\rangle|^2}{E_\alpha - E_\beta} + \langle\alpha|V_2|\alpha\rangle\right] \tag{6.86}$$

In thermal equilibrium, the averaged phonon number is $1/\left(e^{\hbar\omega_k/k_BT} - 1\right)$, and the phonon density is $\rho(\omega_k) = 3V\omega_k^2/2\pi^2 v^3$. In this case, the summation in (6.85) should be replaced by the following integral after using Debye model.

$$\frac{3V}{2\pi^2 v^3}\int_0^{\omega_D}\frac{\omega_k^3 d\omega_k}{(e^{\hbar\omega_k/k_BT} - 1)} = \frac{3V}{2\pi^2 v^3}\left(\frac{k_BT}{\hbar}\right)^4\int_0^{T_D/T}\frac{x^3}{e^x - 1}dx \tag{6.87}$$

The thermal shifting in the unit of wave number can be expressed as

$$\delta E\left(\mathrm{cm}^{-1}\right) = \alpha\left(\frac{T}{T_\mathrm{D}}\right)^4\int_0^{T_\mathrm{D}/T}\frac{x^3}{e^x - 1}dx \tag{6.88}$$

where α has the following relation with a expressed in (6.86) and $\hbar = 1/(2\pi c)$, then one has

$$\alpha = \frac{3D^2}{4\pi^3\rho v^5 c}\left(\frac{k_BT_D}{\hbar}\right)^4\left[\sum_{\beta\neq\alpha}\frac{|\langle\alpha|V_1|\beta\rangle|^2}{E_\alpha - E_\beta} + \langle\alpha|V_2|\alpha\rangle\right] \tag{6.89}$$

For the case of $\left|E_i - E_j\right| \leq \hbar\omega_D$, the second-order term can be neglected, (6.84) can be similarly written as

$$\begin{aligned}\delta E_i &= \frac{\hbar D^2}{2Mv^2}\sum_{\beta\neq\alpha}\frac{3V}{2\pi^2 v^3}|\langle\alpha|V_1|\beta\rangle|^2 P\int_0^{\omega_\mathrm{D}}\frac{2(E_\alpha - E_\beta)\omega_k^3}{\left[(E_\alpha - E_\beta)^2 - (\hbar\omega_k)^2\right](e^{\hbar\omega_k/k_BT} - 1)}d\omega_k \\ &= \frac{\hbar D^2}{Mv^2}\sum_{\beta\neq\alpha}\frac{3V}{2\pi^2 v^3}|\langle\alpha|V_1|\beta\rangle|^2\left(\frac{k_BT}{\hbar}\right)^4\frac{1}{(kT)^2}(E_\alpha - E_\beta)\ P\int_0^{T_\mathrm{D}/T}\frac{x^3}{e^x - 1}\frac{1}{(T_{\alpha\beta}/T)^2 - x^2}dx \\ &= \sum_{\beta\neq\alpha}\frac{3\omega_{\alpha\beta}^3 D^2}{2\pi^2\rho v^5}|\langle\alpha|V_1|\beta\rangle|^2\left(\frac{T}{T_{\alpha\beta}}\right)^2 P\int_0^{T_\mathrm{D}/T}\frac{x^3}{e^x - 1}\frac{1}{(T_{\alpha\beta}/T)^2 - x^2}dx\end{aligned} \tag{6.90}$$

where $T_{\alpha\beta} = (E_\alpha - E_\beta)/k_B$, $\omega_{\alpha\beta} = 2\pi(E_\alpha - E_\beta)/h$. P means the principal value of the integral. By using a formula similar to (6.67), the above formula can be reduced and expressed in the unit of wave number

$$\delta E_i(\mathrm{cm}^{-1}) = \sum_{\beta \neq \alpha} \left(\frac{1}{2\pi^2 c}\right) \beta_{\alpha\beta} \left(\frac{T}{T_{\alpha\beta}}\right)^2 \mathrm{P} \int_0^{T_\mathrm{D}/T} \frac{x^3}{e^x - 1} \cdot \frac{1}{\left(T_{\alpha\beta}/T\right)^2 - x^2} dx \quad (6.91)$$

$$\beta_{\alpha\beta} = \frac{3\omega_{\alpha\beta}^3 D^2}{2\pi\rho\hbar v^5} |\langle\alpha|V_1|\beta\rangle|^2$$

The above calculation does not calculate separately the thermal shifting of the higher and the lower energy levels, it just finds the magnitude of the difference to give the sign of the spectral shifting. We cannot predict whether the thermal spectral line shifting is red or blue one. The experiments show that most of the thermal spectral line shiftings are red one [22]. Therefore, the phenomena of the thermal spectral line shifting should have a more comprehensive study. An important aspect of the research on thermal shifting that should be mentioned is the thermal shifting of the "barycenter" of the spectral terms induced by lattice vibration. The mechanism of this "barycenter" shifting can be the modulation of the spin–orbital coupling by lattice vibration mentioned by Kaminskii [22]. Another mechanism of thermal shifting of the "barycenter" was proposed by the author [23]. The decrease of the Coulomb interaction between electrons caused by the exchange of phonons generates a thermal shifting of the "barycenter" of the spectral term and ^{2S+1}L was expressed as

$$G_T\left(^{2S+1}L\right) = \frac{g\left(^{2S+1}L\right) S_\rho^2}{\omega_\rho\left(e^{\hbar\omega_\rho/k_B T} - 1\right)}$$

where $g(^{2S+1}L)$ denotes the average pair number of the electron satisfying phonon exchange condition, ω_ρ is the frequency of the phonon and S_ρ is a parameter measuring the electron–phonon interaction strength.

The values of $\left(\frac{T}{T_D}\right)^4 \int_0^{T_D/T} \frac{x^3}{e^x - 1} dx$ can be found in the Appendix G. For example, when $T_D/T = 1.2$ ($T_D = 500$ K, $T = 417$ K), it is 1.724×10^{-1}, and when $T_D/T = 0.6$ ($T_D = 500$ K, $T = 833$ K), it is 4.405×10^{-1}.

At the end of this section, it should be pointed out that the mass difference of host ions certainly has important effects on the spectral line thermal broadening and shifting. It can be seen from the fact that D^2 appears as a factor of the parameter α and D^4 appears as a factor of the parameter $\bar{\alpha}$. It should also be mentioned that the thermal broadening and shifting of spectral lines are inversely proportional to v^{10} and v^5, respectively, where v is the sound velocity of the crystal, owing to the fact that sound velocity of the crystal is inversely proportional to the product of crystal density ρ and compressibility κ [24]. If the binding energy of crystal is almost the same, the compressibility of homogeneous solid is increased with mass difference of their ions [25]. It means the larger the D factor, the larger is the compressibility κ. On the other hand, the mass of anion usually is smaller than that of the cation so that the density of the materials with the same anion is increased with the D factor. For example

$Y_3Al_6O_{12}$: $D^2 = 1.35$, $\rho = 4.55$ g/cm^3; $YAlO_3$: $D^2 = 1.44$, $\rho = 5.35$ g/cm^3; $Y_3Ga_6O_{12}$: $D^2 = 1.55$, $\rho = 5.83$ g/cm^3; $Gd_3Ga_6O_{12}$: $D^2 = 1.86$, $\rho = 7.05$/cm^3; $Lu_3Ga_6O_{12}$: $D^2 = 1.95$, $\rho = 7.79$ g/cm^3.

It demonstrates that the mass difference of the host ions affects the thermal broadening and shifting of spectral lines not only directly through the D factor but also indirectly through its variation of material density and compressibility. Therefore, the larger mass difference factor D always corresponds to the wider thermal broadening and larger shifting of the spectral lines. A detailed example will be shown in Sect. 6.7.

6.7 Examples for the Calculation of Thermal Spectral Line Broadening and Shifting

In order to have a perceptual knowledge about thermal broadening and shiftings of the spectral lines, the study of phonon effects on the fluorescence lines of Nd^{3+} ions in crystal $Gd_3Sc_2Ga_3O_{12}$ by Chen and Di Bartolo [26] will be introduced as an example. We discuss only four spectral lines corresponding to the transitions from crystal field energy level R_1 (11,439 cm^{-1}) and R_2 (11,500 cm^{-1}) of multiplet $^4F_{3/2}$ to the lowest energy level Y_1 (1983 cm^{-1}) and the highest energy level Y_6 (2432 cm^{-1}) of multiplet $^4I_{11/2}$ (Fig. 6.3). Other experimental data can be directly referred to their paper [26].

On the thermal broadening of spectral lines, the total broadening of fluorescence lines was written as a sum of the following terms [26]

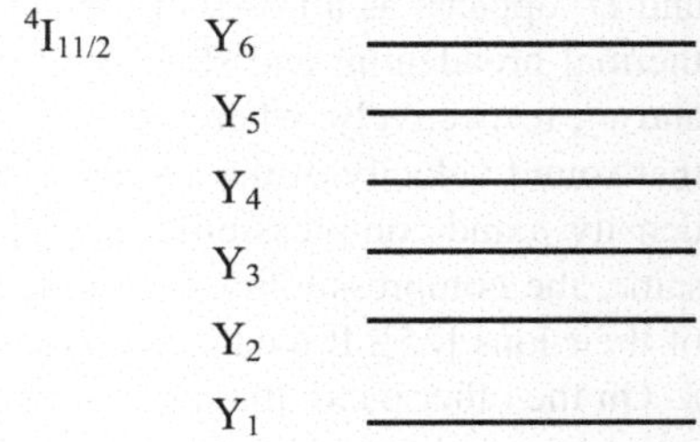

Fig. 6.3 Energy levels of $^4F_{3/2}$ and $^4I_{11/2}$ multiplets of Nd^{3+} ion

$$\Delta E_i(\text{cm}^{-1}) = \Delta E_i^{\text{strain}} + \Delta E_i^{\text{M}} + \sum_{j<i} \bar{\beta}_{ij} + \Delta E_i^{\text{DT}} + \Delta E_i^{\text{R}}$$

The first term $\Delta E_i^{\text{strain}}$ is the broadening introduced by the defects and other inhomogeneous in the crystals and is independent of the temperature. The second term ΔE_i^M is the broadening induced by multi-phonon emission and is essentially independent of the temperature. The third term corresponds to the last term of (6.68), which is the contribution of the spontaneous phonon emission and is also independent of temperature. ΔE_i^{DT} consisting of the first two terms in (6.68) can be neglected at very low temperature. ΔE_i^R is the broadening introduced by phonon Raman scattering. The major contribution to linewidth at low temperature comes from phonon spontaneous emission, that is, the third term of (6.68).
At temperature of 78 K, the spectral line corresponding to $R_1 \rightarrow Y_1$ transition has nearly a Lorentzian line shape (it shows that the crystal is rather perfect, and the influence of defects is small) and has a linewidth of 2.49 cm^{-1} smaller than that of any other spectral lines, because R_1 and Y_1 are the lowest levels at multiplets $^4F_{3/2}$ and $^4I_{11/2}$, respectively, and so there is no spontaneous phonon emission contribution to the linewidth.

Fitting the experimental data of thermal line broadening to theoretical calculation results by taking into account or neglecting the effect of ΔE_i^{DT}, respectively, Chen and Di Bartolo [26] found that the contribution of ΔE_i^{DT} cannot be neglected at temperature 78 K; otherwise the Debye temperature T_D obtained will be different from that obtained by the fitting of thermal line shifts. For the $R_1 \rightarrow Y_1$ emission line, the phonon absorption $Y_1 \rightarrow Y_2$ and phonon absorption $R_1 \rightarrow R_2$ have contributed to the thermal line broadening while for the $R_2 \rightarrow Y_1$ emission line, the phonon absorption $Y_1 \rightarrow Y_2$ and phonon absorption $R_1 \rightarrow R_2$ as well as the spontaneous phonon emission $R_2 \rightarrow R_1$ have contributed to the thermal line broadening. Noting that $\beta_{R1\rightarrow R2} = \beta_{R2\rightarrow R1}$, then the spectral line thermal broadening of the transitions $R_1 \rightarrow Y_1$ and $R_2 \rightarrow Y_1$, except the linewidth contribution of the spontaneous phonon emission $R_2 \rightarrow R_1$ for $R_2 \rightarrow Y_1$ emission line, can be expressed as

$$\Delta E(\text{cm}^{-1}) = \Delta E_0 + \beta_{R_1\rightarrow R_2}\frac{1}{\exp(\Delta E_{R_2\rightarrow R_1}/k_BT) - 1} + \beta_{Y_1\rightarrow Y_2}\frac{1}{\exp(\Delta E_{Y_2\rightarrow Y_1}/k_BT) - 1} + \bar{\alpha}\left(\frac{T}{T_D}\right)^7 \int_0^{T_D/T} \frac{x^6 e^x}{(e^x - 1)^2} dx$$

The last term in the above equation is the contribution of Raman scattering, where $\bar{\alpha}$ is also a fitting parameter like those of $\beta_{Y1\rightarrow Y2}$, $\beta_{R1\rightarrow R2}$, ΔE_0, and T_D. The fitting results are ΔE_0 = 2.1 cm^{-1} for $R_1 \rightarrow Y_1$ emission line and ΔE_0 = 2.5 cm^{-1} for $R_2 \rightarrow Y_1$ emission line. The reason for this difference has been stated previously, that is because the thermal line broadening of $R_2 \rightarrow Y_1$ emission line has more contribution $\beta_{R2\to R1}$ that comes from the phonon spontaneous emission and it shows $\beta_{R1\to R2} = \beta_{R2\to R1} = 0.4\ \text{cm}^{-1}$. Other parameters fitted are: $\beta_{Y1\to Y2} = 0.158\ \text{cm}^{-1}$, $\bar{\alpha} = 34\,\text{cm}^{-1}$, $T_D = 500$ K and $\beta_{Y1\to Y2} = 0.288\ \text{cm}^{-1}$, $\bar{\alpha} = 49\,\text{cm}^{-1}$, T_D= 500 K for $R_1 \to Y_1$ and $R_2 \to Y_1$ transitions, respectively (Fig. 6.4).

The effect of temperature on the position of the lines $R_1 \to Y_1$, $R_2 \to Y_1$, $R_1 \to Y_6$, $R_2 \to Y_6$, and $R_1 \to Z_5$ (also known as X_5, the highest energy level of $^4I_{9/2}$) was also studied by Chen and Di Bartolo [26]. All the measured lines shifted to the red (longer wavelength) except for the $R_1 \to Z_5$ line which shifted to the blue (shorter wavelength) with the increase of temperature. According to (6.88), the fitting parameters are: $R_1 \to Y_1$ and $R_2 \to Y_1$: $\alpha = -61\ \text{cm}^{-1}$, $T_D = 600$ K; $R_1 \to Y_6$: $\alpha = -32\ \text{cm}^{-1}$, $T_D = 760$ K; $R_2 \to Y_6$: $\alpha = -26\ \text{cm}^{-1}$, $T_D = 660$ K. The fitting curve for $R_1 \to Y_1$ transition is shown in Fig. 6.5.

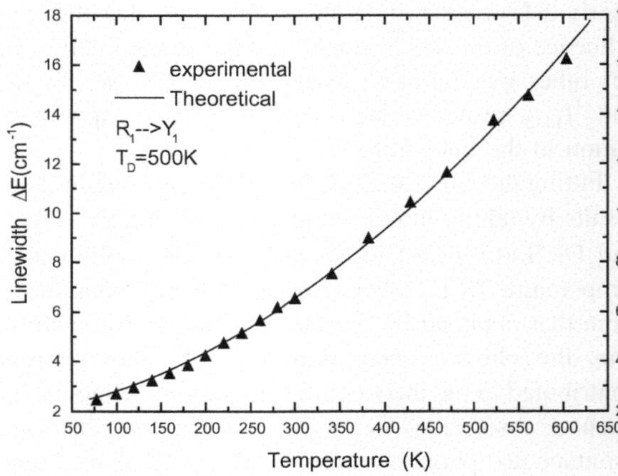

Fig. 6.4 The fitting results of thermal broadening of emission spectral lines for transition $R_1 \to Y_1$ [26]

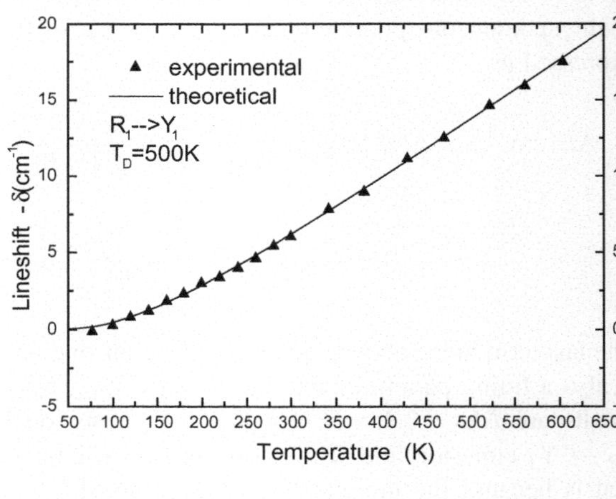

Fig. 6.5 The fitting results of thermal line shifting of emission spectral lines for transitions $R_1 \to Y_1$ [26]

The thermal spectral line broadening and shifting have nearly the same orders of magnitude for the same rare earth ion in different materials or the different rare earth ions in the same material. The following are some other data for comparison. The fitting parameters for 1.23% Er^{3+}: $YAlO_3$ [27] are: $\bar{\alpha} = 45 \pm 5\,\mathrm{cm}^{-1}$, $T_D = 350 \pm 30$ K for thermal broadening; $\alpha = -10 \pm 1\ \mathrm{cm}^{-1}$, $T_D = 300 \pm 30$ K for thermal shifting. The data obtained have no substantial difference for different doping concentration. The fitting parameters for 1% Nd^{3+}: $CaWO_4$ [28] are: $\bar{\alpha} = 70\,\mathrm{cm}^{-1} \pm 0.5\%$, $T_D = 200$ K $\pm$ 0.5% for thermal broadening, and for Nd^{3+}:$RbMnF_3$ [29] are: $\bar{\alpha} = 76 \pm 12\,\mathrm{cm}^{-1}$, $T_D = 200 \pm 20$ K for thermal broadening; $\alpha = -30 \pm 6\ \mathrm{cm}^{-1}$, $T_D = 500 \pm 40$ K for thermal shifting. One of the reasons for obtaining different Debye temperature for thermal broadening and thermal shifting is that the contribution of single phonon absorption and emission has not been taken into account.

It must be pointed out that the thermal spectral line broadenings and shiftings are different for different energy level pairs of the same rare earth ion, or the same rare earth ion and the same energy level pair but in different materials. For example, for Nd^{3+}: LLGG crystal [30] the width of spectral line $R_1 \rightarrow Y_3$ increases by 17.3 cm^{-1} when temperature rises from 10 to 300 K but the width of spectral line $R_1 \rightarrow Z_5$ only increases by 8.1 cm^{-1} for the same temperature increment. The thermal spectral line shiftings of this crystal between different energy level pairs have much difference. The spectral line of $R_1 \rightarrow Z_5$ blue shifts by 16 cm^{-1} but the spectral line of $R_1 \rightarrow Y_3$ red shifts by 3 cm^{-1} when temperature rises from 10 to 300 K. In the case of another garnet crystal Nd^{3+}: CAZGAR [31] spectral line of $R_1 \rightarrow Z_6$ blue shifts by 2.3 cm^{-1} while the longer wavelength spectral line $R_1 \rightarrow Y_1$ red shift by 3.2 cm^{-1} in a temperature range of 10–300 K.

Originally, the thermal spectral line shifting that results from electron–phonon interaction should be red shifting, but why some of the spectral lines have blue shifting? Actually, it should be noted that the lattice expands at higher temperature then the distance between the active ions and their ligands will be lengthened, and the crystal field will be weakened, so in the above example the crystal field splitting of multiplet $^4I_{9/2}$ becomes smaller. The position of the energy level above the "barycenter" of the multiplet such as Z_5 will be lower because the "barycenter" of the multiplet is not affected by the crystal field interaction. Although the crystal field thermal effect is generally relatively small, but for the highest crystal energy levels in the multiplets like Z_5 and Y_6, the quantity of energy level position change is considerable. The aforementioned spectral line $R_1 \rightarrow Z_5$ of Nd^{3+}: GSGG crystal has a thermal blue shifting while all the spectral lines of Nd^{3+}: YAG crystal reported by Kushida [32] have thermal red shifting except the spectral lines of $R \rightarrow Z_5$ and $R \rightarrow Y_6$ have thermal blue shifting. These are the embodiment of the above mechanism. Kushida [32] attempted to explain these phenomena by a pushing effect between multiplets $^4I_{13/2}$ and $^4I_{11/2}$ as well as that between multiplets $^4I_{11/2}$ and $^4I_{9/2}$, but he neither proposed its physical mechanism nor explained if the higher multiplet has pushing effect to lower multiplet and whether the lower multiplet has the same pushing effect to the higher multiplet and how this effect change the position of the "barycenter" of the multiplet. Therefore the pushing effect is difficult to explain the phenomena of thermal blue shifting of the spectral line.

The following examples illustrate the effect of mass difference of host ions on the spectral line thermal broadening of active ions. Take the thermal spectral line broadening of Nd^{3+} ion in garnet crystals LLGG and CAZGAR as examples [30, 31]. By using their $R_1 \rightarrow Z_5$ spectral line thermal broadening data T_D and $\bar{\alpha}$ and the expression of D factor given in Sect. 6.4, the following parameters are calculated: for LLGG crystal D=1.44, $\bar{\alpha}/T_D^7 = 2.5 \times 10^{-17}\,\text{cm}^{-1}/\text{K}^7$ and for CAZGAR crystal D=1.22, $\bar{\alpha}/T_D^7 = 1.65 \times 10^{-18}\,\text{cm}^{-1}/\text{K}^7$. When temperature rises from 10 K to 300 K, the spectral line $R_1 \rightarrow X_5$ thermal broadenings of Nd^{3+}: LLGG crystal and Nd^{3+}: CAZGAR crystal are 8.1 and 4.5 cm^{-1}, respectively. The situation is similar for the transition metal ions. By using T_D and $\bar{\alpha}$ data of R line in Cr^{3+}: YAG and Cr^{3+}: YGG [33], the calculated result for YAG crystal is $D = 1.17$, $\bar{\alpha}/T_D^7 = 1.53 \times 10^{-17}\,\text{cm}^{-1}/\text{K}^7$; for YGG crystal it is $D = 1.23$, $\bar{\alpha}/T_D^7 = 8.77 \times 10^{-17}\,\text{cm}^{-1}/\text{K}^7$. When the temperature rises from 7 K to 240 K, the R line broadenings of Cr^{3+}: YAG crystal and Cr^{3+}: YGG crystal are 5.4 and 25 cm^{-1}, respectively.

Just as pointed out in Sect. 6.4, the spontaneous phonon emission constitutes the main contribution to the spectral line broadening at very low temperature. Therefore, the transition to higher energy states of final multiplet will have larger spectral linewidth. This is a common phenomenon, not only demonstrated on the $R_1 \rightarrow Y_1$ and $R_2 \rightarrow Y_1$ spectral lines but also can be clearly seen in the following low temperature emission and absorption spectra: Fig. 6.6a is the emission spectrum of Pr^{3+}: YAB excited to multiplet $^1D_2$ measured at 10 K [34]. It can be seen that the spectral linewidth increases with wavelength. The emission lines with longer wavelength are those jumping to higher crystal field energy levels of ground multiplet which have more spontaneous phonon emission channels. Figure 6.6b is absorption spectrum of NAB crystal that corresponds to the transition to multiplet

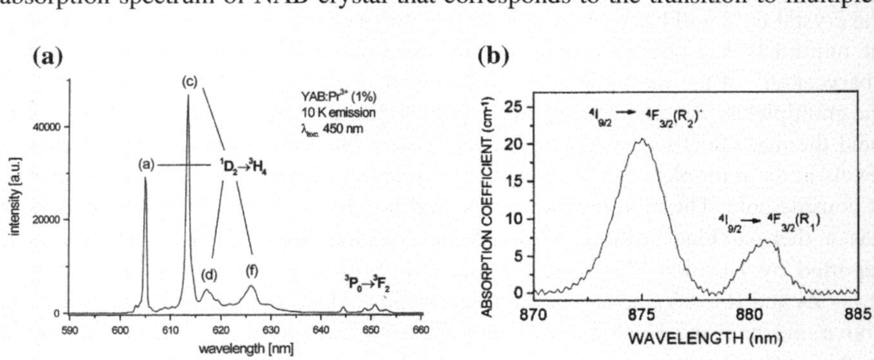

Fig. 6.6 **a** YAB: Pr^{3+} $^1D_2 \rightarrow {}^3H_4$ emission spectrum at 10 K; (a), (c), (d), (f) are transition to crystal energy levels in 0, 231, 332, 560 cm^{-13} of H_4 [34]; **b** $^4I_{9/2} \rightarrow {}^4F_{3/2}$ absorption spectrum of NAB at 10 K [35]

$^4F_{3/2}$ measured at 10 K [35]. The absorption spectral line peak at shorter wavelength corresponds to the transition to a higher crystal field energy level R_2 of multiplet $^4F_{3/2}$ which has the probability to emit phonon so that it has wider spectral linewidth compared to that of the absorption spectral line peak at longer wavelength which is the transition to the lowest crystal field energy level R_1 of multiplet $^4F_{3/2}$ and the level is impossible to emit phonon.

The spectral line thermal broadening of laser materials not only has practical meaning for usually laser application but also is an important subject of the study of coherent optical signal processing, storage and quantum information technology, because in these cases the spectral line thermal broadening appears as optical dephasing and decorrelation in time domain.

References

1. A.S. Davydov, *Quantum Mechanics*, 2nd edn. (Pergamon Press, Oxford, 1976)
2. C. Kittel, *Introduction to Solid State Physics*, 6th edn. (Wiley, New York, 1976)
3. W. Ludwig, in *Phonons and Phonon Interactions*, T.A. Bak (ed.), (W.A. Benjamin, INC, New York, Amsterdam, 1964)
4. P.A.M. Dirac, *The Principles of Quantum Mechanics* (Oxford, 1958)
5. K. Huang, A. Rhys, Proc. Roy. Soc. A **204**, 406 (1950)
6. M. Born, K. Huang, *Dynamics Theory of Crystal Lattice* (Oxford, 1954)
7. W. Magnus, F. Oberhettinger, P.R. Soni, *Formulas and Theorems for the Special Functions of Mathematical Physics* (Springer, New York, 1966)
8. B. Di Bartolo, in *Spectroscopy of Solid-State Laser-Type Materials*, B. Di Bartolo (ed.), (Plenum Press, New York, 1987)
9. K. Huang, Progress in Physics **1**, 31 (1981). (In Chinese)
10. K. Rebane, *Impurity Spectra of Solid* (Plenum Press, New York, 1970)
11. D.E. McCumber, M.D. Sturge, J. Appl. Phys. **34**, 1682 (1963)
12. R.M. Macfarlane, J. Lum. **85**, 181 (2000)
13. D. Hsu, J.L. Skinner, J. Chem. Phys. **81**, 1604–5471 (1984); **83**, 2097–2107 (1985)
14. B. Di Bartolo, *Optical Interactions in Solids* (Wiley, New York, 1968)
15. Z.D. Luo, Y.D. Huang, X.Y. Chen, J. Physics, Condens. Matter. **13**, L447 (2001)
16. A. Ellens, H. Andres, M.L.H ter Heerdt et al., J. Lum., **66** and **67**, 240 (1996)
17. A. Ellens, H. Andres, A. Meijerink et al., Phys. Rev B **55**, 173 (1997)
18. A. Ellens, H. Andres, M.L.H ter Heerdt et al., Phys. Rev B **55**, 180 (1997)
19. K.H. Hellwege, Ann. Phys. **40**, 529 (1941)
20. W.F. Krupke, Phys. Rev. **145**, 325 (1966)
21. D.L. Orth, R.J. Mshl, J.L. Skinner, J. Phys.: Condens. Matter **5**, 2533 (1993)
22. A.A. Kaminskii, *Laser Crystal Physics* (Springer, New York, 1980)
23. Z.D. Luo, Sci. Bull. **17**, 786 (1979). (in Chinese)
24. J, Callaway, *Quantum Theory of the Solid State*, 2nd edn (Academic Press, Inc., 1991)
25. H.G.F. Winkler, *Struktur und Eigenschaften der Kristalle* (Springer, Berlin, 1955)
26. X. Chen, B. Di Bartolo. J. Lumin. **54**, 309 (1993)
27. C. Peistner, P. Albers, W. Luthy et al., Phys. Lett. A **137**, 457 (1989)
28. R. Moncorge, D. Pacheco, B. Di Bartolo, Phys. Stat. Sol. (a) **43**, K45 (1977)
29. J. Dunko, D. Pachcco, B. Di Bartolo, Phys. Stat. Sol. (a) **63**, K31 (1981)
30. D.K. Sardar, R.M. Yow, F.S. Salinas, Opt. Mater. **18**, 301 (2001)
31. D.K. Sardar, R.M. Yow, Opt. Mater. **10**, 191 (1998)

32. A. Kushida, Phys. Rev. **185**, 500 (1969)
33. A.P. Vink, A. Meijerink, J. Lumin. **87–89**, 601 (2000)
34. M.H. Bartl, K. Gatterer, E. Cavalli, et al., Spectrochimica Acta Part A **57**, 1981 (2001)
35. D. Jaque, O. Enguita, G.U. Caldino et al., J. Appl. Phys. **90**, 561 (2001)

Chapter 7
Energy Levels and Spectroscopic Properties of Transition Metal Ions

The ruby crystal used by Maiman in 1960 to observe laser emission first in the world is aluminum oxide crystal doped with trivalent chromium (Cr^{3+}:Al_2O_3). Tunable laser crystals and laser materials for Q switch and mode-locked devices have been widely used in the field of laser technology after 1970. Most of them are doped with transition metal group ions (only the ions of the first transition group, i.e. the ions of iron group are discussed here), including titanium sapphire (Ti^{3+}: Al_2O_3), trivalent chromium-doped lithium strontium aluminum fluoride (Cr^{3+}: $LiSrAlF_6$), and tetravalent chromium-doped magnesium olivine (Cr^{4+}:Mg_2SiO_4) and yttrium aluminum garnet (Cr^{4+}:$Y_3Al_5O1_2$). On the other hand, the crystals doped with bivalent iron group ions like V^{2+}, Co^{2+}, Ni^{2+}, and Fe^{2+} also have laser emission. More than 10 years ago, tunable laser wavelength had been extended to 4.7 μm [1] (shown in Fig. 7.1).

The basic theoretical knowledge of the energy level and spectroscopy characteristics for transition metal ions in crystals will be introduced in this chapter. The method of group chain scheme will be used to calculate matrix elements of crystal field energy level splitting and transition probability. On the other hand, some related laser properties will be briefly mentioned. There are many excellent monographs [2–5] dealing with crystal field theory of transition metal ions and a very extensive volume of literature on the iron group ions doped laser crystals, which will not be introduced in detail, so the readers can directly refer to. The character of the electronic structure of these ions is that they have an unfilled $3d$ shell, that is, their electronic configurations are $3d^q$, where the electron number q can vary from 1 to 9. The $3d^q$ shells of these ions are not shielded by outer filled shells like those of rare earth ions; in this situation the crystal field and lattice vibration have a strong effect on their electronic states and the transitions between these states. The crystal field energy is comparable to the energy of Coulomb interaction between electrons but larger than that of the spin–orbit interaction. Therefore, in order to perform the theoretical calculation of energy levels, the crystal field perturbation of d electron must be considered at first, from high symmetry field to lower symmetry field step by step, then the spin–orbit coupling is

Z. Luo and Y. Huang, *Physics of Solid-State Laser Materials*, Springer Series in Materials Science 289, https://doi.org/10.1007/978-981-32-9668-8_7

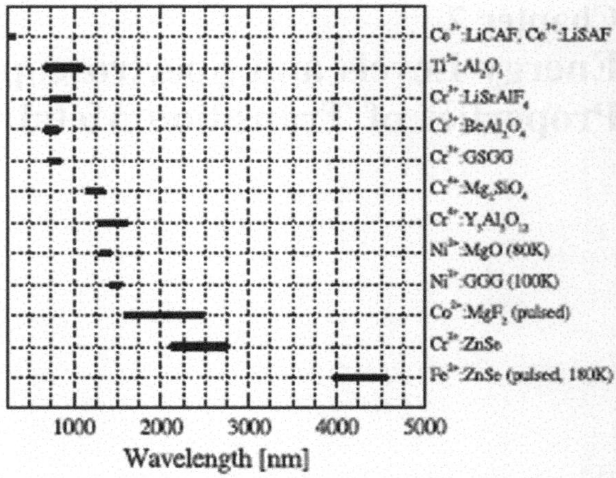

Fig. 7.1 Overview of tunable solid-state lasers [1]

subsequently included. The discussion of Chap. 6 shows that, in this situation, there are larger shift between the zero points of configuration coordinate curve of the excited state and that of the ground state; hence the emission and absorption spectra show the pattern of multi-phonon side band.

Solid-state laser properties of transition metal ions will be discussed as examples. However, the knowledge of energy level structure and transition characteristics of the transition metal ions studied in this chapter can also be used to luminescent materials.

It should be said that in order to accord with energy level symbols usually used in spectroscopy of transition metal ions, the Milliken's symbols of group representation are adopted in this chapter except those used in 3*jm* factors.

7.1 Energy Levels and Spectral Properties of $3d^1$ Electron System

Supposing Ti^{3+} ions in crystal are at the center of a regular octahedron consisting of six anions with charge $-Ze$, let's consider the crystal field effect on the *d* electron. Obviously, the interaction potential of these anions with *d* electron can be, in point charge model, expressed as

$$V_c(\boldsymbol{r}) = \sum_{i=1}^{6} Ze^2/|\boldsymbol{R}_i - \boldsymbol{r}| \tag{7.1}$$

where $\boldsymbol{r} = (r, \theta, \varphi)$ and $\boldsymbol{R_i} = (R, \Theta_i, \Phi_i)$ are the position vectors of the electron and the ligand *i*, respectively. The Schrödinger equation of the electron can be written as

$$\left[-\left(\hbar^2/2m\right)\Delta + U(r) + V_c(\boldsymbol{r})\right]\varphi(\boldsymbol{r}) = E\varphi(\boldsymbol{r}) \tag{7.2}$$

where $U(r)$ is the interaction potential of the electron with central field of the free ions which is independent of the angular variables, thus can be expressed as a function of scalar coordinate r. The perturbation method, to treat $V_c(\boldsymbol{r})$ as a perturbation, should be used to solve (7.2). Crystal field potential $V_c(\boldsymbol{r})$ is generally expanded in terms of spherical harmonic functions in crystal field theory. It takes the form of

$$V_c(\boldsymbol{r}) = \sum_{k=0}^{\infty}\sum_{m=-k}^{k} r^k B_{kq} C_q^{(k)}(\theta,\varphi) \tag{7.3}$$

where

$$C_q^{(k)}(\theta,\varphi) = \left(\frac{4\pi}{2k+1}\right)^{1/2} Y_{kq}(\theta,\varphi) \tag{7.4}$$

$$B_{kq} = \left(\frac{4\pi}{2k+1}\right)^{1/2} \frac{Ze^2}{R^{k+1}} \sum_{i=1}^{6} Y_{km}^*(\Theta_i,\Phi_i) \tag{7.5}$$

Equation (7.3) is a general expression, when it is applied to specific case studied, then some restrictions should be added.

First, according to Wigner–Eckart theorem and the triangle condition of the 3-j symbol introduced in Chap. 2, it can be seen that in case of d electron where $l = l' = 2$, the related matrix element $\langle \varphi_{nlm}|V_c|\varphi_{n'l'm'}\rangle$ is zero unless $k \leq 4$. For the problem of crystal field energy splitting, the wave functions involved belong to the same configuration (i.e. the same parity) and so the crystal field potential should have even parity, that is only the even k value in $k \leq 4$ should be considered. Secondly, in the first-order approximation, the active ion can be seen as in the center of regular octahedron composed by six negative ions with charge $-Ze$. As pointed out in Chap. 2, V_c is invariant under the operations of cubic group operations. Owing to the existence of the fourth-order symmetry axis, in the range of $k \leq 4$, k can only be equal to 0 or 4 and q must also be 0, 4, and −4. If the term with $k = 0$ is omitted, which has only a parallel shift of all the levels, then the crystal field potential will be

$$V_c = \frac{7Ze^2}{2R^5} r^4 \left\{ C_0^{(4)}(\theta,\varphi) + \left(\frac{5}{14}\right)^{1/2} \left[C_4^{(4)}(\theta,\varphi) + C_{-4}^{(4)}(\theta,\varphi)\right] \right\} \tag{7.6}$$

By using the language of group chain scheme introduced in Chap. 3, it means that when the irreducible representation of rotation group SO_3 is reduced to the irreducible representation of cubic group, there is only the term of $k = 4$ that can have a component belonging to the identity representation. If the wave function of the

$3d$ electron expressed as $\varphi_{32m}= R_{32}(r)Y_{2m}(\theta,\varphi)$, then the matrix elements can be calculated as

$$\begin{gathered}
\langle\varphi_{32,+2}|V_c|\varphi_{32,+2}\rangle = \langle\varphi_{32,-2}|V_c|\varphi_{32,-2}\rangle = Dq \\
\langle\varphi_{32,+1}|V_c|\varphi_{32,+1}\rangle = \langle\varphi_{32,-1}|V_c|\varphi_{32,-1}\rangle = -4Dq \\
\langle\varphi_{32,0}|V_c|\varphi_{32,0}\rangle = 6Dq \\
\langle\varphi_{32,+2}|V_c|\varphi_{32,-2}\rangle = \langle\varphi_{32,-2}|V_c|\varphi_{32,+2}\rangle = 5Dq
\end{gathered}$$

where $Dq = \frac{Ze^2}{6R^5}\langle r^4\rangle_{3d}$, $< r^4 >_{3d}$ is the fourth power of the electronic coordinates r averaged over $3d$ orbital. The calculation of the matrix elements can use (2.43a) of the Wigner–Eckart theorem and 3-j symbol introduced in Chap. 2. For example

$$\begin{aligned}
\langle\varphi_{32,+2}|V_c|\varphi_{32,+2}\rangle &= \frac{7Ze^2}{2R^5}\langle r^4\rangle_{3d}(-1)^{2-2}\langle 2\|C^4\|2\rangle \times \begin{pmatrix} 2 & 4 & 2 \\ -2 & 0 & 2 \end{pmatrix} \\
&= \frac{7Ze^2}{2R^5}\langle r^4\rangle_{3d}\times\frac{5\sqrt{2}}{\sqrt{35}}\times\left(\frac{1}{18\times 35}\right)^{1/2} = \frac{Ze^2}{6R^5}\langle r^4\rangle_{3d} \equiv Dq
\end{aligned} \tag{7.7}$$

where 3-j symbol can be found from Rotenberg [6]; at the same time note that the 3-j symbols vanish unless $m_1 + m_2 + m_3 = 0$ and so for the calculation of above three matrix elements only the first term of (7.6) should be considered. On the other hand, (2.53) is used to calculate $\langle 2\|C^{(4)}\|2\rangle$.
Therefore the following secular equation is obtained

$$\begin{vmatrix}
E^0+Dq-E & 0 & 0 & 0 & 5Dq \\
0 & E^0-4Dq-E & 0 & 0 & 0 \\
0 & 0 & E^0+6Dq-E & 0 & 0 \\
0 & 0 & 0 & E^0-4Dq-E & 0 \\
5Dq & 0 & 0 & 0 & E^0+Dq-E
\end{vmatrix} = 0$$

By exchanging the fifth row with the second row and then exchanging the fifth line with the second line, the secular equation changes to

$$\begin{vmatrix}
E^0+Dq-E & 5Dq & 0 & 0 & 0 \\
5Dq & E^0+4Dq-E & 0 & 0 & 0 \\
0 & 0 & E^0+6Dq-E & 0 & 0 \\
0 & 0 & 0 & E^0-4Dq-E & 0 \\
0 & 0 & 0 & 0 & E^0-Dq-E
\end{vmatrix} = 0$$

This determinant equation is equivalent to a three-order one with only diagonal elements and a two-order one. It is easy to find out that its solution can be divided into two groups, one has double degeneracy, the other has triplet degeneracy, that is $(E) = E^0 + 6Dq$ with double degeneracy and $E(T_2) = E^0 - 4Dq$ with triplet degeneracy. The spin degeneracy has not been considered now.

From Chap. 2 we know that the representation D_L ($L = 2$) of the rotation group SO_3 will be reduced to one irreducible representation $\Gamma_3(E)$ (two-dimension) and one irreducible representation $\Gamma_5(T_2)$ (three-dimension) of group O (it can be found in Appendix B, there is $2 \rightarrow 3 \oplus 5$ of the $SO_3 \rightarrow O$ irreducible representation reduction). Only the electronic states of the orbital and their degeneracy are considered here, so the spin degeneracy has not been included. The energy levels are denoted by the symbols of group irreducible representation as shown in Fig. 7.2.

When active ions have the octahedral coordination symmetry, the energy levels of ^2E and $^2T_2$ are shown in Fig. 7.2, in which the energy of ^2E is higher than that of the $^2T_2$. However, if the active ions have tetrahedral or cubic coordination symmetry, the energy of the ^2E is lower than that of the $^2T_2$; besides their crystal field splittings are smaller by a factor of 4/9 and 8/9, respectively. This is because in these three coordination symmetries, the Dq values have the following relation

$$Dq(\text{octahe}) = -\frac{9}{4}Dq(\text{tetrat}) = -\frac{9}{8}Dq(\text{cubic}) \tag{7.8}$$

In titanium sapphire (Ti^{3+}:Al_2O_3), the positions occupied by Ti^{3+} ions have a small trigonal distortion, that is, they have the crystal field components with point group symmetry C_3, which further split the energy levels with symmetry of point group O. Since the crystal field does not change the spin multiplicity, in the following discussion, the sign of the spin multiplicity is omitted for the symbols of all the energy levels. The irreducible representation $\Gamma_5(T_2)$ of group O becomes reducible in group C_3 and can be reduced to the sum of two irreducible representations $A_1 \oplus E$ $(\Gamma_1 \oplus (\Gamma_{2\oplus 3}))$ of group C_3. Correspondingly, the energy level

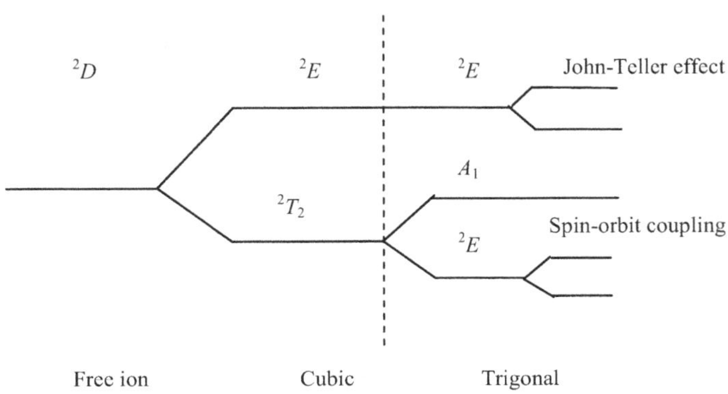

Fig. 7.2 Energy levels of d^1 electron in cubic and trigonal fields

$E(\Gamma_5)$ is split into two energy levels $E(\Gamma_1)$ and $E(\Gamma_{2\oplus3})$ (the Γ_2 and Γ_3 are one-dimensional irreducible representations and conjugate to each other, and together form a two-degree degenerate level). On the other hand, the irreducible representation $E(\Gamma_3)$ of group O corresponds to the irreducible representation E $(\Gamma_{2\oplus3})$ of group C_3. It means that the energy level $E(\Gamma_3)$ of the system with symmetry O will not be further split in the trigonal crystal field. Moreover, this double-electron state has no spin–orbit splitting, as discussed in the following.

For a single electron, the spin–orbit interaction Hamiltonian is expressed as (1.2), where the operator of orbital angular momentum belongs to $T_1(\Gamma_4)$ irreducible representation of the group O. If the orbital wave function belongs to G irreducible representation of group O, then the condition for a non-zero spin–orbit interaction matrix element is that the product $G \otimes T_1(\Gamma_4) \otimes G$ should include $A_1(\Gamma_1)$ representation of group O. Therefore, the matrix element of the spin–orbit interaction between the energy levels with orbital wave functions belonging to $T_1(\Gamma_4)$ and $T_2(\Gamma_5)$ of group O is non-zero, because $T_2(\Gamma_5) \otimes T_1(\Gamma_4) \otimes T_2(\Gamma_5)$ includes $A_1(\Gamma_1)$ representation of group O.

The ground electronic double-state $E(E)$ is one of the energy levels split under trigonal crystal field from energy level $E(T_2)$ belonging to irreducible representation T_2 of group O (see Fig. 7.2). According to the formula given in Chap. 3, the matrix element of spin–orbit interaction of the wave function with irreducible representation E of C_3 symmetry group can be expressed as the spin–orbit interaction matrix element of wave function with T_2 irreducible representation of group O multiplied by corresponding $3jm$ factor. Therefore, the ground electronic double-state E (E) under C_3 symmetry can be split by spin–orbit interaction into two electronic single levels.

The higher electronic double-state $E(E)$ is the energy levels split under trigonal crystal field from energy level $E(E)$ belonging to E irreducible representation of group O, and its wave function does not satisfy the condition $E(\Gamma_3) \otimes T_1(\Gamma_4) \otimes E(\Gamma_3)$ that includes $A_1(\Gamma_1)$ so that it cannot split by the spin–orbit interaction. However, some of the energy levels without spin–orbit splitting can be split by Jahn-Teller effect, that is the degeneracy will be removed by the electron–lattice vibration interaction. Similar to the analysis given by Sugano [3] for the E level with cubic symmetry, the Jahn-Teller splitting of E level with trigonal symmetry can be calculated. If one denotes two coordinates of normal modes by Q_1 and Q_2 belonging to irreducible representation E which have interaction with electron states, introducing the polar angle θ and radius vector ρ, coordinates Q_1 and Q_2 can be expressed as

$$Q_1 = \rho \cos \theta, Q_2 = \rho \sin \theta$$

Therefore, after Jahn-Teller splitting the energy separation between these two levels will be

$$\Delta E_{JT} = 2A\rho$$

where A has the following relation with Q_1 and Q_2

$$A = \frac{1}{2}\langle E \| V_{E_g}(r) \| E \rangle \begin{pmatrix} E & E & E \\ E & E & E \end{pmatrix}^{O}_{C_3}$$

where $V_{E_g}(r)$ is the coefficient of the first-order term of the electron-phonon interaction energy expanding in terms of the normal coordinates, which belongs to the E_g representation. The $3jm$ factor [7] involved in the formula is

$$\begin{pmatrix} E & E & E \\ E & E & E \end{pmatrix}^{O}_{C_3} \equiv \begin{pmatrix} 3 & 3 & 3 \\ 2 & 2 & 2 \end{pmatrix}^{O}_{C_3} = \frac{1}{\sqrt{2}}$$

In the case of titanium sapphire, the value obtained by experiment is $\Delta E_{JT} = 1850\ \text{cm}^{-1}$ [8].

The energy levels of Ti^{3+} in titanium sapphire have been shown in Fig. 7.2. It should be pointed out that the five electronic single-states split by crystal field and spin–orbit interaction or Jahn-Teller effect still have Kramers degeneracy, which can only be removed by magnetic interactions. The absorption band of titanium sapphire is peaked at 490 nm corresponding to the transition from low-energy double-state E (E) to high-energy double-state E(E). The peak of the emission band is at 770 nm and the π polarized spectrum has higher intensity than that of the σ polarized spectrum at 300 K. Its absorption and emission spectra are shown in Fig. 7.3.

It is shown that the titanium sapphire used as tunable laser crystal has the following distinct advantages: First, the separation between the excited state and the ground state is proportional to crystal field strength and so is strongly affected by crystal field and lattice vibration. There are large Stokes shift between the excited

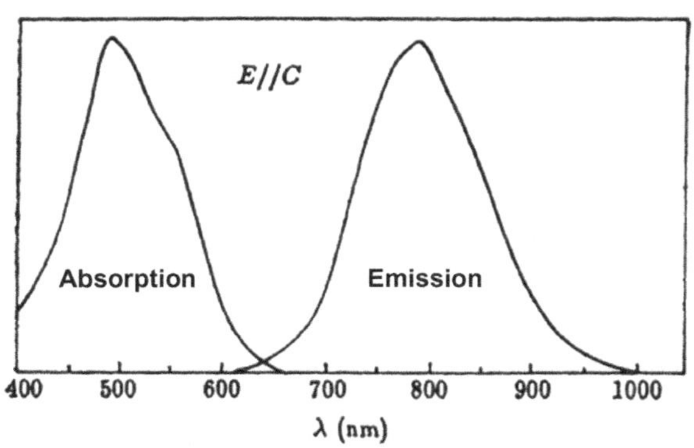

Fig. 7.3 Absorption and emission spectra of titanium sapphire

and ground states, which leads to a very wide phonon side band; hence the titanium sapphire laser has a very wide tunable range from 650 to1060 nm. Secondly, the initial and final states of the transition have the same spin multiplicity, thus the transition is spin allowed and so is strong. Correspondingly, the lifetime of the upper laser state is as short as 3.2 μs, thus it can only be efficiently pumped by laser sources. Thirdly, the aluminum oxide is a material with very high Debye temperature so that the electrons of Ti^{3+} ions have strong interaction with lattice vibration. In this way, the absorption and emission will be intensified. As a result, the pumping efficiency is higher than 25%. Fourthly, as shown in the energy level diagram, above the fluorescent level, there are no other excited states; hence the excited state absorption is non-existent. These are the prominent advantage of titanium sapphire over other tunable laser crystals. Other spectral properties of the titanium sapphire can be referred to the paper published by Moulton [9]. Ti: $BeAl_2O_4$ is also a trivalent titanium tunable laser crystal, and its spectral properties can be found in [10]. It should be pointed out that the disadvantage of Ti^{3+} laser materials is the valence instability of Ti^{3+} ions; they can change to Ti^{2+} or Ti^{4+} and will reduce the laser performance.

7.2 Energy Levels and Spectral Properties of $3d^2$ Electron System

Transition metal ions Cr^{4+}, V^{3+}, Ti^{2+}, and Mn^{5+} have electronic shell $3d^2$; among these the Cr^{4+} ion is more stable. One can investigate the electronic states of d^2 configuration in crystals by two schemes.

First, the weak field scheme: It can be seen from Table 1.1 in Chap. 1 that the terms of the free ions with configuration d^2 are: spin single-state—^1S, ^1D, ^1G and spin triple-state—^3P, ^3F. When the ions are sited in positions with symmetry lower than that of the rotation group SO_3, then the free ion states ^{2S+1}L belonging to irreducible representation D_L of group SO_3 will be split into crystal field states belonging to irreducible representations of group O. The reduction of irreducible representation can be obtained by the theorem introduced in Chap. 2. One can refer to the reduction table in Appendix B to obtain the following results:

$$\begin{aligned}&{}^1S(L=0)-{}^1A_1; D(L=2)-{}^1E,{}^1T_2;{}^1G(L=4)-{}^1A_1,{}^1E,{}^1T_1,{}^1T_2;\\&{}^3P(L=1)-{}^3T_1;{}^3F(L=3)-{}^3A_2,{}^3T_1,{}^3T_2.\end{aligned}$$

Secondly, the strong field scheme: As having been seen in the previous section, single electron in the cubic field will have two energy levels E and T_2. Two d electrons can be: both in E state; both in T_2 state; or one in E state and the other in T_2 state. After introducing the interactions between electrons and between the spin and orbital momentum, the final electronic energy levels usually will be denoted by irreducible representation symbol with lowercase letters. Therefore, the states occupied by single electron are designated by letter e and t_2. The electronic system studied can have the configurations of t_2^2, e^2, or $t_2^1e^1$. By the multiplication table of the irreducible representation of group O, the following can be obtained

$$
\begin{aligned}
T_2(5) \otimes T_2(5) &= A_1(1) \oplus E(3) \oplus T_1(4) \oplus T_2(5) \\
E(3) \otimes E(3) &= A_1(1) \oplus A_2(2) \oplus E(3) \\
T_2(5) \otimes E(3) &= T_1(4) \oplus T_2(5)
\end{aligned}
$$

The figures in the brackets of Muliken symbols are the subscripts i of Bethe symbol Γ_i. The corresponding relationship between the Muliken symbol and the Bethe symbol (and the subscript of Bethe symbol) is listed to facilitate the search for the relevant tables of this book. The following will be described in accordance with the Muliken symbols usually used in the literature of spectroscopy for iron group ions. The relations of two kind symbols for low symmetry group can be found in Appendix A.

According to Pauli exclusion principle, two electrons in the states of T_1 and T_2 in configuration $t_2^1e^1$ can have their spin either parallel (spin triple state) or antiparallel (spin single-state), which results in the states of $^3T_1$ $^1T_1$, $^3T_2$, and $^1T_2$. The spin multiplicity of the configurations of t_2^2 and e^2 can be determined similarly. According to Pauli exclusion principle, the total electron wave functions should be antisymmetry. Thus, if the orbital wave function is symmetry, the spin wave function should be antisymmetry. Contrarily, if the orbital wave function is antisymmetry, the spin wave function should be symmetry. In order to determine the symmetry properties of orbital wave function, the symmetry properties of the irreducible representation reduced by direct product of group representations discussed in Chap. 2 should be utilized.

The vector space of t_2^2 is the direct product of two t_2 vector spaces $t_2 \otimes t_2$. If the basic vector of one of these vector spaces is expressed by $\psi(1,i)$ (ith basic vector of electron 1), the other is $\psi(2,j)$ (jth basic vector of electron 2), then the basic vector of the direct product space will be $\psi(1,i)\psi(2,j)$. This direct product space can be reduced to a sum of a series of subspaces with each subspace corresponding to one irreducible representation reduced from the direct product representation of $T_2 \otimes T_2$. The basic functions of these subspaces are the linear combinations of $\psi(1,i)\psi(2,j)$. As to the symmetry properties of exchange electronic coordinates, the basic vector of the direct product space can be divided into: symmetry type $\psi(1,i)\psi(2,j) + \psi(2,i)\psi(1,j)$ and antisymmetry type $\psi(1,i)\psi(2,j)$-$\psi(2,i)\psi(1,j)$ ($i \neq j$). The subspaces formed by symmetric basic vectors correspond to one or several symmetric irreducible representations and those formed by antisymmetric basic vectors correspond to one or several antisymmetric irreducible representations. The character of the symmetric representation of the square direct product is denoted by $[\chi^2](G)$ and that of the antisymmetric representation is denoted by $\{\chi^2\}(G)$ (in the problem discussed, G is one of the symmetry operation of the cubic group O). Obviously, $[\chi^2](G)$ can be calculated as follows

$$G[\psi(1,i)\psi(2,j)+\psi(2,i)\psi(1,j)]=\sum_{lm}G_{li}G_{mj}[\psi(1,l)\psi(2,m)+\psi(2,l)\psi(1,m)]$$

The indexes l and m are only the symbols for summation; exchanging them will not alter the values of the sum. Therefore, the above expression can be changed into

$$\frac{1}{2}\sum_{lm}(G_{li}G_{mj}+G_{mi}G_{lj})[\psi(1,l)\psi(2,m)+\psi(2,l)\psi(1,m)]$$

The character $[\chi^2](G)$ obviously is to multiply from the left of the wave function $\psi(1,i)\psi(2,j)+\psi(2,i)\psi(1,j)$ then let $l = i$, $m = j$ and sum over i and j as shown in the following

$$\begin{aligned}[\chi^2](G) &= \sum_{ij}[\psi(1,i)\psi(2,j)+\psi(2,i)\psi(1,j)]\mathrm{G}[\psi(1,i)\psi(2,j)+\psi(2,i)\psi(1,j)] \\ &= \sum_{ij}\frac{1}{2}(G_{ii}G_{jj}+G_{ij}G_{ji})=\frac{1}{2}\left[\chi(G)^2+\chi(G^2)\right]\end{aligned} \tag{7.9}$$

because

$$\sum_i G_{ii}=\sum_j G_{jj}=\chi(G),\ \sum_{ij}G_{ij}G_{ji}=\sum_j G_{jj}^2=\chi(G^2)$$

Similarly, one can obtain

$$\{\chi^2\}(G)=\frac{1}{2}\left[\chi(G)^2-\chi(G^2)\right] \tag{7.10}$$

The characters of the symmetry and antisymmetry representations of the cubic group are listed in Table 7.1. According to the reduction formula of irreducible representation ((2.7) in Chap. 2) and the characters given in Table 7.1, it is possible to determine the symmetry and antisymmetry irreducible representations. By using the symmetry representation character shown in Table 7.1 $[\chi^2](T_2) = \chi(A_1) + \chi(E) + \chi(T_2)$ and $\{\chi^2\}(T_2) = \chi(T_1)$. It shows that the direct product $T_2 \otimes T_2$ includes symmetry irreducible representations A_1, E, and T_2, and antisymmetry irreducible representation T_1. The antisymmetry character of the total wave function requires that the symmetry orbital wave function should be coupled with antisymmetry spin wave function. Contrarily, the antisymmetry orbital wave function should be coupled with symmetry spin wave function.

Consequently, t_2^2 configuration generates three-spin single states $^1A_1$, ^1E, and $^1T_2$ as well as one-spin triple state $^3T_1$. Similarly, e^2 configuration includes two-spin single states $^1A_1$, ^1E, and one-spin triple state $^3A_2$. Obviously, the results obtained by the strong field scheme are the same as those of the weak field scheme.

Table 7.1 Symmetry and antisymmetry representations of $T_2 \otimes T_2$ of the cubic group O

O	E	$8C_3$	$3C_2$	$6C_4$	$6C_2'$
$\chi(A_1)$	1	1	1	1	1
$\chi(A_2)$	1	1	1	−1	−1
$\chi(E)$	2	−1	2	0	0
$\chi(E^2)$	2	−1`	2	2	2
$\chi(T_1)$	3	0	−1	1	−1
$\chi(T_2)$	3	0	−1	−1	1
$\chi(T_2^2)$	3	0	3	−1	3
$[\chi^2](E)$	3	0	3	1	1
$\{\chi^2\}(E)$	1	1	1	−1	−1
$[\chi^2](T_2)$	6	0	2	0	2
$\{\chi^2\}(T_2)$	3	0	−1	1	−1

By means of the term energy of the free ions given in Chap. 1, it is convenient to calculate energy levels of the ions in a crystal field by the weak field scheme. The vector space formed by the electronic wave function of the entire $3d^2$ configuration can be reduced to several subspaces; each subspace has the same irreducible representation and spin multiplicity. Owing to the fact that the crystal field Hamiltonian belongs to the identity representation of the point group of active ions and does not include spin variable, the crystal field matrix elements between wave functions belonging to different subspaces are equal to zero. Therefore, the calculation of energy eigenvalue only needs to be performed in the same irreducible representation subspace. As a simple example, one can see the energy calculation of the levels belonging to A_1 representation of cubic group O. Both the free ion terms ^1S and ^1G generate level A_1 in the cubic crystal field. The electrostatic interaction energy between electrons has been given in Chap. 1, i.e.

$$\langle d^2, {}^1S, A_1 | H_0 + H_{ee} | d^2, {}^1S, A_1 \rangle = A + 14B + 7C$$
$$\langle d^2, {}^1G, A_1 | H_0 + H_{ee} | d^2, {}^1G, A_1 \rangle = A + 4B + 2C$$

The calculation of the matrix elements of the crystal field Hamiltonian can utilize directly the $3jm$ factor of the group-chain $SO_3 \supset O$ and (2.53) as well as the formulas similar to (3.18) and (3.19). The reduced matrix elements of tensor $U^{(4)}$ can be found from the published tables [11]. In the case of d electron in cubic field, there is only one term in the crystal field Hamiltonian

$$V_c = C_0^4 b_0^4 \tag{7.11}$$

By using tetragonal coordinate system, that is adopting the tetragonal symmetry axis as z-axis, it can be obtained in the electrostatic model that

$$C_0^4 = 42(3/7)^{1/2}Dq \tag{7.12}$$

The energy of $^3A_2$ level is equal to that of the spectral term ^3F of the free ion, that is A-$8B$ (shown in Chap. 1, Table 1.4) plus the following crystal field energy

$$\begin{aligned}
&\langle d^2, {}^3F, {}^3A_2 | V_c | d^2, {}^3F, {}^3A_2 \rangle \\
&= \begin{pmatrix} 3 & 4 & 3 \\ 2 & 1 & 2 \end{pmatrix}_O^{SO_3} \begin{pmatrix} 2 \\ \gamma \end{pmatrix} \langle d^2, {}^3F \| U^{(4)} \| d^2, {}^3F \rangle \langle d \| C^4 \| d \rangle C_0^4 \\
&= \frac{\sqrt{2}}{\sqrt{3 \times 11}} \times 1 \times \frac{\sqrt{11}}{\sqrt{5}} \times \frac{5\sqrt{2}}{\sqrt{35}} \times \frac{42\sqrt{3}}{\sqrt{7}} Dq = 12Dq
\end{aligned} \tag{7.13}$$

where $2jm$ symbol equals +1, therefore

$$\langle d^2, {}^3F, {}^3A_2 | H_0 + H_{ee} + V_c | d^2, {}^3F, {}^3A_2 \rangle = A - 8B + 12Dq \tag{7.14}$$

Similarly, the energy of $^3T_2$ is equal to the A-$8B$ plus the following crystal field energy

$$\begin{aligned}
&\langle d^2, {}^3F, {}^3T_2\gamma | V_c | d^2, {}^3F, {}^3T_2\gamma \rangle \\
&= \begin{pmatrix} 3 & 4 & 3 \\ 5 & 1 & 5 \end{pmatrix}_O^{SO_3} \begin{pmatrix} 5 \\ \gamma \end{pmatrix} \begin{pmatrix} 5 & 1 & 5 \\ \gamma & 1 & \gamma \end{pmatrix}^O \langle d^2, {}^3F \| U^{(4)} \| d^2, {}^3F \rangle \langle d \| C^4 \| d \rangle C_0^4 \\
&= \frac{-1}{3\sqrt{2} \times 11} \times \frac{1}{\sqrt{3}} \times \frac{-\sqrt{11}}{\sqrt{5}} \times \frac{5\sqrt{2}}{\sqrt{35}} \times \frac{42\sqrt{3}}{\sqrt{7}} Dq = 2Dq
\end{aligned} \tag{7.15}$$

In the above equation γ represents the component of T_2 irreducible representation. The products of $2jm$ symbol with $3jm$ symbol following it for three γ components all equal to $+1/\sqrt{3}$, therefore

$$\langle d^2, {}^3F, {}^3T_2\gamma | H_0 + H_{ee} + V_c | d^2, {}^3F, {}^3T_2\gamma \rangle = A - 8B + 2Dq \tag{7.16}$$

In the strong field scheme, the energy of $^3A_2$ level is $20Dq$ and the energy of $^3T_2$ level is $10Dq$ [5]. The energy difference of the two levels is also $10Dq$.

All the above matrix elements include a common term A, while it has only the effect to shift all the energy levels parallel. If we focus on the separation between different levels, the term A can be omitted. Substituting the detailed value of B and Dq into above secular determinant or matrix elements of Hamiltonian, the eigenvalues of all the energy levels discussed can be obtained. Drawing a curve of energy eigenvalue E/B versus the crystal field strength Dq/B, the Tanabe-Sugano (T-S) energy level diagram shown in Fig. 7.4 can be obtained.

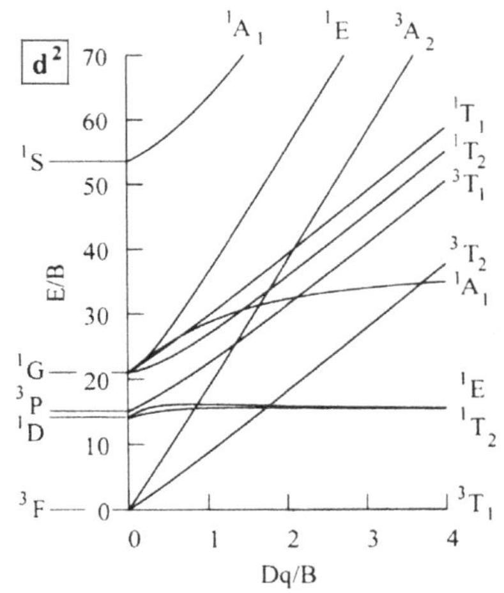

Fig. 7.4 Tanabe-Sugano diagram of d^2 electron configuration in the octahedron position

This T-S energy level diagram is in fact similar to that of the d^8 configuration in the octahedron position. In the latter case, the ground state is $^3A_2$ but not $^3T_1$ and the third excited state with the same spin multiplicity belongs to $^3T_1$.

The relationship of the energy level of the transition metal ions in the octahedron position with that in the tetrahedron position is described below.

The previous calculation shows that if the transition metal ions occupy the octahedron position, the crystal field parameter in the group chain scheme can be expressed as the form of (7.12). The T-S energy level diagram given in many reference books is for this octahedron position. However, when the active ions are sited in tetrahedron position, (7.8) shows that the crystal field parameter Dq is not only different in number but also opposite in sign to that for the octahedron position. Consequently, their energy level diagrams will be different. The following relation between the reduced matrix element of tensors $U^{(k)}(4\,l + 2-q)$ and $U^{(k)}(q)$ was given by Nielson et al. [11]

$$\left\langle \alpha SL \middle\| U^{(k)}(4l+2-q) \middle\| \alpha SL' \right\rangle = -(-1)^k \left\langle \alpha SL \middle\| U^{(k)}(q) \middle\| \alpha SL' \right\rangle$$

Applying it to the case studied $k = 4$, the sign of reduced matrix elements of tensor $U^{(4)}$ for d^q configuration is the opposite of that for d^{10-q} configuration. Therefore, the d^q configuration in octahedron position has a similar T-S energy level diagram as d^{10-q} configuration in tetrahedron position, because the negative sign generated in crystal parameter is cancelled by the negative sign coming from reduced matrix elements of tensor $U^{(4)}$. This is the reason why the T-S energy level diagram for d^8 configuration in octahedron position is similar to that of d^2 configuration in tetrahedron position, if one concerns only the relative position of different energy levels.

As typical examples of the energy levels and spectra of d^2 electronic configuration, the energy levels and spectra of Cr^{4+}:Y_2SiO_5 and Cr^{4+}:Mg_2SiO_4 will be analyzed. The former is a very good example for the study of the spectra of tetravalent chromium-doped crystals, because there is pure Cr^{4+} doping, where Cr^{4+} ions occupy definite C_{3v} positions.

The energy levels of crystal Cr^{4+}:Y_2SiO_5 are shown in Fig. 7.5. Cr^{4+} ions in crystal Cr^{4+}:Mg_2SiO_4 occupy Si positions of the SiO_4 tetrahedron and have approximately the C_{3v} symmetry [12]. Note that the axis a has a threefold symmetry.

Now let's turn to the discussion of the selection rule of absorption transitions. It can be seen from the character table of group C_{3v} that the z component of the electric-dipole moment belongs to representation A_1 of the group C_{3v} and its x and y components belong to representation E. The allowed electric-dipole transitions should satisfy the requirement of $\langle i|\boldsymbol{er}|f\rangle \neq 0$. In the language of group theory, the product of the irreducible representations of the initial and the final states should include the irreducible representations of the electric-dipole moment. Note that the z axis should be the threefold axis of SiO_4, that is the crystal axis a. The transition $^3A_2 \to {}^3A_2$ ($^3T_1$) shown in Fig. 7.5 has polarization of electric field $\boldsymbol{E}$ parallel to axis a, because of the relations $A_2 \otimes A_2 = A_1$ (the z component of the electric-dipole moment belongs to A_1 representation). Similarly, the transition $^3A_2 \to {}^3E$ ($^3T_1$) has polarization of electric field $\boldsymbol{E}$ perpendicular to axis a, because of the relation $A_2 \otimes E = E$. Strictly speaking, the position symmetry of Cr^{4+} ions in crystal Y_2SiO_5 is C_s, Therefore, the transition $^3A_1$ ($^3T_2$)$\to {}^3A_2$ forbidden in the C_{3v} symmetry case is allowed in crystal Cr^{4+}:Y_2SiO_5, which corresponds to a weak absorption at

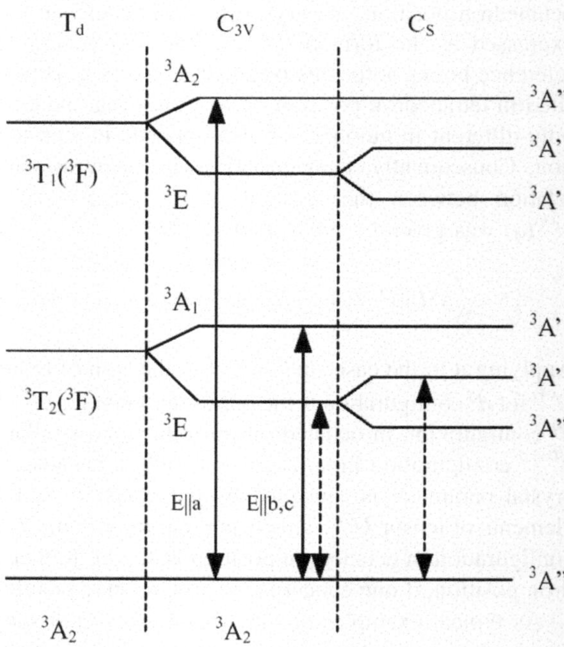

Fig. 7.5 Energy level diagram of Cr^{4+}:Y_2SiO_5

wavelength of 882.6 nm. The weakness of this absorption line shows that the distortion of crystal field to C_s symmetry is small. The polarized absorption spectrum of the Cr^{4+}:Y_2SiO_5 crystal is shown in Fig. 7.6 and the fluorescence spectrum of this crystal is shown in Fig. 7.7.

The laser properties of this crystal at the temperature of liquid nitrogen were studied by Chai et al. [13] Actually there is a laser emission even at temperature of 257 K. It is very interesting to note that pumped at different wavelengths, their laser and tunable performances have a big difference. Pumped by 1064 nm-laser, the emission peak is at 1270 nm and has linewidth of 10 nm. However, the emission peak is at 1225 nm and has linewidth of 23 nm when the pumping wavelength is 532 nm. Finally, pumped at 840 nm, the emission peak is still at 1225 nm while the linewidth changes to 30 nm.

Chromium ions in crystal Mg_2SiO_4 may exist at the same time as trivalent and tetravalent states with trivalent chromium occupies the octahedron Mg position (having symmetry of C_s and C_i) and the tetravalent one occupies the tetrahedron Si position (having approximately C_{3v} symmetry with C_s distortion component). These have been confirmed by detailed structure data, electron paramagnetic resonance and electron nuclear double-resonance experiments [12, 14, 15].

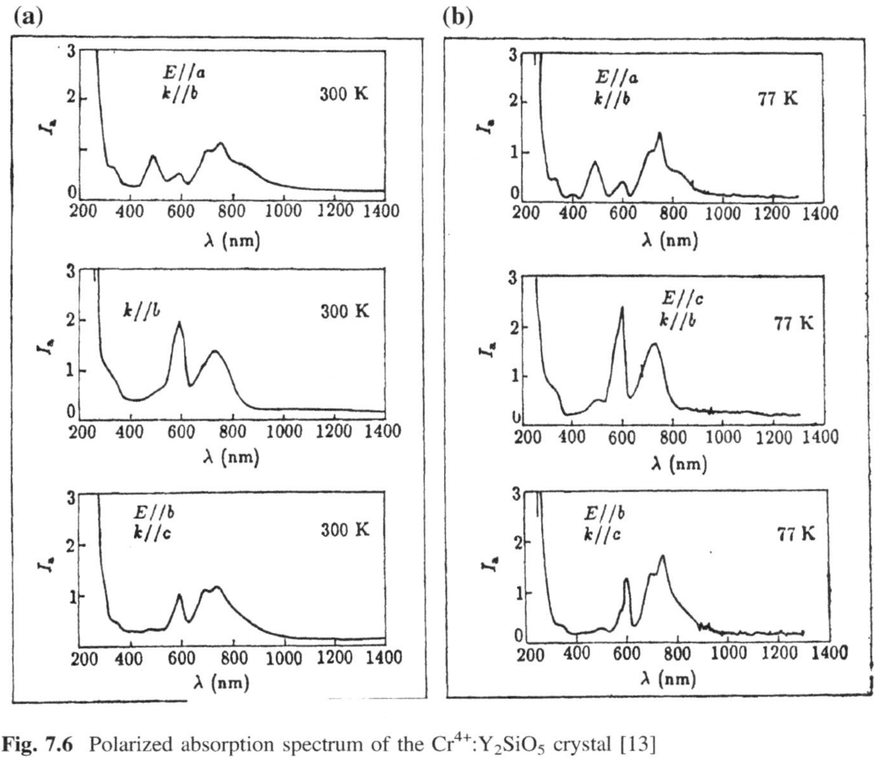

Fig. 7.6 Polarized absorption spectrum of the Cr^{4+}:Y_2SiO_5 crystal [13]

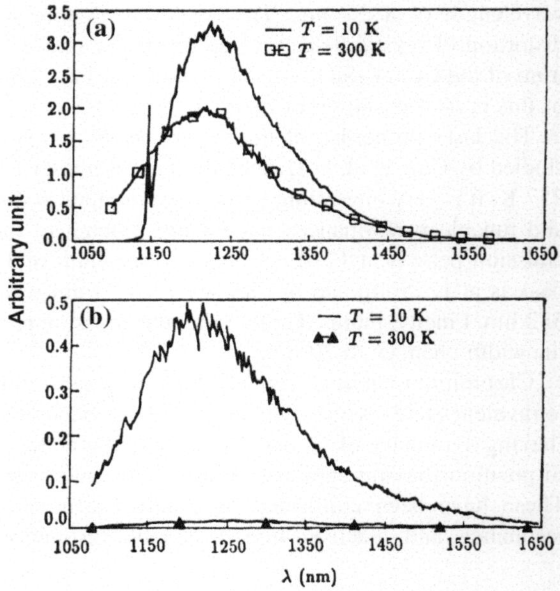

Fig. 7.7 Fluorescence spectrum of Cr^{4+}:Y_2SiO_5 crystal [13]

However, as one of the tunable laser crystal, the Cr^{4+}:Mg_2SiO_4 has a series of advantages, although the spectrum of this crystal is rather complicated. Particularly, its laser emission tunable wavelength range of 850–1400 nm covers the wavelength of 1300 nm, which is very important to the laser optical communication. It also covers the water absorption wavelength range, and by tuning the wavelength of the laser, the penetration depth in the human body can be controlled, hence is very useful in laser medical treatments. The Cr^{4+}:Mg_2SiO_4 laser has become one of the important commercial laser systems at room temperature [16], and it is necessary to know more about its spectral properties.

The ratio of the amount of Cr^{4+} ions to that of Cr^{3+} ions depends on the condition of crystal growth. For the crystals grown in reduction atmosphere, the ratio of the Cr^{4+} will be decreased and that of the Cr^{3+} will be increased. Therefore, in order to obtain crystal with larger amount of Cr^{4+}, the crystal growth should be in standard condition. Their main transition wavelengths are listed in Table 7.2 [16].

Table 7.2 Main transitions and their central wavelengths of Cr^{4+}:Mg_2SiO_4 crystal

Transition channel	Central wavelength (nm)
$^3A_2 \rightarrow {}^3T_2$	1092.9
$^3A_2 \rightarrow {}^3T_1(t_2e)$	648.1
$^3A_2 \rightarrow {}^1E$	629.9
$^3A_2 \rightarrow {}^3T_1(t_2^2)$	373
$^3A_2 \rightarrow {}^1T_2$	348
$^3T_2 \rightarrow {}^3A_2$	1142.9

The spectral and tunable laser properties of Cr^{4+} ions in garnet crystals especially in YAG crystal have been fully studied [17]. Wavelength of these lasers covers the wavelength range of minimum loss fiber communication and eye safe. Among these, Cr^{4+}:YAG has realized 1309–1596 nm tunable laser emission at room temperature with peak emission cross-section of 1.7×10^{-19} cm^2, while Cr^{4+}: YSAG has realized 1446–1604 nm tunable laser emission at room temperature with peak emission cross-section of 1.5×10^{-19} cm^2.

7.3 Energy Levels and Spectral Properties of $3d^3$ Electronic System

Now turn to the energy level structure and spectral properties of trivalent chromium, which have been studied first and intensively among transition metal ions. Compared with $3d^2$ system, $3d^3$ system has more complicated energy level structure. However, using the group chain scheme, the crystal field calculation will be very simple. Taking the upper laser state ${}^4T_2$ and lower laser state ${}^4A_2$ of the Cr^{3+} tunable laser crystal as examples, let us see how to use the method proposed to calculate energy eigenvalues of the crystal field Hamiltonian. Both the states come from free ion's spectral term ${}^4F$ and there are only one ${}^4A_2$ state and one ${}^4T_2$ state in the configuration $3d^3$, so it is only necessary to calculate the diagonal matrix elements for the calculation of crystal field energy levels

$$
\begin{aligned}
&\langle d^3, {}^4F, {}^4T_2\gamma | V_c | d^3, {}^4F, {}^4T_2\gamma \rangle \\
&= \begin{pmatrix} 3 & 4 & 3 \\ 5 & 1 & 5 \end{pmatrix}_O^{SO_3} \begin{pmatrix} 5 \\ \gamma \end{pmatrix} \begin{pmatrix} 5 & 1 & 5 \\ \gamma & 1 & \gamma \end{pmatrix}^O \langle d^3, {}^4F \| U^{(4)} \| d^3, {}^4F \rangle \langle d \| C^4 \| d \rangle C_0^4 \\
&= \frac{-1}{3\sqrt{2} \times 11} \times \frac{1}{\sqrt{3}} \times \frac{\sqrt{11}}{\sqrt{5}} \times \frac{5\sqrt{2}}{\sqrt{35}} \times \frac{42\sqrt{3}}{\sqrt{7}} Dq \\
&= -2Dq
\end{aligned}
\tag{7.17}
$$

It should be noted that for irreducible representation T_2, its three γ components that have the same crystal field matrix elements corresponding to the calculated energy level is a threefold degenerate electronic orbital energy level. On the other hand,

$$\begin{aligned}&\langle d^3, {}^4F, {}^4T_1\gamma|V_c|d^3, {}^4F, {}^4T_1\gamma\rangle\\&= \begin{pmatrix} 3 & 4 & 3 \\ 4 & 1 & 4 \end{pmatrix}_O^{SO_3} \begin{pmatrix} 4 \\ \gamma \end{pmatrix} \begin{pmatrix} 4 & 1 & 1 \\ \gamma & 1 & \gamma \end{pmatrix}^O \left\langle d^3, {}^4F \| U^{(4)} \| d^3, {}^4F \right\rangle \langle d\|C^4\|d\rangle C_0^4\\&= \frac{1}{\sqrt{2\times 11}} \times \frac{1}{\sqrt{3}} \times \frac{\sqrt{11}}{\sqrt{5}} \times \frac{5\sqrt{2}}{\sqrt{35}} \times \frac{42\sqrt{3}}{\sqrt{7}} Dq\\&= 6Dq\end{aligned} \tag{7.18}$$

$$\begin{aligned}&\langle d^3, {}^4F, {}^4A_2|V_c|d^3, {}^4F, {}^4A_2\rangle\\&= \begin{pmatrix} 3 & 4 & 3 \\ 2 & 1 & 2 \end{pmatrix}_O^{SO_3} \langle d^3, {}^4F||U^4||d^3, {}^4F\rangle \langle d||C^4||d\rangle C_0^4\\&= \frac{-\sqrt{2}}{\sqrt{3\times 11}} \times \frac{\sqrt{11}}{\sqrt{5}} \times \frac{5\sqrt{2}}{\sqrt{35}} \times \frac{42\sqrt{3}}{\sqrt{7}} Dq\\&= -12Dq\end{aligned} \tag{7.19}$$

${}^4A_2$ and ${}^4T_2$ levels are both derived from the same free ion spectral term ${}^4F$, and the spectral term near it in energy is ${}^4P$, which can only introduce the energy level ${}^4T_1$ with different irreducible representation from those of ${}^4A_2$ and ${}^4T_2$ levels. The energy operator matrix elements between them is equal to zero. Therefore, the energy eigenvalues of ${}^4A_2$ and ${}^4T_2$ levels can be determined by the diagonal matrix element of the energy operator, then

$$E({}^4T_2) - E({}^4A_2) = 10Dq \tag{7.20}$$

By using strong coupling approximation, it can be shown [18] that $E({}^4A_2) = 3A - 15B$ and $E({}^4T_2) = 10Dq + 3A - 15B$, then the separation between these two levels is the same as (7.20), which is a formula often used in the calculation of spectroscopic parameters of tunable laser crystals. Dq is a parameter to characterize the crystal field strength of transition ions. The value of Dq increases with anions according to the order $Cl^- < F^- < O^{2-} < S^{2-}$.

The energy levels of other states can be simply obtained by the group chain scheme. By the weak field scheme, the energy state ${}^2E$ appears in three spectral terms, that is ${}^2G$, ${}^2H$, and two ${}^2D$, so a 4×4 secular equation should be solved. By the strong field scheme, the calculation of the energy level ${}^2E$ involves matrix elements of crystal field and Coulomb interaction which can be referred to Table 9.3 of [18], but the zero-energy point should be shifted, so the diagonal matrix elements should be subtracted by $12Dq$. Both the state ${}^2E$ and ground state ${}^4A_2$ come from orbit t^3 which has a larger energy interval with the t^2e and e^3 orbits and is less affected by these orbits. Therefore, the relationship of energy level position with the intensity of the crystal field strength is basically the same for the state ${}^2E$ and for the ground state ${}^4A_2$, that is, the separation of the state ${}^2E$ with the ground state $^4A_2$ is independent of the crystal field strength and then the electronic transition of $^2E \rightarrow ^4A_2$ results in a very narrow zero phonon lines. The energy levels involved in the absorption and emission transitions for tunable laser are $^4T_1$ and $^4T_2$. Crystal field state $^4T_1$ comes from spectral terms ^4F and ^4P; its calculation involves a 2×2 secular equation, and one of the diagonal matrix elements has been obtained and shown in (7.18). The non-diagonal matrix elements of static electronic and crystal field interaction Hamiltonian between the wave functions belonging to spectral terms ^4F and ^4P should be calculated. The workload of this calculation is small; the only thing to do for the calculation of the matrix element is to find a $3jm$ factor. The results are shown in Table 7.3.

Trivalent chromium ions in the crystals usually occupy octahedron positions and have the T-S diagram shown in Fig. 7.8.

As has been pointed out in previous section, whether the positions of the ions are octahedron or tetrahedron should be noticed when one uses the T-S energy level diagram. Remember that for the ions of iron group, the T-S diagram for the energy

Table 7.3 Energy matrix elements between spin quartet states of d^3 configuration in cubic crystal field

	$^4T_1$	
	^4F	^4P
^{4}F	$3A-15B+6Dq$	$4Dq$
^{4}P	$4Dq$	$3A$
	$^4T_2$	$^4A_2$
	^4F	^4F
^{4}F	$3A-15B-2Dq$	$3A-15B-12Dq$

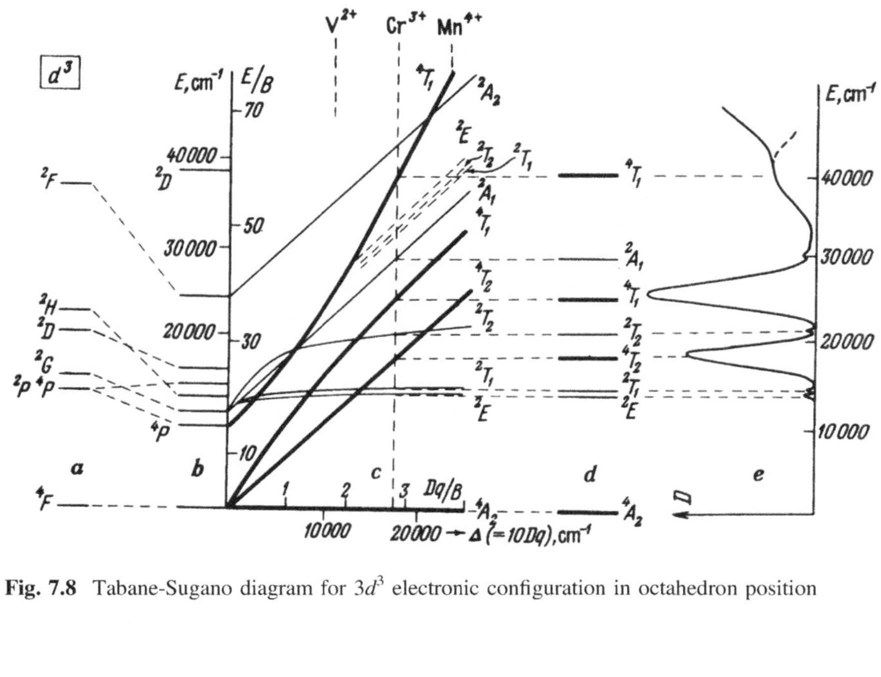

Fig. 7.8 Tabane-Sugano diagram for $3d^3$ electronic configuration in octahedron position

level of $3d^q$ electronic configuration in tetrahedron positions is similar to the T-S diagram for the energy level of $3d^{(4l+1-q)}$ electronic configuration in octahedron positions.

As an example, the energy levels and spectroscopic properties of ruby crystal Cr^{3+}:Al_2O_3 are introduced. It can be seen from Fig. 7.8 that the positions of the energy levels $^2T_1$ and ^2E basically do not change with the strength of crystal field, and hence the effect of the lattice vibration is weak. Moreover, their spin state is different from the ground state, thus limited by the spin forbidden rule. The radiative transition rate is rather weak. The ^2E level is further split into two levels $2\bar{A}$ and $\bar{E}$ with a separation of 29 cm^{-1} by the trigonal crystal field. Laser oscillation usually occurs via the $\bar{E} \rightarrow ^4A_2$ transition corresponding to the zero-phonon line R_1 with spectral linewidth of 11 cm^{-1} at 694.3 nm. When the concentration of Cr^{3+} is high or the sample size is large, the influence of reabsorption makes it have a longer fluorescence lifetime. If it is not affected by the reabsorption, its fluorescence lifetime is 3 ms at very low temperature. At room temperature, some particles will be populated in the $^2T_1$ level and slightly shorten the fluorescence lifetime of the ^2E state. At very high temperature, some particles will also be populated in the $^4T_2$ level and further shorten the fluorescence lifetime. The energy separation between levels $^2T_1$ and ^2E is small and so the active ions in the $^2T_1$ will transit to the ^2E by a strong non-radiative transition as will be pointed out in the next chapter, although the ions in the $^2T_1$ level have only a weak coupling with the lattice vibration. Ruby has a strong crystal field (Dq/B = 2.8 [19]) with absorption bands $^4T_1$ and $^4T_2$ above the energy level of ^2E and the separation between $^4T_2$ and ^2E is 2300 cm^{-1}. In this case, the particle in the absorption band $^4T_2$ has a high non-radiative transition rate to decay to the ^2E, while the particle in the ^2E is impossible to transfer to the $^4T_2$ by thermal population. It is the reason why the ruby has only emission from the ^2E. It should be noticed that the emission cross-section of the R line is even larger than that of the $^4T_2 \rightarrow ^4A_2$ transition due to its narrow linewidth, although the transition of $^2E \rightarrow ^4A_2$ is spin forbidden. Therefore, the ruby crystal becomes the gain medium of the first laser system in the world, although it is a three-level system and usually has a much higher laser threshold than those of the four-level systems.

The trivalent chromium-doped tunable laser crystals have weak crystal field. It can be seen in Fig. 7.8, in this situation, that the position of energy level $^4T_2$ varies greatly with the strength of crystal field. It is close to and usually higher than the level of ^2E, but can also be lower than that of the ^2E in very weak field cases. The level of ^2E forms a "particle reservoir", which transfers the ions coming from absorption band $^4T_1$ to level $^4T_2$ by thermal population, because of its long fluorescence lifetime. The position of the energy level $^4T_2$ changes with the crystal field strength, thus the variation of crystal field by lattice vibration has a strong effect on the potential of ions in the state of $^4T_2$. As a result, the transition of $^4T_2 \rightarrow ^4A_2$ has an emission with broad linewidth. These kinds of tunable laser crystals have their Dq/B value a little higher or even lower than 2.3. The position of the energy level $^4T_2$ is only several hundred wave numbers higher than or even lower than that of the ^2E level. ΔE_{TE} is used to denote the separation between these two levels. A positive or negative ΔE_{TE} represents that the position of the $^4T_2$ is higher or lower than that of the ^2E, respectively. Values of ΔE_{TE}, peak wavelength λ_{peak}, radiative lifetime τ_r, and stimulated emission cross-section σ_{se} for several typical trivalent chromium laser crystals are listed in Table 7.4.

As the crystal field weakens, the fluorescent band emitted by the $^4T_2 \rightarrow {}^4A_2$ transition moves to the long wavelength. Figure 7.9 is a visual picture for this variation.

A series of trivalent chromium tunable laser were developed in 1980s. Cr^{3+}:$BeAl_2O_4$ [20] and Cr^{3+}:$Be_2Al_2(SiO_3)_6$ [21] lasers appeared at first. The former has a threshold of only two-fifths of that for the ruby under the same condition. It can be pumped at high repetition rates and operated at room or even at higher temperature with better laser performance. Its wavelength tunable range is from 701 to 826 nm. Cr^{3+}:Be_2Al_2 $(SiO_3)_6$ crystal has a gain coefficient six times that of the Cr^{3+}:$BeAl_2O_4$ crystal and a wide tunable range of 695–835 nm. In the study of Cr^{3+}-doped tunable laser crystals, in order to reduce the crystal field effect on the Cr^{3+} ions, the host crystals with larger lattice cell constant were adopted [22]. A series of Cr^{3+}-doped garnet crystals were developed in the middle of 1980s, especially, Cr^{3+}:$Gd_3Sc_2Ga_3O_{12}$ crystal [23]. The fluorescence lifetime of the upper laser level for this crystal reaches 120 μs but its tunable range is limited by the excited state absorption on the long wavelength side, while it is limited by the fluorescence reabsorption on the short wavelength side and so has only a tunable range of about 100 nm. It was pointed out by the author [24] that in the oxysalts constituted by anion group with high valence metal ions as its center, the effective charge of the oxygen coordination ions adjacent to the rare earth activator will be relatively small because of the strong polarization of the high valence cation. This chemical factor will also weaken the crystal field strength of the activator. The typical example is Cr^{3+}-doped $ZnWO_4$ with emission phonon side band expanding to 1 μm [24], which has much smaller cell constant compared to Cr^{3+}:$La_3Ga_5SiO_{14}$ with tunable

Table 7.4 Energy separations and spectral parameters of Cr^{3+}-doped tunable laser crystals

Host	$\Delta E_{TE}(cm^{-1})$	$\lambda_{peak}(nm)$	$\tau_r(\mu s)$	$\sigma_{se}(10^{-20}cm^2)$
$BeAl_2O_4$	800	752	240	0.6
$Y_3Ga_5O_{12}$	600	740	241	0.36
$Be_2Al_2(SiO_3)_6$	400	768	60	1.9
$Gd_3Ga_5O_{12}$	300	769	160	0.6
$Y_3Sc_2Ga_3O_{12}$	250	750	139	0.6
$Gd_3Sc_2Ga_3O_{12}$	0	785	115	0.9
$La_3Lu_2Ga_3O_{12}$	<0	830	68	1.6
$LiCaAlF_6$		780	170	1.3(π)
$LiSrAlF_6$		825	67	4.8(π)
$KZnF_3$	−500	820	176	1.3
MgF_2	−2500	~1110		
$ZnWO_4$	−3000	1035	0.5−5.4	43.0

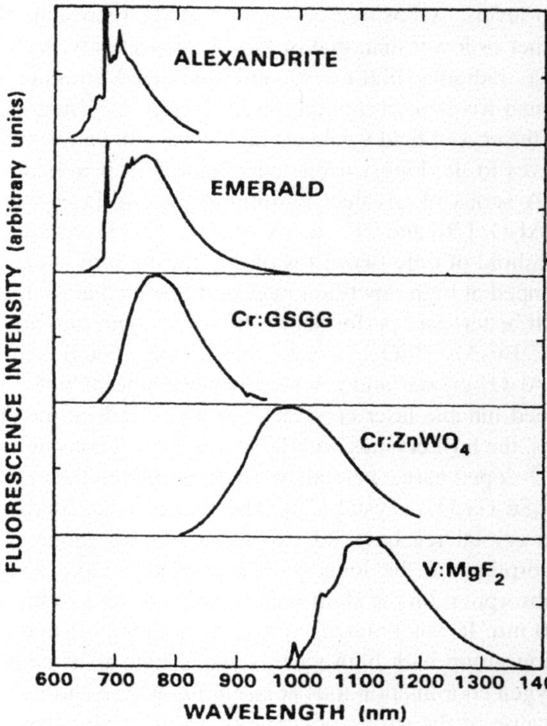

Fig. 7.9 Fluorescence spectra of five laser crystals having d^3 electronic configuration [19]

range of 862–1107 nm [25], the longest wavelength obtained for Cr^{3+}-doped laser crystals.

Cr^{3+}:$LiSrAlF_6$ and Cr^{3+}:$LiCaAlF_6$ are the most important tunable laser crystals in the practical applications. Both of them have large fluorescence linewidths and much longer fluorescence lifetimes compared with titanium sapphire, and thus are suitable for using in diode-pumped tunable femtosecond laser system [26–31].

The energy level character of trivalent chromium (or divalent vanadium) doped tunable laser crystals is that the state of $^4T_2$ is closed to the state of ^2E. The parameter ΔE_{TE} denoting this separation are given in Table 7.4. Owing to the fact that this separation is small, the mixing of the states with different spin multiplicities by spin–orbit interaction relieves partly the forbidden of the $^2E \rightarrow {}^4A_2$ transition. Hence the fluorescence lifetime of the $^4T_2$ state is actually a composite effect of the emission for the transition from the $^4T_2$ state and that from the ^2E state. The effective fluorescence lifetime τ_{eff} satisfies the following formula

$$\frac{1}{\tau_{eff}} = \frac{\frac{1}{\tau_E} + \frac{1}{\tau_T}\exp(-\Delta E/k_B T)}{1 + \exp(-\Delta E/k_B T)} = \frac{1}{\tau_T}\left[\frac{r + \exp(-\Delta E/k_B T)}{1 + \exp(-\Delta E/k_B T)}\right]$$

where τ_E and τ_T refer to $\tau(^2E)$ and $\tau(^4T_2)$, respectively, and $r = \tau(^2E)/\tau(^4T_2)$ is a function of ΔE_{TE}, because the fluorescence lifetime of ^2E depends on the spin–orbit coupling; the smaller the ΔE_{TE}, the larger is the component of the wave function of $^4T_2$ level mixing into that of ^2E level and so the $^2E \rightarrow ^4T_2$ transition has higher transition probability. Anyway, it is a parameter directly related to tunable laser properties and necessary to be further investigated.

Starting from the calculation of transition rate by perturbation method, it is important to calculate the spin–orbit matrix element $\langle \Phi_n(^2\mathrm{E}) | H_{\mathrm{so}} | \Phi_n(^4\mathrm{T}_2) \rangle$. Taking into account the lattice relaxation, this matrix element takes the following form

$$\begin{aligned} \langle \Phi_n(^2E) | H_{\mathrm{so}} | \Phi_m(^4T_2) \rangle &= \langle \varphi_n(^2E) | H_{\mathrm{so}} | \varphi_m(^4T_2) \rangle \langle \chi_n(^2E) | \chi_m(^4T_2) \rangle \\ &\equiv V_{so} \langle \chi_n(^2E) | \chi_m(^4T_2) \rangle \end{aligned}$$

where φ is the wave function of the electron and χ is that of the lattice vibration. The calculation by Struve and Huber in 1985 [32] did not consider overlap integral between lattice wave functions $\chi_n(^2E)$ and $\chi_m(^4T_2)$, so the following result was obtained

$$r = 2\left(1 - \frac{\Delta E_{\mathrm{TE}}}{\sqrt{4|V_{\mathrm{so}}|^2 + \Delta E_{\mathrm{TE}}^2}}\right)^{-1}$$

It differs significantly from the experimental results. After considering overlap integral between lattice wave functions, the calculated result will be determined by Huang–Rhys S factor and phonon energy $\hbar\omega$. It means that the ratio of fluorescence lifetime r will be determined by the matrix element of spin–orbit coupling V_{so}, phonon parameters S and $\hbar\omega$. The crystal field wave functions and the matrix element of spin–orbit coupling V_{so} depend on the site-symmetry of the chromium ions. Therefore, in the lower symmetry situation, it is unreasonable to apply the value of cubic symmetry. The group chain scheme has been used by the authors to calculate this matrix element according to the site-symmetry of the active ions [33] and at the same time to adopt the calculation method for the values of S factor and phonon energy $\hbar\omega$ by a statistical multi-phonon model proposed by K. Huang [34]. Using these results, the ratio of fluorescence lifetime r could be calculated by the DHGIM model [35], modified DHGIM model [36], and frozen lattice model [36], according to the following formula

$$r = \left[|V_{\mathrm{so}}|^2 \times \sum_m \frac{e^{-S} \times S^m}{(E_T^0 + m\hbar\omega - E_E^0) \times m!}\right]^{-1} \left\{1 + \left[|V_{\mathrm{so}}|^2 \times \sum_m \frac{e^{-S} \times S^m}{(E_T^0 + m\hbar\omega - E_E^0) \times m!}\right]\right\}$$

where m refers to phonon number. It is enough to calculate it from $m = 0$ to $m = 10$; the contribution of larger m numbers can be neglected. The obtained results of the matrix element of spin–orbit coupling V_{so}, Huang–Rhys S factor, phonon energy $\hbar\omega$, and the ratio of fluorescence lifetime r are listed in Tables 7.5 and 7.6.

It can be seen that the calculated results are in good agreement with the experimental data, especially those using the modified DHGIM model. In order to obtain better agreement with the experimental data, Wojtowicz [36] introduced a special model in which the lattice is frozen in a position $R_0 = (R_1 + R_2)/2$, where R_1 is the system position when it is in the state of ^2E and R_2 is the system position when it is in the state of $^4T_2$. This is obviously an artificial hypothesis that does not conform to physical reality. The above results show that by using accurate point group to calculate the spin–orbit coupling matrix element of V_{so}, and adopting the correct phonon parameters, it is possible to obtain the results in accordance with the spectroscopic experiments and absolutely unnecessary to introduce an artificial model.

Divalent vanadium has the same electron configuration $3d^3$ as that of the trivalent chromium but with much weaker crystal field. As a result, the working wavelength of tunable laser crystal activated by V^2 is much longer. Typical examples are V^{2+}:MgF_2 [37] and V^{2+}:$CsCaF_3$ [38], which can emit lasers in wavelength ranges of 1.24–1.33 μm and 1.07–1.16 μm, respectively, but both should be operated at 80 K. Owing to their very long fluorescence lifetimes, they are suitable to be pumped by krypton arc lamps and operated continuously. The laser efficiency of this kind of systems is low because of the excited state absorption, especially for the crystal $CsCaF_3$ doped with V^{2+}.

Finally, we would like to mention briefly about Ni^{2+} and Co^{2+}-doped tunable laser crystals with $3d^8$ and $3d^7$ electronic configurations, respectively [9, 39–43]. As has been pointed out previously, the $3d^8$ electronic system has a T-S diagram similar to that of the $3d^2$ system but the crystal field strength parameter has an

Table 7.5 Huang–Rhys *S* factor and phonon energy for four garnet crystals

Crystal	YGG	GGG	YSGG	GSGG
S	5.45	5.04	4.40	4.49
$\hbar\omega$	238.7	252.5	317.8	293.9

Table 7.6 Comparing between experimental and calculated results of lifetime ratio for four garnet crystals

Crystal	ΔE_{TE} (cm^{-1})	Experimental	Calculated $\tau(^2E)/\tau(^4T_2)$		
		$\tau(^2E)/\tau(^4T_2)$	DHGIM	Modified DHGIM	Frozen lattice model
GSGG	50	2.4	3.5	2.6	1.8
YSGG	350	13.7	15.7	15.4	8.7
GGG	380	16.0	16.4	16.2	10.8
YGG	650	26.1	26.5	26.2	21.0

opposite sign. Therefore, the laser transition channel of Ni^{2+}:MgF_2 crystal with crystal field parameter $Dq/B < 1$ ($B = 1030\ cm^{-1}$) is $^3T_2 \rightarrow {}^3A_2$. Ni^{2+}:MgF_2 is a tunable laser crystal studied earliest, which has tunable laser operation at 77 K. This crystal can be pumped by flash lamp or tungsten filament lamp. Certainly, it is better pumped by 1.32 μm laser of Nd^{3+}:YAG crystal. It can be tuned in a range of 1600–1700 nm with central wavelength of 1668 nm. Other Ni^{2+}-doped tunable laser crystal are: Ni^{2+}:$KMgF_3$ (pumped by xenon flash lamp) and Ni^{2+}:MgO (pumped by 1.06 μm laser of Nd^{3+}:YAG crystal) and Ni^{2+}:GGG (pumped by 1.32 μm laser of Nd^{3+}:YAG crystal).

T-S diagram of Co^{2+} ions is shown in Fig. 7.10. The energy levels of Co^{2+} ions are located in the left part of the vertical dotted line corresponding to $D_q/B \cong 1$ in this diagram, because it has a weak crystal field. The ground state is $^4T_1$ split by crystal field from spectral term of ^4F and its excited states are $^4T_2$,$^4T_1$, and ^2E coming from spectral terms ^4F, ^4P, and ^2G, respectively. Co^{2+}:MgF_2 can be tunable lased at either 77 K (fluorescence lifetime of the upper laser level $^4T_2$ is 1.3 ms) or 299 K (fluorescence lifetime of the $^4T_2$ is 36 μs) with tunable wavelength range of 1750–2500 or 1510–2280 nm, respectively [17, 40, 41]. This crystal has good mechanical and thermal properties, pumped by 1.32 μm laser of Nd^{3+}:YAG crystal. The optical-optical efficiency of 2050 nm laser at a repetition rate of 2 Hz reaches 33% at temperature of 282 K [44]. A tunable laser operation of Co^{2+}:$KZnF_3$ in the range of 1650–2070 nm has been realized by argon ion laser pumping at low temperature of 80 K [45, 46].

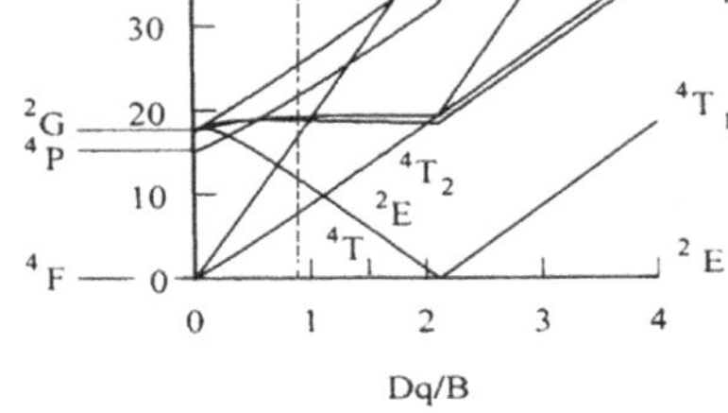

Fig. 7.10 Tanabe-Sugano diagram for d^7 electronic configuration in octahedron position

7.4 Relative Intensity Analysis of *R* Line in Ruby Polarized Absorption Spectrum

Figure 7.11 is a figure of *R* absorption line of Cr^{3+}:Al_2O_3 crystal corresponding to the $^4A_2 \rightarrow {}^2E$ transition. It can be seen that the σ spectrum is obviously stronger than the π spectrum.

Theoretical explanation should start with the structure of ruby crystal. Figure 7.12 is the distribution of Cr^{3+} ions and their nearest neighbor O^{2-} ions in the crystal. The left picture shows that two of the Cr^{3+} ions situate between two layers of O^{2-} ions but there is a certain deviation from the midpoint of the two O^{2-} ion planes. The right picture is the distribution of O^{2-} and Cr^{3+} ions looking down from the top along the C_3 axis of the crystal. It is shown that the vertices of the two triangles above and below the Cr^{3+} ion are the six nearest neighbor coordination O^{2-} anions. The sizes of these two triangles have a little difference and a 4.3° deflection angle, making the position symmetry of Cr^{3+} ion decreasing from C_{3v} to C_3.
The deviation of Cr^{3+} ions from the midpoint of the two O^{2-} planes produces $T_{1u}a_0$ odd crystal field potential, while the 4.3° deviation angle of two O^{2-} triangles results in $T_{2u}x_0$ odd crystal field potential. Owing to the fact that the 4.3° deviation angle is very small so that the strength of $T_{2u}x_0$ odd crystal field potential is much weaker than that of the $T_{1u}a_0$ crystal field potential.

Energy levels ^2E and $^4A_2$ are different in spin multiplicity and so the electric-dipole transition between these two levels is spin forbidden in the first-order

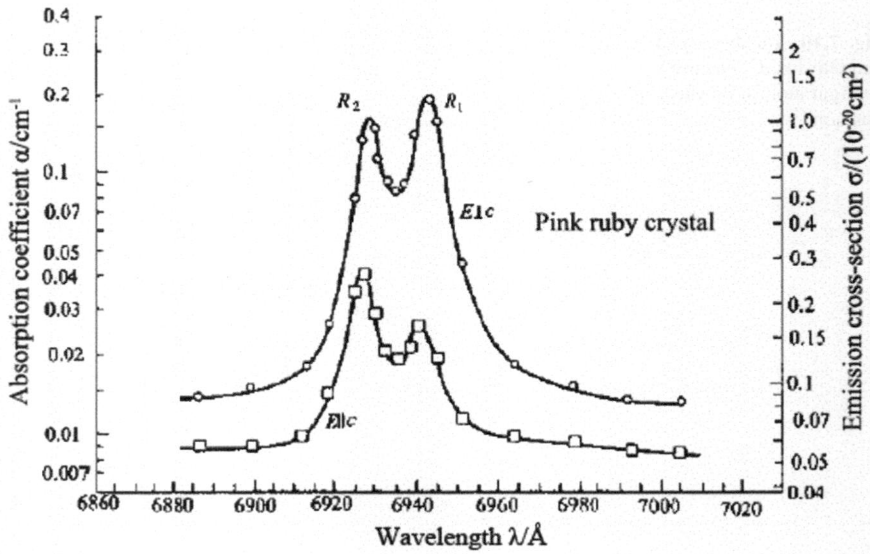

Fig. 7.11 Polarized absorption spectra of the $^4A_2 \rightarrow {}^2E$ transition for Cr^{3+}:Al_2O_3 crystal [47]

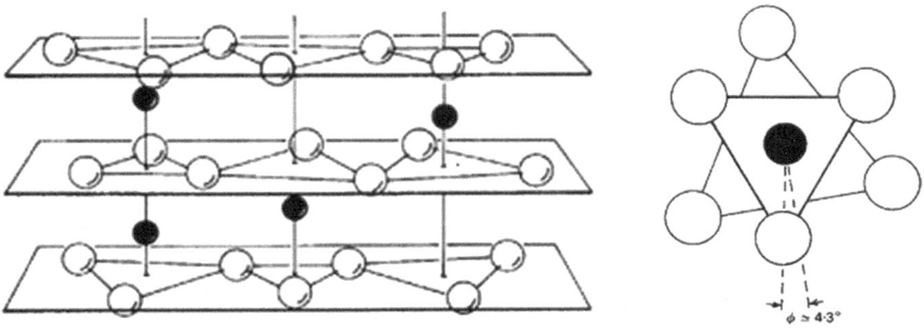

Fig. 7.12 The distribution of Cr^{3+} ions (solid spheres) and O^{2-} ions (hollow spheres) in ruby crystal

approximation. However, owing to the spin–orbit coupling, some components of $^4T_2$ and $^4T_1$ wave functions will be mixed into ^2E wave function and result in the ^2E to $^4A_2$ electric-dipole transition. To investigate the polarization character of this transition, the following perturbation matrix elements should be studied

$$\frac{\langle ^2E|H_{SO}|^4T_1\rangle\langle ^4T_1|\boldsymbol{\mu}|^4A_2\rangle}{E(^4T_1)-E(^2E)} \text{ and } \frac{\langle ^2E|H_{SO}|^4T_2\rangle\langle ^4T_2|\boldsymbol{\mu}|^4A_2\rangle}{E(^4T_2)-E(^2E)} \tag{7.21}$$

where H_{SO} is the spin–orbit coupling Hamiltonian and $\boldsymbol{\mu}$ denotes the operator of electric-dipole moment. It can be seen from the above matrix elements that the transition probability is inversely proportional to the square of energy separation between ^2E and $^4T_1$ or between ^2E and $^4T_2$. Owing to the fact that the separation between ^2E and $^4T_1$ is four times that between ^2E and $^4T_2$, besides the matrix element of $\langle ^4T_1|\boldsymbol{\mu}|^4A_2\rangle$ is equal to zero; therefore, in the comparison of R line absorption intensity in different polarization direction, the contribution of the first matrix element in (7.21) can be neglected. To compare the π and σ polarized intensities of the $^2E{\rightarrow}^4A_2$ transition induced by the ^2E and $^4T_2$ spin–orbit coupling, it is enough to compare the squares of matrix elements $|\langle ^4T_2|\mu_z|^4A_2\rangle|^2$ and $|\langle ^4T_2|\mu_{x,y}|^4A_2\rangle|^2$.

Energy levels $^4T_2$ and $^4A_2$ belong to even parity states. The matrix element of electric-dipole moment $\boldsymbol{\mu}$ between them is equal to zero unless their wave functions are mixed by the wave functions of the electronic configuration $3d^24p$ introduced by odd crystal field potential V_u. In this case, the electric-dipole matrix element between $^4A_2$ and $^4T_2$ is

$$M \equiv \langle ^4T_2|\boldsymbol{\mu}|^4A_2\rangle = \sum_k \frac{\langle ^4T_2|\boldsymbol{\mu}|k\rangle\langle k|V_u|^4A_2\rangle}{E_k - E(^4T_2)} + \sum_k \frac{\langle ^4A_2|\boldsymbol{\mu}|k\rangle\langle k|V_u|^4T_2\rangle}{E_k - E(^4A_2)} \tag{7.22}$$

k denotes the energy levels of odd parity configuration $3d^2 4p$. The energy separation between these states and ${}^4A_2$ or ${}^4T_1$ is much larger than that between ${}^4A_2$ and ${}^4T_2$. However, the energy difference in the denominator of (7.22) can be replaced by the average separation ΔE between energy levels of configuration $3d^2 4p$ and ${}^4A_2$ or ${}^4T_2$. By using $\sum_k |k\rangle\langle k| = 1$, it can be obtained

$$M^2 = 4\left|\left\langle {}^4A_2\right|\boldsymbol{\mu}\otimes V_u\left|{}^4T_2\right\rangle\right|^2/(\Delta E)^2 \quad (7.23)$$

Obviously, the intensity of R line of the transition ${}^2E\rightarrow{}^4A_2$ is proportional to M^2.

The contribution of odd crystal field potential $T_{1u}a_0$ to the square of π polarized matrix element $\left|\langle A_2|\mu_z\otimes V_u|T_2\rangle\right|^2$ can be calculated as follows:

Because π polarization is generated by the z component of the electric-dipole moment, while the function z transform as $T_{1u}a_0$ under cubic group operation, by using following trigonal basis Clebsch-Gordan coefficient (C-G coefficient) of cubic group (it can be found in Appendix F)

$$\langle T_1a_0T_1a_0|A_1e_1\rangle = -1/\sqrt{3}, \langle T_1a_0T_1a_0|T_2x_0\rangle = -\sqrt{2}/\sqrt{3}$$

The expression of the product of electric-dipole operator and the odd crystal field potential operator can be obtained as follows

$$\begin{aligned}\mu_z\otimes V(T_{1u}a_0) &\Rightarrow z\otimes V(T_{1u}a_0) \Rightarrow (T_{1u}a_0)\otimes T_{1u}a_0\\ &= \left(-1/\sqrt{3}\right)A_{1g}e_1 \oplus \left(-\sqrt{2}/\sqrt{3}\right)T_{2g}x_0\end{aligned}$$

According to Wigner–Eckart theorem for point group

$$\langle \Gamma_1\gamma_1\left|O_\gamma^\Gamma\right|\Gamma_2\gamma_2\rangle = (\Gamma)^{-1/2}\langle\Gamma_1\gamma_1|\Gamma_2\gamma_2\Gamma\gamma\rangle\langle\Gamma_1\left\|O_\gamma^\Gamma\right\|\Gamma_2\rangle$$

where (Γ) is the dimension of irreducible representation Γ. Substituting C-G coefficients $\langle A_2e_2|T_2x_0T_1a_0\rangle = 0$ and $\langle A_2e_2|T_2x_0A_1e_1\rangle = 0$ into (7.23), we can obtain

$$\left(M_\pi^{T_{1u}}\left({}^4A_2\rightarrow{}^4T_2\right)\right)^2 = 0 \quad (7.24)$$

It is shown that the intensity of π polarized absorption line is zero.

The calculation of the matrix element for the σ polarized transition involves the electronic radius vector in x direction (it can be shown that if the electronic diameter vector is in y direction, its contribution to the matrix element is zero), its irreducible component can be expressed as

$$(T_{1u}a_- - T_{1u}a_+)/\sqrt{2}$$

Using the related C-G coefficients, the product of the electric-dipole operator and the odd crystal field potential operator can be expressed as

$$T_{1u}a_0 \otimes (T_{1u}a_- - T_{1u}a_+)/\sqrt{2} = \left(1/\sqrt{6}\right)E_g u_+ \oplus \left(-1/\sqrt{6}\right)E_g u_- \oplus (-i/2) \times T_{1g}a_+ \oplus (-i/2)T_{1g}a_- \oplus \left(-1/\sqrt{12}\right)T_{2g}x_+ \oplus \left(1/\sqrt{12}\right)T_{2g}x_-$$

C-G coefficient $\langle A_2e_2|T_2x_{\mp}T_1a_{\pm}\rangle = 1/\sqrt{3}$ can be found in Appendix F, then Wigner–Eckart theorem shows

$$\left(M_\sigma^{T_{1u}}(^4A_2 \to {}^4T_2)\right)^2 = (1/3)\left|\langle{}^4A_2\|\mu V(T_{1u})\|{}^4T_2\rangle\right|^2/(\Delta E)^2 \tag{7.25}$$

then to see the contributions of odd crystal field potential $T_{2u}x_0$ to π and σ polarized transition matrix elements. For the π polarization, the product of electric-dipole operator and odd crystal field potential operator can be expressed as

$$T_{1u}a_0 \otimes T_{2u}x_0 = -1/\sqrt{3}A_{2g}e_2 \oplus \sqrt{2}/\sqrt{3}T_{1g}a_0$$

By using Wigner–Eckart theorem and related C-G coefficient $\langle A_2e_2|T_2x_0T_1a_0\rangle = 1/\sqrt{3}$, the square of transition matrix element is

$$\left(M_\pi^{T_{2u}}(^4A_2 \to {}^4T_2)\right)^2 = (2/9)\left|\langle{}^4A_2\|\mu V(T_{2u})\|{}^4T_2\rangle\right|^2/(\Delta E)^2 \tag{7.26}$$

To calculate the contribution of odd crystal field potential $T_{2u}x_0$ to σ polarized transition matrix element, the y direction electronic radius vector should be concerned. Its irreducible component can be expressed as

$$i(T_{1u}a_- + T_{1u}a_+)/\sqrt{2}$$

Then the product of electric-dipole operator and odd crystal field potential operator can be expressed using the C-G coefficient

$$T_{2u}x_0 \otimes i(T_{1u}a_+ + T_{1u}a_-)/\sqrt{2} = \left(-1/\sqrt{6}\right)E_g u_+ \oplus \left(1/\sqrt{6}\right)E_g u_- \oplus \left(-i/\sqrt{12}\right) T_{1g}a_+ \oplus \left(-i/\sqrt{12}\right)T_{1g}a_- \oplus \left(-1/\sqrt{4}\right)T_{2g}x_+ \oplus \left(1/\sqrt{4}\right)T_{2g}x_-$$

Similarly, by using Wigner–Eckart theorem and related C-G coefficient $\langle A_2e_2|T_2x_{\mp}T_1a_{\pm}\rangle = 1/\sqrt{3}$, one obtains

$$\left(M_{\sigma}^{T_{2u}}\left({}^{4}A_{2} \rightarrow {}^{4}T_{2}\right)\right)^{2} = (1/9)\left|\left\langle {}^{4}A_{2}\|\mu V(T_{2u})\|{}^{4}T_{2}\right\rangle\right|^{2}/(\Delta E)^{2} \tag{7.27}$$

Summarize the results obtained above

$$\left[M_{\sigma}\left({}^{4}A_{2} \rightarrow {}^{4}T_{2}\right)\right]^{2} = \left[1/3\left|\left\langle {}^{4}A_{2}\|\mu V(T_{1u})\|{}^{4}T_{2}\right\rangle\right|^{2} + 1/9\left|\left\langle {}^{4}A_{2}\|\mu V(T_{2u})\|{}^{4}T_{2}\right\rangle\right|^{2}\right]/(\Delta E)^{2} \tag{7.28}$$

$$\left(M_{\pi}\left({}^{4}A_{2} \rightarrow {}^{4}T_{2}\right)\right)^{2} = (2/9)\left|\left\langle {}^{4}A_{2}\|\mu V(T_{2u})\|{}^{4}T_{2}\right\rangle\right|^{2}/(\Delta E)^{2} \tag{7.29}$$

Consider that $V(\mathrm{T}_{1u}a_0) > V(\mathrm{T}_{2u}x_0)$, the following conclusion is obvious

$$\left(M_{\sigma}\left({}^{4}A_{2} \rightarrow {}^{4}T_{2}\right)\right)^{2} > \left(M_{\pi}\left({}^{4}A_{2} \rightarrow {}^{4}T_{2}\right)\right)^{2} \tag{7.30}$$

Therefore, the experimental fact that the intensity of σ polarized absorption *R* line is stronger than that of the π polarized absorption *R* line is completely demonstrated. In fact, this is also a good explanation of the experimental fact that the intensity of σ polarized ${}^{4}A_{2}\rightarrow{}^{4}T_{2}$ absorption line is stronger than that of the π polarized ${}^{4}A_{2}\rightarrow{}^{4}T_{2}$ absorption line. Clearly, this phenomenon comes from the fact that the main odd crystal field potential of Cr^{3+} ion in ruby crystal is $V(\mathrm{T}_{1u}a_0)$. If deviation angle of two ligand O^{2-} triangles above and below the Cr^{3+} ion increases, its π polarized absorption intensity will be stronger.

7.5 Estimation of Trivalent Chromium Ion Spectral Parameters in Solid-State Laser Materials

In the quality test and evaluation of laser crystals, it is often necessary to estimate the spectral parameters by the experimental spectra measurement. In these estimations, it is often possible to use relationship between some parameters obtained by simple spectroscopic measurements to calculate parameters which are originally difficult to determine or can only be determined by complicated measurements and calculations. This section will introduce two direct and simple methods for the calculation of the spectral parameters of the trivalent chromium ion doped in crystals which have been proposed by the authors [34, 48].

The trivalent chromium ion-doped crystals with weak crystal field are used as tunable laser crystals; their laser properties to a large extent depend on the crystal field parameter $10Dq$, effective phonon energy $\hbar\omega$, and Huang–Rhys S factor. For this ion, the energy $E_a({}^{4}A_{2}\rightarrow{}^{4}T_{2})$ corresponding to the ${}^{4}A_{2}\rightarrow{}^{4}T_{2}$ absorption peak is $10Dq$ (see (7.20)), therefore, crystal field strength Dq can be obtained directly by spectral measurement. On the other hand, the calculation of effective phonon energy $\hbar\omega$ and Huang–Rhys S factor is much more troublesome. It can be obtained either by using the measurement of line shape curve to fit with the theoretical line shape function or by comparing the temperature dependence of the linewidth obtained in experiment with the expression of (6.55). Anyway, the results obtained are only some kinds of approximate phenomenological parameters, because the formulas used are introduced by phenomenological model. Actually, by almost the same phenomenological model and the same order of approximation, the above results can be obtained in a more direct and simple way. This is the main idea of the method introduced in this section.

Under point-charge approximation, crystal field parameter $10Dq$ can be expressed as [3]

$$10Dq = \frac{10}{6}\left(Ze^2\langle r^4\rangle_{3d}/R^5\right) \tag{7.31}$$

Owing to the fact that the amplitude of lattice vibration δR satisfies the relation of $\delta R/R \approx 10^{-2}$, it can be assumed that Z and $< r^4 >_{3d}$ have not any substantial variation in the process of lattice vibration. It means that the crystal field energy E depends on the fifth power of R. Consider all the interactions including point charge Coulomb interaction, the general dependent relation of the crystal field energy will take the following form:

$$E = AR^{-n} \tag{7.32}$$

where A is a constant, n assumes the value of 5 in point-charge approximation and a value between 4.5 and 5 in general. By (7.32) it is obvious

$$\frac{dE}{dR} = -nE/R \tag{7.33}$$

The energy variation δE in the process of electronic transition also satisfies the same relation

$$\frac{d\delta E}{dR} = -n\delta E/R \tag{7.34}$$

On the other hand, by assuming a single-mode model as that adopted by the configuration coordinate model, we discuss in fact the interaction of the electrons in the central active ion with a "breathing mode" vibration of six ligand ions, which has A_{1g} symmetry. This local structure consists of six linear oscillators; each has an energy change of $\delta E/6$ during the electronic transition. If the variation of configuration coordinates of any of these oscillators during the electronic transition is denoted by Δ_{ji}, then the interaction force generated by the transition for each oscillator can be expressed as

$$F = -\frac{1}{6}\frac{d\delta E}{dR} = M\bar{\omega}^2\Delta_{ji} \tag{7.35}$$

where M refers to as the mass of the ligand ion. According to the definition of Huang–Rhys factor and the fact that there are six oscillators in the model discussed, noting that the mass of oscillator has been dealt with as a unit in the formula of S in Chap. 6, it is obvious

$$S = 3M\bar{\omega}\Delta_{ji}^2/\hbar \tag{7.36}$$

Therefore, the following relation will be obtained by (7.34), (7.35), and (7.36)

$$S = \frac{n^2(\delta E)^2}{12\hbar MR^2\bar{\omega}^3} \tag{7.37a}$$

or

$$S\hbar\bar{\omega} = \frac{n^2\hbar^2}{12MR^2}\left(\frac{\delta E}{\hbar\bar{\omega}}\right)^2 \tag{7.37b}$$

Equation (6.54) shows that the Stocks shift corresponding to the difference of the peak of the absorption spectrum E_a and that of the emission spectrum E_e is equal to the double of $S\hbar\bar{\omega}$, and δE in the above equation is the peak of the absorption spectrum E_a, then

$$\hbar\bar{\omega} = E_a\left(\frac{n}{R}\right)\left[\frac{\hbar^2}{6M(E_a - E_e)}\right]^{1/2} \tag{7.38}$$

Associating with (6.55), the full-width at half-maximum of the emission line can be expressed as

$$W = [4(\ln 2)(E_a - E_e)\hbar\bar{\omega}\coth(\hbar\bar{\omega}/2k_B T)]^{1/2} \tag{7.39}$$

Equations (7.32)–(7.39) are really introduced by simple single-mode approximation, which is the same as that has been adopted in the performance investigation of tunable laser crystals. Therefore, one can use these formulas to describe performances of tunable laser crystals without considering whether they have real microcosmic meaning, but remembering that they are only phenomenological parameters. Actually, electron–phonon interaction in laser materials is indeed multi-mode and multi-frequency, and the phonon state density of different mode is always different. They are still phenomenological and even two or three modes are used to describe the spectral properties.

Ducilos et al. [49] have shown that for $Cr^{3+}:Al_2O_3$ at a pressure as high as 35 GPa, n in (7.32) increases with R. Therefore they have assumed that n/R is a constant within a certain range of pressure. For oxides and fluorides, it can be assumed that $n/R = 2.25 \times 10^8\ \text{cm}^{-1}$. The mass of oxygen ion is $M_O = 2.7 \times 10^{-23}$ g and that of fluorine ion is $M_F = 3.2 \times 10^{-23}$ g, then the phonon energy of effective single mode for the transition between electronic levels $^4A_2$ and $^4T_2$ can be expressed as, for oxides

$$\hbar\bar{\omega} \approx 2.25E_a[0.3456/(E_a - E_e)]^{1/2} \tag{7.40}$$

for fluorides

$$\hbar\bar{\omega} \approx 2.25E_a[0.2910/(E_a - E_e)]^{1/2} \tag{7.41}$$

By means of (7.39), (7.40), and (7.41), the effective phonon energy and width of the spectral line in single-mode approximation can be obtained by the data of the peak positions of the absorption and emission spectra. By using (6.54), (7.40), and (7.41), the effective phonon energy and Huang–Rhys factor can be calculated. Compared with the results obtained by other methods, it is evident that the approach proposed is suitable for the investigation of the performances of laser crystal. This comparison is shown in three tables followed.

The calculated results and the comparison with the experimental data of the linewidth are given in Tables 7.7 and 7.8. The relative errors for oxide and fluoride crystals are all less than 10% in the 23 examples given, except for those cases in which disorders or second centers appear in the crystals. Comparing the effective phonon energy and Huang–Rhys factor calculated by the direct method proposed with those obtained by other methods published in the literature, it can be seen that the direct method proposed also lead to very good results, which are given in Table 7.9.

It is simple to measure the width of the spectral line. By using the direct method proposed to calculate linewidth and comparing with the experimental result, it is possible to find if there is disorder structure or multi-center in crystals. In this case, the calculated linewidth will be much smaller than the measured value obtained by the experiment. Certainly, to obtain a correct experimental data of linewidth, the response curve of the spectrometer should certainly be corrected.

Table 7.7 Comparison of calculated results with experimental data of emission linewidth for the transition $^4T_2 \rightarrow {}^4A_2$ of Cr^{3+} ions in fluoride crystals

Crystal	T (K)	E_a (cm^{-1})	E_e (cm^{-1})	W Cal. (cm^{-1})	W Exp. (cm^{-1})	Ref.	Error (%)
$KZnF_3$	80	14858	12934	1482	1521	[50]	2.6
ScF_3	14	14280	12028	1510	1495	[51]	1.0
$LiSrAlF_6$	20	15676	12821	1679	1690	[52]	1.2
$LiSrGaF_6$	295	15873	12054	2237	2438	[53]	8.2
$Na_3Ga_2Li_3F_{12}$	18	16069	13313	1685	1740	[54]	3.2

Table 7.8 Comparison of calculated results with experimental data of emission linewidth for the transition $^4T_2 \rightarrow ^4A_2$ of Cr^{3+} ions in oxide crystals

Crystal	T (K)	E_a (cm^{-1})	E_e (cm^{-1})	W Cal. (cm^{-1})	W ExP. (cm^{-1})	Ref.	Error (%)
$Al_2(WO_4)_2$	300	15229	12157	2073	2100	[24, 55]	1.3
$ZnWO_4$	300	13766	10278	2117	2105	[36]	0.6
$ZnWO_4$	9	13500	10145	1693	1588	[56]	6.6
$YAl_3(BO_3)_4$	300	16949	13513	2215	2127	[57]	4.1
$La_3Ga_5SiO_4$	300	15748	σ:11395	2406	2127	[26]	1.3
	300	15748	π:11549	2370	2525	[26]	6.1
YGG	300	16293	13691	1971	1900	[32]	1.5
GGG	300	15963	13617	1886	1900	[32]	0.7
YSGG	300	16128	13332	2022	2100	[32]	3.7
LGG	300	14794	11758	2051	2100	[32	2.3
GSGG	300	15631	12993	1963	2000	[32]	1.9
	177	15430	13506	1606	1558	[58, 59]	3.1
	89	15430	13624	1538	1429	[58, 59]	7.6
GAGG	300	15600	13444	1817	1732	[58, 59]	4.9
	100	15600	13717	1576	1514	[58, 59]	4.1
$Ca_3Ga_2Ge_4O_{14}$	300	16000	10989	2560	3645	[50]	29.7[a]
$Sr_3Ga_2Ge_4O_{14}$	300	16260	11905	2420	4085	[50]	40.8[a]
Sc_2O_3	14	15200	11875	1792	2329	[51]	23.0[b]

[a]There is disorder for the crystal field of chromium ions, [b]there is two chromium centers

Table 7.9 Comparison of the results of the method proposed with those obtained by other methods for effective phonon energy and Huang–Rhys factor of the transition $^4T_2 \rightarrow ^4A_2$ of Cr^{3+} ions in fluoride crystals

Crystal	T	E_a	E_e	Results of this method		Results by other method		Ref.
	(K)	(cm^{-1})	(cm^{-1})	S	$\hbar\bar{\omega}$ (cm^{-1})	S	$\hbar\bar{\omega}$ (cm^{-1})	
K_2NaScF_6	295	15600	13100	3.30	379	3.95	380	[60]
K_2NaGaF_6	295	16000	13600	3.03	396	3.98	378	[60]
$Na_3Ga_2Li_3F_{12}$	18	16069	13313	3.7	372	3.5	400	[54]
ScF_3	14	14280	12028	3.0	367	3.0	380	[51]

There is another simple method to calculate effective phonon energy and Huang–Rhys factor if one has correct linewidth data. In the following, the calculation method without simplified supposition of (7.35) will be introduced.

The detailed description of this method can be referred to [34]. The main basis of this method is (6.54) and (6.55), but it technically introduces a ratio γ

$$\gamma = \frac{W^2}{\Delta E_s/2} \tag{7.42}$$

By means of (6.54) and (6.55), we have

$$\gamma = 8(\ln 2)\hbar\bar{\omega}\coth\left(\frac{\hbar\bar{\omega}}{2k_B T}\right) \tag{7.43}$$

This method requires the experimental data of the peak positions of the absorption and emission spectra as well as the full-width at half-maximum of the emission line W. From the experimental data mentioned, γ can be calculated by (7.42). As shown in (7.43), the phonon energy $\hbar\bar{\omega}$ has a one-to-one corresponding relation between γ and phonon energy $\hbar\bar{\omega}$. It is easy to make a curve or to list a table by this equation and then from the γ value one can determine phonon energy $\hbar\bar{\omega}$, although it is difficult to express $\hbar\bar{\omega}$ as an analytic function of γ. Then S factor can be found by (6.54). It has indicated that $\hbar\bar{\omega}$ and S factor obtained by these two methods introduced are almost identical.

The calculation in this section shows that for the strong electron–phonon system, the single-mode approximation is suitable to describe shape of spectral line including phonon side band, although it is a rough model. In view of the fact that the phonons have different frequencies, Huang Kun has proposed a statistical multi-phonon model [61]. According to this model, the author designed a method for measuring the phonon energy and Huang–Rhys factor by using measured spectral data [34]. However, from the practical point of view, the single-mode approximation is enough for the study of laser crystal.

Of course, the single-phonon model has a lot of limitations. For example, when the spectral measurement can resolve zero-phonon line and a certain order of phonon side band, their intensity ratios cannot be always predicted by the single mode. When active ions situated in the positions with central symmetry, the electric-dipole transition within the same configuration is forbidden but the lattice vibration breaks this central symmetry and introduces forced electric-dipole transition. In this case, the single phonon line but not the zero-phonon line is induced by electric-dipole transition, and so the zero-phonon line is much weaker.

Further investigation shows that the intensity of electron–phonon coupling depends on the concentration of the active ions. Auzel [62] has reported the increase of Huang–Rhys S factor along with the concentration of the active ions. Physically, it can be interpreted as the range of electron interaction increases with the concentration of the active ion; it's like the wave function of electrons with a larger range. When the concentrations of Ni^{2+} are 2 and 10%, the S factor will be 0.5 and 2.25, respectively, for Ni^{2+}:MgF_2 crystal.

References

1. S. Kück, Appl. Phys. B **72**, 515 (2001)
2. J.S. Griffith, *The Theory of Transition-Metal Ions* (Cambridge University Press, Cambridge, 1961)
3. S. Sugano, Y. Tanabe, H. Kamimura, *Multiplets of Transition-Metal Ions in Crystals* (Academic Press, New York, 1970)
4. C.A. Morrison, *Crystal Fields for Transition-Metal Ions in Laser Host Materials* (Springer, Berlin, 1992)
5. B. Henderson, R.H. Bartram, *Crystal-Field Engineering of Solid- State Laser Materials* (Cambridge University Press, Cambridge, 2000)
6. M. Rotenberg, R. Bivins, N. Metropolis et al., *The 3-j and 6-j Symbols* (MIT Press, Massachusetts, 1959)
7. P.H. Butler, *Pont Group Symmetry Application: Method and Tables* (Plenum, New York, 1981)
8. E.D. Nelson, J.Y. Wong, A.L. Schawlow, Phys. Rev. **156**, 298 (1967)
9. P.F. Moulton, in *Laser Handbook*, vol. 5, ed. by M. Bass, M.L. Stitch (North-Holland Publishers, Amsterdam, 1985), pp. P203–P289
10. A. Suginoto, Y. Segawa, P.H Kim., et al., J. Opt. Soc. Am. B, **6**, 2334 (1989)
11. C.W. Nielson, G.F. Koster, *Spectroscopic Coefficients for the* p^n, d^n, f^n *Configurations* (MIT Press, Massachusetts, 1959)
12. R. Verdun, L.M Thomas, D.M. Andrauskas, et al., in ed. by M.L. Shand, H.P. Jenssen, *OAS Proceedings on Tunable Solid State Lasers*, vol. 5, (1989), p. 85
13. B.H.T. Chai, Y. Shimony, C. Deka, et al., in ed. by L.L. Chase, A.A. Pinto, *OAS Proceedings on Advanced Solid State Lasers*, vol. 13 (1992)
14. H. Rager, Phys. Chem. Miner. **1**, 371 (1977)
15. L.H. Bershov, J.M. Gaite, S.S. Hafner, et al., Phys. Chem. Miner. **9**, 95 (1983)
16. V. Petricevic, S.K. Gayen, R.R. Alfano, in ed. by M L. Shand, H.P. Jenssen, *OAS Proceedings on Tunable Solid State Lasers*, vol. 5, p. 77 (1989)
17. A. Sennaroglu, Prog. Quant. Elect. **26**, 287 (2002)
18. B. Henderson, G.F. Imbusch, *Optical Spectroscopy of Inorganic Solids* (Clarendon Press, Oxford, 1989)
19. P.F. Moulton, Proc. IEEE **80**, 348 (1992)
20. J.C. Walling, O.G. Peterson, H.P. Jenssen, et al., IEEE J. Quant. Elect. **QE-16**, 1302 (1980)
21. M.L. Shand, S.T. Lai, IEEE J. Quant. Elect. QE-20, 105 (1984)
22. P.T. Kenyon, L. Andrews, B. McCollum, et al., IEEE J. Quant. Elect. **QE-18**, 1189 (1982)
23. B. Struve, G.J. Huber, Appl. Phys. **57**, 45 (1985)
24. Z.D. Luo, J.M. Chen, et al., Chin. Phys. **6**, 991 (1986)
25. W. Kolbe, K. Petermann, G. Huber, IEEE J. Quant. Elect. **QE-21**, 1596 (1985)
26. S.T. Lai, B.H.T. Chai, M. Long, et al., IEEE J. Quant. Elect. **QE-24**, 1922 (1988)
27. S.A. Payne, L.L. Chase., H.W. Newkirk, et al., IEEE J. Quant. Elect. **QE-24**, 2243 (1988)
28. S.A Payne, L.L. Chase., L.K. Smith, et al., J. Appl. Phys. **66**, 1051 (1988)
29. R. Scheps, Opt. Mater. **1**, 1 (1992)
30. M.D. Perry, Laser Focus World **27**, 69 (1991)
31. P. Beaud, M. Richardson, E.J. Miesak et al., Opt. Lett. **18**, 1550 (1993)
32. B. Struve, G.J. Huber., Appl. Phys. B 36, 195 (1988)
33. Z.D. Luo, Y.D. Huang, J. Phys.: Condens. Matter **4**, 9751 (1992)
34. Z.D. Luo, J.M. Chen, Chin. Phys. **11**, 399 (1991)
35. C.J. Donnelly, S.M. Healy, T.J. Glynn et al., J. Lumin. **42**, 119 (1988)
36. A.J. Wojtowicz, M. Grinberg, A. Lempicki, J. Lumin., **50**, 231 (1991)
37. L.F. Johnson, H.J. Guggenheim, J. Appl. Phys. **38**, 4837 (1967)
38. W. Knierim, A. Honold, U. Brauch et al., J. Opt. Soc. Am. B **3**, 119 (1986)

39. L.F. Johnson, R.E. Dietz, H.J. Guggenheim, Phys. Rev. Lett. **11**, 318 (1963); Phys. Rev. **149**, 179 (1966)
40. P.F. Moulton, A. Mooradian, Appl. Phys. Lett. **35**, 838 (1979)
41. B.C. Johnson, P.F. Moulton, A. Mooradian, Opt. Lett. **9**, 116 (1984)
42. D. Welford, P.F. Moulton, Opt. Lett. **13**, 975 (1988)
43. V. Petricevic, S.K. Gayen, R.R. Alfano, Appl. Opt. **27**, 4162 (1988)
44. D.M. Rines, G.A. Rines, D. Welford, et al., in ed. by L.L. Chase, A.A. Pinto, *OAS Proceedings on Advanced Solid State Lasers*, vol. 13. p. 161 (1992)
45. W. Kunzel, U. Durr, Opt. Comm. **36**, 383 (1981)
46. W. Kunzel, U. Durr, Appl. Phys. B **28**, 233 (1982)
47. W. Koechner, *Solid-State Laser Engineering*, (Springer, Berlin, 1976)
48. Luo Zundu, Huang Yidong, J. Phys.: Condens. Matter **5**, 9411 (1993)
49. S.J. Ducilos, Y.K. Vohra, A.L. Ruoff, Phys. Rev. B **41**, 5372 (1990)
50. A.A. Kaminskii, A.V. Butashin, A.A. Demidovich et al., Phys. Status. Solidi. (a) **112**, 197 (1989)
51. G.J. Huber, S.A. Payne, L.L. Chase et al., J. Lumin. **39**, 259 (1988)
52. S.A. Payne, L.L. Chase, G.D. Whike, J. Lumin. **44**, 167 (1989)
53. L.K. Smith, S.A. Payne, W.L Kway, et al., IEEE J. Quant. Elect. **QE-28**, 2612 (1992)
54. J.A. Caird, S.A. Payne, P.R. Stave, et al., IEEE J. Quant. Elect. **QE-24**, 1077 (1988)
55. K. Petermann, G.J. Huber, J. Lumin. **31–32**, 71 (1984)
56. M. Yamaga, A. Marshall, K.P. O'Donnell et al., J. Lumin. **47**, 65 (1990)
57. J.M. Chen, Z.D. Luo, A.D. Jiang, Chinese J. Lumin. **11**, 104 (1990)
58. B. Henderson, A. Marshall, M. Yamaga, K.P. O'Donnell et al., J. Phys.: Solid State Phys. **21**, 6187 (1988)
59. K.P. O'Donnell, A. Marshall, M. Yamaga, et al. J. Lumin. 1989, **42**(365) (1990)
60. L.J. Andrews, A. Lempicki, B.C. McCollum et al., Phys. Rev. B **34**, 2735 (1986)
61. K. Huang, J. Lumin. **31–32**, 738 (1984)
62. F. Auzel, in *Proceedings of the First International School on Excited States of Transition Elements* (World Scientific, Singapore, 1989)

Chapter 8
Non-radiative Transition Inside Ions

One of the important effects of lattice vibration on spectral properties of active ions is the emission or absorption of phonon between the electronic energy levels without the emission or absorption of photon. This is the so-called non-radiative transition. In this chapter, the non-radiative transition process discussed will not involve the energy transfer between ions which will be discussed in the next chapter. In solid-state laser materials, one of the non-radiative transition process is the transfer of an active ion energy to lattice vibration; thus its fluorescence efficiency is reduced, which can be expressed as

$$\eta_f = \frac{W_r}{W_r + W_{nr}}$$

where W_{nr} refers to radiative transition probability and W_r denotes non-radiative transition probability. Obviously, a high non-radiative transition probability leads to low fluorescence efficiency. On the other hand, non-radiative transition is also a useful process for laser materials, because the active ions that transit from pump band to upper laser level and from lower laser level decay to ground state in a four-level system should rely on this process. There are extensive volume of papers and many monographs published dealing with this process since the 1950s, for example, the papers of Huang and Rhys [1], Perlin [2], Riseberg and Weber [3], Perlin and Kaminskii [4], Schuurmans and Dijk [5], Auzel [6], Hagston and Lowther [7], and Huang [8, 9] as well as monographs of Englman [10], Stoneham [11], and Fong [12]. Although many works in this area have been reported, the study of this high-order quantum process is still immature. We cannot give a detailed theoretical description on this process but can introduce some knowledge related to the spectroscopic properties of solid-state laser materials and focus on the study of weak electron–phonon coupling system such as rare earth ions in the materials.

In principle, there is a process where the electron from a lower energy level transits to a higher energy level and absorbs energy from lattice system, but the

Z. Luo and Y. Huang, *Physics of Solid-State Laser Materials*, Springer Series in Materials Science 289, https://doi.org/10.1007/978-981-32-9668-8_8

probability is very low especially when the energy separation between high and low energy levels is larger than that of a few phonons, so the probability can be neglected. In the language of thermodynamics, the process of energy transfer from electronic system with high order to lattice vibration system with low order is irreversible. The irreversibility is related to the increase of system entropy. In order to have a deep understanding about this problem, readers can refer to the book published by Fong [12], in which the relaxation processes including multi-phonon non-radiative process are dealt with by statistical theory. The book written by Englman [10] also discussed this problem.

The process mentioned above is a multi-phonon one and will be dealt with by the same theoretical model as that adopted in Chap. 6, considering it as a lattice relaxation process excited by non-adiabatic interaction.

8.1 Introduction of Non-radiative Transition Matrix Elements

In the development process of non-radiative transition theory, three different methods have been adopted to calculate the matrix element of non-adiabatic operator, that is, Condon approximation, non-Condon approximation, and static coupling approximation. The calculation of some authors in earlier time has shown that the calculation result of non-Condon approximation is higher by 2–3 orders of magnitude than that of the Condon approximation. For the situation of single-frequency mode, the transition probabilities calculated by these two approximations differ by p^2 times (p is the phonon number emitted in the transition). Thereafter, the calculation of static coupling approximation has also shown that its result is higher than that of the Condon approximation by several orders of magnitude. In 1980s, K. Huang [8, 9] analyzed different approximations in the calculation of non-radiative transition probability. By using only the non-diagonal part of electron-phonon interaction $H'_{\mathrm{eL}} = \sum\sum_{i\neq j} |j\rangle\langle j|H_{\mathrm{eL}}|i\rangle\langle i|$ while assuming a special adiabatic electronic wave function

$$\varphi_i(x) = \varphi_i^0(x) + \sum_i \frac{\langle j|H_{\mathrm{eL}}|i\rangle}{W_i^0 - W_j^0 + (\langle i|H_{\mathrm{eL}}|i\rangle - \langle j|H_{\mathrm{eL}}|j\rangle)}\varphi_j^0(x)$$

a non-radiative transition matrix element the same as that of the static coupling theory was obtained [8].

Commutation rule of quantum mechanics had been used by the author [13] to transform the matrix elements of non-adiabatic operator in Condon approximation. In this way, a non-radiative transition matrix element the same as that of the static coupling theory also was obtained without assuming an explicit function expression of the electronic wave function at the beginning of discussion. This method will be described in the following.

The Hamiltonian of the electronic wave function can be expressed as follows

$$H_e = T_e + V_{ee} + V_{en} \tag{8.1}$$

In the adiabatic approximation, the wave function of electron-phonon system can be expressed as $\psi(r, R) = \Phi(r, R)\varphi(R)$, where $\Phi(r, R)$ is the electron wave function and $\varphi(R)$ is the phonon wave function. By using the Condon approximation, after expanding the electronic wave function to a power series of the amplitude R of lattice vibration, the terms of R^n with $n \geq 2$ in the expansion expression can be neglected. Therefore, the first term of the non-adiabatic operator of (6.6) is equal to zero. If H_{nr} denotes the non-adiabatic interaction, then its effect on the wave function can be written as

$$H_{nr}\Psi(r, R) = -\sum_{\alpha} \frac{\hbar^2}{M_\alpha} \frac{\partial}{\partial R_\alpha} \Phi(r, R) \frac{\partial}{\partial R_\alpha} \varphi(R)$$

If one uses normal coordinates Q as that used in the literature [8, 9], then

$$H_{nr}\Psi(r, Q) = -\sum_{k} \hbar^2 \left(\frac{\partial}{\partial Q_k} \Phi(r, Q) \right) \left(\frac{\partial}{\partial Q_k} \varphi(Q) \right) \tag{8.2}$$

where $\Psi(r, Q) = \Phi(r, Q)\varphi(Q)$, $\varphi(Q)$ is a continued product of vibrational wave functions corresponding to a series of normal vibrational modes.

According to the golden rule in quantum mechanics, the transition probability can be expressed as

$$W_{nr} = \frac{2\pi}{\hbar} \underset{m'}{\mathrm{Av}} \sum_{m} |\langle fm|H_{na}|im'\rangle|^2 \delta(E_{im'} - E_{fm}) \tag{8.3}$$

The above formula has been summed over different final vibrational states and averaged over different initial vibrational states. For the sake of simplicity, it is assumed that in the electronic initial state, the ions are populated in the zero vibrational state $m' = 0$. If the final state density complying with the law of energy conservation is $\rho(E)$, then the non-radiative transition probability can be expressed as

$$W_{nr} = \frac{2\pi}{\hbar} |\langle fm|H_{na}|i0\rangle|^2 \rho(E) \tag{8.4}$$

It is necessary to further investigate the coordinate derivative of the electronic wave function before substituting the Hamiltonian of (8.2) into (8.4) and calculating the transition probability. Let

$$\nabla_{Q_k}\Phi(r,Q) = \frac{\partial}{\partial Q_k}\Phi(r,Q)$$

The electronic wave function $\Phi(r, Q)$ is an eigenfunction of H_e expressed by (8.1). The first two terms of H_e involves only electronic coordinates and commute with ∇_{Q_k}, therefore

$$\begin{aligned}\lfloor\nabla_{Q_k}, H_e\rfloor\Phi_i &= \nabla_{Q_k}H_e\Phi_i(r,Q) - H_e\nabla_{Q_k}\Phi_i(r,Q)\\ &= \nabla_{Q_k}V_{en}\Phi_i(r,Q) - V_{en}\nabla_{Q_k}\Phi_i(r,Q)\\ &= (\nabla_{Q_k}V_{en})\Phi_i(r,Q)\end{aligned}$$

On the other hand

$$H_e\Phi_f(r,Q) = E_f(e)\Phi_f(r,Q),\ H_e\Phi_i(r,Q) = E_i(e)\Phi_i(r,Q) \tag{8.5}$$

where $E_f(e)$ and $E_i(e)$ stand for the energy eigenvalues of the final and the initial electronic states, respectively. Therefore

$$\langle f|[\nabla_{Q_k}, H_e]|i\rangle = [E_i(e) - E_f(e)]\langle f|\nabla_{Q_k}|i\rangle \equiv \Delta E(e)\left\langle f\left|\frac{\partial}{\partial Q_k}\right|i\right\rangle = \langle f|\nabla_{Q_k}V_{en}|i\rangle$$

and so

$$\left\langle f\left|\frac{\partial}{\partial Q_k}\right|i\right\rangle = \frac{1}{\Delta E(e)}\left\langle f\left|\frac{\partial}{\partial Q_k}V_{en}\right|i\right\rangle \tag{8.6}$$

where $|i\rangle$ and $\langle f|$ are the bra and ket of the initial and final electronic wave functions, respectively and $\Delta E(e) = E_i(e) - E_f(e)$. The lattice wave functions satisfy the following equations

$$H_L\varphi_{im'}(Q) = E_{im'}(p)\varphi_{im'}(Q);\ H_L\varphi_{fm}(Q) = E_{fm}(p)\varphi_{fm}(Q) \tag{8.7}$$

where the lattice Hamiltonian is

$$H_L = \sum_k \frac{1}{2}\left(-\hbar^2\frac{\partial^2}{\partial Q_k^2} + \omega_k^2 Q_k^2\right) \equiv \sum_k \frac{1}{2}(P_k^2 + \omega_k^2 Q_k^2)$$

and $E_{im'}(p)$ and $E_{fm}(p)$ are the eigenenergy of H_L, that is, the energy of lattice system in the states *im′* and *fm*. One can find the following commutation relation in quantum mechanics [14]

$$f(P)Q - Qf(P) = -i\hbar\frac{\partial}{\partial P}f(P) \tag{8.8}$$

If

$$f(P) \equiv T_{\rm L} = \frac{\sum_k P_k^2}{2}$$

Except $T_{\rm L}$, other terms in the lattice Hamiltonian $H_{\rm L}$ commute with Q and so we have

$$H_{\rm L}Q_k - Q_k H_{\rm L} = -i\hbar \frac{\partial}{\partial P_k} f = -i\hbar P_k = -\hbar^2 \frac{\partial}{\partial Q_k} \tag{8.9}$$

Calculating the matrix element of operator (8.9) between initial and final vibrational states and using (8.7), then

$$\left\langle \varphi_{fm}(Q) \left| \frac{\partial}{\partial Q_k} \right| \varphi_{im'}(Q) \right\rangle = -\frac{1}{\hbar^2} \Delta E(p) \langle \varphi_{fm}(Q) | Q_k | \varphi_{im'}(Q) \rangle \tag{8.10}$$

where $\Delta E(p) = E_{fm}(p) - E_{im'}(p)$ is the energy difference between final and initial vibrational states. The δ function in (8.3) requires that the process should keep the energy conservation, that is

$$E_f(e) + E_{fm}(p) = E_i(e) + E_{im'}(p)$$

or

$$\Delta E(p) = \Delta E(e)$$

Consequently, after substituting (8.6) and (8.10) into (8.2), one can obtain the following relation

$$\langle \Psi_{fm}(r,Q) | H_{\rm na} | \Psi_{im'}(r,Q) \rangle = \sum_k \langle \Psi_{fm}(r,Q) | \nabla_{Q_k} V_{\rm en} \times Q_k | \Psi_{im'}(r,Q) \rangle \tag{8.11}$$

where $\Psi_{jn}(r,Q) = \Phi_j(r,Q)\varphi_{jn}(Q)$. Expanding $V_{\rm en}$ in terms of $Q_{\rm k}$, it can be obtained

$$V_{\rm en} = V_{\rm en}^0 + \sum_k \left. \frac{\partial V_{\rm en}}{\partial Q_k} \right|_{Q_k=0} Q_k + \sum_k \left. \frac{\partial^2 V_{\rm en}}{\partial Q_k^2} \right|_{Q_k=0} Q_k^2 + \cdots \tag{8.12}$$

Obviously, $\nabla_{Q_k} V_{\rm en} = \partial V_{\rm en}/\partial Q_k|_{Q_k=0}$, by (8.11) one has

$$\begin{aligned} \langle \Psi_{fm}(r,Q) | H_{\rm na} | \Psi_{im'}(r,Q) \rangle &= \left\langle \Psi_{fm}(r,Q) \left| \sum_k \partial V_{\rm en}/\partial Q_k|_{Q_k=0} Q_k \right| \Psi_{im'}(r,Q) \right\rangle \\ &= \langle \Psi_{fm}(r,Q) | H_{\rm eL} | \Psi_{im'}(r,Q) \rangle \end{aligned}$$

In the above matrix element, $H_{eL} = \sum_k \partial V_{en}/\partial Q_k|_{Q_k=0} Q_k$, it has the same expression as the H_{eL} expression given by Kun Huang, in which $\partial V_{en}/\partial Q_k|_{Q_k=0}$ is equivalent to $u_k(x)$ in (3.2.1) of reference [9]. It is a function that depends only on the electronic coordinates.

If the electronic wave function takes the zero approximation term in its expansion in terms of the vibrational coordinates, that is, it depends only on electronic coordinates, and then the result obtained is the same as that of the static coupling approximation

$$\langle \Phi_f(r)\varphi_{fin}(Q)|H_{eL}(Q)|\Phi_i(r)\varphi_{im'}(Q)\rangle$$

It shows that the simple way in the calculation of the non-radiative transition probability is to use commutation rule in quantum mechanics to transform the non-adiabatic operator at first and then substitute the electronic wave function and electron–phonon interaction potential. In this procedure, it is impossible to underestimate the non-radiative transition probability, although the Condon approximation is adopted for the electronic wave functions. Peuker et al. [15] also showed that the non-radiative transition matrix element obtained by non-Condon approximation and static coupling approximation can also be obtained by Condon approximation.

8.2 Promoting Mode and Accepting Mode in Non-radiative Transition Process

A further expression for non-radiative transition probability can be obtained by (8.2) and (8.4)

$$W_{nr} = \frac{2\pi}{\hbar}\left|\sum_k \hbar^2 L_{fi}^k \langle \varphi_{fm}|\frac{\partial}{\partial Q_k}|\varphi_{i0}\rangle G\right|^2 \rho(E) \tag{8.13}$$

The concepts of accepting mode and promoting mode are introduced to describe different effects of the vibrational modes. The product of matrix elements $L_{fi}^k = \langle \Phi_f(r, Q_k)|\partial/\partial Q_k|\langle \Phi_i(r, Q_k)|\rangle$ and $\langle \varphi_{fm}|\partial/\partial Q_k|\varphi_{i0}\rangle$ is not equal to zero in promoting mode and so it can stimulate the transition, which are labeled by index k. On the other hand, accepting mode cannot stimulate the transition but can receive energy in the non-radiative transition process. By the discussion in Chap. 6, because of the lattice relaxation, the product of the overlap integral $G = \prod_\lambda \langle \varphi_{fm\lambda}|\varphi_{i0\lambda}\rangle$ is non-zero. The factor G is called accepting mode factor and λ is the index of accepting mode.

In (8.13), there is a transformation between normal coordinates Q_k and lattice coordinates R_n

$$L_f^k = \left\langle f \left| \frac{\partial}{\partial Q_k} \right| i \right\rangle = \sum_n C_{nk} \langle f | \nabla_{R_n} | i \rangle \tag{8.14}$$

where C_{nk} is the transformation coefficients between the matrix elements of lattice vibrational coordinates and those of normal coordinates. It should be noted that although (8.14) is a first-order perturbation expression, the non-radiative transition probability is a small quantity of $2p$-order

$$W_{\text{nr}} = O\left(|H_{\text{nr}}|^{2p} \right)$$

This is due to the fact that, as pointed out by Englman [10], the wave functions have been "renormalized" by the lattice vibration and so

$$G = \prod_\lambda \langle \varphi_{fn\lambda} | \varphi_{i0\lambda} \rangle = O\left(|H_{\text{nr}}|^{p-1} \right)$$

Therefore, p phonon non-radiative transition is still a p-order quantum process.

The vibrational wave function φ_n of the nth excited state with phonon frequency ω_k (6.42) can be rewritten as

$$\varphi_n = \left(\frac{\omega_k}{\pi \hbar} \right)^{1/4} \frac{1}{\sqrt{2^n n!}} e^{-\frac{1}{2}\frac{\omega_k}{\hbar} Q_k^2} H_n \left(\sqrt{\frac{\omega_k}{\hbar}} Q_k \right) \tag{8.15}$$

By using (6.44), it is easy to show that in the condition of $m = 1$, $\varDelta_k \to 0$ $(S_k \to 0)$, $\left| \left\langle \phi_{fm} | \partial / \partial Q_k | \phi_{i0} \right\rangle \right|^2$ reaches its maximum value at $\omega_k / 2\hbar$. It shows that the promoting modes are those with Huang–Rhys factor $S_k \to 0$. However, $S_k \to 0$ means that the corresponding modes have a minimum value of $\partial V_{\text{en}} / \partial Q_k$. The optical modes correspond to the vibrations in which the movement directions of opposite charge particles are opposite and the acoustic modes correspond to the vibrations in which the movement directions of opposite charge particles are the same. Obviously, the acoustic modes have the minimum variation of V_{en} and so the promoting modes should be acoustic modes. The non-radiative transition probability can be written as

$$W_{\text{nr}} = \pi \hbar^2 \sum_k \left| L_{fi}^k \right|^2 \omega_k G^2 \rho(E) \tag{8.16}$$

The summation is over all the promoting modes.

8.3 Non-radiative Transition Probability for Weak Coupling Systems

By using overlap integral of lattice vibrational functions obtained in Chap. 6, it is easy to show that the accepting mode factor is $G^2 = S^{p_1}e^{-S}/p_1!$, where S is Huang-Rhys factor and p_1 is the total phonon number that the accepting mode consumed, while the phonon number emitted in the non-radiative transition process is $p = p_1 + 1$. If the frequencies of the promoting modes and the accepting modes are ω_k and ω_m, respectively, and the separation between two electronic energy level is ΔE, then $p_1 = (\Delta E - h\omega_k)/h\omega_m$. For multi-phonon non-radiative transition, p_1 is a larger number and so by the Stirling approximation formula $p_1! = p_1^{p_1}(2\pi p_1)^{1/2}\exp(-p_1)$, the following expression can be obtained

$$W_{\mathrm{nr}} = \left[\frac{\pi\hbar^2}{\sqrt{2\pi}}\sum_k \left|L_{fi}^k\right|^2 \omega_k e^{-S}\rho(E)\right]\exp(-p_1\beta) \tag{8.17}$$

where

$$\beta = \left(\frac{p_1 + 1/2}{p_1}\right)\ln p_1 - \ln S - 1,\ p_1 = \frac{\Delta E - \hbar\omega_k}{\hbar\omega_m}$$

Substituting the expression of β and p_1 to (8.17), the expression known as exponential energy gap law of the non-radiative transition probability resulted as follows

$$W_{\mathrm{nr}} = W\exp(-\alpha\Delta E) \tag{8.18}$$

where

$$\alpha = \beta/\hbar\omega_m,\ W = \left[\frac{\pi\hbar^2}{\sqrt{2\pi}}\sum_k \left|L_{fi}^k\right|^2 \omega_k e^{-S}\rho(E)\right]\exp\left(\frac{\omega_k}{\omega_{m,}}\beta\right)$$

β in the above formula is in fact a slow variation function of p_1, and W and α are also not the constants. Therefore, (8.18) is only an approximate energy gap exponential relation of the transition probability. The relation between β and p_1 can be tabulated as:

$S = 0.1$, $\beta(10)/\beta(2) = 1.71$									
P_1	2	3	4	5	6	7	8	9	10
β	2.17	2.58	2.86	3.07	3.2	3.39	3.51	3.62	3.72
$S = 0.2$, $\beta(10)/\beta(2) = 2.05$									
P_1	2	3	4	5	6	7	8	9	10
β	1.48	1.89	2.17	2.38	2.5	2.76	2.82	2.93	3.03

It can be seen that only for the system with smaller value of S factor (weak coupling system), the assumption of constant β is reasonable. Therefore the exponential energy gap relation of non-radiative transition probability is only valid for the weak electron–phonon coupling system.

Schuurmans et al. [5] modified the above relation to put all the variables depending on p_1 into exponential function. Writing the total product of all the terms depending on p_1 in (8.17) as

$$\exp[-(p_1 - 1)\beta']$$

then

$$\beta' = \frac{2p_1 + 1}{2(p_1 - 1)} \ln p_1 - \frac{p_1}{p_1 - 1} \ln S - \frac{p_1}{p_1 - 1} \tag{8.19}$$

The detailed calculation shows that compared with the variation of β, β' is a much slower function of p_1

$S = 0.1$									
p_1	2	3	4	5	6	7	8	9	10
β'	4.34	3.88	3.83	3.84	3.89	3.96	4.01	4.07	4.15
$S = 0.2$									
p_1	2	3	4	5	6	7	8	9	10
β'	2.95	2.84	2.90	2.98	3.06	3.14	3.22	3.30	3.36

Note that β and β' increase with p_1. It means that the decrease of non-radiative transition probability is faster than that predicted by the exponential energy gap law.

Suppose $\alpha' = \beta'/\hbar\omega_m$, then non-radiative transition probability has the following form

$$W_{\mathrm{nr}} = W' \exp[-\alpha'(\Delta E - \hbar\omega_k - \hbar\omega_{\mathrm{m}})] \tag{8.20}$$

where

$$W' = \left[\frac{\pi\hbar^2}{\sqrt{2\pi}} \sum_k \left|L_{fi}^k\right|^2 \omega_k e^{-S} \rho(E)\right]$$

W' is a quantity independent of p_1 and ΔE, and α' is substantially independent of ΔE. Therefore, (8.20) is a more accurate exponential relation of the non-radiative transition probability with the energy gap. It should be pointed out that the above formula has a bit difference from that given by Schuurmans et al. [5]. The effective energy gap in the above formula is ΔE minus one phonon energy of accepting mode and another of promoting mode instead of two phonon energies of accepting modes. Generally, the phonon energy of accepting mode is different from that of promoting mode. However, as shown in (8.20), the high-energy acoustic phonons and the high-energy optical phonons both have greater contribution to the non-radiative transition, so long as L_{fi}^{k} is non-zero.

Equation (8.20) can be written in the form of (8.18), if one puts the factor of $\exp[\alpha'(\hbar\omega_k + \hbar\omega_m)]$ into W'. Nevertheless, the pre-exponential factor in (8.18) has a strong dependence on the host crystals. As pointed out by Schuurmans et al. [5], the pre-exponential factor in (8.18) can have a variation of 4–5 orders of magnitude for different host crystals or glasses doped with the same rare earth ions but the variation in the pre-exponential factor in (8.20) is within one order of magnitude. It should be pointed out that the Huang–Rhys factor S is dependent on the kinds of rare earth ions, although the difference of S factors is within a factor of two. Consequently, strictly speaking, the values of β' and α' for different kinds of rare earth ion in the same host are different.

The above exponential energy gap law can be used to describe the transition probability of multi-phonon non-radiative process involving more than five phonons, although the demonstration is limited to the processes of 3–5 phonons. The reader can refer to the article published by Riserberg and Weber [3] to know more about this problem. It should be admitted that the multi-phonon non-radiative process is a complicated high-order quantum process and it is difficult to describe all its characteristics by any one approximate theoretical method. In the description of the exponential energy gap law of non-radiative transition probability, the perturbation theory introduced by Hagston and Lowther [7] should be mentioned. According to their analysis, non-radiative transition resulted from the following electron–phonon interaction potential

$$V_{\mathrm{eL}} = V_1\varepsilon + V_2\varepsilon^2 + \cdots V_n\varepsilon^n + \cdots \tag{8.21}$$

where

$$V_n = \frac{\partial^n V_{\mathrm{eL}}}{\partial Q^n}$$

It should be noted that in this theory the zero-order electronic wave functions are the eigenfunctions of Hamiltonian that do not include V_{eL} and the phonon wave functions are independent of the electronic states, that is, without lattice relaxation and non-adiabatic interaction. The contributions of p-order perturbation of first-order term V_1 and the first-order perturbation of p-order term V_p to the probability of p phonon transition are compared [7]. Their results showed that the major contribution is due to the first-order perturbation of p-order term and the transition probability takes the following form

$$W_{\rm nr}^{p} = \frac{2\pi}{\hbar}|\langle f|V_p|i\rangle|^2|\langle m|\varepsilon^p|n\rangle|^2\delta(E_i - E_f - p\hbar\omega_{\rm eff}) \tag{8.22}$$

In this formula, $\hbar\omega_{\rm eff}$ is the effective phonon energy. Their calculation showed that the ratio of interaction matrix element of a p-phonon transition process to that of a $(p-1)$-phonon transition process is varied in a small range, that is

$$2 \leq \left|\frac{\langle a|V_p|b\rangle}{\langle a|V_{p-1}|b\rangle}\right|^2 \leq 4$$

In the following calculation this ratio is approximately assumed to be 3. Therefore, the ratio of transition probability of p-phonon process $W_{\rm nr}^{p}$ to that of a $(p-1)$-phonon transition process $W_{\rm nr}^{p-1}$ can be approximately considered as a constant

$$R_n = \frac{W_{\rm nr}^{p}}{W_{\rm nr}^{p-1}} = \left|\frac{\langle a|V_p|b\rangle}{\langle a|V_{p-1}|b\rangle}\right|^2\langle\varepsilon^2\rangle \approx 3\frac{\hbar\omega_{\rm D}}{50\rho v^5} \tag{8.23}$$

In the calculation of above formula, (6.29) for phonon state density (neglecting the difference between the velocities of transverse and longitude acoustic waves) and the strain expression of the linear combination of creation operator and annihilation operator have been used. The integral over phonon frequency is from 0 to Debye frequency $\omega_{\rm D}$. Obviously, the magnitude of R_n depends only on the variety of host materials. Consequently

$$W_{\rm nr}^{p} = W_{\rm nr}^{0}R_n = W_{\rm nr}^{0}\exp(-\alpha\Delta E) \tag{8.24}$$

If one takes $\alpha = -\ln R/\hbar\omega_{\rm eff}$, then α depends only on the type of host material.

It must be pointed out that it is hard to see the approximate degree of the exponential law by this analysis; the problem is it is difficulty to estimate the error of transition probability introduced by assuming a constant ratio of $|\langle a|V_p|b\rangle/\langle a|V_{p-1}|b\rangle|^2$ that is difficult to be estimated. Using the density matrix method of statistical theory, Fong [12] introduced, in weak coupling condition, a formula for non-radiative transition probability as follows

$$W_{\rm nr} = A\exp(-p\gamma_p) \tag{8.25}$$

where $p = \Delta E/\hbar\omega_k$. The pre-exponential factor A is composed of the square of matrix element $|\langle i|\partial V/\partial Q_k|f\rangle|^2$ and the averaged phonon number factor $(1+\bar{n})^p$ while γ_p is a slow variation function of p

$$\gamma_p = \ln\left(\frac{f}{2L_k g_k^2}\right) + \frac{1}{2p}\ln\left\{(f/2)\left[1+(1+f)^{-2}\right]\right\} - \frac{f}{2p} - \frac{1}{p}\ln\left[\frac{f(2+f)}{4L_k g_k^2}\right] \tag{8.26}$$

where

$$f = (p-3) + \left[(p-1)^2 + 4\right]^{1/2}$$

Similar to Huang–Rhys factor, $L_k g_k^2$ is a quantity representing the strength of electron–phonon coupling. The strict energy gap exponential law requires that γ_p should be a constant. However, it can be seen from (8.26) and in the following table that γ_p is a slow variation function of p.

$L_k g_k^2 = 0.04$									
p	3	4	5	6	7	8	9	10	15
γ	1.68	2.27	2.80	3.02	3.29	3.51	3.69	3.85	4.44
$L_k g_k^2 = 0.02$									
p	3	4	5	6	7	8	9	10	15
γ	2.14	2.79	3.25	3.60	3.88	4.11	4.31	4.48	5.09

Obviously, the result is the same as presented previously that the non-radiative transition probability increased faster than that predicted by the exponential law, when the energy gap narrows down. However, it has been pointed out by many review articles, for example Riseberg and Weber [3], that the exponential law is useful at least for the estimation of the orders of magnitude of non-radiative transition probability. Figure 8.1 is a typical example showing the relation of non-radiative transition probabilities versus energy gap for five laser crystals.

The comparisons between experimental data and those of the calculated results given by Perlin and Kaminskii [4] for different rare earth ions in crystal YAP and YAG are shown in Fig. 8.2a, b.

The following is some reference values for the parameters in (8.24) [16].

Host	W^0 (sec^{-1})	α (cm)	$\hbar\omega_{\mathrm{eff}}$ (cm^{-1})
YAG	9.7×10^{7}	3.1×10^{-3}	≈700
	2.235×10^{8}	3.5×10^{-3}	≈700
YAP	5.0×10^{9}	4.6×10^{-3}	550–600
	6.425×10^{9}	4.69×10^{-3}	≈600
Y_2O_3	2.7×10^{8}	3.8×10^{-3}	430–550
	1.204×10^{8}	3.53×10^{-3}	≈600
$LiYF_4$	3.5×10^{7}	3.8×10^{-3}	≈400
	6.4×10^{7}	3.6×10^{-3}	≈560
LaF_3	6.6×10^{8}	5.6×10^{-3}	≈350
	3.966×10^{9}	6.45×10^{-3}	≈305
SrF_2	3.1×10^{8}	4.5×10^{-3}	≈360
	3.935×10^{8}	4.6×10^{-3}	≈350
$LaCl_3$	1.5×10^{10}	1.3×10^{-2}	≈260
	3.008×10^{10}	1.37×10^{-2}	≈250

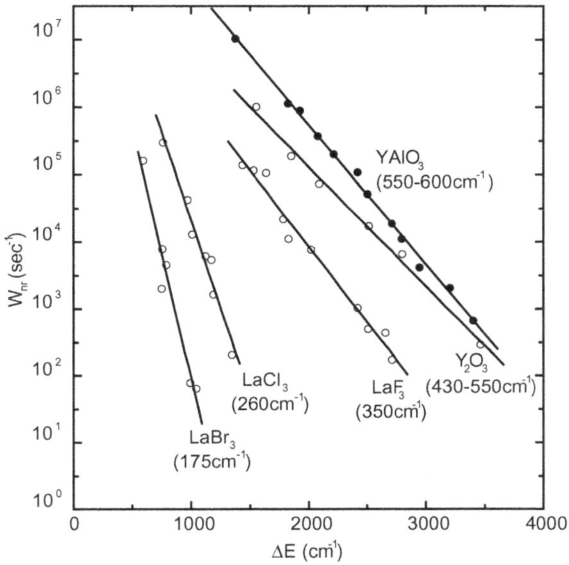

Fig. 8.1 Non-radiative transition probability versus energy gap for laser crystals of $YAlO_3$, Y2O3, LaF_3, $LaCl_3$, and $LaBr_3$ [3]

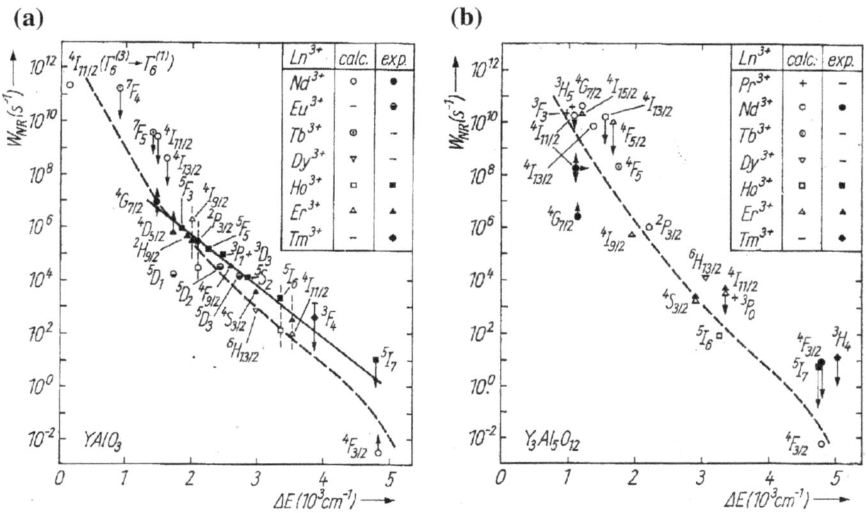

Fig. 8.2 **a** Non-radiative transition probabilities of a series of rare earth ions in crystal YAP at $T \to 0$ K; **b** non-radiative transition probabilities of a series of rare earth ions in crystal YAG at $T \to 0$ K

The phonon energy of borate crystals or glasses reaches 1400 cm^{-1}. Their non-radiative transition probabilities are much higher than those of other crystals and glasses. As an example, for $YAl_3(BO_3)_4$ (YAB) crystal, $W^0 = 8.0 \times 10^8\ s^{-1}$, $\alpha = 2.1\times10^{-3}$ cm, $\hbar\omega_{\text{eff}} = 1400\,\text{cm}^{-1}$. These three parameters for four kinds of glasses are shown as follows [16]:

Host	W^0 (sec^{-1})	α (cm)	$\approx$1200 $\hbar\ \omega_{\text{eff}}$ (cm^{-1})
Phosphate glass	5.4×10^{12}	4.7×10^{-3}	
Silicate glass	1.4×10^{12}	4.7×10^{-3}	$\approx$1100
Germanate glass	3.4×10^{9}	4.9×10^{-3}	$\approx$900
Fluoride glass	1.6×10^{10}	5.2×10^{-3}	$\approx$500

8.4 Parallelism Between Non-radiative Transition Probability and Radiative Transition Probability

By using commutation formula equivalent to (8.8)

$$f(\boldsymbol{p})\boldsymbol{r} - \boldsymbol{r}f(\boldsymbol{p}) = -i\hbar\frac{\partial}{\partial \boldsymbol{p}}f$$

where $f(\boldsymbol{p}) = T_e = \frac{p^2}{2m}$, and the operator form of electronic momentum $\boldsymbol{p}$, the following relation is obtained

$$T_e\boldsymbol{r} - \boldsymbol{r}T_e = -\frac{\hbar^2}{m}\nabla_{r_e} \tag{8.27}$$

In the Hamiltonian H_e of (8.1), except T_e, the other two terms are commuted with electronic coordinates. It is obvious from (8.27) that

$$H_e\sum_i \boldsymbol{r}_{ei} - \sum_i \boldsymbol{r}_{ei}H_e = -\frac{\hbar^2}{m}\sum_i \nabla_{r_{ei}} \tag{8.28}$$

and by using (8.5) and (8.28), the following formula resulted

$$\left\langle f\left|\sum_i \boldsymbol{r}_{ei}\right|i\right\rangle = \frac{\hbar^2}{m\Delta E(e)}\left\langle f\left|\sum_i \nabla_{r_{ei}}\right|i\right\rangle \tag{8.29}$$

∇_{r_e} and T_e are differential operators and commute with each other. On the other hand, V_{ee} has the following form

$$V_{ee} = \sum_{e \neq e'} V(r_e - r_{e'})$$

and commutes with $\sum_e \nabla_{r_e}$. Therefore, $\sum_e \nabla_{r_e}$ commutes with the first two terms in the expression of H_e. Similar to the introduction of (8.6), one has

$$\left\langle f \left| \sum_i \nabla_{r_{ei}} \right| i \right\rangle = \frac{1}{\Delta E(e)} \left\langle f \left| \sum_i \nabla_{r_{ei}} V_{en} \right| i \right\rangle \tag{8.30}$$

The interaction potential between any electron and any lattice ion can be expressed in the following general form

$$V_{en} = f(\boldsymbol{r}_{ei} - \boldsymbol{R}_n) \tag{8.31}$$

The potential function in static electric model is the well-known Coulomb potential. By (8.31)

$$\nabla_{r_{ei}} V_{en} = -\nabla_{R_n} V_{en}$$

Therefore, according to (8.6) in the form of lattice coordinate system and (8.30), we have

$$\left\langle f \left| \sum_i \nabla_{r_{ei}} \right| i \right\rangle = -\left\langle f \left| \sum_n \nabla_{R_n} \right| i \right\rangle \tag{8.32}$$

By using (8.29) and (8.32), it can be shown that

$$\langle f | \sum_n \nabla_{R_n} | i \rangle = -\frac{m\Delta E(e)}{\hbar^2} \langle f | \sum_i \boldsymbol{r}_{ei} | i \rangle = -\frac{m\Delta E(e)}{\hbar^2 e} \langle f | \sum_i \boldsymbol{m}_{ei} | i \rangle \tag{8.33}$$

where $\boldsymbol{m}_{ei} = e\boldsymbol{r}_{ei}$ is the dipole moment of electron i. The above formula can be written as

$$\langle f | \sum_i \nabla_{R_n} | i \rangle = -\frac{m\Delta E(e)}{\hbar^2 e} \boldsymbol{M}_{fi} \tag{8.34}$$

In the zero-order approximation it is assumed that the matrix element $\boldsymbol{M}_{fi}$ is independent of lattice coordinates. By (8.14) and (8.34), we obtain

$$L_{fi}^{k} = -\sum_{n} \frac{m\Delta E}{\hbar^{2} e} C_{nk} \boldsymbol{M}_{fi} \tag{8.35}$$

Equation (8.35) shows that L_{fi}^{k} is a linear combination of the dipole matrix element $\boldsymbol{M}_{fi}$, therefore the relation between the non-radiative transition probability and the electric-dipole transition probability, which is proportional to the square of the $\boldsymbol{M}_{fi}$, can be clearly obtained from (8.20) and (8.35). The ratio between these two transition probabilities will be given in the following.

By means of (8.35), the pre-exponential factor in (8.20) can be related to radiative transition probability W_{r} [5]

$$W' \approx 40 \frac{c^{3}}{M_{1} \Delta E} W_{\mathrm{r}} \tag{8.36}$$

where M_1 is the mass of ligand ions of the rare earth ion. This relationship is sometimes called parallelism between the non-radiative and radiative transition probabilities.

As early as the beginning of the 1980s, Ray and Chowdhury [17] had actually discussed this parallelism in a special problem of non-radiative transition excited by the ligands. They mentioned about one interesting situation that the rare earth ions are sited in the positions with central symmetry. In this situation the electric-dipole transition is forbidden and the above parallelism undoubtedly requires that the probability of non-radiative transition is also very small, thus ensuring that the related electronic states have a very long fluorescence lifetime. The energy transfer efficiency between Ho^{3+} ions and that between Ho^{3+} and Yb^{3+} or Sm^{3+} ions sited at central symmetry positions is as high as 99%. This is an indirect evidence of this parallelism. If the non-radiative transition probability is still high, it is impossible to provide the long-lived intermediate states to make an energy transfer with so high efficiency. This parallelism is also introduced in 1983 from another analysis by the author [18].

8.5 Temperature Dependence of Non-radiative Transition Probability in Weak Coupling Systems

8.5.1 Experimental

Many luminescent materials can only emit at lower temperature but not at higher temperature. This is known as the thermal quenching of luminescence. The thermal quenching can be caused by the increase of the non-radiative transition probability of the electron in the active ion with the increase of temperature, and also can be caused by the increase of the electronic transition probability between the electronic state of active ion and the charge transfer state of the ligand after the temperature rises. Following this charge transfer, the electron will undergo a strong coupling non-radiative transition to another electronic state of active ion. The first and the second kinds of thermal quenching involve the temperature dependence of non-radiative transition probability in weak coupling and strong coupling systems, respectively. The mechanism of the first kind of thermal quenching will be addressed.

Simply speaking, when temperature is raised, the electron–phonon interaction becomes stronger, thus the non-radiative transition probability W_{nr} is higher. For a definite luminescence system, the radiative transition probability W_r is independent of the temperature basically, therefore the fluorescence efficiency will become lower and even tend to zero when the temperature is raised. From the expression of overlapping integral of phonon wave function in Chap. 6, it can be seen that for rare earth ions, $S \ll 1$. The overlapping integral of $p - 1$ accepting modes has the following relation

$$G \propto \frac{S^{p-1}(1+\bar{n})^{p-1}}{(p-1)!}$$

The square mean of the matrix element $\langle \varphi_{fm}|\partial/\partial Q_k|\varphi_{i0}\rangle$ of promoting mode includes a factor $(1+\bar{n})$, therefore the non-radiative transition probability includes a factor $(1+\bar{n})^p$. It can also be seen from (8.25) that the pre-exponential factor A includes a factor of $(1+\bar{n})^p$. Obviously, the relation of the non-radiative transition probability with the temperature can be written as

$$W_{nr}(T) = W_{nr}(0)(1+\bar{n})^p \tag{8.37}$$

where

$$\bar{n} = \frac{1}{\exp(\hbar\omega_{eff}/kT) - 1} \tag{8.38}$$

This relation can be explained by the example of Nd^{3+}:YAG system. The energy separation between $^4F_{3/2}$ state and $^4I_{15/2}$ state is about 5000 cm^{-1} and this crystal has effective phonon energy of about 700 cm^{-1}, so the non-radiative transition is a 7-phonon process. According to (8.37), it is easy to obtain the following results: the non-radiative transition probability at 300 and 600 °C are three times and 11 times of that at the room temperature, respectively. At very high temperature, the

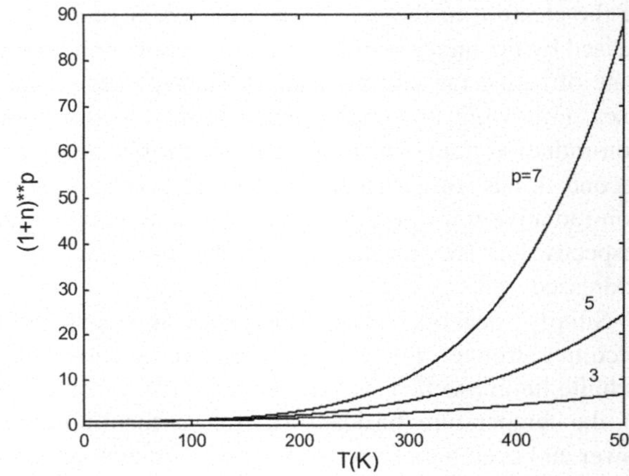

Fig. 8.3 Transition probability versus temperature for three different orders of multi-phonon process

non-radiative transition probability is proportional to seven powers of the temperature. The thermal quenching is obvious. Figure 8.3 shows this relation for a series of different orders of multi-phonon process. The effective phonon energy $h\omega_{\mathrm{eff}}$ is assumed as 250 cm^{-1}.

8.6 Non-radiative Transition in Strong Coupling Systems

Strong electron-phonon coupling means Huang–Rhys factor $S > 1$. The transition metal group ions in materials belong to this case. The formulas introduced in previous sections with the assumption of $S \ll 1$ are invalid in this situation. In order to obtain calculation formulas of the non-radiative transition probability, one can start with a phenomenological discussion by using configuration coordinate curve. Because in the strong electron–phonon coupling case $S > 1$, the origins of the configuration coordinates of potential curves for the excited state and that of the ground state have a considerable shift. Consequently, the two potential curves will have an intersect point (see Fig. 8.4). Suppose the energy difference between this intersect point and the zero-potential point of the excited state is ΔE_{e}. When electronic energy reaches or exceeds ΔE_{e}, the active ions will go from the excited state to the ground state. This is the most important non-radiative pathway in such a strong coupling system. The electrons populated in the excited state according to Boltzmann distribution law and the non-radiative transition probability can be expressed as

$$W_{\mathrm{nr}} = W_{\mathrm{nr}}^{0} \exp(-\Delta E_{\mathrm{e}}/k_{\mathrm{B}}T) \tag{8.39}$$

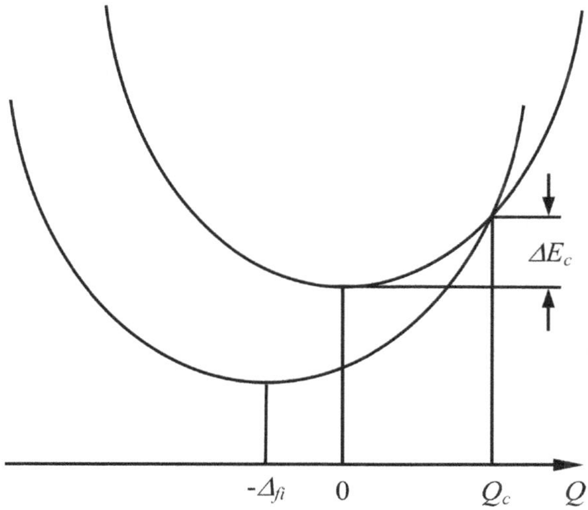

Fig. 8.4 Configuration coordinate diagram for non-radiative transition

The process of having this form of transition probability expression is usually called a thermal activated process and ΔE_e denotes the activation energy, which can be found in the following way.

If the lowest energy point of the potential energy curve for the initial state i is at the configuration coordinate $Q = 0$, assuming a single-phonon mode with phonon frequency $<\omega>$, then the potential energy for the initial state can be written as

$$E_i + \frac{1}{2}\langle\omega\rangle^2 Q^2 \tag{8.40}$$

The corresponding potential energy of the final state f is

$$E_f + \frac{1}{2}\langle\omega\rangle^2 (Q + \Delta_{fi})^2 \tag{8.41}$$

The above two energy curves intersect at the point with configuration coordinates $Q = Q_c$, then

$$Q_c = \frac{E_{if} - \frac{1}{2}\langle\omega\rangle^2 \Delta_{fi}^2}{\langle\omega\rangle^2 \Delta_{fi}} \tag{8.42}$$

Obviously, the activation energy ΔE_e is

$$\Delta E_e = \frac{1}{2}\langle\omega\rangle^2 Q_c^2 = \frac{\left(E_{if} - \frac{1}{2}\langle\omega\rangle^2 \Delta_{fi}^2\right)^2}{2\langle\omega\rangle^2 \Delta_{fi}^2} \tag{8.43}$$

According to the formula introduced in Chap. 6, in the case of single mode, we have

$$S\hbar\langle\omega\rangle = \frac{1}{2}\langle\omega\rangle^2 \Delta_{fi}^2$$

Therefore, the activation energy can be expressed as

$$\Delta E_e = \frac{\left(E_{if} - S\hbar\langle\omega\rangle\right)^2}{4S\hbar\langle\omega\rangle} \tag{8.44}$$

By the above simple discussion, the relation between the non-radiative transition probability and the energy gap as well as the temperature can be obtained, although it is impossible to know the detail about pre-exponential factor and in what condition this expression can be used.

For a more sophisticated discussion, one should start with the general results given in the classical paper of Huang and Rhys [1]. When the phonon number in the initial state is non-zero, the overlap integral G consists of five terms: G_{p+2}, G_{p+1}, G_p, G_{p-1}, and G_{p-2}. Considering non-radiative transition between large energy gap in condition of larger Huang-Rhys factor, it can be assumed that the phonon-order in the above five terms are the same. Therefore

$$W_{nr} \cong -\hbar^2 \sum_l \left(L_{fi}^l\right)^2 \left(\bar{n} + \frac{1}{2}\right) \exp[-S(1+2\bar{n})] \left(\frac{1+\bar{n}}{\bar{n}}\right)^{p/2} I_p\left[2S\sqrt{\bar{n}(\bar{n}+1)}\right] \tag{8.45}$$

Expanding the above Bessel function, then

$$\begin{aligned} W_{nr} &= -\hbar^2 \sum_l \left(L_{fi}^l\right)^2 \left(\bar{n} + \frac{1}{2}\right) \exp[-S(1+2\bar{n})] \sum_{k=0}^{\infty} \frac{[S(1+\bar{n})]^{k+p}}{k!(k+p)!} (S\bar{n})^k \\ &= -\hbar^2 \sum_l \left(L_{fi}^l\right)^2 \left(\bar{n} + \frac{1}{2}\right) \exp[-S(1+2\bar{n})] \frac{[S(1+\bar{n})]^p}{p!} \left[1 + \frac{S^2\bar{n}(\bar{n}+1)}{p+1} + \cdots\right] \end{aligned} \tag{8.46}$$

The relation of the non-radiative transition probability with the energy gap cannot be seen from this formula. The saddle point approximation will be used to introduce this relation in strong electron–phonon coupling situation. Generally, the non-radiative transition probability can be expressed as

$$W_{\mathrm{nr}} = \frac{1}{\hbar} \int_{-\infty}^{+\infty} f(z) e^{g(z)} \mathrm{d}z \tag{8.47}$$

where

$$f(z) = \left(\frac{1}{2} \sum_q L_{fi}^q \left\{ \left(\Delta_{fq} + \Delta_{iq} \right) + \Delta_{ifq} \left[\cos\left(z\hbar\omega_q \right) + i\coth\left(\hbar\omega_q / 2k_B T \right) \sin\left(z\hbar\omega_q \right) \right] \right\} \right)^2 + \frac{1}{2} \sum_q \left(L_{fi}^q \right)^2 \left(\frac{\hbar}{\omega_q} \right) \left[\coth\left(\hbar\omega_q / 2k_{\mathrm{B}} T \right) \cos\left(z\hbar\omega_q \right) + i\sin\left(z\hbar\omega_q \right) \right] \tag{8.48}$$

and

$$g(z) = -iz\Delta E_{\pm q} + \sum_k \frac{\omega_k}{2\hbar} \Delta_{ifk}^2 \{ \coth(\hbar\omega_k / 2k_B T)[\cos(z\hbar\omega_k) - 1] + i\sin(z\hbar\omega_k) \} \tag{8.49}$$

where

$$\Delta E_{\mp q} = \begin{cases} \Delta E - \hbar\omega_q \\ \Delta E + \hbar\omega_q \end{cases}$$

where ΔE_{-q} corresponds to the emission of a promoting phonon, while ΔE_{+q} corresponds to the absorption of a promoting phonon. The right-hand side of (8.49) includes the factor of phonon frequency, which is obviously a large positive real number. Therefore, the right-hand side of (8.49) can be written as a large positive real number t multiplied by a function $J(z)$, that is

$$g(z) = tJ(z)$$

Equation (8.47) is generally an integral of complex function and can be calculated approximately by saddle point method. It can be seen from the form of the function that (8.48) is a slowly varied function of z. By using saddle point method introduced by Irving and Mullineux [19] and taking into account that the probability is a positive number as well as the phase factor can be assumed a value of 1, the result of the integration is

$$W_{\mathrm{nr}} \approx \frac{1}{\hbar} \left[\frac{2\pi}{g''(z_0)} \right]^{1/2} f(z_0) \exp[g(z_0)] \tag{8.50}$$

where z_0 is the so-called saddle point, and it satisfies the condition of the first-order derivative with respect to z to be zero

$$g'(z_0) = 0 \tag{8.51}$$

The problem now is how to calculate saddle point z_0 according to (8.51). By considering the case of strong electron–phonon coupling and expanding the cos(zhω_k) and sin(zhω_k) functions of (8.49) of z, the second term in the right-hand side can be written as

$$S(T)\sum_{n\geq 1}\frac{(-1)^n}{(2n)!}\left\langle(\hbar\omega)^{2n}\right\rangle_T z^{2n} + iS\sum_{n\geq 0}\frac{(-1)^n}{(2n+1)!}\left\langle(\hbar\omega)^{2n+1}\right\rangle z^{2n+1} \tag{8.52}$$

where

$$S(T) = \sum_k \Delta_{ifk}{}^2\left(\frac{\omega_k}{2\hbar}\right)\coth(\hbar\omega_k/2k_{B'}T) \tag{8.53}$$

$$\left\langle(\hbar\omega)^{2n}\right\rangle_T = \frac{1}{S(T)}\sum_k\left[\Delta_{ifk}^2\left(\frac{\omega_k}{2\hbar}\right)\coth(\hbar\omega_k/2k_BT)\right](\hbar\omega_k)^{2n} \tag{8.54}$$

$$\left\langle(\hbar\omega)^{2n+1}\right\rangle = \frac{1}{S}\sum_k\left[\Delta_{ifk}^2\left(\frac{\omega_k}{2\hbar}\right)\right](\hbar\omega_k)^{2n+1} \tag{8.55}$$

It can be shown [9] in high temperature strong coupling approximation, the terms with powers higher than z^2 can be neglected, then the saddle point z_0 calculated according to (8.51) leads to the following result

$$z_0 = -i\frac{(\Delta E_{\mp q} - S\hbar\langle\omega\rangle)}{S(T)\left\langle(\hbar\omega)^2\right\rangle_T}$$

Obviously, at saddle point we have

$$\left\langle(\hbar\omega)^2\right\rangle_T z_0^2 \approx \left\{\frac{\left(\frac{\Delta E_{\mp q}}{\hbar\langle\omega\rangle} - S\right)}{\left[S\coth\left(\frac{\hbar\langle\omega\rangle}{2k_BT}\right)\right]}\right\}^2$$

Therefore, when

$$\frac{\left|\frac{\Delta E_{\mp q}}{\hbar\langle\omega\rangle} - S\right|}{\left(S\coth\frac{\hbar\langle\omega\rangle}{2k_BT}\right)} \ll 1 \tag{8.56}$$

it is reasonable to neglect the terms higher than z^2 in the expansion of (8.52). Inequality (8.56) corresponds to the criterion introduced by Lin et al. [20].

At high temperature

$$k_B T \gg \hbar\omega_q \tag{8.57}$$

therefore

$$\coth \hbar\omega_q/2k_B T \approx 2kT/\hbar\omega_q \tag{8.58}$$

Substituting it into (8.48) and (8.49) to obtain $f(z_0)$ and $g(z_0)$, and expanding $f(z_0)$ as a power series of T and keeping only linear term of T, then the results for $f(z_0)$ and $g(z_0)$ are

$$f(z_0) \approx \left(\sum_q \frac{\left(L_{fi}^q\right)^2}{\omega_q^2} \right) k_B T \tag{8.59}$$

$$g(z_0) \approx -\frac{\left(\Delta E_{\mp q} - S\hbar\langle\omega\rangle\right)^2}{4k_B T S\langle\omega\rangle} \tag{8.60}$$

In the introduction of (8.59) and (8.60), the terms with negative and zero powers of T are neglected, because $T \gg 1$. According to (8.50), the non-radiative transition probability can be obtained

$$W_{nr} \approx \frac{1}{\hbar}\left(\frac{\pi kT}{S\hbar\langle\omega\rangle}\right)^{1/2} \left(\sum_q \frac{\left(L_{fi}^q\right)^2}{\omega_q^2} \right) \exp(-\Delta E_e/k_B T) \tag{8.61}$$

Transition probability can be written in this form, which means that the process is a thermal activation process with activation energy

$$\Delta E_e = \frac{\left(\Delta E_{\mp q} - S\hbar\langle\omega\rangle\right)^2}{4S\hbar\langle\omega\rangle} \tag{8.62}$$

Note that (8.61) has the same form as that of (8.39), while (8.62) is the same as (8.44). It means that the non-radiative transition in strong coupling system is a thermal activation process. However, it should be noted that the inequality (8.56) has to be satisfied.

In multi-mode approximation, a similar formula can also be introduced. Freed and Jortner [21] introduced a formula for the calculation of the non-radiative transition probability; in this case

$$W_{nr} = \sqrt{\frac{\pi}{4S\hbar\langle\omega\rangle k_B T^*}} \sum_q \left|C_{ee'}^q\right| \omega_q \coth\left(\hbar\omega_q/2k_B T\right) \exp(-\Delta E_e/k_B T^*) \tag{8.63}$$

where C_{ij}^q denotes the matrix element between electronic states i and j

$$C_{ij}^q = i\hbar\langle i|\frac{\partial}{\partial Q_q}|j\rangle \tag{8.64}$$

The summation in (8.63) should be over all the promoting mode q. The relation between effective temperature T^* and the environment temperature T is

$$T^* = \frac{\hbar\langle\omega\rangle}{2k_B}\coth(\hbar\langle\omega\rangle/2k_B T) \tag{8.65}$$

where $\langle\omega\rangle$ denotes the averaged promoting mode frequency.
At very high temperature, that is, $T \gg \hbar\langle\omega\rangle/2k_B$

$$\coth(\hbar\langle\omega\rangle/2k_B T) \to 2k_B T/\hbar\langle\omega\rangle$$

then

$$T^* \approx T$$

Niznan et al. [22] pointed out that this kind of Gaussian function relation between the non-radiative transition probability and the energy gap is only valid if the following condition is satisfied.

$$\frac{\Delta E/\hbar\langle\omega\rangle - S}{S\coth^2(\hbar\langle\omega\rangle/2k_B T)} \ll 1 \tag{8.66}$$

The meaning of high temperature and strong coupling can be seen from (8.58) and (8.56). Equation (8.56) shows that for different magnitude of energy gap, the requirement of strong coupling for Huang–Rhys factor is different. By the calculation of the Cr^{3+}:Al_2O_3 system, this can be clearly seen. Assuming $h <\omega> =$ 250 cm^{-1}, $S = 7$ [23], at a temperature higher than room temperature $kT > 200\ cm^{-1}$, (8.58) is roughly satisfied. For an energy gap of $\Delta E = 2500\ cm^{-1}$, the left side of (8.56) is smaller than 0.18. The criterion for strong coupling is satisfied, but for a larger energy gap, for example $\Delta E = 5000\ cm^{-1}$, the above criterion cannot be satisfied. Therefore, for the ${}^4T_2 \to {}^4A_2$ transition in ruby, the thermal activation relation cannot be satisfied at room temperature. Only at much higher temperature it is valid. However, the situations for the ${}^4T_2 \to {}^2E$ and ${}^4T_2 \to {}^2T_1$ transitions are different. The experimental results obtained by Misu [24] showed the following relation at high temperature:

$$W_{nr}({}^4T_2 \to^2 T_1) = 2.7 \times 10^8 \times \exp(-\Delta E_e/kT),\ \Delta E_e = 2200\,\mathrm{cm}^{-1}$$
$$W_{nr}({}^4T_2 \to^2 E) = 4.4 \times 10^8 \times \exp(-\Delta E_e/kT),\ \Delta E_e = 2600\,\mathrm{cm}^{-1}$$

For the $^2E \rightarrow {}^4A_2$ transition in ruby, because the potential energy curves of the two correlated electronic levels are basically parallel, the non-radiative transition between them cannot be described by the above strong coupling approximation.

8.7 Nonlinear Theory of Non-radiative Transition

The theory introduced previously is mainly based on the supposition that the expansion of the electron–phonon interaction Hamiltonian in terms of the strain keeping only the linear term is enough. This is the so-called linear theory of non-radiative transition. Perlin and Kaminskii [4] pointed out that this linear theory gives a too low transition probability for 3–6 multi-phonon non-radiative transition while by the nonlinear theory they obtained the results well compared with the experimental data. Orlovskii and Basiev et al. [25–28] applied nonlinear theory to the non-radiative transition of many laser crystals and obtained some valuable conclusions. Some of their conclusions on the relationship of the non-radiative transition rate with the structure and chemical composition of the host materials will be introduced in the following [28].

Their study shows that the non-radiative transition rate can be expressed as

$$W_{\mathrm{NR}}(J \rightarrow J') = \frac{1}{137}(2J+1)^{-1}(2J'+1)^{-1}\sum_{k=2,4,5} \Xi_k(p)\left|\left\langle LSJ\|U^{(k)}\|L'S'J'\right\rangle\right|^2 \tag{8.67}$$

$$\Xi_k^{pc}(p) = \eta^p\left(\bar{\xi}^k/R_0^k\right)^2[(2p+2k)!/(2k+1)!p!](2l+1)^2\begin{pmatrix} l & l & k \\ 0 & 0 & 0 \end{pmatrix}^2\frac{Ze_s^2}{\bar{v}\hbar R_0^2} \tag{8.68}$$

$$\Xi_k^{ex}(p) = \eta^p\left(\frac{1}{137}\right)\left(G_s|S_s|^2 + G_\sigma|S_\sigma|^2 + G_\pi\gamma_k|S_\pi|^2\right)4(2l+1)2^p\Phi_{kp}\begin{pmatrix} l & l & k \\ 0 & 0 & 0 \end{pmatrix}^2\frac{Zc}{(\bar{v}R_0)^2} \tag{8.69}$$

$$\eta \approx \frac{\hbar}{8\pi c\bar{v}MR_0^2} \tag{8.70}$$

Equations (8.68) and (8.69) denote the contributions of the point charge Coulomb interaction and the exchange interaction to the non-radiative transition rates, respectively. In the above four formulas, p is the phonon number in non-radiative transition process, R_0 denotes the separation between the active ion and its nearest neighbor anion and $l = 3$ for $4f$ electrons, $\bar{v}$ is the effective phonon frequency in the unit of wave number, ξ is the radius of rare earth ion, Z is the number of the anions nearest to the RE ion, and M is the reduced mass of the anion and cation of host material.

The contribution of Coulomb interaction to the non-radiative transition rate is proportional to Z, e_s^2, and $\left(\overline{\xi^k}\right)^2$. On the other hand, it is inversely proportional to $\overline{\nu}^{p+1}$, M^p, and $R_0^{2k+2p+2}$.

The contribution of exchange interaction is composed of the overlap effect of 4f electronic wave function with the ligand wave function, the effect of ligand charge distribution on the Coulomb interaction, as well as the effect of charge transfer state. The non-radiative transition rate of exchange interaction is proportional to Z and the summation of the product G and S^2, while inversely proportional to R_0^{2p+2}, M^p, and $\overline{\nu}^{2p+2}$. Crystal field fitting parameters G of fluoride and oxide crystals are $G_s = G_\sigma = G_\pi \cong 8$–10.

Lattice dynamics parameter η is affected greatly by the mass of lattice ions, and the orders of magnitude of η is in the range of 10^{-3}–10^{-4} for fluoride and oxide crystals. The larger the reduced mass M of the lattice ions, the lower is the non-radiative transition rate W_{NR}. The shorter the distance between the active ion and its nearest neighbor anion and the larger the radius of nearest neighbor anion leads to the higher the W_{NR}. On the other hand, the higher the degree of electronic cloud overlapping, that is the larger covalent bond proportion, the higher is the W_{NR}. Furthermore, by comparing the W_{NR} expression given above and the expression of electric-dipole transition probability A_R obtained by the J–O theory, it can be seen that both W_{NR} and A_R are proportional to the square of reduced matrix element of unit operator and it also demonstrates the parallelism between W_{NR} and A_R.

The contribution of Coulomb interaction to the non-radiative transition rate depends largely on the radius of active ion. Some data of rare earth ions usually used in laser crystals are listed as follows (in atomic unit).

Ion	$\langle \xi^2 \rangle$	$\langle \xi^4 \rangle$	$\langle \xi^6 \rangle$
Pr^{3+}	1.09	2.82	15.7
Nd^{3+}	1.01	2.40	12.4
Ho^{3+}	0.69	1.22	4.5
Er^{3+}	0.67	1.13	3.98
Tm^{3+}	0.63	1.07	3.65

Nonlinear theory shows that the non-radiative transition rate W_{NR} is different for the same active ion in different crystals as well as different active ions in the same crystal. In all the following transitions discussed, the contribution of $k = 4$ and $k = 6$ terms in (8.67) is much smaller than that of the $k = 2$ term because Ξ_4 and Ξ_6 are much smaller than Ξ_2 [28], so that it is enough only to compare the values of the reduced matrix element of $U^{(2)}$ for different transitions. The radius of Nd^{3+} ion is larger than that of Er^{3+} ion. The reduced matrix element $U^{(2)}$ for the ${}^4G_{7/2} \rightarrow {}^4G_{5/2} + {}^2G_{8/2}$ transition of Nd^{3+} ion and that for the ${}^4F_{5/2} \rightarrow {}^4F_{7/2}$ transition of Er^{3+} ion is nearly the same. In LaF_3 crystal both the transitions belong to a 4-phonon process but the non-radiative transition rate W_{NR} of the former is about six times that of the latter. The radius of Pr^{3+} ion is larger than that of Ho^{3+} ion. The reduced matrix element $U^{(2)}$ for the $^1G_4 \rightarrow {}^3F_4$ transition of Pr^{3+} ion is nearly the same as that of the $^5I_5 \rightarrow {}^5I_6$ transition of Ho^{3+} ion. In YLF crystal both the transitions belong to a 5-phonon process but the non-radiative transition rate W_{NR} of the former is about 2.6 times of the latter. The radius of Er^{3+} ion is a little larger than that of Tm^{3+} ion. The reduced matrix element of $U^{(2)}$ for the $^3F_{5/2} \rightarrow {}^3F_{7/2}$ transition of Er^{3+} ion is equal to that for the $^3F_3 \rightarrow {}^3H_4$ transition of Tm^{3+} ion, but the non-radiative transition rate W_{NR} of the former is about 1.5 times of the latter. The effect of the radius of active ion on the non-radiative transition rate is mainly due to the fact that Coulomb and exchange interactions are stronger for larger active ion radius. The effect of Coulomb interaction is reflected by the average square of ξ^k in (8.68), while the effect of exchange interaction is reflected by the overlap parameters.

If the lattice dynamics parameter η is larger, the combined effect of η and the radius ξ makes the difference of non-radiative transition rate W_{NR} more obvious. For the LaF_3 crystal, $\bar{\nu}$ = 400 cm^{-1}, R_0 = 2.5 Å, $M = 2.28 \times 10^{-23}$g while for the YLF, $\bar{\nu}$ = 560 cm^{-1}, R_0 = 2.27 Å, $M = 2.60 \times 10^{-23}$g. Calculating by (8.70), the η value of the former is larger than the latter by a ratio of 2.1, in addition that the radius of Nd^{3+} ion is larger than that of the Tm^{3+} ion. Therefore, for the same 4-phonon transition process, the W_{NR} of the $^4G_{7/2} \rightarrow {}^4G_{5/2} + {}^2G_{7/2}$ transition of Nd^{3+}:LaF_3 crystal is 4.5 times that of the $^3H_5 \rightarrow {}^3F_4$ transition of Tm^{3+}:YLF crystal, although the ratio of the square reduced matrix element $U^{(2)}$ for the former to that of the latter is 6/9. The square reduced matrix element of $U^{(2)}$ for the $^4G_{7/2} + {}^2K_{13/2} \rightarrow {}^4G_{5/2} + {}^2G_{7/2}$ transition of Nd^{3+} ion is 0.0583 while that for the $^4D_{3/2} \rightarrow {}^2P_{3/2}$ transition is 0.0118, that is, the former is 4.9 times of the latter. In Nd^{3+}:YAG crystal the non-radiative transition rate of the former is 5.9 times that of the latter.

The lattice dynamics parameter η is proportional to the mean square of lattice vibration amplitude; therefore if the melting point of the crystal is lower or the hardness of the crystal is lower, the lattice vibration amplitude will be larger at the same temperature, then the non-radiative transition rate of active ion in this crystal generally will be higher, when the other factors are similar.

Non-radiative transition rate decreases with the increase of the distance between the active ion and its ligands. Following is the non-radiative transition rate W_{NR} for the $^3P_0 \rightarrow {}^1D_2$ transition of Pr^{3+} ion in four different crystals. Comparing the Pr^{3+} ion substituting for La^{3+} ion with that substituting for Y^{3+} ion, the radius of La^{3+} ion larger than that of the Y^{3+} ion results in the distance between the Pr^{3+} ion and its nearest neighbor O^{2+} ion larger for the former than that for the latter. Therefore in the following table, the non-radiative transition rate of the first crystal is lower than that of the second crystal, while the third crystal is lower than that of the fourth crystal.

Crystal	ΔE (cm^{-1})	$\hbar\omega_{max}$ (cm^{-1})	n	W_{NR} (s^{-1})
$LaNbO_4$	3500	810	4.32	4.5×10^4
$YNbO_4$	3500	830	4.16	2.2×10^6
$Na_5La(WO_4)_4$	3740	960	3.896	6.0×10^4
$Na_5Y(WO_4)_4$	3720	960	3.875	1.0×10^5

For the non-radiative transition process emitting phonon number $p > 6$, the non-radiative transition rate obtained by nonlinear theory is still lower than the experimental result; moreover, the deviation between the calculated values and the experimental data will increase with p.

8.8 Stimulated Non-radiative Transition

According to the aforementioned theory of non-radiative transition, by a simple calculation, it can be seen that in the four-level high power continuous laser, the stimulated radiative transition (SRT) probability between upper and lower laser levels is much higher than the decay rate from the lower laser level to the ground state. The continuous Nd^{3+}:YAG laser can be used as an example to illustrate this problem.

It is well known that the photon number in the laser cavity can be calculated by the following expression

$$n_k = \frac{2L_c P_{out}}{c\gamma_{out} h\nu_k} \tag{8.71}$$

where P_{out} is laser output power, L_c is the length of laser cavity, and γ_{out} is loss per pass due to the mirror transmission. The laser wavelength of Nd^{3+}:YAG crystal is λ_k =1.064 μm (emission generated by the transition of $R_2 \rightarrow Y_3$, that is, from energy level R_2 of multiplet $^4F_{3/2}$ to energy level Y_3 of multiplet $^4I_{11/2}$) and the frequency of laser is $\nu_k \approx 2.8\times10^{14}$ s^{-1}. Its multi-mode oscillation linewidth is in the range of 15–30 K GHz [29]. Suppose it is 20 K GHz, then $\Delta\nu_k \approx 2\times10^{10}$ s^{-1}. When laser output power reaches 10 W $\approx 5\times10^{23}$ $cm^{-1}s^{-1}$ (energy in the unit of wave number), and suppose the length of cavity is L_c = 20 cm, the loss per pass γ_{out} = 0.2 and that the photon number is $n_k \approx 3.6\times10^{11}$. Suppose the shape of laser spectral line is Lorenz type, then the relation of the spontaneous transition probability A_{R2Y3} with the stimulated emission cross-section σ_{em} is [30]

$$A_{R2Y3} = \frac{4\pi^2 n^2 \Delta\nu\sigma_{em}}{\lambda^2} \tag{8.72}$$

Using the data of stimulated emission cross-section $\sigma_{em} = 6.5 \times 10^{-19}$ cm^2 for laser transition $R_2 \rightarrow Y_3$ [30], $A_{R2Y3} \approx 1.5\times10^2$ s^{-1} can be obtained. The SRT probability of $R_2 \rightarrow Y_3$ transition will be $n_k A_{R2Y3}$, and it reaches 5.4×10^{13} s^{-1}. However, according to the experimental result [16], the non-radiative transition probability of $^4I_{11/2} \rightarrow {}^4I_{9/2}$ at room temperature is 10^{10} s^{-1} orders of magnitude at most (taking account of the nonlinear mechanism, the theoretical calculation value can only reach 1.6×10^{10} s^{-1} [16]). It means that when the output power of Nd^{3+}: YAG laser reaches $P = 10$ W, the SRT probability of $^4F_{3/2} \rightarrow {}^4I_{11/2}$ transition is far outweighing the non-radiative transition probability of $^4I_{11/2} \rightarrow {}^4I_{9/2}$. In this case, the particles in the low laser level cannot be removed in time, so that the population inversion of the upper and the lower laser levels cannot be maintained and the bottleneck effect is inevitable. However, as it is well known, the output power of continuous Nd^{3+}:YAG laser has reached the scale of kilowatt nowadays, but no bottleneck effect appeared. Obviously, the experimental fact can only be explained by a new non-radiative transition mechanism excited by the laser itself. In the following it will be demonstrated that the mechanism proposed by the author [31, 34] referred to as stimulated non-radiative transition (SNT), can well explain the related phenomenon.

The starting point to solve this problem is to deal with the photon–electron–phonon system as a whole and write the Hamiltonian of the system as

$$H = H_e + H_r + H_l + H_{er} + H_{el} \tag{8.73}$$

where H_e, H_r, and H_l express the Hamiltonians of electron, photon, and phonon, respectively, H_{er} is the Hamiltonian of photon–electron interaction, and H_{el} is the Hamiltonian of electron–phonon interaction. As being shown in Chap. 4, in dipole approximation, H_{er} can be written as

$$H_{er} = \frac{e}{mc}\boldsymbol{p} \cdot \boldsymbol{A} = \frac{e}{m}\sum_{\mathbf{k}\alpha}\left(\frac{2\pi\hbar}{V\omega_{\mathbf{k}}\varepsilon}\right)^{1/2} \boldsymbol{e}_\alpha(\boldsymbol{k}) \cdot \boldsymbol{p}\left[a_{k\alpha}^{-}(t)e^{i\boldsymbol{k}\cdot\boldsymbol{r}} + a_{\mathbf{k}\alpha}^{+}(t)e^{-i\boldsymbol{k}\cdot\boldsymbol{r}}\right] \tag{8.74}$$

Suppose there are N phonon modes Q_i, so that H_{el} is

$$H_{el} = \frac{1}{\sqrt{N}}\sum_i \frac{\partial V}{\partial Q_i} Q_i \tag{8.75}$$

The transition between electronic energy levels corresponds to the absorption or emission in the wavelength range of ultraviolet–visible–near infrared. Suppose the photon and the phonon has no direct interaction, so that the term to describe its interaction is not included in (8.73).

If there are n_k photons in the electromagnetic field, the wave function of electron–photon system can be written as

$$|\phi_{ik}\rangle = |\psi_i\rangle|n_k\rangle \tag{8.76}$$

where $|\psi>$ is electronic wave function and $|n_k>$ is photon wave function. Phonon system can be regarded as slow system compared with photon-electron system.

Similar to the adiabatic approximation in lattice relaxation theory, the initial state wave function of photon-electron-phonon system can be expressed as

$$|\Phi_{ik\lambda}\rangle = |\psi_i\rangle|n_k\rangle|\varphi_{i\lambda}\rangle \tag{8.77}$$

By the single-mode approximation of the radiation field and phonon field, the zero-order eigenvalue of initial photon–electron–phonon system is expressed as

$$E_{ik\lambda} = E_i + \left(n_k + \frac{1}{2}\right)\hbar\omega_k + \left(n_\lambda + \frac{1}{2}\right)\hbar\omega_\lambda \tag{8.78}$$

The perturbation of non-diagonal matrix elements of interaction Hamiltonian $H_{\rm er}$ and $H_{\rm el}$ induce the radiative transition, non-radiative transition, and light scattering processes. SNT is a process induced by the following second-order perturbation matrix element of $H_{\rm er}$

$$M_{if}^{(2)} = \sum_{vk_1\lambda} \frac{\langle\Phi_{ik\lambda}|H_{\rm er}|\Phi_{vk_1\lambda}\rangle\langle\Phi_{vk_1\lambda}|H_{\rm er}|\Phi_{fk_1\lambda_1}\rangle}{E_{ik\lambda} - E_{vk_1\lambda} + i\delta} \tag{8.79}$$

In this process, the electron excited by the photon transits from the most probable phonon state λ of the initial electronic state i to the same phonon state λ of the intermediate electronic state v and stimulated by the photon transiting simultaneously to the final phonon excited state λ_1 of final electronic state f. In this process, the electronic system undergoes a non-radiative transition of $i \to f$, the phonon system absorbs the energy emitted by the electron, while the photon system completed a virtual process without energy variation. The initial electronic state i, the intermediate electronic state v, and final the electronic state f discussed here are the crystal field energy levels $^4I_{11/2}$, $^4F_{3/2}$, and $^4I_{9/2}$ multiplets of Nd^{3+} ion, respectively. The energy of electron and phonon satisfies the following relations

$$\Delta E_{if} = E_i - E_f > 0,\ n_{\lambda_1} > n_\lambda,\ E_i - E_f = (n_{\lambda_1} - n_\lambda)\hbar\omega_\lambda \equiv p\hbar\omega_\lambda$$

In (8.79) δ is the linewidth of laser beam in the unit of energy. When the photon energy $\hbar\omega_k$ satisfies the resonance condition $E_{ik\lambda} - E_{vk_1\lambda} = E_i + \hbar\omega_k - E_v = 0$, this process has practical physical meaning.

In order to introduce expression of the transition probability, suppose the N phonon modes which interacted with electronic system have the same frequency ω_λ and the radiation field also has one mode ω_k, the initial photon wave function can be written as

$$|n_i\rangle = |n_1, n_2, \ldots, n_k, \ldots\rangle \tag{8.80}$$

On the other hand, the photon wave functions of the intermediate and the final states are

$$|n_v\rangle = |n_1, n_2, \ldots, n_k - 1, \ldots\rangle, \quad |n_f\rangle = |n_1, n_2, \ldots, n_k, \ldots\rangle \tag{8.81}$$

According to the golden rule of quantum mechanics, similar to the deducing method of Heitler [32], the transition probability of this process can be expressed as

$$W^{\mathrm{SNT}} = \frac{2\pi}{\hbar}\int \left|M_{if}^{(2)}\right|^2 \rho(E_f)\mathrm{d}E_f \tag{8.82}$$

Considering that the final state is an electronic discrete stationary state but a photon–phonon composed continuum, the product of final state density $\rho(E_f)\mathrm{d}E_f$ can be expressed as

$$\begin{aligned}\rho(E_f)\mathrm{d}E_f &= \rho(E_k)\rho(E_{\lambda'})\delta(E_{if} - p\hbar\omega_{\lambda'})\mathrm{d}E_k\mathrm{d}E_{\lambda'} \\ &= \frac{1}{\hbar}\rho(\omega_k)\rho(\omega_{\lambda'})\delta(\omega_{\lambda'} - \omega_k/p)\mathrm{d}\omega_k\mathrm{d}\omega_{\lambda'}\end{aligned}$$

In the above deduction, the following relations are used

$$\omega_k = E_{if}/\hbar,\ \omega_\lambda = \omega_k/p\ \delta(E_{if} - p\hbar\omega_{\lambda'}) = \frac{1}{\hbar}\delta(\omega_{\lambda'} - \omega_k/p)$$

Substituting these expressions to (8.82) and integrating over $\omega_{\lambda'}$, the probability of SNT can be expressed as follows

$$W^{\mathrm{SNT}} = \frac{2\pi}{\hbar^2}\rho(\omega_\lambda)\int \left|M_{if}^{(2)}\right|^2 \rho(\omega_k)\mathrm{d}\omega_k \tag{8.83}$$

In virtue of linewidth $\Delta\omega_k$ of the laser beam is very narrow, the above expression is approximately equal to

$$W^{\mathrm{SNT}} \approx \frac{2\pi}{\hbar^2}\left|M_{if}^{(2)}\right|^2 \rho(\omega_k)\Delta\omega_k\rho(\omega_\lambda) \tag{8.84}$$

The state density $\rho(\omega_k)$ of final photon state can be calculated by (4.15) of Chap. 4, which is rewritten in the following

$$\rho(\omega_k) = \frac{V\omega_k^2}{8\pi^3c^3}\mathrm{d}\Omega_k \tag{8.85}$$

The accepting mode of the non-radiative transition is optical mode, $\rho(\omega_\lambda)$ is the state density of optical phonon mode, which is a quantity related to the crystal structure and the magnitude of wave vector $\boldsymbol{k}$ and has not a simple analytic expression. The matrix element of (8.79) can be expressed as

$$\langle \Phi_{ik\lambda} | H_{\text{er}} | \Phi_{vk_1\lambda} \rangle = \langle \psi_i | \langle n_k | H_{\text{er}} | n_k - 1 \rangle | \psi_v \rangle \langle \varphi_{i\lambda} | \varphi_{v\lambda} \rangle \tag{8.86a}$$

$$\langle \Phi_{vk_1\lambda} | H_{\text{er}} | \Phi_{fk\lambda_1} \rangle = \langle \psi_v | \langle n_k - 1 | H_{\text{er}} | n_k \rangle | \psi_f \rangle \langle \varphi_{v\lambda} | \varphi_{f\lambda_1} \rangle \tag{8.86b}$$

Integrating the above transition matrix elements over the photon wave function at first and then referring to the (4.16) and (4.17) and the Poisson equation of quantum mechanics used in the derivation of (4.19) and (4.20), the transition matrix element for the photon-stimulated absorption process in electric dipole approximation can be obtained as

$$\langle \psi_i | H_{\text{er}} | \psi_v \rangle \langle \varphi_{i\lambda} | \varphi_{v\lambda_1} \rangle = \left[\frac{2\pi\hbar\omega_k n_k}{V\varepsilon} \right]^{1/2} \left\langle \psi_i \middle| e \sum_j r_j \middle| \psi_v \right\rangle \langle \varphi_{i\lambda} | \varphi_{v\lambda} \rangle \tag{8.87a}$$

For photon-stimulated emission the transition matrix element is

$$\langle \psi_v | H_{\text{er}} | \psi_f \rangle \langle \varphi_{v\lambda} | \varphi_{f\lambda_1} \rangle = \left[\frac{2\pi\hbar\omega_k n_k}{V\varepsilon} \right]^{1/2} \left\langle \psi_v \middle| e \sum_j r_j \middle| \psi_f \right\rangle \langle \varphi_{v\lambda} | \varphi_{f\lambda_1} \rangle \tag{8.87b}$$

For the special process considered here, only one intermediate state satisfy the condition of $E_{ik\lambda} - E_{vk_1\lambda} = 0$ and remember that the initial phonon state is the same as the phonon state of the intermediate electronic state, $|\varphi_{i\lambda}\rangle = |\varphi_{v\lambda}\rangle$, $\langle \varphi_{i\lambda} | \varphi_{v\lambda} \rangle = 1$ and so the square of matrix element in (8.79) can be expressed as

$$\left(M_{if}^{(2)} \right)^2 \approx M_{iv}^2 M_{vf}^2 \left| \langle \varphi_{i\lambda} | \varphi_{f\lambda_1} \rangle \right|^2 / \delta^2$$

Because

$$\varphi_\lambda = \prod_s \varphi_{\lambda_s}$$

$\left| \langle \varphi_{i\lambda} | \varphi_{f\lambda_1} \rangle \right|^2$ can be referred to as lattice relaxation factor f_{lr} having an expression the same as (6.47) of Chap. 6.

$$f_{\text{lr}} = \left| \langle \varphi_{i\lambda} | \varphi_{f\lambda_1} \rangle \right|^2 = e^{-S} \frac{S^{-p}}{p!}$$

where S is Huang–Rhys factor of the crystal, so that the probability of SNT is

$$W^{\mathrm{SNT}} \approx \frac{2\pi}{\hbar^2} M_{iv}^2 M_{vf}^2 \rho(\omega_k)\rho(\omega_\lambda)\Delta\omega_k f_{\mathrm{lr}}/\delta^2$$

$$M_{iv}^2 = \left[\frac{2\pi\hbar\omega_k n_k}{V\varepsilon}\right]\left|\left\langle \psi_i \middle| e\sum_j r_j \middle| \psi_v \right\rangle\right|^2$$

$$M_{vf}^2 = \left[\frac{2\pi\hbar\omega_k n_k}{V\varepsilon}\right]\left|\left\langle \psi_v \middle| e\sum_j r_j \middle| \psi_f \right\rangle\right|^2$$

By using the stimulated absorption transition probability from initial state to intermediate state

$$W_{iv}^{\mathrm{SRT}} = \frac{2\pi}{\hbar^2}|M_{iv}|^2 \rho(\omega_k)$$

and the spontaneous emission transition probability from intermediate state to final state is

$$A_{vf} = \frac{2\pi}{\hbar^2}|M_{vf}|^2 \rho(\omega_k)/n_k$$

Therefore, the SNT probability is expressed as

$$\begin{aligned} W^{\mathrm{SNT}} &\approx n_k A_{vf} W_{iv}^{SRT}\left(\frac{\hbar^2\Delta\omega_k\rho(\omega_\lambda)}{2\pi\delta^2\rho(\omega_k)}\right) f_{\mathrm{lr}} \equiv r n_k A_{vf} W_{iv}^{srt} \\ r &= \left(\frac{\hbar^2\Delta\omega_k\rho(\omega_\lambda)}{2\pi\delta^2\rho(\omega_k)}\right) f_{\mathrm{lr}} = \left(\frac{\hbar}{2\pi\delta}\right)\left(\frac{\rho(\omega_\lambda)}{\rho(\omega_k)}\right) f_{\mathrm{lr}} \end{aligned} \tag{8.88}$$

In the above expression $\delta = \hbar\Delta\omega_k$ has been used. The Huang–Rhys factor S is generally in the order of magnitude 10^{-2}–10^{-1} for rare earth ions in crystals, and in YAG crystal, it can supposed to be 0.05. At room temperature, the minimum energy separation between the crystal energy level Y_3 of $^4I_{11/2}$ and the crystal energy level X_3 of $^4I_{9/2}$ is about 2000 cm^{-1} [16] and the effective phonon energy of YAG crystal is about 700 cm^{-1}. The phonon number in (8.80) can be taken as $p = 3$, therefore, $f_{\mathrm{lr}} = 2 \times 10^{-5}$. On the other hand, by using $\Delta\nu_k \approx 2\times10^{10}$ s^{-1}, $\delta \approx 0.7$ cm^{-1} and $\hbar/2\pi\delta = 1.2 \times 10^{-12}$ s, it can be obtained that $r = 2.6\times10^{-17}\times(\rho(\omega_\lambda)/\rho(\omega_k))$ s. According to the ratio of fluorescence branch ratios of $R_2 \rightarrow Y_3$ transition to $R_2 \rightarrow X_3$ (Z_3) transition being 0.125/0.054 and $A(R_2 \rightarrow Y_3) \approx 1.5\times10^2$ s^{-1}, it can be estimated that $A(R_2 \rightarrow X_3) \approx 65$ s^{-1}. Substituting it into (8.88), the following expression resulted

$$W^{SNT}\left({}^4I_{11/2} \to {}^4I_{9/2}\right) \approx 6 \times 10^{-5}(\rho(\omega_\lambda)/\rho(\omega_k))W^{SRT}(R_2 \to Y) \qquad (8.89)$$

The calculation of the density $\rho(\omega_\lambda)$ of optical phonon mode for YAG crystal is a problem to be studied, however, owing to the fact that the phonon velocity is much less than that of the photon, without doubt the density $\rho(\omega_\lambda)$ of optical phonon mode has more than 5 orders of magnitude higher than that of the photon mode density $\rho(\omega_k)$. Therefore, when laser power reaches 10 W, it can be seen that

$$W^{SNT}\left({}^4I_{11/2} \to {}^4I_{9/2}\right) \gg W^{SRT}(R_2 \to Y_3)$$

That is, the probability of SNT from the lower laser level to the ground state is much higher than that of SRT from upper laser level to lower laser level. For lasers with higher output power, the multiple of the SNT probability higher than that of the SRT will be increased, and the bottleneck effect will never happen.

References

1. K. Huang, A. Rhys, Proc. Roy. Soc. **A204**, 406 (1950)
2. Yu.E. Perlin, Uspekli fiz. Nauk **80**, 553 (1963)
3. L.A. Riserberg, M.J. Wiber, in *Progress in Optics*, ed. by E. Wolf, vol. XIV (North- Holland Publ. Co., Amsterdam, 1976)
4. Yu.E. Perlin, A.A. Kaminskii, Phys. Stat. Sol. (B) **132**, 11 (1985)
5. M.F.H. Schuumams, J.M.F. van Dijk, Phys. B **123**, 131 (1984)
6. F. Auzel, in *Advances in Nonradiative Processes in Solids*, ed. by B. Di Bartolo (Plenum Press, New York, 1991)
7. W.E. Hagston, J.E. Lowther, Physica **70**, 40 (1973)
8. H. Kun, Scientia Sinica, **XXIV**, 27 (1981)
9. Huang Kun, Phys. Prog. **1**, 31 (1981). (in Chinese)
10. R. Englman, in *Radiationless Decay of Ions and Molecules in Solids* (North-Holland, 1979)
11. A.M. Stoneham, *The Theory of Defects in Solids* (Clarendon Press, Oxford, 1975)
12. F.K. Hong, *Theory of Molecular Relaxation: Application in Chemistry and Biology* (Wiley-Interscience, New York, 1975)
13. Z.D. Luo, Chin. J. Lumin. **20**, 1 (1999)
14. L.D. Landau, E.M. Lifshitz, *Quantum Mechanism* (Pergamon Press, Oxford, 1958)
15. K. Peuker, R. Enderlein, A. Schenk, E. Gutsche, Phys. Stat. Sol. B **109**, 599 (1982)
16. A. A. Kaminskii, *Crystalline Lasers: Physical Processes and Operating Schemes* (CRC Press, Boca Raton, 1996)
17. D.S. Ray, M. Chowdhury, Chem. Phys. **61**, 157 (1981)
18. Z.D. Luo, Chem. Phys. Lett. **94**, 498 (1983)
19. J. Irving, N. Mullineux, *Mathematics in Physics and Engineering* (Academic Press Inc., New York, 1959)
20. S.H. Lin, H.P. Lin, D. Knittel, J. Chem. Phys. **64**, 441 (1975)
21. K.F. Freed, J. Jortner, J. Chem. Phys. **52**, 6272 (1970)
22. B. Nitznan, S. Mukamel, J. Jortner, J. Chem. Phys. **63**, 200 (1975)
23. B. Henderson, G.F. Imbusch, *Optical Spectroscopy of Inorganic Solids* (Clarendon Press, Oxford, 1989)
24. Misu, J. Phys. Soc. Japan, 19, 2260 (1964)

25. YuV Orlovskii, K.K. Pukhov, T.T. Basiev et al., Opt. Mater. **4**, 583 (1995)
26. T.T. Basiev, YuV Orlovskii, K.K. Pukhov et al., Laser Phys. **8**, 1139 (1998)
27. K.K. Pukhov, T.T. Basiev, Yu.V. Orlovskii, et al., J. Lumin. **76/77**, 586 (1998)
28. T.T. Basiev, Yu.V. Orlovskii, K.K. Pukhov et al., J. Lumin. **68**, 241 (1996)
29. J.E. Geusic, IEEE J. Quant. Elect. 2: Digest of Tech. Paper, in *1966 International Quantum Electronics Conference*, Apr 1966
30. W. Koechner, in *Solid-State Laser Engineering* (Springer, 1976)
31. Z.D. Luo, in *Proceedings of the First International School on Excited States of Transition Elements*, ed. by B. Jezowska-Trzebiatowska, J. Legendziewicz, W. Strek (World Scientific, Singapore, 1989)
32. W. Heitler, *The Quantum Theory of Radiation* (Clarendon Press, Oxford, 1954)
33. J. Calloway, *Quantum Theory of the Solid State* (Academic Press, New York, 1976)
34. Z.D. Luo, Phys. Stat. Sol. b **133**, K77 (1986)

Chapter 9
Energy Transfer and Migration Between Ions

In the previous discussion on the absorption and emission of light, all processes take place in a single active center. In fact, other centers in the medium contribution to the absorption and emission of a single active center. It involves the physical processes of energy transfer and migration, which are very important in the study of laser materials. After absorbing excitation energy, the electrons in luminescent center transit to excited states, then the excitation energy can be consumed usually by light emission or transferred to thermal energy of lattice vibration. However, part of the excitation energy can be transferred from one center to other centers. Basically, there are three kinds of centers in laser materials, that is, active center (activator), sensitized center (sensitizer), and quenching center. The activator is luminescent center in the materials. The sensitizer is defined as the center used to absorb excitation energy then transfer it to the activator, whereas the quenching center only consumes the excitation energy being transferred to it. Certainly, ion, group, or molecule of host material can also act as a sensitizer to transfer absorbed excitation energy to activator. This is called as host sensitization. Usually, the energy exchange between sensitizers or activators is defined as energy migration, while the energy exchange between sensitizer and activator is defined as energy transfer. We will discuss these two processes but does not involve electric charge migration.

The energy transfer and migration discussed are the results of electric multi-pole, magnetic dipole, or exchange interactions. As early as 1940s, Forster [1, 2] developed the theory of dipole-electric interaction and Dexter [3] further generalized it to include high-order multi-pole and exchange interactions in 1950s. Orbach [4], Miyakawa and Dexter [5] extended the theory to include phonon-assisted transfer. With the rapid development in the field of laser and fluorescent materials, especially the development of powerful laser spectroscopy (time-resolved spectroscopy, selection excitation spectroscopy, fluorescence narrowing spectroscopy, and so on.), many micro processes including many ultrafast processes in the material can be observed in depth and concretely. All of these stimulate the research interest in the area of energy transfer and make it has a great progress. In this chapter, some basic

Z. Luo and Y. Huang, *Physics of Solid-State Laser Materials*, Springer Series in Materials Science 289, https://doi.org/10.1007/978-981-32-9668-8_9

theoretical knowledge of this field closely related to the spectroscopic properties of solid-state laser material will be introduced. There is a very extensive volume of literature published, which the reader can directly refer to. Among these, the review articles written by Auzel [6–8], Blasse [9, 10], Yen [11], and Holstein [12] are worthy to be read to obtain a general idea about the development in this field.

9.1 Theory of Resonant Energy Transfer

There are several types of energy transfer between active ions: Radiation reabsorption (the radiation trapping effect mentioned earlier), photoconductivity, exciton migration, and non-radiative resonance energy transfer. The following discussion is about non-radiative resonance energy transfer, in which the light transition is a virtual process. Only two-body interaction is to be dealt with, although there are three-body and even many-body interactions in fact. Phonon-assisted energy transfer and energy migration will be introduced later.

Assume that in the initial state *i*, there is a sensitizer (S) in excited state while a nearby activator (A) is in ground state. In the final state *f*, S goes down to ground state, while A goes up to excited state (as shown in Fig. 9.1). This process is introduced by a series of interactions between ions. One should first examine their interaction Hamiltonian.

Hamiltonian H of the system can be divided into two parts, H_0 of the single ion and H_{sa} of the interaction between two ions, expressed as

$$H = H_0 + H_{sa} \tag{9.1}$$

H_{sa} (excluding exchange interaction temporarily) is

$$H_{sa} = \frac{e^2}{\varepsilon|\boldsymbol{R}_{sa} - \boldsymbol{r}_s + \boldsymbol{r}_a|} \tag{9.2}$$

where R_{sa} is the distance vector between two nuclei, r_s and r_a are the electronic coordinates of S and A, respectively. ε is the dielectric constant of the material. Expanding (9.2) in terms of R_{sa}, then

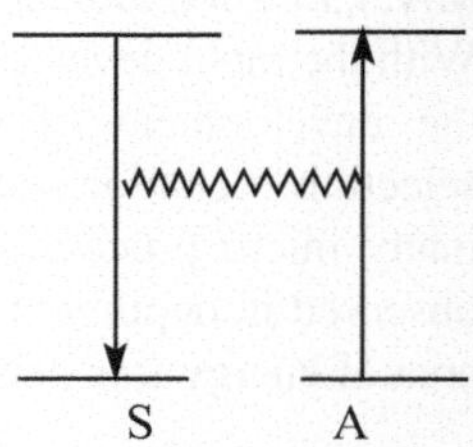

Fig. 9.1 Schematic diagram for resonant energy transfer between ions

$$H_{\mathrm{sa}} = H_{\mathrm{sa}}^{\mathrm{edd}} + H_{\mathrm{sa}}^{\mathrm{edq}} + H_{\mathrm{sa}}^{\mathrm{eqd}} + H_{\mathrm{sa}}^{\mathrm{eqq}} + H_{\mathrm{sa}}^{\mathrm{mdd}} + \cdots \tag{9.3}$$

where the first term is the electric dipole–dipole interaction expressed as

$$H_{\mathrm{sa}}^{\mathrm{edd}} = \frac{1}{\varepsilon R_{\mathrm{sa}}^{3}} \left[\boldsymbol{m}_{\mathrm{ed}}^{\mathrm{s}} \cdot \boldsymbol{m}_{\mathrm{ed}}^{\mathrm{a}} - \frac{3}{R_{\mathrm{sa}}^{2}} \left(\boldsymbol{m}_{\mathrm{ed}}^{\mathrm{s}} \cdot \boldsymbol{R}_{\mathrm{sa}} \right) \left(\boldsymbol{m}_{\mathrm{ed}}^{\mathrm{a}} \cdot \boldsymbol{R}_{\mathrm{sa}} \right) \right] \tag{9.4}$$

$\boldsymbol{m}_{\mathrm{ed}}^{\mathrm{s}}$ and $\boldsymbol{m}_{\mathrm{ed}}^{\mathrm{a}}$ denote the electric dipole of S and A, respectively. The other terms in (9.3) are the Hamiltonians of electric dipole–quadrupole, electric quadrupole–quadrupole, and magnetic dipole–dipole interactions. H_{sa} usually can be expanded as

$$H_{\mathrm{sa}} = \sum_{k_1 k_2 q_1 q_2} \left(\frac{e^2}{\varepsilon R^{k_1 + k_2 + 1}} \right) C_{q_1 q_2}^{(k_1 k_2)} D_{q_1}^{(k_1)}(r_{\mathrm{s}}) D_{q_2}^{(k_2)}(r_{\mathrm{a}}) \tag{9.5}$$

Multi-pole moment in the above expression is

$$D_q^{(k)}(r) = \sum_i \left(\frac{4\pi}{2k+1} \right)^{1/2} r_m^{(k)} Y_{kq}(\theta_i, \varphi_i) \tag{9.6}$$

It is summed over all the electrons in S or A and

$$C_{q_1 q_2}^{(k_1 k_2)} = (-1)^{k_1} \left[\frac{4\pi(2k_1 + 2k_2)!}{(2k_1)!(2k_2)!} \right]^{1/2} \begin{pmatrix} k_1 & k_2 & k_1 + k_2 \\ q_1 & q_2 & -q_1 - q_2 \end{pmatrix} \left[Y_{k_1 + k_2 q_1 + q_2}(\theta, \varphi) \right]^* \tag{9.7}$$

The first term in (9.3), that is, the (9.4), corresponds to the case $k_1 = k_2 = 1$. The second term corresponds to $k_1 = 1$, $k_2 = 2$, the third term corresponds to $k_1 = 2$, $k_2 = 1$, the fourth term corresponds to $k_1 = 2$, $k_2 = 2$, and so on. The wave function of the initial state is $|\Psi(r_{\mathrm{s}}, \alpha_{\mathrm{e}}; r_{\mathrm{a}}, \alpha_{\mathrm{g}})\rangle$ and that for the final state is $|\Psi(r_{\mathrm{s}}, \alpha_{\mathrm{g}}; r_{\mathrm{a}}, \alpha_{\mathrm{e}})\rangle$. α refers to the electronic states. Consequently, the probability of energy transfer becomes

$$W = \frac{2\pi}{\hbar} \left| \left\langle \Psi(\boldsymbol{r}_{\mathrm{s}}, \alpha_{\mathrm{e}}; \boldsymbol{r}_{\mathrm{a}}, \alpha_{\mathrm{g}}) \right| H_{\mathrm{sa}} \left| \Psi(\boldsymbol{r}_{\mathrm{s}}, \alpha_{\mathrm{g}}; \boldsymbol{r}_{\mathrm{a}}, \alpha_{\mathrm{e}}) \right\rangle \right|^2 \rho(E_f) \delta(E_f - E_i) \tag{9.8}$$

A detail expression for the energy transfer probability of electric dipole–dipole interaction can be introduced as follows. Let's focus on the energy transfer between one S and one A. By excluding the exchange interaction between these two ions, the zero-order wave function can be expressed as a product of the wave functions of these two ions, that is

$$\Psi(r_s, \alpha_i; r_a, \alpha_j) = \Psi(r_s, \alpha_i)\Psi(r_a, \alpha_j) \tag{9.9}$$

Owing to the fact that the direction of electric dipole is arbitrary in the space, substituting (9.4) into (9.8), the following scalar products should be averaged over γ, γ_1, and γ_2, that is, the angles between $\boldsymbol{r}_a$ and $\boldsymbol{r}_s$, $\boldsymbol{r}_s$ and $\boldsymbol{R}_{sa}$, and $\boldsymbol{r}_a$ and $\boldsymbol{R}_{sa}$, respectively

$$\boldsymbol{m}^{s}_{ed} \cdot \boldsymbol{m}^{a}_{ed} = m^{s}_{ed} m^{a}_{ed} \cos\gamma, \boldsymbol{m}^{s}_{ed} \cdot \boldsymbol{R}_{sa} = m^{s}_{ed} R_{sa} \cos\gamma_1, \boldsymbol{m}^{a}_{ed} \cdot \boldsymbol{R}_{sa} = m^{a}_{ed} R_{sa} \cos\gamma_2$$

It is easy to obtain

$$\frac{1}{4\pi}\int \cos^2\gamma \mathrm{d}\Omega = \frac{1}{3}, \frac{1}{4\pi}\int \cos\gamma \mathrm{d}\Omega = 0, \frac{1}{4\pi}\int \sin\gamma \mathrm{d}\Omega = 0 \tag{9.10}$$

The results for the average over γ_1 and γ_2 are the same. The average square of (9.4) consists of three terms and by using (9.10) it can be shown that the sum of the first and the third term is equal to $(4e^4/3)|\langle r_s\rangle|^2|\langle r_a\rangle|^2$. The calculation of the second term involves the relationship between three vectors and the angles γ_1, γ_2, and γ. The scalar products of $\boldsymbol{r}_a$ with $\boldsymbol{r}_s$, $\boldsymbol{r}_s$ with $\boldsymbol{R}_{sa}$ and $\boldsymbol{r}_a$ with $\boldsymbol{R}_{sa}$ can be calculated by using a spherical coordinate system with $\boldsymbol{R}_{sa}$ acting as a polar axis. In this way one has $\gamma = \gamma_1 - \gamma_2$. The calculation result of this second term is $-(2e^4/3)|\langle r_s\rangle|^2|\langle r_a\rangle|^2$. Therefore

$$\left\langle \left| \boldsymbol{m}^{s}_{ed} \cdot \boldsymbol{m}^{a}_{ed} - \frac{3}{R^2_{sa}}\left(\boldsymbol{m}^{s}_{ed} \cdot \boldsymbol{R}_{sa}\right)\left(\boldsymbol{m}^{a}_{ed} \cdot \boldsymbol{R}_{sa}\right) \right|^2 \right\rangle_{av} = \left(\frac{2e^4}{3}\right)|\langle r_s\rangle|^2|\langle r_a\rangle|^2 \tag{9.11}$$

where $|\langle r_s\rangle|^2$ denotes the matrix element $|\langle\Psi(r_s, \alpha_e)|r_s|\Psi(r_s, \alpha_e)\rangle|^2$, and similar for $|\langle r_a\rangle|^2$. The transition matrix element multiplied by state density $\rho(E_f)$ in (9.8) can be written as the transition matrix element multiplied by the product of line shape factor $g_s(E)G_a(E)$ and then integrate over the energy E, where

$$\int g_s(E)\mathrm{d}E = \int G_a(E)\mathrm{d}E = 1$$

For the sake of simplicity, the degeneracy of the energy levels has not been taken into account and by using (9.4) and (9.11), we obtained

$$W_{dd} = \frac{4\pi}{3\hbar\varepsilon^2}\frac{1}{R^6_{sa}}\int |e\langle r_s\rangle|^2|e\langle r_a\rangle|^2 g_s(E)G_a(E)\mathrm{d}E \tag{9.12}$$

If the probability A_s of S adopts the expression of (5.2a) and denote the factor $\left[3n/(n^2+2)\right]^2$ as β, then $|e\langle r_s\rangle|^2 = 3\hbar c^3\beta A_s/4\omega^3 n^3$. On the other hand, the square of dipole matrix element can be expressed by oscillator strength of (5.18a) as $|e\langle r_a\rangle|^2 = 3\mathrm{h}e^2\beta f_a/2mn\omega_s$. Transform the variable ω to wave number $\bar{\nu}$ by using

$g(E) = ng(\bar{\nu})/hc$ and $\omega n = 2\pi c\bar{\nu}$; substitute $A_s = \eta_s/\tau_{sf}$. The energy transfer probability of the electric dipole–dipole interaction finally takes the form

$$W_{dd} = \frac{1}{R_{sa}^6}\frac{3f_a\eta_s}{64\pi^4 n^3\tau_{sf}}\frac{e^2}{mc^2}\beta^2\int\frac{g_a(\bar{\nu})G_s(\bar{\nu})}{\bar{\nu}^4}d\bar{\nu} \tag{9.13}$$

where $\varepsilon = n^2$ has been used. The factor β approximately equals to 1 and so will be omitted in the following formulas. Using $\bar{\nu}_{av}$ to represent the average wave number in the overlap region of emission spectrum of S and absorption spectrum of A, then the transfer probability can be expressed approximately as

$$W_{dd} \approx \frac{1.36\times10^{-16}\text{cm}}{R_{sa}^6}\frac{f_a\eta_s}{n^3\tau_{sf}}\frac{O_{in}}{\bar{\nu}_{av}^4} \tag{9.14}$$

where the diameter of the electron $e^2/mc^2 = 2.82\times10^{-13}$ cm has been substituted and $O_{in} = \int g_a(\bar{\nu})G_s(\bar{\nu})d\bar{\nu}$ is the overlap integral between emission spectrum of S and absorption spectrum of A. If one uses both the oscillator strengths of S and A, then (9.12) takes the form

$$W_{dd} = \frac{1}{R_{sa}^6}\frac{3\pi\hbar e^4 f_a f_s}{n^6 m^2\omega^2}\int g_s(E)G_a(E)dE \tag{9.15}$$

Considering the difference of one factor $(1/4\pi\varepsilon_0)^2$ in different unit system, the difference in the definition of oscillator strength used and the approximation of $3n/(n^2+2) \approx 1$ adopted here, it can be seen that formula (9.15) is consistent with the formula (10.12) in the book of Henderson and Imbusch [13].

The conceptions of critical interaction distance and critical concentration have been used in the literature to deal with energy transfer between ions. When the separation between S and A equals to the critical distance, the energy transfer probability will be equal to the fluorescence decay rate $1/\tau_{sf}$ of S. By (9.14), it is obvious that the critical interaction distance takes the following form

$$R_c = 2.27\times10^{-3}\left(\frac{f_a\eta_s O_{in}}{n^3\bar{\nu}^4}\right)^{1/6}\text{cm} \tag{9.16}$$

The critical concentration C_c is defined as a concentration in which there is an interaction ion in the sphere with the radius of critical interaction distance; obviously

$$C_c = \left(\frac{4\pi}{3}R_c^3\right)^{-1} \tag{9.17}$$

By means of critical interaction distance, the energy transfer probability of dipole–dipole resonance can be written as

$$W_{dd} = \frac{R_C^6}{R_{sa}^6 \tau_{sf}} \tag{9.18}$$

The critical interaction distance and critical concentration can be roughly estimated for the energy transfer of rare earth ions. A typical value of 10^{-6} for the oscillator strength of rare earth ions can be assumed, and the wave number of the radiation is supposed to be 2×10^4 cm^{-1} and $\eta_s \approx 1$, $n = 1.7$. The overlap integral O_{in} is estimated to be 5×10^{-3} cm. Substituting all these data into (9.16) results in $R_c \approx 4.2 \times 10^{-8}$ cm. By (9.17) one can obtain $C_c = 3.2 \times 10^{21}$ cm^{-3}. It indicates that the energy is easy to be transferred by the interaction of ions.

The expressions of energy transfer probability by high-order multi-pole interaction will not be introduced in detail here. The relations between them can be written as

$$W_{dq} = (\lambda/R_{sa})^2 (f_Q/f_D) W_{dd} \tag{9.19}$$

$$W_{qq} = (\lambda/R_{sa})^4 (f_Q/f_D)^2 W_{dd} \tag{9.20}$$

where λ is the wavelength of the transition and f_D and f_Q are the oscillator strengths of electric-dipole and electric-quadrupole transitions, respectively. From (9.5), the energy transfer probability by multi-pole interactions is closely related to the distance between two kinds of ions:

Dipole-Dipole	$(1/R_{sa})^6$
Dipole-Quadrupole	$(1/R_{sa})^8$
Quadrupole-Quadrupole	$(1/R_{sa})^{10}$

Besides, energy transfer induced by the magnetic dipole–dipole interaction has similar distance dependence as that induced by the electric dipole–dipole interaction. For 3*d* electrons, the additional exchange interaction among them has to be taken into account. The energy transfer probability depends on both coordination number Z and wave function overlap integral S

$$W_{EX} \propto \exp[-2Z \ln S] \tag{9.21}$$

9.2 Phonon-Assisted Energy Transfer Between Ions

The energy transfer related energy level separation between S and A in the material is not well matched in many cases. Therefore, there is an energy mismatch between the photon energy emitted by S and that absorbed by A. Thus, the overlap integral O_{in} in (9.14) is equal to zero generally. The energy transfer in this situation makes

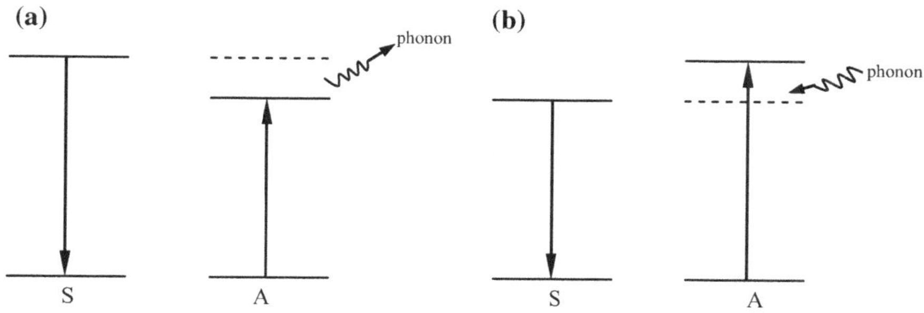

Fig. 9.2 Schematic diagram of phonon-assisted energy transfer, S: sensitizer, A: activator. **a** Energy mismatch ΔE_{sa} compensated by phonon emission; **b** energy mismatch ΔE_{sa} compensated by phonon absorption

using phonon as an intermediate. This process is called phonon-assisted energy transfer, and it is shown schematically in Fig. 9.2.

In this kind of energy transfer process, the energy mismatch can be compensated by absorbing or emission one or two phonons, so as to maintain the energy conservation of the whole process. However, the multi-phonon-assisted energy transfer has a very low probability. As will be discussed later, there is also an exponential energy gap law here.

Let's first introduce the expression of transfer probability for single-phonon-assisted energy transfer. It can also be calculated by the golden rule in quantum mechanics, that is

$$W_{1-\mathrm{p}} = \frac{2\pi}{\hbar} |M_{\mathrm{if}}|^2 \delta(\hbar\omega_k \pm \Delta E_{\mathrm{sa}}) \tag{9.22}$$

where the matrix element M_{if} is a second-order matrix element because it involves multi-pole interaction between ions as well as electron–phonon interaction, and can be expressed as the product of the matrix elements of ion–ion interaction and that of electron–phonon interaction

$$\begin{aligned} M_{\mathrm{if}} = & \sum_{m=\mathrm{s,a}} \frac{\langle a^*, s, n_k \pm 1|H_{\mathrm{sa}}|a, s^*, n_k \pm 1\rangle\langle a, s^*, n_k \pm 1|H_{\mathrm{ep}}(m)|a, s^*, n_k\rangle}{E_s - (E_s \pm \hbar\omega_k)} \\ & + \sum_{m=\mathrm{s,a}} \frac{\langle a^*, s, n_k \pm 1|H_{\mathrm{ep}}(m)|a^*, s, n_k\rangle\langle a^*, s, n_k|H_{\mathrm{sa}}|a, s^*, n_k\rangle}{E_{\mathrm{s}} - E_{\mathrm{a}}} \end{aligned} \tag{9.23}$$

where a and s denote the ground states (a^* and s^* denote excited states) of activator (A) and sensitizer (S), respectively. As introduced previously, the first-order electron–phonon interaction can be expressed in terms of the strain ε as

$$H_{\mathrm{ep}} = V_1 \varepsilon$$

In the above expression, strain tensor ε has been averaged over the solid angle in the space. By using formula in Chap. 6, ε can be expressed as linear combination of phonon creation and phonon annihilation operators and an angle averaged factor α is introduced, which has the order of magnitude 1.

$$\varepsilon = i \sum_{\boldsymbol{k}} \left(\frac{\hbar \omega_{\boldsymbol{k}} \alpha}{2Mv^2} \right)^{1/2} (b_{\boldsymbol{k}} - b_{\boldsymbol{k}}^{+}) \tag{9.24}$$

This expression has the similar form as that introduced in Sect. 6.5; the difference is that an angle factor α with order of magnitude 1 is introduced here.

On the other hand, if we use the notations $\langle \varphi | V_1 | \varphi \rangle = f$ and $\langle \varphi^* | V_1 | \varphi^* \rangle = g$ to denote the matrix elements of V_1 in low energy state φ and high energy state φ^*, respectively, where $\varphi = a$ or s, then

$$\begin{aligned}
\langle a, s^*, n_k \pm 1 | H_{ep}(s) | a, s^*, n_k \rangle &= g(s) \langle n_k \pm 1 | \varepsilon(s) | n_k \rangle \\
&= g(s) \langle n_k \pm 1 | \varepsilon | n_k \rangle \exp(\mp i\boldsymbol{k} \cdot \boldsymbol{r}_s) \\
\langle a, s^*, n_{jk} \pm 1 | H_{ep}(a) | a, s^*, n_{jk} \rangle &= f(a) \langle n_k \pm 1 | \varepsilon(a) | n_k \rangle \\
&= f(a) \langle n_k \pm 1 | \varepsilon | n_k \rangle \exp(\mp i\boldsymbol{k} \cdot \boldsymbol{r}_a) \\
\langle a^*, s, n_k \pm 1 | H_{ep}(s) | a^*, s, n_k \rangle &= f(s) \langle n_k \pm 1 | \varepsilon(s) | n_k \rangle \\
&= f(s) \langle n_k \pm 1 | \varepsilon | n_k \rangle \exp(\mp i\boldsymbol{k} \cdot \boldsymbol{r}_s) \\
\langle a^*, s, n_k \pm 1 | H_{ep}(a) | a^*, s, n_k \rangle &= g(a) \langle n_k \pm 1 | \varepsilon(a) | n_k \rangle \\
&= g(a) \langle n_k \pm 1 | \varepsilon | n_k \rangle \exp(\mp i\boldsymbol{k} \cdot \boldsymbol{r}_a)
\end{aligned} \tag{9.25}$$

The relation of strain ε with wave vector $\boldsymbol{k}$ and position vector $\boldsymbol{r}$ can be referred to Chap. 6. Owing to the fact that the ion–ion multi-pole interaction does not involve phonon coordinates and the phonon wave functions are normalized, we have

$$\langle a^*, s, n_k \pm 1 | H_{\mathrm{sa}} | a, s^*, n_k \pm 1 \rangle = \langle a^*, s, n_k | H_{\mathrm{sa}} | a, s^*, n_k \rangle = \langle a^*, s | H_{\mathrm{sa}} | a, s^* \rangle$$

If J_{sa} is used to represent the above matrix element, then it is easy to obtain

$$M_{\mathrm{if}} = \mp J_{\mathrm{sa}} \frac{\langle n_k \pm 1 | \varepsilon | n_k \rangle}{\Delta E_{\mathrm{sa}}} \exp(\mp i\boldsymbol{k} \cdot \boldsymbol{r}_{\mathrm{a}}) [(f(s) - g(s)) \exp(\mp i\boldsymbol{k} \cdot \boldsymbol{r}_{\mathrm{sa}}) - (f(a) - g(a))] \tag{9.26}$$

where $\boldsymbol{r}_{\mathrm{sa}} = \boldsymbol{r}_{\mathrm{s}} - \boldsymbol{r}_{\mathrm{a}}$ and ΔE_{sa} is absolute value of energy difference between the electronic excited state of S and A. For the single-phonon process, we have $\Delta E_{\mathrm{sa}} = |E_{\mathrm{s}} - E_{\mathrm{a}}| = \hbar\omega_k$. In the introduction of formula (9.26), it should be noted that when $E_{\mathrm{s}} > E_{\mathrm{a}}$, the electronic system will emit one phonon in the energy transfer process, thus the phonon state becomes $n_k + 1$, and the energy difference in the denominator of the first term of (9.23) is $-\hbar\omega_k = -\Delta E_{sa}$; when $E_s < E_a$, the electronic system will absorb one phonon in the energy transfer process, thus the phonon state become $n_k - 1$, and the energy difference in the denominator of the first term of (9.23) is $\hbar\omega_k = \Delta E_{sa}$. In the case of S and A are the same or similar kind of ion, $f(a) - g(a) \approx f(s) - g(s)$. By using (9.22), the single-phonon-assisted energy transfer probability takes the form

$$W_{1-p} = \frac{2\pi}{\hbar} \frac{J_{sa}^2 (f-g)^2}{\Delta E_{sa}^2} \sum_k |\langle n_k \pm 1|\varepsilon|n_k\rangle|^2 h(\boldsymbol{k}, \boldsymbol{r}_{sa}) \delta(\hbar\omega_k \pm \Delta E_{sa}) \tag{9.27}$$

$$h(\boldsymbol{k}, \boldsymbol{r}_{sa}) = |\exp(-i\boldsymbol{k} \cdot \boldsymbol{r}_{sa}) - 1|^2$$

On the other hand, by using (9.24), for a particular phonon mode one has

$$|\langle n_k \pm 1|\varepsilon|n_k\rangle|^2 = \left(\frac{\hbar\omega_k \alpha}{2Mv^2}\right) \begin{Bmatrix} n_k + 1 \\ n_k \end{Bmatrix} \tag{9.28}$$

The summation over k in (9.27) can be changed to integral according to the following way

$$\sum_k |\langle n_k \pm 1|\varepsilon|n_k\rangle|^2 = \frac{V}{8\pi^3} \int |\langle n_k \pm 1|\varepsilon|n_k\rangle|^2 4\pi k^2 dk \tag{9.29}$$

where $\frac{V}{8\pi^3}$ is the state density in k space, and it can be changed to an integral over ω_k by $\omega_k = kv$

$$\frac{V}{2\pi^2 v^3} \int |\langle n_k \pm 1|\varepsilon|n_k\rangle|^2 \omega_k^2 d\omega_k \tag{9.30}$$

The factor $h(\boldsymbol{k}, \boldsymbol{r}_{sa})$ has different values in two different cases. When energy mismatch ΔE_{sa} is large, $k \cdot r_{sa} > 1$. The wavelength of the phonon is in the same order of magnitude as the distance between S and A. In this case the space average of $h(k, r_{sa})$ is about 2. By using the following relation

$$\delta(\hbar\omega_k \pm \Delta E_{sa}) = \frac{d\omega_k}{dE} \delta(\omega_k \pm \Delta E_{sa}/\hbar) = \frac{1}{\hbar} \delta(\omega_k \pm \Delta E_{sa}/\hbar)$$

the single-phonon-assisted energy transfer probability can be calculated according to (9.27) to obtain the result as follows

$$W_{1-\mathrm{p}} = \frac{J_{\mathrm{sa}}^2 (f-g)^2 \Delta E_{\mathrm{sa}}}{6\pi\hbar^4\rho} \frac{\alpha}{v^5} \begin{Bmatrix} n_k + 1 \\ n_k \end{Bmatrix} \tag{9.31}$$

where ρ denotes the density of the material. When energy mismatch ΔE_{sa} is small, $k \cdot r_{\mathrm{sa}} \ll 1$, and the space average of $h(k, r_{\mathrm{sa}})$ is approximately $(kr_{\mathrm{sa}})^2/3$. After the similar calculation, the following expression for single-phonon-assisted energy transfer probability can be obtained

$$W_{1-\mathrm{p}} = \frac{J_{\mathrm{sa}}^2 (f-g)^2 \Delta E_{\mathrm{sa}}^3 r_{\mathrm{sa}}^2}{6\pi\hbar^6\rho} \frac{\alpha}{v^7} \begin{Bmatrix} n_k + 1 \\ n_k \end{Bmatrix} \tag{9.32}$$

In thermal equilibrium, n_k in (9.31) and (9.32) should be replaced by the following average value

$$\bar{n} = \frac{1}{\exp(\Delta E_{\mathrm{sa}}/k_{\mathrm{B}}T) - 1} \tag{9.33}$$

Two-phonon-assisted energy transfer probability can be introduced by third-order perturbation method in the similar way. For the case of small energy mismatch, two-phonon inelastic scattering may result in an efficient energy transfer, and in this case $\Delta E_{\mathrm{sa}} = \hbar\omega_{k'} - \hbar\omega_k$. If energy mismatch is larger than the energy of one phonon, in the third-order perturbation, the absorption and emission of two phonons satisfy $\Delta E_{\mathrm{sa}} = \hbar\omega_{k'} + \hbar\omega_k$.

The energy gap law of phonon-assisted energy transfer should be simply mentioned. It involves a system with strong electron phonon coupling. Although there is a certain amount of mismatch between the related energy gaps of S and A, but the overlapping integral of the phonon sideband of the emission spectrum for S and that of the absorption spectrum for A is not equal to zero. Miyakawa and Dexter [5] demonstrated by using generating function method that the relation of energy transfer probability for p-phonon-assisted process with that for the zero-phonon energy transfer process can be expressed as

$$W(p) = W_{\mathrm{sa}}(0)\exp(-\beta\Delta E) \tag{9.34a}$$

$W_{\mathrm{sa}}(0)$ and β is expressed as

$$W_{\mathrm{sa}}(0) = \frac{2\pi J_{\mathrm{sa}}^2 \int g_{\mathrm{a}}(E) G_{\mathrm{s}}(E) \mathrm{d}E}{\hbar} e^{-(\bar{g}_{\mathrm{s}} + \bar{g}_{\mathrm{a}})} \tag{9.34b}$$

$$\beta = (\hbar\omega_k)^{-1} \left\{ \ln \left[\frac{p}{(\bar{g}_{\mathrm{s}} + \bar{g}_{\mathrm{a}})(\bar{n}+1)} \right] - 1 \right\} \tag{9.34c}$$

In the above expression, $\bar{g}_{\mathrm{s}}$ and $\bar{g}_{\mathrm{a}}$ are the coupling coefficients of electron–phonon coupling (written as g_{a} and g_{b} in [5]). By using these expressions, the temperature-dependent relation can be deduced. Substituting $p = \Delta E_{sa}/\hbar\omega_k$ and (9.34c) into (9.34a), it can be seen that the p-phonon energy transfer probability is proportional to the following factor

$$[\bar{n}+1]^p = \left[\frac{\exp(\hbar\omega_k/k_B T)}{\exp(\hbar\omega_k/k_B T) - 1}\right]^p$$

If one uses the average phonon energy $\hbar\omega$ to express the compensation of the energy gap mismatch, $\hbar\omega = \Delta E_{sa}/p$. At low temperature $\exp(\hbar\omega/k_B T) \gg 1$, $\exp(\hbar\omega/k_B T)/(\exp(\hbar\omega/k_B T) - 1) \approx 1$, then the energy transfer probability is independent of the temperature. However, at a higher temperature, the transfer probability will be increased. For the transfer processes involving 4–10 phonons, Yamada [14] fitted the calculation values to the experimental data according to the above-mentioned factor and the result was satisfactory. For the transfer processes involving 1–2 phonons, Colins and Di Bartolo [15] observed that the transfer probability decreases with temperature. By taking into account the particle distribution in the initial and final states, the abnormal temperature dependence can be well explained. It demonstrates, on the other hand, that (9.34a) is suitable only for the transfer processes involving large number of phonons.

Auzel and Chen [16] pointed out that the phonon energy in energy transfer and phonon side band processes is actually lower than that in non-radiative transition process. They calculated this difference for trivalent erbium in glass ZBLAN and silicon glass. For trivalent erbium in glass ZBLAN, in the probability calculation of non-radiative transition, phonon energy of 580 cm^{-1} should be used while that of phonon side band or energy transfer, phonon energy can only be 325–350 cm^{-1}. For trivalent erbium in silicon glass, the former is $\hbar\omega = 1100\,cm^{-1}$ while for the latter it is $\hbar\omega = 500\,cm^{-1}$.

9.3 Statistical Theory of Energy Transfer Between Ions

The distance between S and A is different in different locations of the media. To obtain a formula which can be practically used in the calculation of transfer probability, it should be based on statistical theory. A detailed procedure to do this is as follows: First, investigate the rate equation of particle number for a certain S–A distance to establish an expression of time-dependent particle number; then statistically average the particle number variation to find out the average expression of the variation of particle number. Suppose S can be classified into several classes, in the case that same class S has the same A surrounding. By using (9.18), the k class S has the following expression of energy transfer probability

$$W_{\mathrm{dd}}^{k} = \sum_{i=1}^{N_{\mathrm{a}}} \frac{R_{\mathrm{c}}^{6}}{R_{ik}^{6}\tau_{\mathrm{sf}}} \tag{9.35}$$

The summation is over all the A, where R_{ik} is the distance between the S of the k class and the A of the i class. Assuming no energy migration among S, the particle number of S in k class will satisfy the following rate equation:

$$\frac{\mathrm{d}N_{sk}}{\mathrm{d}t} = -\frac{1}{\tau_{\mathrm{sf}}}\left(1 + \sum_{i=1}^{N_{\mathrm{a}}} \frac{R_{\mathrm{c}}^{6}}{R_{ik}^{6}}\right)N_{sk} \tag{9.36}$$

where the first term accounts for the decay of S by spontaneous emission and the second term corresponds to the decay produced by energy transfer. Integrating (9.36), then

$$N_{sk} = N_{sk}(0)\exp\left\{-(\tau_{\mathrm{sf}})^{-1}t\left[1 + \sum_{i=1}^{N_{\mathrm{a}}}(R_{\mathrm{c}}/R_{ik})^{6}\right]\right\} \tag{9.37}$$

For the random distribution of A, the probability that the S belongs to the k class is

$$P_{k}\mathrm{d}k = \prod_{i=1}^{N_{a}} \frac{4\pi R_{ik}^{2}\mathrm{d}R_{ik}}{V} \tag{9.38}$$

The statistical average of N_{sk} denoted by $\bar{N}_{sk}$ is

$$\bar{N}_{sk} = \int N_{sk}P_{k}\mathrm{d}k \tag{9.39}$$

In the problem discussed, the condition $(\tau_{\mathrm{sf}})^{-1}t(R_{\mathrm{c}}/R_{ik})^{6} \ll 1$ is always satisfied. In this situation, Forster [1] did the integration of (9.39) and obtained the following result

$$\bar{N}_{sk} = N_{sk}(0)\exp\left[-(\tau_{\mathrm{sf}})^{-1}t - (C_{\mathrm{a}}/C_{\mathrm{c}})(\pi t/\tau_{\mathrm{sf}})^{1/2}\right] \tag{9.40}$$

where C_{a} denotes the concentration of A, and C_{c} is the critical concentration.

On the basis of the above theory, Inokuti and Hirayama [17] assumed a random spatial distribution of A around S and which is not affected by the distribution of S. Without considering the energy migration between S, a general number decay expression of S in fluorescence level for multi-pole interactions (n = 6, 8, 10) was derived

$$\bar{N}_{sk} = N_{sk}(0)\exp\left[-(\tau_{sf})^{-1}t - \Gamma\left(1-\frac{3}{n}\right)(C_a/C_c)(t/\tau_{sf})^{3/n}\right] \tag{9.41}$$

The time-variation of S fluorescence intensity, therefore, is

$$I(t) = I(0)\exp\left[-(\tau_{sf})^{-1}t - \Gamma\left(1-\frac{3}{n}\right)(C_a/C_c)(t/\tau_{sf})^{3/n}\right] \tag{9.41a}$$

This expression is also written as

$$I(t) = I(0)\exp\left[-\frac{t}{\tau_{sf}} - \gamma t^{3/n}\right] \tag{9.41b}$$

where

$$\gamma = \Gamma\left(1-\frac{3}{n}\right)(C_a/C_c)\tau_{sf}^{-3/n} \tag{9.41c}$$

In these expressions, $n = 6$ for dipole–dipole interaction, $n = 8$ for dipole–quadrupole interaction, and $n = 10$ for quadrupole–quadrupole interaction. The gamma functions in (9.41) are:

$$\Gamma(1-3/6) = \Gamma(1/2) = \pi^{1/2} \approx 1.77,\ \Gamma(1-3/8) = \Gamma(5/8) \approx 1.43, \Gamma(1-3/10) = \Gamma(7/10) \approx 1.30.$$

Note that (9.40) and (9.41) are obtained in the condition of no energy migration among S.

The characteristic of this energy transfer pattern is: in the initial stage of the luminescence decay curve of S (i.e. $t \ll \tau_{sf}$), $\ln[I(t)/I(0)]$ is proportional to $t^{3/s}$ but with the increase of time, the number of non-excited A around excited S is decreased. Owing to the fact that the energy transfer probability between S and A is greatly decreased with the increase of distance between these two ions so that more S will decay by their luminescent emission and their luminescence decay curve will become approximately simple exponential decay curve. Here, simple exponential decay curve means that the exponent of exponential function only involves a linear term of time, while the non-simple exponential decay curve refers to the nonlinear term of time appearing in the exponent .That is to say, the total fluorescence decay curve is divided into two segments of the initial non-simple exponential decay and the subsequent approximately simple exponential decay. It should be pointed out that I-H model was obtained under the assumption of without energy migration between S; it is applicable for the case of low concentration of S and A.

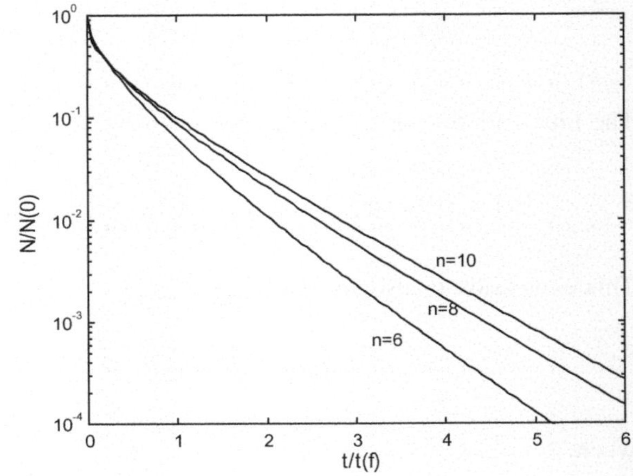

Fig. 9.3 Schematic diagram of the decay of particle number of sensitizer for different types of multi-pole interaction under the assumption of $C_a/C_0 = 1$. Abscissa is in the unit of the lifetime of S ion

The decay rates for different interactions are different. The dipole–dipole interaction produces the highest decay rate, the dipole–quadrupole interaction has a lower decay rate, and the quadrupole–quadrupole interaction has the lowest decay rate. These relations are shown in Fig. 9.3 with exponential coordinate of particle number versus t/τ_{sf}.

9.4 Energy Migration Between Ions

When A and S are coexistent in the sample and the concentration of S is high enough, the excited energy can migrate among S before transfer to A. This is a multi-step process. The transfer of excitation energy can be seen as the migration of exciton among S and the trapping of the exciton at site of A will modify the decay rule of the excited state population of S. In this section, the main results of the related theory will be briefly introduced. Generally speaking, there are two kinds of method to deal with this problem, namely random walk theory and diffusion theory.

First let us discuss the simplest case, in which the S-A interaction is equivalent to the S-S interaction and belong to the electric dipole–dipole interaction. During the energy transfer of S $\rightarrow$ A, S can also emit photon directly. The energy transfer to A will be described as exciton random walk to the positions of A (trap). Suppose the propagating time of the exciton from one ion to another ion is t_h. The emission probability of S per second is $1/\tau_s$. On the other hand, the possibility of the exciton fall into the traps for walking to a new lattice position is C_a (the concentration of A). When the exciton walks to a new lattice position, the probability for falling into the traps per second will be C_a/t_h, so that the lifetime of the exciton in this walking process is t_h/C_a. Therefore, the number N_0 of original exciton will be reduced to $N_0\exp\ [-t_h/(t_h/C_a)] = N_0\exp\ (-C_a)$; in other words, the decay rate of the exciton number by walking to a new lattice position is $\exp(-C_a)$. According to mathematical theory of random walk [18], the probability $W(0, n)$ for the n-step random walk can be expressed as the product of the corresponding probability of random walk at successive steps

$$W(0,n) = W(0,1)W(1,2)W(2,3)\ldots W(n-1,n) \tag{9.42}$$

The mathematical theory of random walk in lattice has been studied by Montroll et al. [19] According to their result, the number of lattice positions visited by the exciton after its n-step random walk (n is a very large number) can be expressed as $S_n n$ ($S_n = 0.66$ for simple cubic lattice, $S_n = 0.74$ for face-centered cubic lattice, and $S_n = 0.72$ for body-centered cubic lattice). After n-step random walk the decay rate of the exciton will be

$$[\exp(-C_a)]^{S_n n} = \exp(-nS_n C_a)$$

Supposing there are N_0 excitons at $t = 0$, then after time t_1 by the trapping effect, the exciton number will become

$$N_0 \exp(-nS_n C_a) \tag{9.43}$$

where $n = t_1/t_h$. On the other hand, after n-step walk, by the direct emission, the decay factor of exciton is obviously

$$\exp(-t_1/\tau_s) \tag{9.44}$$

The exciton number without both trapping and emission will be

$$N_0 \exp(-nS_n C_a)\exp[-t_1/\tau_s] = N_0 \exp[-t_1(S_n C_a/t_h + 1/\tau_s)] \equiv N_0 \exp(-t_1/\tau_s') \tag{9.45}$$

Therefore the fluorescence lifetime of S affected by both energy transfer and radiation transition will be

$$\tau_s' = \frac{1}{1/\tau_s + C_a S_n/t_h} \tag{9.46}$$

Suppose W_{rw} is the energy transfer probability generated by random walk; W_0 and W denote the transition probabilities without A and the total transition probability of S in case the coexistence of A and S, respectively, then

$$W = W_0 + W_{rw} \tag{9.47}$$

Obviously

$$W_{rw} = S_n C_a / t_h \tag{9.48}$$

By the formula introduced previously, the jumping time of exciton is the reciprocal of energy transfer

$$t_h = \tau_s (R_{ss}/R_c)^6 \tag{9.49}$$

where R_c is the critical interaction distance defined in (9.16) and R_{ss} denotes the distance between S and its nearest neighbor S. It is easy to realize that $(R_{ss}/R_c)^6 = (C_c/C_s)^2$, where C_s and C_c are the concentration and critical concentration of S.

If the fluorescence intensities of S without and with the presence of A denoted by I_{s0} and I_s, respectively, it can be proved that

$$I_s = I_{s0}/(1 + \tau_s W_{rw}) = I_{s0}/[1 + S_n C_a (C_s/C_c)^2] \tag{9.50}$$

By using the conceptions of exciton jumping and average jumping time, Brushtein introduced another luminescence decay formula of S in the case of S-S energy migration probability is higher than S-A energy transfer probability, that is $\alpha_{ss} > \alpha_{sa}$. This is called hopping model [20]

$$I(t) = I(0)\exp\left(-\frac{t}{\tau_{sf}} - \xi\sqrt{t} - K't\right) \tag{9.51}$$

where K' is energy transfer probability assisted by energy migration and the values of ξ and K' by the electric-dipole interaction can be expressed as

$$\xi = \frac{4}{3}\pi^{3/2} C_a \sqrt{\alpha_{sa}} \tag{9.52}$$

$$K' = \pi(2\pi/3)^{5/2} C_a C_s \sqrt{\alpha_{sa}\alpha_{ss}} \tag{9.53}$$

On the other hand, according to (9.14), by taking into account degeneracy g_e of the excited level and degeneracy g_g of the ground level of S, the energy transfer parameter α_{sa} and the energy migration parameter α_{ss} for the electric-dipole interaction can be expressed as

$$\alpha_{sa}^{dd} \approx 1.36 \times 10^{-16}\text{cm}\frac{f_a \eta_s}{n^3}\left(\frac{g_g}{g_e}\right)\frac{O_{in}}{\bar{\nu}^4}, \alpha_{ss}^{dd} \approx 1.36 \times 10^{-16}\text{cm}\frac{f_s \eta_s}{n^3}\left(\frac{g_g}{g_e}\right)\frac{O_{in}}{\bar{\nu}^4} \tag{9.54}$$

These two parameters also denote as C_{sa} and C_{ss} and can also be calculated by emission and absorption cross-section as follows

$$\alpha_{sa} \equiv C_{sa} = \frac{3c}{8\pi^4 n'^2}\left(\frac{g_g}{g_e}\right)\int \sigma_{em}^{(s)}(\lambda)\sigma_{ab}^{(a)}(\lambda)d\lambda,$$
$$\alpha_{ss} \equiv C_{ss} = \frac{3c}{8\pi^4 n'^2}\left(\frac{g_g}{g_e}\right)\int \sigma_{em}^{(s)}(\lambda)\sigma_{ab}^{(s)}(\lambda)d\lambda \quad (9.55)$$

In the case of energy migration among S has high probability; one can also treat the energy migration process as diffusion process. Martin et al. [21] used diffusion model to study energy migration and energy transfer processes for random distribution of ion. The density $\rho(\boldsymbol{R}, t)$ of excited S satisfy the following equation

$$\frac{\partial}{\partial t}\rho(\boldsymbol{R},t) = \left[D_s\Delta - \frac{1}{\tau_{sf}} - \sum_{j=1}^{N_A} W_{sa}(\boldsymbol{R}_s - \boldsymbol{R}_{aj})\right]\rho(\boldsymbol{R},t) \quad (9.56)$$

where D_s is a diffusion parameter which characterizes the migration processes among S, τ_{sf} is the lifetime of S and $W_{sa}(\boldsymbol{R}_s - \boldsymbol{R}_{aj})$ is the energy transfer probability from S at position $\boldsymbol{R}_s$ to A at position $\boldsymbol{R}_{aj}$. The first term in the right-hand side of this equation expresses the diffusion effect; it has not appeared in (9.36). By using Padé approximation to solve (9.56), we can obtain

$$I(t) = I(0)\exp\left[-\frac{t}{\tau_{sf}} - \frac{4\pi C_A}{3}\Gamma\left(1-\frac{3}{n}\right)\left(\alpha_{sa}^{(n)}t\right)^{3/n}\left(\frac{1+a_1x+a_2x^2}{1+b_1x}\right)^{(n-3)/(n-2)}\right] \quad (9.57)$$

where $x = D_s\left(\alpha_{sa}^{(n)}\right)^{-2/n} t^{1-2/n}$ and the expression of D_s as well as the values of a_1, a_2, and b_1, for dipole–dipole interaction $n = 6$, dipole–quadrupole interaction $n = 8$, and quadrupole–quadrupole interaction $n = 10$ respectively are shown as follows:

n	D_s	a_1	a_2	b_1
6	$D_s = 3.376C_s^{4/3}\alpha_{ss}^{(6)}$	10.866	15.500	8.743
8	$D_s = 2.924C_s^2\alpha_{ss}^{(8)}$	17.072	35.860	13.882
10	$D_s = 4.559C_s^{8/3}\alpha_{ss}^{(10)}$	24.524	67.909	20.290

Obviously, when $D_s = 0$, (9.57) is identical to (9.41) because in this case (9.56) is actually the same as (9.36).

Yokota and Tanimoto [22] solved (9.56) before Martin for the special case of electric-dipole interaction and obtained the result of (9.57) for $n = 6$

$$I(t) = I(0)\exp\left[-\frac{t}{\tau_{sf}} - \frac{4\pi C_a}{3}\pi^{1/2}\left(\alpha_{sa}^{(6)}t\right)^{1/2}\left(\frac{1+10.87y+15.50y^2}{1+8.743y}\right)^{3/4}\right] \quad (9.58)$$

$$y = D_s \left(\alpha_{sa}^{(6)}\right)^{-1/3} t^{2/3}$$

Equations (9.57) and (9.58) are all referred to as Yokota–Tanimoto model (Y-H model) in the literature.

In the above expressions, the diffusion parameter D_s is dependent on the S-S energy migration probability and the concentration of S. In the case of low S concentration Huber [23] obtained its expression for multi-pole interaction as follows

$$D_s = \frac{1}{2}(n-5)^{-1}\left(\frac{4}{3}\pi C_s\right)^{\frac{n-2}{3}} \alpha_{ss}^{(n)} \tag{9.59}$$

For dipole–dipole interaction, this diffusion parameter is

$$D_s = \frac{1}{2}\left(\frac{4}{3}\pi C_s\right)^{\frac{4}{3}} \alpha_{ss}^{(6)} \tag{9.60}$$

After a long enough time t, the fluorescence decay of S will be asymptotically simple exponential and if S-S energy migration is dipole–dipole process then the decay rate τ^{-1} follows the following equation

$$\frac{1}{\tau} = \frac{1}{\tau_{sf}} + W_{sa}^{(\mathrm{diff})} \tag{9.61}$$

Accordant to the calculation of Huber [23]

$$W_{sa}^{(\mathrm{diff})} = 8.5 C_s \left(\alpha_{sa}^{(6)}\right)^{1/4} D_s^{3/4} = 21 C_s C_a \left(\alpha_{sa}^{(6)}\right)^{1/4} \left(\alpha_{ss}^{dd}\right)^{3/4} \tag{9.62}$$

It can be demonstrated that when $\alpha_{ss} \gg \alpha_{sa}$ the hopping model is suitable, while when $\alpha_{ss} \ll \alpha_{sa}$ diffusion model is suitable.

The interaction between S increases with the temperature and is proportional to T^3 [24]. Therefore, the diffusion model can be used to describe the energy transfer at relatively low temperature, whereas at higher temperature the hopping model should be used. A typical example is presented in the following.

Hegarty et al. [24] studied the energy transfer rule of Nd^{3+}:PrF_3. Nd^{3+} concentration in this sample is 5 at.%. The emission of Pr^{3+} is observed under pulse laser excitation. The investigation of logarithm relation between W_{sa} (W_{DA}) and $\alpha_{ss}^{(6)}/\alpha_{sa}^{(6)}\left(\alpha_{DD}^{(6)}/\alpha_{DA}^{(6)}\right)$ in the temperature range of 10–13 K showed that the slope of $\left(\alpha_{ss}^{(6)}\right)^{3/4}$ is in agreement with the result obtained by diffusion model, but at the temperature higher than 15 K, it should be described by exciton hopping model. This is shown in Fig. 9.4.

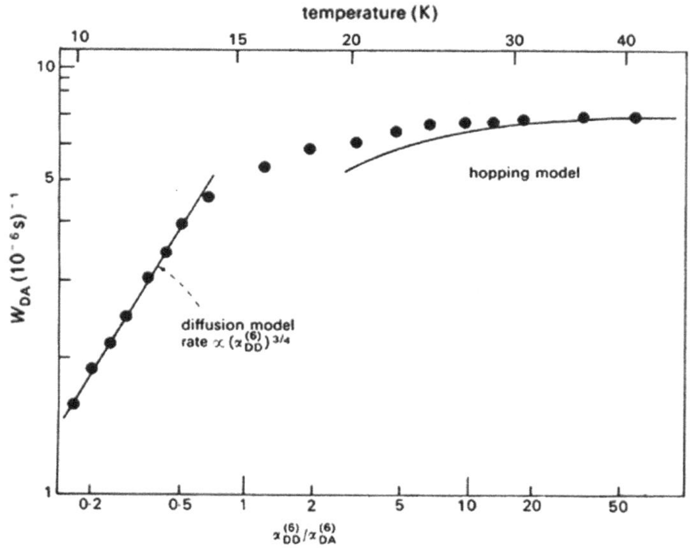

Fig. 9.4 The variation of exciton transfer versus temperature and the ratio of $\alpha_{DD}^{(6)}/\alpha_{DA}^{(6)}$ in Nd^{3+}: PrF_3 [24]

In some cases, the following rate equations can be used to calculate the energy transfer probability by means of the data of fluorescence intensity and fluorescence lifetime

$$\frac{dN_s}{dt} = P_s - \frac{N_s}{\tau_{sf}^0} - K_t N_s, \frac{dN_a}{dt} = -\frac{N_a}{\tau_{af}^0} + K_t N_s \tag{9.63}$$

where N_s and N_a are the particle numbers of S and A in excited state, while τ_{sf}^0 and τ_{af}^0 denote the lifetime of S and A, respectively. K_t is the transfer rate between S and A, while P_s is the exciting rate of S. Obviously, in the case of steady state (P_s is constant), the solutions of N_s and N_a are

$$N_s = \frac{P_s \tau_{sf}^0}{1 + K_t \tau_{sf}^0}, N_a = \frac{K_t P_s \tau_{sf}^0 \tau_{af}^0}{1 + K_t \tau_{sf}^0} \tag{9.64}$$

while in the case of pulse excitation, the solutions are

$$N_S(t) = N_S(0) \exp\left[-(1 + K_t \tau_{sf}) \frac{t}{\tau_{sf}}\right] \tag{9.65a}$$

$$N_A(t) = \frac{K_t N_S(0) \tau_{sf}^0 \tau_{af}^0}{\tau_{sf}^0 - \tau_{af}^0 - K_t \tau_{sf}^0 \tau_{af}^0} \left\{ \exp\left[-\left(1 + K_t \tau_{sf}^0\right) \frac{t}{\tau_{sf}^0}\right] - \exp\left[-\frac{t}{\tau_{af}^0}\right] \right\} \tag{9.65b}$$

The decay of fluorescence intensity with time written by (9.65a) will be

$$I(t) = I(0)\exp\left(-\frac{t}{\tau_{sf}^{0}} - K_t t\right) \tag{9.65c}$$

This expression is the same as (9.51) by neglected $\xi(t)^{1/2}$; however, the condition of obtaining (9.51) is $\alpha_{ss} > \alpha_{sa}$. Owing to the fact that $\xi = 7.42C_a(\alpha_{sa})^{1/2}$ and $K' = 19.9C_sC_a(\alpha_{ss}\alpha_{sa})^{1/2}$, if $C_s \gg C_a$ the relation $K't \gg \xi(t)^{1/2}$ is certainly satisfied. Therefore, when S-S energy migration probability is much higher than S-A energy transfer probability and the concentration of S is much higher than that of A, the decay of S fluorescence intensity after doping of A is still a simple exponential type; the only difference is the decay rate is increased. On the other hand, a conclusion can be drawn: the premise conditions of using the above rate equation to analyze energy transfer process are the S-S energy migration probability is much higher than that of the S-A energy transfer and the concentration of S is much higher than that of the A. The laser material of Yb^{3+} ions sensitize Tm^{3+} ions is one of the examples which satisfy this condition, because the energy migration between Yb^{3+} ions is a resonance process; what is more, the overlapping between the emission spectrum and the absorption spectrum of Yb^{3+} ion is great, while the energy transfer between Yb^{3+} ion and Tm^{3+} ion is a phonon-assisted process so that $\alpha_{ss} > \alpha_{sa}$. On the other hand, in this kind of laser material, in order to increase the pumping light absorption efficiency, the concentration of sensitized Yb^{3+} ion is much higher than that of the activated Tm^{3+} ion. Therefore, the above premise conditions for using rate equation to analyze energy transfer process are satisfied.

Energy transfer rate K_t can also be obtained by fluorescence lifetime measurement. If the fluorescence lifetime of S is τ_{sf} with A in the sample, while the fluorescence lifetime of S is τ_{sf}^{0} without A in the sample, then the transfer probability will be

$$K_t = \frac{\tau_{sf}^{0} - \tau_{sf}}{\tau_{sf}\tau_{sf}^{0}} \tag{9.66}$$

By using this formula, it is only necessary to obtain the experimental data of fluorescence lifetime for S with and without A in the sample, respectively. On the other hand, one can use integrated fluorescence intensities I_s and I_a of S and A when two kinds of ions coexist in the sample. It can be shown that in this case one has the following expression

$$\frac{I_a}{I_s} = \frac{A_a N_a \nu_a}{A_s N_s \nu_s} = \frac{\tau_{sr} N_a \lambda_s}{\tau_{ar} N_s \lambda_a} \tag{9.67}$$

where N_a and N_s are concentration, while A_s and A_a are spontaneously transition probabilities of S and A, respectively; meanwhile, τ_{sr} and τ_{ar} are corresponding radiative lifetimes. By using measured integrated luminescence intensities I_s and I_a, when two kinds of ions are present in the sample simultaneously and the solutions of (9.64) and (9.67), the energy transfer rate in steady-state condition can be obtained as

$$K_t = \frac{\tau_{ar} I_a \lambda_a}{\tau_{af} \tau_{sr} I_s \lambda_s} \tag{9.68}$$

To analyze the relation between the Yb^{3+} fluorescence lifetime and the concentration of Yb^{3+} and Tm^{3+} ions, one can also use the following simple model. To facilitate the discussion of the problem, the Yb^{3+} ion is denoted by S while the Tm^{3+} ion is denoted by A. First, let us discuss the effect of the presence of A on the number of S in excited state. Suppose the atomic percentage concentrations of S is C_s and that of A is C_a and S–S energy migration probability is W_{ss} while S $\rightarrow$ A energy transfer probability is W_{sa}.

Under certain excitation conditions, the S $\rightarrow$ A energy transfer rate is proportional not only to transfer probability W_{sa} but also to concentration C_a. Suppose the proportional constant is g, then by the fluorescence decay of S and S $\rightarrow$ A energy transfer, the change rate of S particle in excited state will be

$$\frac{dN_S}{dt} = -\frac{N_S}{\tau_{sf}} - gC_a W_{sa} N_S \tag{9.69}$$

where τ_{sf} is fluorescence lifetime of S. In the calculation of the number decrease of excited state S caused by fluorescence decay, the effect of radiation trapping should be deducted. Suppose the reabsorption probability of S is W_{reab}, the number of excited state S generated by the reabsorption is proportional to the product of $C_s W_{reab}$ and so can be written as $g' C_s W_{reab}$. Therefore, the rate equation of S should be

$$\frac{dN_S}{dt} = -\left(\frac{1}{\tau_{sf}} + gC_a W_{sa} - g' C_s W_{reab}\right) N_S \tag{9.70}$$

Integrating the above equation, we obtain

$$N_S(t) = N_S(0) \exp\left[-\left(\frac{1}{\tau_{sf}} + gC_a W_{sa} - g' C_s W_{reab}\right) t\right] \tag{9.71}$$

From the formula (9.13), the ratio of W_{sa} and W_{ss} is expressed as

$$\frac{W_{sa}}{W_{ss}} = s\left(\frac{R_{ss}}{R_{sa}}\right)^6 = s\left(\frac{C_s + C_a}{C_s}\right)^2 \tag{9.72}$$

where R_{ss} is the average separation between different S, R_{sa} is that between A and S, and the parameter s only depends on the spectral properties of A and S but independent of their concentrations. If R_{sc} represents the average separation of S at critical concentration C_{sc}, then by using (9.18), W_{ss} can be expressed as

$$W_{ss} = \tau_{sf}^{-1}\left(\frac{R_{sc}}{R_{ss}}\right)^6 = \tau_{sf}^{-1}\left(\frac{C_s}{C_{sc}}\right)^2 \tag{9.73}$$

Therefore

$$W_{sa} = \frac{s}{\tau_{sf}}\left(\frac{C_s + C_a}{C_{sc}}\right)^2 \tag{9.74}$$

In addition, the probability of radiation trapping is proportional to the fluorescence emission probability of S and inversely proportional to the square of ions separation [25]; therefore it can be written as

$$W_{reab} \propto \tau_{sf}^{-1}\left(\frac{1}{R_{ss}}\right)^2 = r\tau_{sf}^{-1}(C_s)^{2/3} \tag{9.75}$$

Integrating the effects of the above three processes, the S number in excited state at time t is

$$N_S(t) = N_S(0)\exp\left\{-\frac{t}{\tau_{sf}}\left[1 + sgC_{sc}^{-2}C_a(C_a + C_s)^2 - g'rC_s^{5/3}\right]\right\} \tag{9.76}$$

That is, after considering the fluorescence emission of S, the radiation trapping and the energy transfer to A, the fluorescence lifetime of S can be expressed as

$$\tau'_{sf} = \tau_{sf}/\left[1 - g'rC_s^{5/3} + sgC_{sc}^{-2}C_a(C_a + C_s)^2\right] \tag{9.77}$$

The second term in the denominator of the above expression takes into account the radiation trapping effect while the third term is the contribution of sensitizing effect on the fluorescence lifetime.

If one discusses not the S $\rightarrow$ A energy transfer but the concentration dependent fluorescent quenching of S, the same deduction method can be used. However, the above S $\rightarrow$ A energy transfer should be replaced by A $\rightarrow$ Q energy transfer (Q denotes the quenching center). To discuss the fluorescence lifetime of A varied with the concentration of A and Q, the above parameters r, g, and g' should be replaced by r_1, g_1, and g_1', then

$$\tau'_{af} = \tau_{af}/\left(1 - r_1g'_1C_a^{5/3} + s_1g_1C_{ac}^{-2}C_q\left(C_q + C_a\right)^2\right) \tag{9.78}$$

whereτ_{af} is fluorescence lifetime of A, C_q is the atomic percentage concentrations of quenching center Q, and C_{ac} is the critical concentration of A. The second term on the denominator of the above expression takes into account the radiation trapping effect, while the third term is the contribution of quenching effect on the fluorescence lifetime.

Equation (9.78) can be used to analyze the concentration dependent fluorescent quenching and radiation trapping effect of Yb^{3+}: YAB crystal [26]. Fluorescence concentration quenching of Yb^{3+} ion in this crystal is due to the impurity ion in the crystal, because the purities of Yb_2O_3 and Y_2O_3 raw materials used for growing crystals were 99.99 and 99.999% respectively. The percent content of impurity ions in the crystal, especially that of the rare earth impurities, which are difficult to separate in the raw material, can be seen as proportional to the percentage of Yb_2O_3 in the crystal. On the other hand, from the crystal growth point of view, it can be assumed that the point defects which may form the energy traps in the crystal are also related to the concentration of doped Yb^{3+} ion. At lower Yb^{3+} concentration (e.g. lower than 50%), the concentration of point defect can be seen as proportional to the Yb^{3+} concentration while at the Yb^{3+} concentration higher than 50%, the concentration of point defect can be seen as decreased with Yb^{3+} concentration, because the increase of Yb^{3+} concentration is equivalent to the reduction of impurity ion Y^{3+} from the perfect crystal of YbAB. If the atomic percentage concentration of Yb^{3+} ion is C_{Yb} then the atomic percentage of fluorescence quenching center can be assumed as αC_{Yb} for the Yb^{3+} concentration lower than 50%. In (9.78) let $C_{\mathrm{a}} = C_{\mathrm{Yb}}$ and $C_{\mathrm{q}} = \alpha C_{\mathrm{Yb}}$, then the following can be obtained

$$\tau'_{\mathrm{af}} = \tau_{\mathrm{af}} / \left(1 - r_1 g'_1 C_{\mathrm{Yb}}^{5/3} + s_1 g_1 \alpha (1+\alpha)^2 C_{\mathrm{ac}}^{-2} C_{\mathrm{Yb}}^3\right) \equiv \tau_{\mathrm{af}} / \left(1 - A C_{\mathrm{Yb}}^{5/3} + B C_{\mathrm{Yb}}^3\right) \tag{9.79}$$

$$A = r_1 g'_1, B = s g_1 \alpha (1+\alpha)^2 C_{\mathrm{ac}}^{-2}$$

The measured values of the fluorescence lifetime for Yb^{3+}: YAB powder sample placed in index matching fluid with different Yb^{3+} concentrations are listed in Table 9.1. The experimental data can be fitted to the (9.79) very well. The fitting parameters are $\tau_{\mathrm{af}} = 510$ μs, $A = 1.89$ and $B = 95.79$. Figure 9.5 is the fitting result

Table 9.1 Yb: YAB crystal fluorescence lifetimes at different Yb^{3+} atomic percentage concentrations [26]

C_{Yb} (%)	τ_{bf} (μs)	τ_{pf} (μs)	τ_{df} (μs)	τ_{r} (μs)
5.6	680	/	/	600
6.5	960	557	513	606
17.3	910	449	480	652
29.2	237	141	490	620
40.7	194	122	509	616
65.5	20	23	541	575
75.0	77	26	520	613
100	17	23	535	672

τ_{bf}, τ_{pf}, τ_{df}, andτ_{r} are the Yb^{3+} fluorescence lifetime measured values in bulk sample, powder sample, and diluted powder sample, respectively and radiative lifetime calculated values of $^2F_{5/2}$ multiplet. The data of τ_{bf} are taken from reference [27]

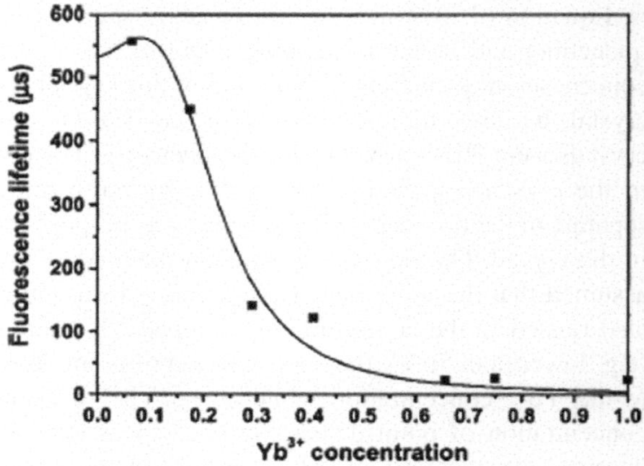

Fig. 9.5 Room temperature fluorescence lifetime of Yb^{3+}: YAB crystal in powder samples at 1038 nm under excitation at 930 nm versus Yb^{3+} concentration. The points represent the experimental data and the solid curve is the fitting result from (9.79) [26]

of the relation of the fluorescence lifetimes at 1038 nm with the Yb^{3+} atomic percentage concentration for Yb^{3+}: YAB crystal powder sample at room temperature.

The above analysis of the effects of concentration dependent fluorescent quenching and radiation trapping is only fitted to the data of powder sample placed in index matching fluid. However, due to the fact that the refractive index of the matching fluid still has some difference from that of the YAB crystal powder, the internal reflection is still existence, so that $A \neq 0$. Nevertheless, the particle size of the samples with different Yb^{3+} concentration is the same; the values of A for different sample are approximately the same, so the experimental data can be fitted. The result obtained by this method can also easily explain the data of diluted powder sample in Table 9.1. All these powder samples are diluted by pure YAB powder to an average Yb^{3+} concentration 0.5%. Although the difference of the distribution of Yb^{3+} ions in the sample causing the difference in parameter r results in several times difference of parameter A, but because the Yb^{3+} concentration is very low, $F = 1 - AC_{\mathrm{Yb}}^{5/3} + BC_{\mathrm{Yb}}^{3}$ and is always very close to 1 ($A = 2.0$, $F = 0.9989$; $A = 20.0$, $F = 0.9959$). This is the reason that the fluorescence lifetimes of all these diluted samples are about 510 μs.

In Table 9.1, the fluorescence lifetime of bulk sample certainly cannot be explained by (9.79). One of the important reasons is in these samples, the defects, bubbles, and dislocations, which can cause the internal reflection of fluorescence emission, are different for different crystal grown batches, and even different for the different parts of the same piece of grown crystal. This complexity makes the theoretical model not correctly describe the contribution of radiation trapping effect on the fluorescence lifetime. However, it can be seen from this table that the fluorescence lifetime of bulk sample with low Yb^{3+} concentration is more significantly longer than that of the powder sample in index matching fluid with same Yb^{3+} concentration. The reason for this phenomenon is that in low Yb^{3+} ion concentration bulk sample the internal reflection radiation can maintain at a higher intensity for a longer time in spite of the weak reabsorption of radiation in a single pass.

With the increasing of Yb^{3+} concentration, the fluorescence lifetime of diluted samples has a change from longer to shorter and then from shorter to longer. The reason for this change can be interpreted as: when the concentration of Yb^{3+} ion is lower than 50%, the increase of Yb^{3+} ion concentration corresponds to the increase of crystal defects; while when the concentration of Yb^{3+} ion is higher than 50%, the increase of Yb^{3+} ion concentration corresponds to the reduction of crystal defects as has been explained.

The study of energy transfer always involves the calculation of fluorescence lifetime. It is necessary to discuss the problem of how to calculate this parameter correctly.

It can be seen from the above discussion that in the case of isolated ions or energy migration probability between ions is higher than that of energy transfer probability, the fluorescence decay is described by a simple exponential function. However, in the general case of energy transfer between ions, such as that shown in (9.41), (9.51), and (9.58), in addition to a linear term of time, a nonlinear term of time appears in the exponent of the function of fluorescence intensity. In this case, the attenuation curve of the fluorescence intensity in the logarithmic coordinate is not a straight line so that the calculation method of fluorescence lifetime should be studied. Some one used the following formula to calculate this kind of non-simple fluorescence decay

$$\tau_{\mathrm{f}} = \int_0^{\infty} I(t)\mathrm{d}t/I_0 \tag{9.80a}$$

It can be shown that this formula is applicable only to the case of simple fluorescence decay and can be demonstrated from the definition of fluorescence lifetime.

Fluorescence lifetime is the time in which the fluorescent particle stays in the fluorescent level. Obviously, the particles transition at different moments has different lifetimes, so that only the average lifetime of the particle in the fluorescent level has physical meaning. If one divides the fluorescence process into infinite small time intervals Δt, at moment t the particle number in the fluorescent level is $N(t)$ and the transition probability for the downward transition is A, then the particle number downward transition from the moment t to the moment $t + \Delta t$ will be $N(t)A\Delta t$. The fraction of particle having lifetime t relative to the total particle number of fluorescent level is

$$N(t)A\Delta t/\sum_{t=0}^{t=\infty} N(t)A\Delta t$$

Therefore, the average fluorescence lifetime is

$$\tau_{\mathrm{f}} = \lim_{\Delta t \to 0}\left(\sum_{t=0}^{t\to\infty} t\left[\frac{N(t)A\Delta t}{\sum_{t=0}^{t\to\infty} N(t)A\Delta t}\right]\right) = \frac{\int_0^\infty tN(t)A\mathrm{d}t}{\int_0^\infty N(t)A\mathrm{d}t} = \frac{\int_0^\infty tI(t)\mathrm{d}t}{\int_0^\infty I(t)\mathrm{d}t} \tag{9.80b}$$

This formula is the correct one for the fluorescence lifetime calculation.

The calculation formula (9.80a) is practically the result of (9.80b) in a special case when the fluorescence decay satisfy the following equation

$$I(t) = I_0 e^{-At}, \text{ where } \tau_f = A^{-1}$$

According to (9.80a), the fluorescence lifetime will be

$$\tau_{\mathrm{f}} = \int_0^\infty I(t)\mathrm{d}t/I_0 = \int_0^\infty e^{-At}\mathrm{d}t = -A^{-1}\int_0^\infty \mathrm{d}e^{-At} = A^{-1} = \tau_{\mathrm{f}}$$

Substitute this simple exponential expression of $I(t)$ into the right side of (9.80b), then

$$\begin{aligned}\int_0^\infty tI(t)\mathrm{d}t &= I_0\int_0^\infty te^{-At}\mathrm{d}t = -I_0A^{-1}\int_0^\infty t\mathrm{d}e^{-At} = -I_0A^{-1}te^{-At}\Big|_0^\infty + I_0A^{-1}\int_0^\infty e^{-At}\mathrm{d}t \\ &= \tau_{\mathrm{f}}\int_0^\infty I(t)\mathrm{d}t\end{aligned}$$

that is, it agrees with (9.80b). However, if the fluorescence decay satisfy the following equation

$$I(t) = I_0 e^{-At-Bt^q}\,(q<1)$$

the results obtained by (9.80a) and (9.80b) are different. It means that in this case the lifetime obtained by (9.80a) is not consistent with the correct definition of fluorescence lifetime.

9.5 Characteristics of Concentration Dependent Fluorescence Quenching for Self-activated Laser Crystals

In this section we would like to take the self-activated laser crystals as an example to introduce the research result of the author to the relation of energy transfer probability with the chemical composition of the crystals. The so-called self-activated laser crystal is one kind of crystals in which the active ion (A) is the constituent of the crystal but not doped impurity. It is well known that for many laser crystals such as YAG and YAP, the concentration of A is usually a few percent in atomic percentage. For the crystals with higher concentration of A, the quenching effect has to be considered. A general viewpoint about the requirement for a crystal without quenching effect is as follows: the shortest separation between the positions which can be occupied by A should be longer than 60 nm. However, among the crystals satisfying this geometric condition, what is the relation between the quenching effect and the characteristics of the host crystals? Auzel [8] reported that for the weak quenching crystals, the crystal field should be weak enough to satisfy the following inequality

$$N_v = \left(\sum_{k,q\neq 0} B_{kq}^2\right)^{1/2} \leq 1800\,\text{cm}^{-1} \tag{9.81}$$

The investigation of emission intensity and fluorescence lifetime has shown that the strength of quenching effect is proportional to the square of the concentration of A for the crystals like YAG and YAP, but it is proportional to the concentration of A for self-activated laser crystals [28]. Obviously, the fluorescence quenching in the crystals like YAG and YAP is generated by the electric dipole–dipole cross-relaxation process $^4F_{3/2}$ + $^4I_{9/2}$→$^4I_{15/2}$ + $^4I_{15/2}$, because the probability of this process is inversely proportional to the six powers of the separation between A, as shown in (9.14); that is, proportional to the square of the concentration of A.

Auzel [29] discussed this problem by the overlap integral $\int g_a(E)G_s(E)\mathrm{d}E$ in the expression of resonant energy transfer probability of (9.12). In order to have a non-zero value of this integral, the energy released by the Nd^{3+} ions in excited state should be equal to the energy absorbed by the neighbor Nd^{3+} ions in the ground state. It requires that the related pair of energy levels satisfy the matching condition. It can be clearly seen in Fig. 9.6.

Figure 9.6 shows that the resonant energy transfer condition is $E_{1M} = E_{2M}$. e_1 is the separation between barycenters of $^4I_{15/2}$ and $^4I_{9/2}$ while e_2 is the separation between barycenters of $^4F_{3/2}$ and $^4I_{15/2}$. The resonant energy transfer condition can be expressed as

$$\Delta E_2 - \Delta E_1/2 + \Delta E_3/2 = e_1 - e_2 \tag{9.82}$$

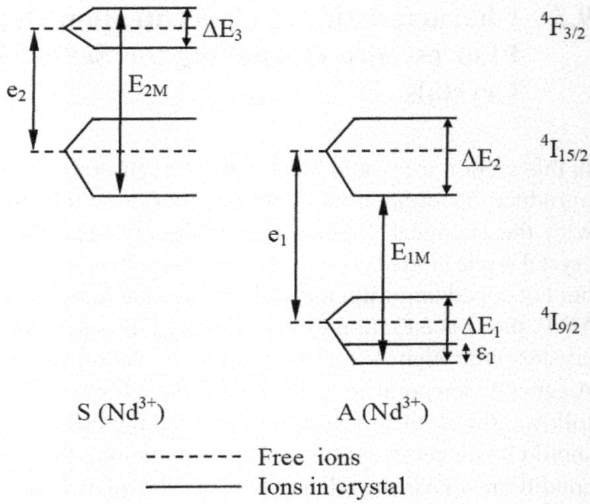

Fig. 9.6 Schematic diagram for the energy levels related to the concentration quenching effect of fluorescence level $^4F_{3/2}$ of Nd^{3+} ion

where ΔE_1, ΔE_2, and ΔE_3 are crystal field splitting of $^4I_{9/2}$, $^4I_{15/2}$, and $^4F_{3/2}$, respectively. The following scalar crystal field parameter is a very interesting quantity

$$N_v = \left[\left(\frac{4\pi}{2k+1}\right)\sum_{k,q\neq 0} B_{kq}^2\right]^{1/2} \approx \left[\sum_{k,q\neq 0} B_{kq}^2\right]^{1/2} \tag{9.83}$$

It was found experimentally [28] that no matter the Nd^{3+} ion sited in what kind of position, the crystal field splitting of the above spectral terms approximately have the following relations

$$\Delta E_1 \approx 0.23N_v, \Delta E_2 \approx 0.27N_v, \Delta E_3 \approx 0.026N_v \tag{9.84}$$

Substituting the above relations to (9.82), it is easy to find

$$N_v = (e_1 - e_2)/0.168$$

Therefore, if the crystal field is weak enough to satisfy the condition $N_v < (e_1 - e_2)/0.168$, then the resonant energy transfer condition cannot be satisfied. The fluorescence energy cannot transfer from multiplet $^4F_{3/2}$ of one Nd^{3+} ion to multiplet $^4I_{15/2}$ of another Nd^{3+} ion by cross-relaxation, and the fluorescent quenching will not occur. This is the reason why the strength of quenching effect of self-activated laser crystal is not proportional to the square of the concentration of A. It should be seen that apart from the above-mentioned cross-relaxation, the fluorescence quenching effect of Nd^{3+}-doped laser crystal may also be the energy migration between $^4F_{3/2}$ multiplet of different Nd^{3+} ions and finally fall into the "trap" of impurity ions or crystal defects that existed in the crystal. This is the reason why the strength of quenching effect is proportional to the concentration of A in self-activated laser crystal, because the concentration of the "trap" is usually proportional to the concentration of doping ions.

However, there is still no answer to the relation of quenching effect with chemical composition of crystals. The author [30] pointed out that the fluorescence quenching effect is closely related to the chemical characteristics of ligand anions or ligand anion groups by using a parameter R, or parameter $L = 1/R = r_-/(r_+\Delta x)$ [31] to measure the strength of quenching effect, where r_+ is the radius of the cation directly connected to the coordination anion, r_- is the radius of the anion, and Δx is the difference of the electronegativity between cation and anion. Taking seven Nd^{3+}-doped laser crystals as an example, the relation of parameter L with Nd^{3+} concentration of unit quenching ratio is listed in Table 9.2. If τ_r indicates the radiative lifetime of Nd^{3+} ion in the crystal, τ_f is its fluorescence lifetime. Quenching rate of fluorescence is defined as $K_q = (\tau_r - \tau_f)/\tau_f\tau_r$. The quenching rate of fluorescence lifetime reduced to half of the radiation lifetime is obviously $K_{1/2} = 1/\tau_r$. Quenching ratio is defined as $R_q = K_q/K_{1/2} = (\tau_r - \tau_f)/\tau_f$. It can be clearly seen that the crystals with larger L value have higher Nd^{3+} concentration of unit quenching ratio. $NdAl_3(BO_3)_4$ crystal has a large value of L and so the quenching effect is very weak. It has been demonstrated by experiment that this crystal is a good material for microchip laser [32–36].

To study the relationship between the concentration quenching strength with the chemical composition of the crystals, the situation when the resonant energy transfer condition cannot be satisfied should also be discussed. The single-phonon-assisted energy transfer probability can be calculated by (9.31) and (9.32) for the cases of the energy mismatch is large and small, respectively. It can be seen from these formulas that the key quantities are

$$f = \left\langle k \left| \frac{\partial V}{\partial Q} \right| k \right\rangle, \; g = \left\langle k^* \left| \frac{\partial V}{\partial Q} \right| k^* \right\rangle$$

and phonon velocity v. The probability of energy transfer is proportional to $(f - g)^2$ and inversely proportional to v^5 or v^7. In order to have a weaker cross-relaxation, the magnitude of f and g should be small and the phonon velocity v should be high. It has been pointed out [30] that the weak crystal field leads to a small magnitude of f and g as well as the difference between them. Consequently, it is important to investigate the chemical composition of Nd^{3+}-doped laser crystals in which Nd^{3+} ions experience a weak crystal field as well as a high phonon velocity. What are the relations of the parameter L with crystal field strength and the velocity of phonon? If the radius r_+ of the ligand directly connected to Nd^{3+} ion is smaller but r_- of the cations is larger, and the difference of the electronegativity Δx between the cation and the anion is smaller, then the parameter $L = r_-/(r_+\Delta x)$ will assume a larger value. In this case, the polarization effect of the cations on the ligand will be stronger. It means that the ligands of the Nd^{3+} ion will have a smaller effective charge. It is easy to realize that it corresponds to a weak crystal field as well as a smaller variation of the crystal field potential with the movement of normal

Table 9.2 Relation of parameter L with Nd^{3+}concentration of unit quenching ratio R_q

Crystal	Space group	$L = R_-/R_+\Delta x$	Nd^{3+} number of unit quenching ratio ($10^{21}/cm^3$)
$NdAl_3(BO_3)_4$ (NAB)	D_3^7-R32	4.3	2.17
$LiNd(PO_3)_4$	C_{2h}^6-C2/c	2.7	1.62
$NaNd(PO_3)_4$	C_{2h}^6-C2/c	2.7	1.55
NdP_5O_{14}	C_{2h}^5-P2/c	2.7	1.56
$NaNd(WO_4)_4$	C_{4h}^6-I4$_1$/*a*	1.5	0.95
$K_5Nd(MoO_4)_4$	C_{3v}^5-R3m	1.5	0.75
$NdMgAl_{11}O_{19}$	D_{6h}^4-6$_3$/mmc	1.2	0.25

coordinates so that a smaller value of $f - g$. On the other hand, the phonon velocity is inversely proportional to the square root of the compressibility of material while the compressibility of the ionic crystal is decreased with the decrease of ion–ion spacing and the increase of electrovalence [37] (the increase of cation valence corresponds to its electronegativity closer to the electronegativity of anions [38]). Therefore, for the same oxide (the same r_-), the smaller cation radius r_+ and smaller electronegativity difference (a larger parameter L) corresponds to a weak crystal field (a smaller $f - g$) and a smaller a compressibility (a higher phonon velocity) of the host material, that is, a lower phonon-assisted energy transfer probability as well as a lower cross-relaxation probability and therefore a weak fluorescence concentration quenching effect of Nd^{3+} ions. Of course, the concentration dependent fluorescence quenching is an effect conditioned by many factors; a qualitative parameter cannot be used to describe all the aspects. However, when we discuss the influence of the chemical composition on the effect of fluorescence quenching, parameter L is a useful qualitative parameter.

References

1. T. Forster, Ann. Phys. 255 (1948)
2. T. Forster, Z. Naturforsch. **A4**, 321 (1949)
3. D.L. Dexter, J. Chem. Phys. **21**, 836 (1953)
4. R. Obach, in *Optical Properties of Ions in Crystals*, ed. by H.W. Moos, H.M. Crosswhite (Wiley-Interscience, New York, 1967)
5. T. Miyakawa, D.L. Dexter, Phys. Rev. B **1**, 2961 (1970)
6. F. Auzel, in *Luminescence of Inorganic Solids*, ed. by B. DI Barstool (Plenum, New York, 1978)
7. F. Auzel, in *Radiationless Processes,* ed. by B. Di Bartolo, V. Golberg (Plenum, New York, 1980)
8. F. Auzel, in *Spectroscopy of Solid-State Laser-Type Materials*, ed. by B. Di Bartolo (Plenum, New York, 1987)
9. G. Blasse, in *Radiationless Processes,* ed. by B. Di Bartolo, V. Golberg (Plenum, New York, 1980)
10. G. Blasse, in *Spectroscopy of Solid-State Laser-Type Materials*, ed. by B. Di Bartolo (Plenum, New York, 1987)

11. W.M. Yen, in *Spectroscopy of Solids Containing Rare Earth Ions*, ed. by A.A. Kaplyanskii, R.M. Macfarlane (North-Holland, Amsterdam, 1987)
12. T. Holstein, S.K. Lyo, R. Orbach, in *Laser Spectroscopy of Solids*, ed. by W.M. Yen, P.M. Selzer (Springer, Berlin, 1981)
13. B. Henderson, G.F. Imbusch, *Optical Spectroscopy of Inorganic Solids* (Clarendon Press, Oxford, 1989)
14. M. Yamada, S. Shionoya, T. Kushida, J. Phys. Soc. Jpn. **32**, 1577 (1972)
15. J.M. Colins, B. Di Bartolo, J. Lumin. **69**, 335 (1996)
16. F. Auzel, Y.H. Chen, J. Lumin. **66&67**, 224 (1996)
17. M. Inokuti, F. Hirayama, J. Chem. Phys. **43**, 1978 (1965)
18. F. Spitzer, *Principles of Random Walk* (D. Van Nostrand, Princeton, 1964)
19. E.W. Montroll, J. Math. Phys. **6**, 167 (1965)
20. A.I. Burshtein, Sov. Phys. JEIP **35**, 882 (1972)
21. I.R. Martin, V.D. Rodriguez, U.R. Rodriguez-Mendoza et al., J. Chem. Phys. **111**, 1191 (1999)
22. M. Yokota, O. Tanimoto, J. Phys. Soc. Jpn. **22**, 779 (1967)
23. D.L. Huber, Phys. Rev. B **20**, 2307 (1979)
24. J. Hegarty, D.L. Huber, W.M. Yen, Phys. Rev. B **23**, 6271 (1981)
25. F. Auzel, in *Spectroscopic Properties of Rare Earths in Optical Materials*, ed. by G. Liu, B. Jacquier (Springer, Berlin, 2005), p. 267
26. J.S. Liao, Y.F. Lin, Y.J. Chen et al., J. Opt. Soc. Am. B **23**, 2572 (2006)
27. P. Wang, J.M. Dawes, P. Dekker, et al., J. Opt. Soc. Am. B **16**, 63 (1999)
28. A.A. Kaminskii, *Laser Crystal* (Springer, Berlin, 1980)
29. F. Auzel, Mat. Res. Bull. **14**, 223 (1979)
30. Z.D. Luo, Commun. Fijian Inst. Res. Struct. Matter (1), 1 (1976) (in Chinese)
31. Z.D. Luo, A.D. Jiang, Y.C. Huang et al., Sci. China (Ser. A) **34**, 762 (1991)
32. D. Jaque, O. Enquitra, S.J. Gacia, et al., App. Phys. Lett. **76**, 2176 (2000)
33. D. Jaque, O. Enguita, U. Caldino, et al., J. App Phys. **90**, 561 (2001)
34. Z.D. Luo, Y.D. Huang, M. Montes et al., App. Phys. Lett. **85**, 715 (2004)
35. E. Bovero, Z.D. Luo, Y.D Huang., et. al. App. Phys. Lett. **87**, 211108 (2005)
36. M. Montes, D. Jaque, Luo Zundu. Opt. Lett. **30**, 397 (2005)
37. H.G.F. Winkler, *Struktur und Eigenschaften der Kristalle* (Springer, Berlin, 1955)
38. C.A. Coulson, *Valence* (Oxford University Press, London, 1961)

Chapter 10
Laser and Physical Properties of Materials

After the study of basis rules of radiative and non-radiative transition and main spectral properties of the materials, it is necessary to investigate the relationship between the laser performance and the physical properties of the materials and to realize the requirements of laser devices on the material properties. To this end, it is useful to give an outline of fundamentals of solid-state laser, although the detailed discussion is not in the scope of this book.

10.1 Brief Introduction of Solid-State Laser Principle

There are two types of energy level system of solid-state laser that have been generally used, that is, the so-called three-level system and four-level system. Figure 10.1 is a schematic diagram of their energy levels.

In the three-level system, the energy levels of laser operation are the absorption band, the upper laser level, and the lower laser level. The active ions absorb the pump light and transit to absorption band then non-radiative decay to the upper laser level. The radiative transition between the upper laser level and the lower laser level emits the radiation at laser wavelength. In order to generate laser emission, the laser material must be placed in a laser resonator and the intensity of pump light should be strong enough to produce enough population inversion between the upper and the lower laser levels, so that the gain of the medium at laser wavelength is enough to compensate for its losses. The probability of the stimulated radiative transition is proportional to the photon number of the same resonator mode in the resonator. Therefore, the avalanche-like increase of the photon number will generate radiation with high intensity, high monochromaticity, high coherence, and high directionality, that is, the laser emission. The difference between the four-level system and the three-level system is in the lower laser level. Different from the three-level system, the low laser level of four-level system is not its ground state. Typical example of the three-level system is ruby $Cr^{3+}:Al_2O_3$; its energy level and

Z. Luo and Y. Huang, *Physics of Solid-State Laser Materials*, Springer Series in Materials Science 289, https://doi.org/10.1007/978-981-32-9668-8_10

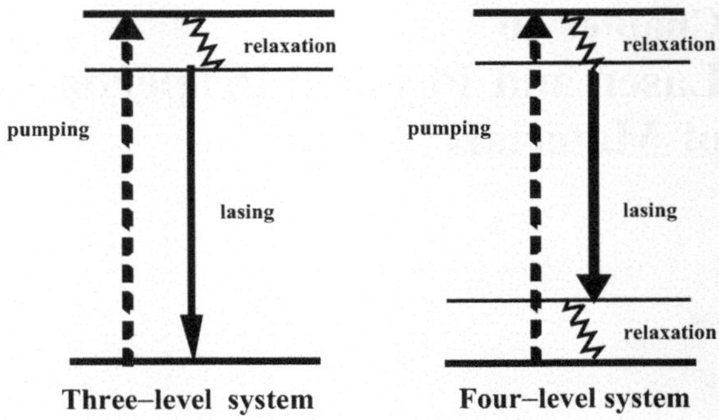

Fig. 10.1 Schematic diagram of three-level system and four-level system of laser

spectral properties have been discussed in Chap. 7. On the other hand, there are many examples of four-level system. The neodymium-doped yttrium-aluminum garnet Nd^{3+}:$Y_3Al_5O_{12}$(Nd^{3+}:YAG) is a typical crystal which has been widely used in laser technique. Whether the three-level systems or the four-level systems, the first condition for the laser emission is the pumping should be strong enough to make the laser material become an optical gain medium.

The early solid-state lasers usually used rare gas discharge lamps as pump sources (flash lamp is used in the pulse excitation while in the continuous excitation the arc lamp should be used). Laser material and straight tubular pumping lamp are usually placed in two focus lines of the elliptical cylindrical light condenser having inner surface coated with gold film or silver film of high reflectivity. Owing to the fact that the emission spectra of the pump light and the absorption spectra of the laser material cannot match well and the gathering efficiency of light is difficult to reach 100%, the pumping efficiency is always too low. On the other hand, the unwanted radiation of pumping lamp always rise the temperature of the laser material.

Figure 10.2 is a schematic diagram of a solid-state laser. The pump source used is a semiconductor laser (LD), which is an efficient pumping method not producing too much heat and can keep the good performance of laser material at a lower temperature.

The rear mirror in Fig. 10.2 is usually in the form of dielectric film deposited directly on the surface of the laser material. It has high transmission at the

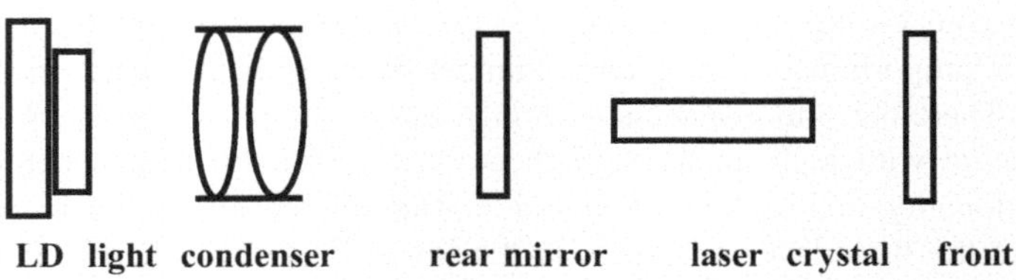

Fig. 10.2 Schematic diagram of a solid-state laser pumped by LD

wavelength of pumping light but high reflectivity at the laser wavelength. The output mirror (front mirror) should have certain amount of transmission at the laser wavelength to output laser emission. If the reflectivity of the rear mirror and front mirror at the laser wavelength is R_1 and R_2, respectively, then the laser light with intensity I_0 will become $I_0R_1R_2 = I\mathrm{e}^{-2(-1/2\ln R1R2)}$ after back and forth reflection between the rear mirror and the front mirror. By using loss coefficient $\gamma' = -\ln(R_1R_2)^{1/2}$ the laser intensity can be expressed as $I = I_0e^{-\gamma'}$. If R is written as $R \equiv (R_1R_2)^{1/2}$, in the case of small output coupling $-\ln R \approx 1-R$, that is, the output loss $\gamma' = 1-R$. Besides the scattering loss of the optical resonator, there are the absorption and scattering of the impurities and defects that existed in the materials. All these losses can be represented by β. Therefore, the total loss is

$$\gamma = 1 - R + \beta \tag{10.1}$$

Photon lifetime in a laser cavity is a conception commonly used in laser science and technology, and it can be measured by the time the laser intensity in the cavity decay to e^{-1} of its original intensity. By the above discussion it is obvious that passing through the resonant cavity for one time, the laser light intensity I_0 will be attenuated from I_0 to $I_0e^{-\gamma}$. After a period of time t, the distance traveled by the light is ct/n (n is refractive index of laser material). If the length of the laser material is L, then in the period of time t the laser light pass through the resonant cavity ct/nL times, so that the laser light decays into

$$I = I_0(e^{-\gamma})^{ct/nL} = I_0\exp\left(-\frac{ct\gamma}{nL}\right) \tag{10.2}$$

Obviously, the photon lifetime in the laser cavity is

$$\tau_c = \frac{nL}{c\gamma} \tag{10.3}$$

The quality factor Q_c of resonator is defined as the 2π times of the cycle number in which the energy in the cavity is consumed, that is, $Q_c = 2\pi\nu_0 I/(\mathrm{d}I/\mathrm{d}t)$. By using (10.2) and (10.3), it is obvious that

$$Q_c = 2\pi\nu_0\frac{nL}{c\gamma} = 2\pi\nu_0\tau_c \tag{10.4}$$

then

$$\gamma = \frac{2\pi\nu_0 nL}{cQ_c} \tag{10.5}$$

Under the excitation of pump source, the inversion concentration between the upper and the lower laser levels is ΔN and so the laser material becomes a gain medium with its gain coefficient α_g proportional to ΔN

$$\alpha_g(\nu) = \alpha'_g(\nu)\Delta N \equiv \sigma_{em}(\nu)\Delta N \tag{10.6}$$

where $\alpha'_g(\nu)$ is the gain coefficient for each inversion particle, that is, the emission cross-section σ_{em} of the laser material and by (5.66) of Chap. 5, it is

$$\sigma_{em}(\nu) = \frac{c^2}{8\pi\nu^2 n^2} A g(\nu) = \frac{c^2}{8\pi\nu^2 n^2 \tau_r} g(\nu) \tag{10.7}$$

The condition of laser generation is obviously the product of the gain coefficient α_g multiplied by the material length L must be equal to or greater than the loss coefficient γ. By using (10.3), (10.6), and (10.7), one can obtain

$$\frac{c^3 \Delta N \tau_c}{8\pi n^3 \nu^2 \tau_r} g(\nu) \geq 1 \tag{10.8}$$

Among the various parameters of the above formula, from the angle of pump intensity, what can be changed only is ΔN. When ΔN reaches a value which can satisfy (10.8), it is said that the laser oscillation threshold is achieved, then the number of inversion particle per unit volume is called the threshold inversion concentration ΔN_{th}, that is

$$\Delta N_{th} = \frac{8\pi n^3 \nu^2 \tau_r}{c^3 \tau_c g(\nu)} \tag{10.9}$$

In virtue of $\int g(\nu)d\nu = 1$, approximately one has $g(\nu) = 1/\Delta\nu$, therefore

$$\Delta N_{th} = \frac{8\pi n^3 \nu^2 \Delta\nu \tau_r}{c^3 \tau_c} = p\frac{\tau_r}{\tau_c} = p\frac{\tau_f}{\tau_c \eta_f} \tag{10.10}$$

where $p = 8\pi n^3 \nu^2 \Delta\nu / c^3$ gives the mode density, τ_r is the radiative lifetime, τ_f is the fluorescence lifetime, and η_f is the fluorescence quantum efficiency.

The threshold pump energy and pump power for three-level system and four-level system will be studied in the following.

For three-level systems, because the lower laser level is ground state, the particle number per unit volume on the upper laser level must be more than half of the total particle concentration, that is

$$N_u = N_1/2 + \Delta N$$

where N_1 is the active ion concentration in the ground state. Owing to the fact that N_1 is several orders of magnitude greater than ΔN, one can approximately adopt $N_u \approx N_1/2$ and N_1 simply is the doping concentration of active ion.

For a four-level system, the lower laser level is not the ground state, at general temperature, its particle concentration can be neglected compared with the total particle concentration in the medium; therefore, the particle concentration difference between the upper and the lower laser levels is that pumped to the upper laser level $N_u = \Delta N$.

In the case of pulse pumping, if the bandwidth of the pump pulse is shorter than the fluorescence lifetime then the threshold energy ε_{th} required by unit volume of the laser material is equal to the photon energy of the pumping light multiplied by the particle concentration that must be pumped. Owing to the fact that the active ions in absorption band and fluorescent energy level have both downward radiative and non-radiative transitions, therefore, if the particle number involved in the radiative transition of the upper laser level is n, then particles of n/η should be pumped into the upper laser level. Here η is the total quantum efficiency. $\eta = \eta_1 \eta_2$, η_1 denotes the quantum efficiency of absorption band and η_2 is the quantum efficiency of the upper laser level, that is, the fluorescence quantum efficiency. Because usually one has $\eta_1 \approx 1$, generally the approximation $\eta \approx \eta_2 \equiv \eta_f$ can be adopted, therefore the threshold energy is expressed as

$$E_{th} = \varepsilon_{th} V = \frac{N_1 \hbar\omega_p V}{2\eta_f} \quad \text{three-level system} \tag{10.11}$$

$$E_{th} = \varepsilon_{th} V = \frac{\Delta N_{th} \hbar\omega_p V}{\eta_f} \quad \text{four-level system} \tag{10.12}$$

Substitute the expression (10.10) of ΔN_{th} into (10.12), the relationship between the threshold energy and the spectral parameters for four-level system can be obtained as

$$E_{th} = \frac{p\tau_f \hbar\omega_p V}{\tau_c \eta_f^2} \tag{10.13}$$

In the case of continuous pumping, below the threshold, the particle energy in the upper laser level is only consumed by spontaneous radiative transition and non-radiative decay in the time interval of fluorescence lifetimeτ_f, therefore the threshold pumping power $P_{th} = E_{th}/\tau_f$ can be expressed as following for three-level system and four-level system, respectively

$$P_{th} = \frac{N_1 \hbar\omega_p V}{2\eta_f \tau_f} \quad \text{three-level system} \tag{10.14}$$

$$P_{th} = \frac{\Delta N_{th} \hbar\omega_p V}{\eta_f \tau_f} = \frac{p\hbar\omega_p V}{\tau_c \eta_f^2} \quad \text{four-level system} \tag{10.15}$$

Most of the pump energy (power) in the three-level systems is consumed in exciting half of the active ions in ground state to the upper laser level. Only beyond this threshold the excitation energy can be converted into laser energy. Comparing with $N_1/2$, ΔN_{th} can be completely neglected; therefore, the laser threshold of the four-level systems is much lower than that of the three-level systems. Moreover, the four-level systems can operate continuously, but the three-level systems can only operate in the pulse mode.

The expressions of laser energy, power, and efficiency of four-level system are derived as follows.

Assuming the energy for exciting the active ions to the upper laser level is E, then after deduction the energy E_{th} consumed in the generation of threshold particles, energy E–E_{th} will be transferred into laser energy. However, the energy consumed in the non-radiative decay also should be taken off; that is, the laser energy obtained should be multiplied by the luminescent efficiency η_f. Therefore

$$E_L = \frac{p\tau_f \hbar\omega_p V}{\tau_c \eta_f}\left(\frac{E}{E_{th}} - 1\right) \tag{10.16}$$

Similarly, it can be concluded that the laser power output in the case of continuous pumping is

$$P_L = \frac{p\hbar\omega_p V}{\tau_c \eta_f}\left(\frac{P}{P_{th}} - 1\right) \tag{10.17}$$

Owing to the fact that reflectivity of the rear mirror and the front mirror at laser wavelength is R_1 and R_2, the output laser energy should be

$$E_{out} = E_L(1 - R_2)\left[1 + R_1R_2 + (R_1R_2)^2 + (R_1R_2)^3 + \cdots\right] = E_L\frac{1 - R_2}{1 - R^2} \tag{10.18}$$

where $R \equiv (R_1R_2)^{1/2}$. Similarly, the output laser power is

$$P_{out} = P_L\frac{1 - R_2}{1 - R^2}$$

In order to have a higher laser output energy or power, obviously E_L or P_L has to be as high as possible. From the point of view of device design, it is necessary to reduce cavity loss and making R_1 as close to 1 as possible, while in the case of high pumping energy and power, R_2 has to be appropriately increased. On the other hand, E_{th} and P_{th} should be as low as possible for laser material. Section 10.2 will start from this point to discuss the quality factor of laser material.

In recent years, a new pumping method has been adopted for some Nd^{3+} laser crystals, such as Nd^{3+}:YVO_4 and Nd^{3+}:YAG; that is directly pump up the active ions to the upper laser level $^4F_{3/2}$ of Nd^{3+}ion [1, 2]. The main advantage is that it can reduce the heat generated during the operation of laser. This problem will be further analyzed in Sect. 10.4. It should be pointed out that because the energy level of absorption band is the upper laser level, the number of energy levels involved in laser operation is less than that of the original four-level system, whether it is a 1.06 nm laser or a 946 nm laser. However, from the angle of the population inversion and the threshold calculation formula, the former is identical to the four-level system which has been analyzed above, but the latter must be analyzed and calculated by using the quasi three-level system.

A typical example of a quasi three-level system is Yb^{3+} laser material. It has become a hot spot in the development and application study of solid laser materials after the development of semiconductor laser. Although this material has a higher threshold than that of the Nd^{3+} laser material but Yb^{3+} ion has simple energy level structure. Theoretically, there is no up-conversion and excited-state absorption and if there are no other rare earth impurities in the host, there will be no concentration quenching and its long fluorescent lifetime is beneficial to energy storage. Moreover, the pump wavelength is close to the wavelength of laser emission, so that the heat loss of pump energy is very low. On the other hand, the broad emission and absorption spectra produced by the strong electron–phonon coupling between the Yb^{3+} ion and the medium are beneficial to LD pumping, and the laser emission tuned in a wider wavelength range, which is conducive to the realization of ultrashort pulse laser output by mode-locked method to realize the laser output of ultrashort pulse. In recent years, the continuous laser power of Yb^{3+} laser has reached kilowatt level and its ultrashort pulse output to reach dozens femtosecond level is not difficult. Their laser performance in many aspects has been reported in many literatures. Here it is not necessary to go into details. However, it should be pointed out that Yb^{3+} ion is a quasi three-level system. The high proportion distribution of particle in the lower laser level results in a high laser oscillation threshold; therefore, search for low threshold host material has become an important problem in material research.

From the perspective of energy level structure, this requires a greater separation between the lower laser level (high crystal field energy level of multiplet $^2F_{7/2}$) and the ground state (lowest crystal field energy level of multiplet $^2F_{7/2}$); consequently, Yb^{3+} ion must be doped in the host material with strong crystal field. The problem of how to find the host material with strong crystal field has been discussed by many authors from different angles. Here just want to point out that Nd^{3+} laser materials have a very rich data of crystal field energy level, so one can use relationship between the total crystal field splitting of the multiplet $^4I_{9/2}$ of Nd^{3+} ion and the multiplet $^2F_{7/2}$ of Yb^{3+} ion obtained by Auzel [3]: $\Delta E(^2F_{7/2}) \approx 1.6\Delta(^4I_{9/2})$ to find the Yb^{3+} laser material with strong crystal field from the Nd^{3+} laser material having strong crystal field.

In the Yb^{3+} ion laser system, the lower laser level is one crystal field energy level of ground multiplet. There is also a certain absorption coefficient (absorption cross-section) at the wavelength of laser emission. Obviously, laser gain cross-section should be deducted by the loss from the absorption at the same wavelength. If the particle number inversion ratio is $f(f = \Delta N/N_0)$, the ratio of particle number remaining in the lower laser level to the particle number in the ground state is $(N_0 - \Delta N)/N_0 = 1 - f$, then the gain cross-section for each inversion particle deducted by ground state absorption cross-section should be

$$\sigma_g(\nu) = f\sigma_{em}(\nu) - (1 - f)\sigma_{ab}(\nu)$$

The inversion population ΔN_{th} at laser oscillation threshold in this situation will be greater than that expressed in (10.10). The general method in this case is to calculate different σ_g values for different f numbers according to measured emission and absorption cross-sections and to investigate in what range of wavelength and what f value the inequality $\sigma_g > 0$ can be satisfied, as well as at what wavelength σ_g reach the maximum, from these data the wavelength of laser emission can be determined.

Solid-state lasers emitting in the spectral range of 1.5 μm are very promising for eye-safe laser, range finding, ophthalmology, fiber-optic communication systems, and optical location. One of the effective ways to realize the laser emission of 1.5 μm band is using Yb^{3+} ion to sensitize Er^{3+}ion. After Yb^{3+} ions efficiently absorb pumping energy at 967 nm and excited to multiplet $^2F_{5/2}$, by the energy transfer of $Yb^{3+} \rightarrow Er^{3+}$, Er^{3+}ions are excited to their multiplet $^4I_{11/2}$, then the upper laser level $^4I_{13/2}$ is populated by $^4I_{11/2} \rightarrow {}^4I_{13/2}$ non-radiative decay. Because Er^{3+} ion has many energy levels in the $4f$ configuration, so that the existent excited-state absorption, up-conversion, and cross-relaxation as well as $Er^{3+} \rightarrow Yb^{3+}$ back energy transfer processes will produce the deexcitation of $^4I_{11/2}$ and $^4I_{13/2}$ multiplets. Therefore, ideal 1.5 μm laser material must be the Yb^{3+}+ Er^{3+}-doped crystals with short $^4I_{11/2}$ fluorescence lifetime and higher thermal conductivity. This is the reason why the Yb^{3+} + Er^{3+}:RAl_3 $(BO_3)_4$(R = Y, Gd, Lu) crystals with high phonon energy have been realized as high efficiency 1.5 μm laser crystal [4–14]. The 1.5 μm continuous laser output with power up to 1 W and low noise laser of power 100 mW has been realized in this type of crystal. The nanosecond, picosecond, and femtosecond short pulse laser output are also realized by Q switch and mode lock method [15–18].

10.2 Quality Factor of Solid-State Laser Materials

It can be seen from previous discussion that to evaluate the performance of a laser material, it should be judged that when the cavity loss is minimum whether it has low threshold power P_{th} or low threshold energy E_{th}, and whether it can absorb more energy or power from pumping sources. Therefore, one should start with threshold optical power flux ϕ_{th} to investigate which material's physical properties make pump threshold optical power flux ϕ_{th} smaller, The power absorbed by the laser material per unit volume should be equal to the power consumed at threshold $\Delta N_{th}\hbar\omega_p/\tau_f$, that is

$$\phi_{th}\sigma_p N_1 = \frac{\Delta N_{th}\hbar\omega_p}{\tau_f}$$

where σ_p is absorption cross-section, then

$$\phi_{\text{th}} = \frac{\Delta N_{\text{th}} \hbar \omega_{\text{p}}}{\tau_{\text{f}} \sigma_{\text{p}} N_1}$$

By using (10.10), the following can be obtained

$$\phi_{\text{th}} = \frac{\Delta N_{\text{th}} \hbar \omega_{\text{p}}}{\tau_{\text{f}} \sigma_{\text{p}} N_1} = \frac{p \hbar \omega_{\text{p}}}{\tau_{\text{c}} \sigma_{\text{p}} \eta_{\text{f}} N_1} \tag{10.19}$$

By (10.7) and the formula of mode density given above, the emission cross-section can transform into

$$\sigma_{em} = \frac{c^2}{8\,\pi\, v^2 n^2} A g(v) = \frac{n}{c p \tau_{\text{r}}}$$

Then the mode density is expressed as

$$p = \frac{n}{c \sigma_{em} \tau_{\text{r}}}$$

In this way, the threshold optical power flux is

$$\phi_{\text{th}} = \frac{\hbar \omega_{\text{p}}}{c \tau_{\text{c}}} \frac{n}{\sigma_{\text{p}} N_1 \tau_{\text{f}} \sigma_{em}} \tag{10.20}$$

The second factor $n/\sigma_{\text{p}} N_1 \tau_{\text{f}} \sigma_{\text{em}}$ in the above formula only relates to the inherent properties of laser materials, and so the quality factor M of laser material can be defined as

$$M = \frac{\sigma_{\text{p}} N_1 \tau_{\text{f}} \sigma_{em}}{n} \tag{10.21}$$

Obviously, the laser material having large M value has a lower laser threshold. The laser technology generally uses the product $\tau_{\text{f}} \sigma_{\text{em}}$ to evaluate the laser threshold; for example, the comparison of related properties of five laser crystals was listed by Loutts [19], as shown in Table 10.1.

Table 10.1 Comparison of spectral properties of several laser crystals

Crystal (cm^2s)	Emission cross-section σ_{em} (cm^2)	Fluorescence lifetime τ_f (s)	Product lose $\tau_f \sigma_{em}$ (cm^2s)
Nd:BFAP	$>5 \times 10^{-19}$	3.7×10^{-4}	$>1.85 \times 10^{-22}$
Nd:SFAP	5.4×10^{-19}	2.98×10^{-4}	1.61×10^{-22}
Nd:YVO_4	10.5×10^{-19}	9.8×10^{-5}	1.03×10^{-22}
Nd:YAG	3.04×10^{-19}	2.7×10^{-4}	8.3×10^{-23}
Nd:YAB	10×10^{-19}	2.0×10^{-5}	2.0×10^{-23}

If the concentration N_1 of doping ion and the absorption cross-section σ_p have not much difference, it is possible to use the above product to compare the performance of different laser materials. Some authors [20] used $\sigma_p N_1 \tau_f \sigma_{em}$ as quality factor of laser performance. Because most laser materials have little difference in refractive index, the quality factor without refractive index can also be used. To compare laser performance under the excitation of semiconductor laser, the absorption cross-section σ_p at pumping wavelength can also be replaced by the absorption peak area A_p at pumping wavelength, it is due to the fact that the emission wavelength of semiconductor laser will change with the temperature, so that the wider linewidth of absorption peak at pumping wavelength is to help the stability of laser output. Therefore, the quality factor of the laser performance of the material can be written as

$$M = \frac{A_p N_1 \tau_f \sigma_{em}}{n}, \text{ or } M = A_p N_1 \tau_f \sigma_{em} \tag{10.22}$$

Although the product $\tau_f \sigma_{em}$ of some laser materials is not very large, but their laser performances are still better than those of the materials with larger product of $\tau_f \sigma_{em}$. For example, for the Nd^{3+}:YAG crystal, the $\tau_f \sigma_{em}$ is 1.63 times that of the Nd^{3+}:KGd $(WO_4)_2$ (Nd^{3+}:KGW) crystal, but its Nd^{3+} ion concentration is only 1/3 of that of the Nd^{3+}:KGW crystal. On the other hand, the pumping light absorption performance of the Nd^{3+}:KGW crystal is better than that of the Nd^{3+}:YAG crystal. The linewidth of absorption peak at 809 nm of the former is three times that of the latter, although the height of absorption peak of the latter is 1.1 times that of the former [21]. From these data it can be predicted that the laser performance of the Nd^{3+}:KGW crystal is better than that of the Nd^{3+}:YAG crystal. To confirm this statement, many experiments [22–27] have shown that the laser performance of Nd^{3+}:KGW crystal is better than that of the Nd^{3+}:YAG crystal, whether it is pumped by LD or flash lamp and whether the laser output wavelength is 1067 or 1350 nm. Certainly, owing to the fact that Nd^{3+}:YAG crystal has far superior thermal performance and so can keep the same performance in high energy or power level but the laser performance of Nd^{3+}:KGW crystal will be deteriorated rapidly with the increase of power or energy and can only be applied to the range of medium and low power. Thermal–optical properties of laser crystals will be discussed in Sect. 10.4.

10.3 Relationship Between Laser Threshold and Chemical Composition of Host Materials

In recent years, the output power of semiconductor laser (LD) is increasing rapidly, they generate laser emission with many different wavelengths, and the price is dropping rapidly. It promotes the rapid development of LD pumped solid-state lasers and puts forward the new requirements of relevant laser materials.

From the point of view of material science, undoubtedly, it is necessary to understand the special requirements of the structure and chemical composition of laser material. Nevertheless, each parameter of laser or spectral performance is related to the static spatial structure of the crystal as well as the lattice vibration and the electron motion of the crystal; therefore it is very difficult to obtain a quantitative relation theoretically. A practical and feasible method is to analyze the relationship of the laser-related performance of the material with their atomic and bonding parameters, in order to build a bridge between the important parameters of laser performance and the chemical composition of laser materials. A qualitative empirical relationship of the laser threshold with the atomic parameters of host crystal which was obtained by the authors [28] will be described in this section.

It can be seen from (10.20) that continuous laser threshold is inversely proportional to the product of stimulated emission cross-section σ_{em} and fluorescence lifetimeτ_f. The relationship of the product $\sigma_{em}\tau_f$ with crystal refractive index and fluorescence spectral linewidth will be analyzed in the following, to find the relations of optical and spectral parameters with the atomic parameters of host crystal and then to discuss the possible effect of the chemical composition of host crystal on the threshold of laser oscillation.

From (5.66) of Chap. 5 and the relationship between spontaneous emission probability A, fluorescence lifetime τ_f, and fluorescence quantum efficiency η_q; that is, $A = \eta_q/\tau_f$, it can be shown that

$$\sigma_{em}\tau_f = \frac{\eta_q c^2}{8\pi v^2 n^2} g(v)$$

Substituting the formula of Gauss or Lorentz line shape function $g(v)$ given in Chap. 6, then

$$\sigma_{em}\tau_f \propto \frac{\eta_q}{n^2 \Delta v} \tag{10.23}$$

From this expression it can be seen that large $\sigma_{em}\tau_f$ value is related to high fluorescence quantum efficiency and narrow emission linewidth as well as small refractive index of host material. In order to have high fluorescence quantum efficiency, electron–phonon interaction should not be strong and the effective phonon energy should not be high. On the other hand, the contribution to the emission linewidth comes from the effects of lattice vibration and the inhomogeneous of crystal field. Except for the group consisting of the atoms with very small mass, such as the borate group BO_3 having especially high vibrational frequency, the difference of phonon frequency for different oxide materials is within two times while the different fluorides have a slightly different but much lower phonon frequency. Weaker crystal field will generally lead to weaker electron–phonon interaction and higher fluorescence quantum efficiency. If crystals have roughly the same integrity, it also corresponds to a narrower inhomogeneous line broadening. At the temperature of laser operation, the contribution to the thermal line broadening mainly comes from Raman process; its thermal linewidth can be expressed by (6.77) and (6.78) of Chap. 6. It can be seen that the thermal linewidth is proportional to the fourth power of the mass difference factor D.

It is difficult to find out the quantitative relationship between the structure and the refractive index of material; however, if one can discover some qualitative relationship between refractive index and such as the parameters of ionic mass, ionic radius and electrical negativity, it will be helpful to the research of new materials. As is well known, the refractive index of material is the square root of the following dielectric tensor

$$\varepsilon_{ik} = \varepsilon_0 \delta_{ik} + a_1 u_{ik} + a_2 u_{ll} \delta_{ik} \tag{10.24}$$

where ε_0 is the dielectric constant without strain. The second and the third terms in the above equation are the dielectric tensors related to the strain of material, while u_{ik} is defined as

$$u_{ik} = \frac{1}{2}\left(\frac{\partial u_i}{\partial X_k} + \frac{\partial u_k}{\partial X_i}\right) \tag{10.25}$$

where u_i and u_k are different components of particle displacement vector. Refractive index can be expressed as the following expression if one does not concern its anisotropy

$$\varepsilon = \varepsilon_0 + a\left(\frac{\partial u}{\partial R}\right)\bigg|_{R=0} \tag{10.26}$$

The first term in the above expression is the polarizability contribution of the original dipole of material which is proportional to original dipole moment. With the increase of the cationic radius r^+ the polarizability of the cation increases and with the increase of atomic number in the same family of periodic table, the polarizability of cations also increases. (Note that with the increase of the atomic number in the same family of periodic table, the electrical negativity difference Δx between cation and anion is increased.) On the other hand, the polarizability of chemical bond is an increasing function of the difference Δx. Therefore, parameter $G = \Delta x r^+$ can be used to measure the dielectric constant ε_0, that is ε_0 is an increasing function of G.

The effect of the difference of host ion mass will be considered as follows. If the host material is composed of the ions with the same mass, then

$$\left(\frac{\partial u}{\partial R}\right)\bigg|_{R=0} = \frac{i}{\sqrt{M}} \sum_{k,} kQ(k) \tag{10.27}$$

where $M = Nm$ is the total mass of host material, while N, k, R, and $Q(k)$ are total particle number, phonon wave vector, coordinate, and canonical coordinate of lattice point, respectively. The material, in fact, is generally composed of two or more than two kinds of ions with different masses. If there are N molecule formula units in the material and each molecule formula unit has $n = n_\alpha + n_\beta$ ions, in which n_α ions with mass of m_α and n_β ions with mass of m_β while the total mass of the material is $M = N\ (n_\alpha m_\alpha + n_\beta m_\beta)$. Referring to (10.27), the average strain of the material as that taken in Chap. 6 will be

$$\left(\frac{\partial \bar{u}}{\partial R}\right)\bigg|_{R=0} = \frac{i}{n}\left[\frac{n_\alpha}{\sqrt{Nnm_\alpha}} + \frac{n_\beta}{\sqrt{Nnm_\beta}}\right]\sum_k kQ(k) \tag{10.28}$$

The above expression can also be written as

$$\left(\frac{\partial \bar{u}}{\partial R}\right)\bigg|_{R=0} = \frac{iD}{\sqrt{M}}\sum_k kQ(k) \tag{10.29}$$

where D is the mass difference factor of host material given in Chap. 6. Obviously, the second term of the expression (10.26) of dielectric constant is proportional to D. In summary, the dielectric constant ε_0 increases with the parameter G, while the second term of (10.26) increases with the increase of parameter D. The dielectric constant ε and the refractive index of the material with the same anion will be decreased with the increase of parameter $P = r^-/\mathrm{GD} = r^-/(r^+\Delta x D) = L/D$ (where $L = r^-/(r^+\Delta x)$ is given in Chap. 9). The refractive index data of oxide compound shown in Fig. 10.3 demonstrate that the refractive index is a decreasing function of P.

From the previous discussion, a conclusion can be drawn that a larger P value generally corresponds to a lower refractive index and a narrower fluorescence linewidth. Then by (10.23) it can be seen that it results in a larger product of fluorescence lifetime and emission cross-section. This is confirmed by the data of 12 laser crystals doped with Nd^{3+} ion shown in Fig. 10.4. In the calculation of parameter P, the value of $r^-/\mathrm{G} = r^-/(r^+\Delta x)$ (i.e. the parameter L) must take the

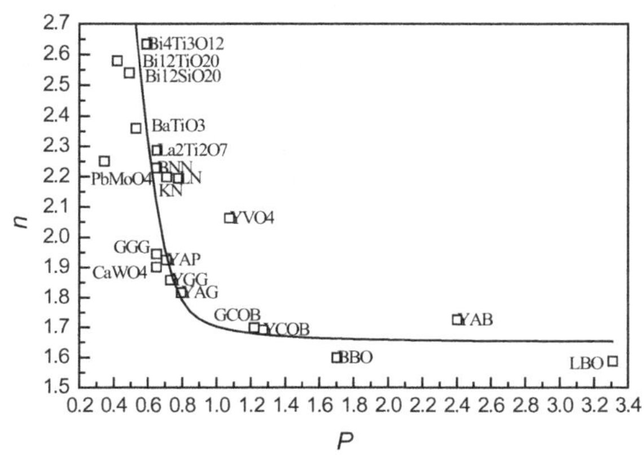

Fig. 10.3 Refractive index as a function of parameter P. The curve in the figure is a fitting curve

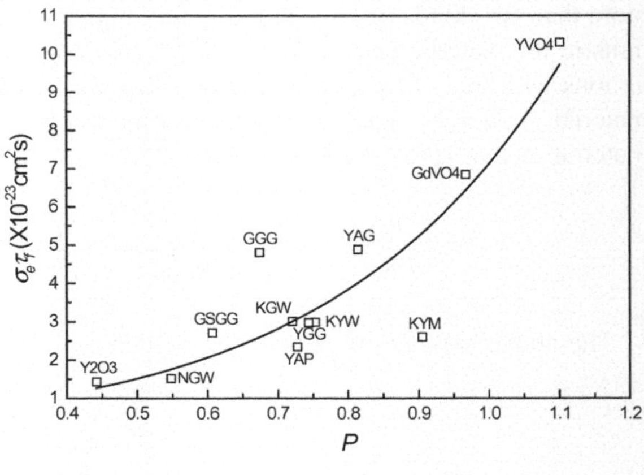

Fig. 10.4 The product of $\tau_f\sigma_{em}$ as a function of the parameter *P*. The curve in the figure is a fitting curve

average value of the simple molecule formula units contained; for example, in the $Y_3Al_5O_{12}$ crystal, its molecular unit can be considered as a complex oxide molecular unit consisting of 3/2 Y_2O_3 unit and 5/2 Al_2O_3 unit, altogether four simple oxide molecular units, so the average of *L* is

$$L(Y_3Al_5O_{12}) = [1.5 \times L(Y_2O_3) + 2.5 \times L(Al_2O_3)] \div 4$$

10.4 Thermo-Mechanical and Thermo-Optical Properties of Solid-State Laser Materials

It has been mentioned in previous section that in high power or high energy application, the laser performance of YAG crystal is better than that of the KGW crystal. This is because the laser performance of KGW crystal but not the YAG crystal is rapidly deteriorated when the temperature is increased. In order to find out its reason, the effect of thermal properties of materials on the laser performance should be studied.

In general, the going up of temperature will widen the fluorescence linewidth and shorten the fluorescence lifetime, and thus increase the laser threshold and decreases the laser output efficiency. On the other hand, the non-uniform temperature distribution in laser crystal will produce thermal stress, thermal birefringence, and thermal lens effects.

First, see the heat produced in the laser operation. Take four-level system as an example; transition from absorption band to the upper laser level is a non-radiative decay process, and part of excited electronic energy is transferred to the heat of host material. There are radiative and non-radiative ways for the transition from the upper laser level to the lower laser level and the proportion of non-radiative decay is calculated by the following formula

$$\frac{W_{\mathrm{nr}}}{W_{\mathrm{r}}+W_{\mathrm{nr}}}=1-\eta_q \tag{10.30}$$

where W_{nr}, W_{r}, and η_{q} are non-radiative decay rate, radiative transition rate, and fluorescence quantum efficiency, respectively. Obviously, the higher the fluorescence quantum efficiency, the less the amount of heat generated by non-radiative decay. The transition from the lower laser level to ground state is also a non-radiative decay process; the excited energy will all be changed to heat energy. If the pump frequency is ω_{p} (wavelengthλ_{p}), laser frequency is ω_{l} (wavelengthλ_{l}), fluorescence quantum efficiency is η_{q}, neglecting the spontaneous emission light energy scattered out of the laser cavity, then the percentage of pump energy converted to the heat energy (thermal loading) is

$$p_t=\frac{\omega_{\mathrm{p}}-\eta_q\omega_{\mathrm{l}}}{\omega_{\mathrm{p}}}=1-\frac{\eta_q\lambda_{\mathrm{p}}}{\lambda_{\mathrm{l}}} \tag{10.31}$$

This formula is only applicable to the operation level near threshold. When operation level much higher than the threshold, the radiative transition rate from the upper laser level to the lower laser level is much higher than the non-radiative decay rate between these two levels. The heat generated by non-radiative decay can be ignored. However, the volume of the pumping area in the laser material is often larger than that of the laser oscillation. In this case to calculate the overall thermal loading, the fluorescence quantum efficiency η_{q} in (10.31) should be replaced by the ratio of the volume of laser oscillation to the volume of pump area. If these two volumes are well coincident, a value close to 1 can be adopted, even if its fluorescence quantum efficiency η_{q} is low. The percentage of pump energy converted to heat energy by different pump methods can be estimated. For example, the Nd^{3+} ion activated crystal or glass pumped by 805 nm semiconductor laser. Suppose in (10.31) $\eta_{\mathrm{q}}=90\%$, then for the laser emission at 1064 nm, it can be calculated that the heat energy generated is 32% of the pump energy. If the same laser material pumped by rare gas discharge lamp, assuming that the average pumping wavelength is 590 nm, then the similar calculation shows the heat energy generated is 50% of the pump energy. If directly pumped to the upper laser level $^4F_{3/2}$ is adopted, and supposing the pumping wavelength is 880 nm, then the thermal loading is reduced to 25%. Obviously, the higher the fluorescence quantum efficiency and the closer the pump band to the upper laser level, the lower is the thermal loss. Moreover, the lower thermal load of laser material will reduce the limitation of laser output by thermal lens effect, thermal broadening of emission spectral line, and shortening of fluorescence lifetime. For example, the NAB crystal, pumped by laser at wavelength 750, 808 nm and directly pump to the upper laser level $^4F_{3/2}$ at 882 nm, the slope efficiency of laser output are 34, 50, and 70%, respectively [29]. For the 946 nm laser operation of Nd^{3+}:YAG crystal, the advantage of direct pump to the upper laser level is even greater. It has been mentioned that the pump wavelength is close to the laser wavelength for Yb^{3+}-doped laser material, and so the thermal loss of pumping light is low. Yb^{3+}:YAG crystal lasers at 1030 nm and diode-pumped at 943 nm, the theoretical thermal loading is $p_t = 0.09$, its thermal loading is lower than half of the thermal loading of the directly pumped 1064 nm laser of Nd^{3+}:YAG. The non-radiative center in laser material and the concentration quenching of active ion will consume the excitation energy and transfer the energy to the heat of host material. Moreover, because of the presence of unwanted impurity ions or other absorption centers in the materials, some of the laser energy can also be converted into heat; all these will further increase the thermal loading of laser material.

Another way to generate heat in laser operation is the up-conversion of laser upper level. When pump power is low, the probability of up-conversion is very low and the effect of up-conversion on fluorescence quantum efficiency and heat production can be neglected, but at a high pumping level, this effect must be considered. Jacinto et al. [30] used experiment of thermal lens to study the effect of energy transfer-assisted up-conversion (ETU) process on the fluorescence quantum efficiency of NAB crystal and confirmed that the effect of up-conversion on fluorescence quantum efficiency and thermal loading cannot be neglected at high pumping level. However, it should be pointed out that there are still some problems in the calculation of that paper and it should be reconsidered. For example, in their formula (7) the fluorescence quantum efficiency η_q at high power pumping level was expressed as

$$\eta_q = \frac{\eta_0}{1 + \beta n_e}$$

where η_0 and βn_e are the fluorescence quantum efficiency at low power pumping level and the effect of up-conversion on this efficiency, respectively. However, as has been mentioned, at a high power laser operation level, because of the stimulated emission transition probability is proportional to laser intensity, so the actual quantum efficiency is much higher than the fluorescence quantum efficiency measured by spectroscopic measurements and also related to the pumping power level, but this fact has not been taken into account. On the other hand, a portion of the excitation energy carried away by the up-conversion process is changed to the light of other wavelengths but not the thermal energy. To correctly calculate η_q, the spectra emitted by the up-conversion process must be measured. Of course, the effect of this process on the fluorescence quantum efficiency and the heat generation is really a problem that should be studied.

Anyhow, the thermal problems in laser operation cannot be ignored. In addition to a considerable part of pump energy change into thermal energy, there is some radiation losses; that is the spontaneous emission loss outside divergence angle of laser emission and the diffraction losses of laser cavity. The proportional relation between the thermal power generated and the output laser power can be expressed as follows [31]

$$P_Q = \chi_q P_{out} = p_q V \tag{10.32}$$

where p_q is the thermal power production per unit volume. The typical value of χ_q is between 1 and 4.

Thermal energy generated in the process of laser operation will rise the temperature of laser material. According to thermodynamic formula, the thermal energy generated and the rise of temperature per second can be expressed as the following expression

$$\partial Q = C_p \rho V \partial T \tag{10.33}$$

where Q, C_p, and ρ denote the thermal energy, the specific heat at constant pressure, and the density of a material, respectively. If there is a temperature gradient in x direction, according to the continuity equation of heat flow, the following equation can be obtained

$$\partial Q = \left(p_q V + KA_c \frac{\partial T}{\partial x} \right) \partial t \tag{10.34}$$

In the right-hand side of the above equation, the first term is the thermal power generated during the laser operation in the laser medium with volume V and sectional area A_c and the second term is the thermal power transferred into the volume V which is the product of the thermal conductivity K, the sectional area A_c, and the temperature gradient $\partial T/\partial x$. Remembering that $\partial V/\partial x = A_c$, deriving the formulas (10.33) and (10.34) with respect to x and equating the two expressions obtained, then the following partial differential equation is resulted

$$\frac{\partial T}{\partial t} = \frac{\chi P_{out}}{C_p \rho V} + \frac{K}{C_p \rho} \frac{\partial^2 T}{\partial x^2} \tag{10.35}$$

In general, $\partial^2 T/\partial x^2$ should be replaced by spatial temperature gradient $\nabla^2 T = (\partial^2/\partial x^2 + \partial^2/\partial y^2 + \partial^2/\partial z^2)T$, so the following equation is obtained

$$\frac{\partial T}{\partial t} = \frac{\chi P_{out}}{C_p \rho V} + \kappa \nabla^2 T \tag{10.36}$$

where thermal diffusion coefficient κ is expressed as

$$\kappa = \frac{K}{C_p \rho} \tag{10.37}$$

Let's discuss the steady temperature distribution in x direction of a slab-shape laser crystal. The sectional area of the crystal perpendicular to the direction x is A_c, the slab has a length l in z direction, width w in y direction, and thickness $2d$ in x direction, that is, the coordinates of x direction of the slab center, the upper interface, and the lower interface are 0, d, and $-d$, respectively. The upper and the lower sides of the slab are cooled by flowing cooling liquid or gas. The pump light uniform illuminates the slab along x direction from the upper and the lower sides.

By solving differential (10.15) with boundary condition $T = T_c$ for $|x| > d + \delta l$, the following temperature distribution in x direction of the laser slab can be obtained [31]

$$T(x) = \begin{cases} T_c + \Delta T_s + \Delta T_0\left(1 - \frac{x^2}{d^2}\right) & x \le d \\ T_c + \Delta T_s\left(1 - \frac{|x-d|}{\delta l}\right) & d \le |x| \le d + \delta l \\ T_c & |x| > d + \delta l \end{cases} \tag{10.38}$$

Between the slab surface and cooling medium, there is a temperature linear transition layer of thickness δl, which depends on surface parameters, coolant parameters, and flow condition. By using the above equations, in a continuous pumping case, it can be shown that the temperature difference between the slab surface and the cooling medium is $\Delta T_s = \chi P_{out}(\delta l)^2/VK$ and the steady-state temperature increase at $x = 0$ is $\Delta T_0 = \chi P_{out} d/4wKl$.

If the laser material in the cooling medium at temperature T_c is a cylinder with radius r_0 and length l, it is cooled by a liquid flowing along the cylinder surface, then the equation of heat flow will be

$$\frac{\partial T}{\partial t} = \frac{\chi P_{out}}{C_p \rho V} + \kappa\left(\frac{\partial^2 T}{\partial r^2} + \frac{1}{r}\frac{\partial T}{\partial r}\right) \tag{10.39}$$

Solving this equation, the temperature distribution in the laser material can be obtained [31]

$$T(r) = \begin{cases} T_c + \Delta T_s + \Delta T_0\left(1 - \frac{r^2}{r_0^2}\right) & r \le r_0 \\ T_c + \Delta T_s\left(1 - \frac{r - r_0}{\delta l}\right) & r_0 \le r \le r_0 + \delta l \\ T_c & r > r_0 + \delta l \end{cases} \tag{10.40}$$

The meaning of ΔT_s is the same as the above and ΔT_0 is the steady temperature increase at $r = 0$. This function is shown in Fig. 10.5. The temperature is highest at the center of the rod, then gradually decreased toward the edge of the rod; it is a parabolic-type function.

In the calculation of formula (10.40), the positive temperature value is adopted for the two opposite direction of the diameter. Let $\nabla^2 T = -\Delta T_s/\delta l^2$ in (10.36), then

$$\Delta T_s = \frac{\chi P_{out} \delta l^2}{\pi r_0^2 l K} \tag{10.41}$$

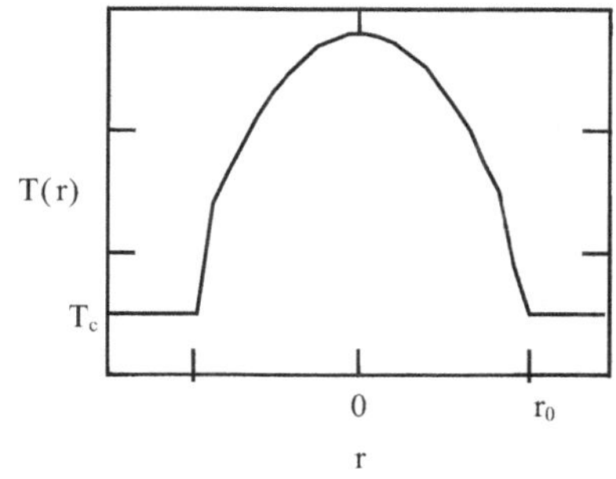

Fig. 10.5 Temperature distribution in the laser rod

Substitute (10.40) into (10.36) and consider that $\nabla^2 T = \frac{\partial^2 T}{\partial r^2} + \frac{1}{r}\frac{\partial T}{\partial r}$ obtained by (10.36) and (10.39), it can be shown that

$$\Delta T_0 = \frac{\chi P_{out} r_0^2}{4KV} = \frac{\chi P_{out}}{4\pi l K} \quad (10.42)$$

In the case of general pulse pumping, by solving time-dependent thermal conduction equation, it can be shown that after switching the pump light, $\Delta T_0(t)$ will decrease with time according to the following formula

$$\Delta T_0(t) = \Delta T_0(0)\exp\left(-\frac{t}{\tau_0}\right) \quad (10.43)$$

where

$$\tau_0 = \frac{r_0^2}{4\kappa} \quad (10.44)$$

Consider that $\partial^2 T/\partial r^2 + r^{-1}\partial T/\partial r \approx -\Delta T_s/\delta l^2$ and $\partial T/\partial t = \partial \Delta T_s/\partial T$. It can be shown that after switching the pump light, that is, the first term in the right-hand side of (10.35) equal to zero, $\Delta T_s(t)$ will be decreased with time according to the following formula

$$\Delta T_s(t) = \Delta T_s(0)\exp\left(-\frac{t}{\tau_s}\right) \quad (10.45)$$

where

$$\tau_s = \frac{\delta l^2}{\kappa} \quad (10.46)$$

It can be seen that the temperature increase ΔT_0 (0) and ΔT_s (0) as well as their decay times τ_0 and τ_s are inversely proportional to thermal conductivity K. For the solid-state laser materials,τ_0 andτ_s are generally of the order of second. Obviously, for the high energy or high power laser applications, the laser materials should have high thermal conductivity.

The period Δt_p of pulsed laser pumping is generally shorter than 5 ms, that is, $\Delta t_p < <\tau_0$ andτ_s, therefore the thermal conduction in the period of pump pulse can be neglected. If the distribution of pump energy is not uniform, the transient thermal stress, thermal birefringence, and thermal lens effects will be generated. Therefore the pump uniformity is very important. Continuous pumping will form a stable thermal distribution, and its temperature distribution is shown in Fig. 10.5. Repeated pumping can be divided into two cases: for the case the time interval of two successive pump pulse is longer than τ_0, then the temperature distribution and thermal effect are similar to those of the single-pulse situation; in case of the pump repetition rate is very high and the time interval of two successive pump pulse is shorter than τ_0, the thermal distribution is similar to that of the continuous pumping.

The above equations are not applicable for the end pump technology commonly used in solid-state laser. In this case, the spatial distribution of pump beam is axisymmetric, and it is generally a Gauss beam near the diffraction limit (single-mode laser beam) or the "top-hat beam" of optical fiber-coupled laser diode. If pump beam is incident to the laser rod with length l and diameter r_0 along the z direction and κ_{ab}, P_{in}, w_p, and p_t are the absorption coefficient of crystal at pump wavelength, input power, pump beam diameter, and thermal loading, respectively, when the pump light is a Gauss beam, the thermal energy as a function of r and z can be expressed as

$$Q(r,z) = \frac{2p_t P_{in}\kappa_{ab}e^{-\kappa_{ab}z-2r/w_p^2}}{\pi w_p^2(1-e^{-\kappa_{ab}l})} \tag{10.47}$$

In case the pump light is top-hat beam, the thermal energy as a function of r and z can be expressed as

$$Q(r,z) = \frac{p_t P_{in}\kappa_{ab}e^{-\kappa_{ab}z}}{\pi w_p^2(1-e^{-\kappa_{ab}l})} \tag{10.48}$$

for $r \le w_p$

$$Q(r,z) = 0$$

for $r > w_p$

The temperature distribution in the laser material can be expressed as [32]

$$T(r,z) - T(r_0,z) = \frac{p_t P_{in}\kappa_{ab}e^{-\kappa_{ab}z}}{4\pi K(1-e^{-\kappa_{ab}l})}\begin{cases}\ln\left(\frac{r_0^2}{w_p^2}\right)+1-\frac{r^2}{w_p^2}, r \le w_p\\ \ln\left(\frac{r_0^2}{r^2}\right), r > w_p\end{cases} \tag{10.49}$$

The temperature distribution along the radius is shown in Fig. 10.6.

The thermo-optic effects on the laser properties of materials should be discussed next.

First, the quantitative description of thermal stress and its physical effects must be made; this involves the concepts of stress and strain in physics. If the plane with normal direction x of an object is subjected to the action of force along direction x, then the stress of the plane per unit area is referred to as σ_{xx} and it will result in the elongation of the object in the x direction. The strain ε_{xx} is defined as normalized length variation $\delta l/l$. According to Hooke's law, the stress σ_{xx} is proportional to the strain ε_{xx}, that is

$$\sigma_{xx} = E_{\mathrm{Y}}\varepsilon_{xx}$$

or

$$\varepsilon_{xx} = \frac{\sigma_{xx}}{E_{\mathrm{Y}}} \tag{10.50}$$

The proportion constant E_{Y} in the formula is the Young's modulus. This stress will lengthen the object in the direction x, while shorten the object in the transverse direction y. The transverse strain ε_{xy} is defined as normalized length variation -$\delta w/w$, and the relation between ε_{xy} and ε_{xx} is

$$\varepsilon_{xy} = -\nu_{\mathrm{Po}}\frac{\sigma_{xx}}{E_{\mathrm{Y}}} \tag{10.51}$$

where ν_{Po} is the Poisson's ratio with a value usually between 0.3 and 0.25 ($\nu_{\mathrm{Po}} = 0.28$ for Nd^{3+}:YAG, $\nu_{\mathrm{Po}} = 0.25$ for ruby, $\nu_{\mathrm{Po}} = 0.33$ for Nd^{3+}:YLF, and $\nu_{\mathrm{Po}} = 0.28$ for Yb^{3+}:YAB). Another important concept is shear stress: force in the

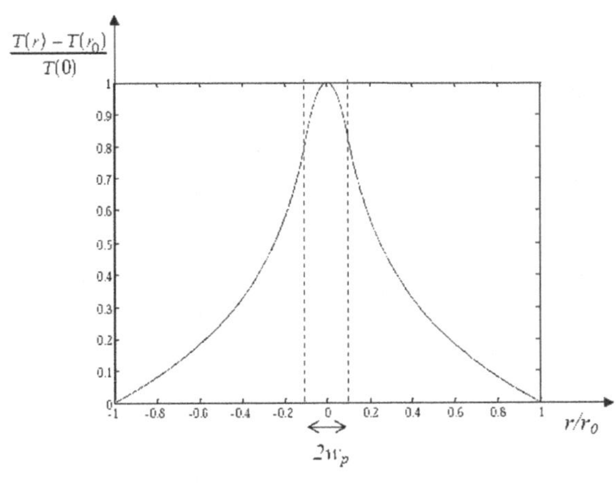

Fig. 10.6 Radial temperature distribution of laser rod pumped by optical fiber-coupled semiconductor laser

y direction applied to the yz plane will cause a shear stress σ_{xy}, which has the following relation with shear strain γ_{xy}

$$\gamma_{xy} = \frac{\sigma_{xy}}{G_R} \tag{10.52}$$

where G_R is called shear modulus of elasticity and has a relation with Young's modulus E_Y and Poisson's ratio ν_{Po} as follows

$$G_R = \frac{E_Y}{2(1+\nu_{Po})} \tag{10.53}$$

The material has a maximum limit stress, and exceeding this limit the material will produce thermal stress damage. In the case of continuous wave pumping, the maximum limit stress occurs at the rod surface produced by temperature distribution in a cylindrical laser rod. As shown in Fig. 10.5, it is [31]

$$\sigma_{max} = \frac{\sqrt{2}}{8\pi} \frac{\alpha \chi E_Y P_{out}}{K(1-\nu_{Po})l} \tag{10.54}$$

Most of the parameters in the above formula have been described, while α is the coefficient of thermal expansion. By (10.54), the maximum laser power output allowed before thermal stress damage of the laser rod can be obtained [31]

$$P_{max} = \frac{8\pi}{\sqrt{2}} \frac{l R_M}{\chi_q} \tag{10.55}$$

where $R_M = \sigma_t K\ (1-\nu_{Po})/\alpha E_Y$ is a measure of the thermal shock resistance of materials and is referred to as thermal shock resistance parameter. The reference value of the thermal shock resistance parameter R_M for several kinds of laser materials are: [33] R_M(glass) = 1, R_M(GSGG) = 6.5, R_M(YAG) = 7.9 and $R_M(Al_2O_3)$ = 100, in the unit of W cm^{-1}.

Obviously, in order to increase the laser output power, the materials with higher heat shock resistance R_M and smaller heat production parameter χ_q should be selected and the length l of laser rod better to be longer.

Actually, due to the thermo-optic effect, the laser output power cannot reach the above maximum value. This is mainly because of that after temperature rising, spectral linewidth becomes wider and fluorescence lifetime becomes shorter. According to (10.21), laser material quality factor will be decreased and laser threshold will be elevated.

On the other hand, it can be seen from (10.40), the temperature of the central region of laser rod is higher than that of its edge; therefore thermal expansion will also being larger in the central region and entire laser rod is changed into a convex lens. In the simple isotropic cases discussed, refractive index change with temperature can be expressed as

$$\Delta n(T) = \frac{dn}{dT}(T - T_0)$$

Referring to (10.40), the refractive index at the radius r can be expressed as

$$n(r) = n(0) + \frac{\mathrm{d}n}{\mathrm{d}T}[T(r) - T(0)] = n(0) - \frac{\mathrm{d}n}{\mathrm{d}T}\Delta T_0\left(\frac{r}{r_0}\right)^2 \tag{10.56}$$

If the temperature coefficient of refractive index is positive (e.g. in YAG crystal), the refractive index increases with the increase of temperature; therefore the refractive index of central region of the rod is higher than that of the rod edge. This will create a convex thermal lens effect. Following is an analysis of this problem.

Suppose that the optical path variation of the light passing the laser rod includes the optical path change $\delta n_{\mathrm{T}} l$ via temperature variation, the optical path change $\delta n_{\mathrm{P}} l$ via photoelastic effect, and the optical path change $n\delta l$ via thermal expansion. Because these three kinds of focusing effect are different in calculation, the focal length f_{T} via temperature variation, the focal length f_{P} via photoelastic effect, and the focal length f_{α} via thermal expansion should be calculated separately, and then the total focal length f_{th} by thermal lensing will be

$$\frac{1}{f_{\mathrm{th}}} = \frac{1}{f_{\mathrm{T}}} + \frac{1}{f_{\mathrm{P}}} + \frac{1}{f_{\alpha}} \tag{10.57}$$

First to calculate δnl, according to (10.56), at the edge of laser rod $r = r_0$ the variation of refractive index is

$$n(r)_{\mathrm{T}} - n(0)_{\mathrm{T}} = -\frac{\mathrm{d}n}{\mathrm{d}T}\Delta T_0\left(\frac{r}{r_0}\right)^2 \tag{10.58a}$$

It can be seen from the above equation that the refractive index in this laser rod shows a quadratic variation with radius r; therefore this laser rod is equivalent to a spherical lens. However, the focal length of a lens-like medium with length l, radius r, and refractive index n_0 can be expressed as follows [33]

$$f \approx -r^2/2l[n(r) - n(0)] \tag{10.59}$$

Comparing the equation of (10.59) with (10.58a) and using (10.42) of ΔT_0, it can be shown that

$$f_{\mathrm{T}} = \frac{\pi r_0^2 K}{\frac{1}{2}\frac{dn}{dT}\chi P_{\mathrm{out}}} \tag{10.60a}$$

δn_{P} via photoelastic effect can be expressed as [33]

$$n(r)_{\mathrm{P}}-n(0)_{\mathrm{P}}=-\frac{1}{2}n_0^3\alpha C_{r,\phi}\chi P_{out}r^2/\pi r_0^2 Kl \tag{10.58b}$$

where α and $C_{\mathrm{r},\phi}$ are the thermal expansion coefficient and the photoelastic coefficient of laser rod, respectively. $C_{\mathrm{r},\phi}$ has two values, one for the radial component and one for the tangential component of polarized light. By using (10.59) and (10.58b), the focal length f_{P} via photoelastic effect can be expressed as

$$f_{\mathrm{P}}=\frac{\pi r_0^2 K}{n_0^3\alpha C_{r,\phi}\chi P_{out}} \tag{10.60b}$$

On the other hand, due to thermal expansion, the laser rod change to a biconvex lens, the focal length f of lens can be usually expressed as [34]

$$f=\frac{1}{(n_0-1)(1/R_1-1/R_2)} \tag{10.61}$$

where R_1 and R_2 are the radius of the front and the rear end face curvature. For the biconvex lens-like laser rod produced by thermal expansion, $R_2=-R_1=-R$ and $R^{-1}=-\mathrm{d}^2l(r)/\mathrm{d}r^2$, by using (10.40), the length $l(r)$ of laser rod varying with respect to r is expressed as

$$l(r)=\alpha l[T(r)-T(0)]=-\alpha l\Delta T_0 r^2/r_0^2$$

The curvature of end face will be equal to

$$R^{-1}=2\alpha l\Delta T_0/r_0^2$$

Therefore, the focal length f_α via thermal expansion can be obtained as follows

$$f_\alpha=\frac{1}{2(n_0-1)R^{-1}}=\frac{r_0^2}{4(n_0-1)\alpha l\Delta T_0}=\frac{\pi r_0^2 K}{(n_0-1)\alpha\chi P_{out}} \tag{10.60c}$$

In introducing the above equation, (10.42) has been used. Then by using (10.60a, 10.60b, 10.60c) and (10.57), the total focal length f_{th} of thermal lensing can be expressed as follows

$$f_{\mathrm{th}}=\frac{\pi r_0^2 K}{\left[\frac{1}{2}\frac{dn}{dT}+\alpha(n_0-1)+\alpha n_0^3 C_{r,\phi}\right]\chi P_{\mathrm{out}}} \tag{10.62}$$

Equation (10.62) is applicable for side pumping laser; this equation is in good agreement with the experiment of side-pumped Nd^{3+}:YAG rod laser [35]. The following expression of the focal length of thermal lens for the laser diode end pumping is derived by Innocenzi [36]. A Gaussian pump beam incident into the crystal was assumed as follows

$$I(r,z) = I_0 \exp\left(-2r^2/w_p^2\right)\exp(-\kappa_p z) \tag{10.63}$$

where κ_p is the absorption coefficient and w_p is the $(1/e^2)$ Gaussian radius of pump beam. With χP_{out}, the thermal power generated in laser operation, the effective focal length for entire laser rod can be expressed as

$$f_{th} = \frac{K\pi w_p^2}{\chi P_{out}(dn/dT)}\left(\frac{1}{1-\exp(-\kappa_p l)}\right) \tag{10.62b}$$

The thermal lens effect of laser material will decrease laser output and laser beam quality, especially the TEM_{00} mode output. In the design of laser devices, some measures can be taken to reduce the influence of thermal lens effect [37–39].

In the case of crystal, the thermal stress causes the refractive index change with temperature by photoelastic effect. The anisotropy of refractive index is described by optical indicatrix in crystal optics. The second-order optical indicatrix tensor $\boldsymbol{N}$ can be expressed as [40]

$$\boldsymbol{N} = \boldsymbol{PE} \tag{10.64}$$

where $\boldsymbol{P}$ is the photoelastic tensor; generally, it is a fourth-order tensor and $\boldsymbol{E}$ is a second-order stress tensor. For different crystal systems, the above tensors have different forms (the symmetry of the crystal results in the vanishing of some components of tensor [40]). The thermal conduction equation of anisotropic crystals will also be anisotropic. When discussing the lens effect, it is necessary to calculate the effect for different crystal orientations, respectively. In addition, the thermal effect of crystal can also cause thermal birefringence [40] and it affects the polarization output of laser. These involved in the design of laser device and can refer to the literatures and books of laser device.

It is necessary to talk about some microscopic relations of the thermal properties before the end of this section. Because of laser material is insulator (except for the semiconductor laser material not discussed in this book), its main mechanism of thermal conduction is phonon conduction, and the microscopic process of thermal resistance is derived by the anharmonic interaction of lattice vibration. It can be demonstrated that the thermal conductivity is proportional to the cubic of Debye temperature T_D and inverse proportional to absolute temperature T and the square of Grüneisen constant γ_G [41]

$$K \propto \frac{T_D^3}{T\gamma_G^2} \tag{10.65}$$

Therefore, the material with a higher maximum phonon energy will have a higher thermal conductivity, if it has the same or almost the same anharmonic, because $\hbar\omega_{\max} \approx kT_D$. On the other hand, the thermal conductivity can be expressed as

$$K = \frac{1}{3} C_v v l_p \tag{10.66}$$

where C_v, v, and l_p are the specific heat, the phonon velocity, and the phonon mean free path of the material, respectively. Considering the Debye temperature and phonon velocity and remembering that the phonon velocity is inversely proportional to the square root of compressibility of materials mentioned in Chap. 9, it can also be seen that the more compact the material structure, the higher is the thermal conductivity. It is also true for the same anisotropic crystal, the thermal conductivity in the direction with compact structure is higher than that in the direction with looser structure; especially for the material with laminated structure, the thermal conductivity along the direction of the layer is significantly higher than that in the direction perpendicular to the layer. On the other hand, thermal expansion of the crystal also is one of the anharmonic effects and the coefficient α of thermal expansion can be expressed as [42, 43]

$$\alpha = \frac{\gamma_G C_v}{\lambda} \tag{10.67}$$

The thermal expansion coefficient is proportional to Grüneisen constant γ_G and inversely proportional to the compressibility λ of crystal. It is also closely related to the structure of the material. The more compact the structure, the higher the binding energy, the smaller the coefficient of thermal expansion. Because the compressibility λ is often different in different directions; in the same anisotropic crystal, there is little difference in Grüneisen constants in different directions. Therefore, the difference in the coefficient of thermal expansion in different directions often comes from the different compressibility. The compressibility is increased with the increase of the distance between ions, therefore in the direction of greater lattice constant; the coefficient of thermal expansion is often relatively smaller. For example, the lattice constants of YVO_4 crystal are: $a = b$=7.12Å, c = 6.29 Å, its thermal expansion coefficients are: $\alpha_a = 4.4\times10^{-6}$/K and $\alpha_c = 11\times10^{-6}$/K;the Nd^{3+}: KGW crystal has lattice constants: a = 8.10Å, b = 10.43Å and c = 7.60 Å, its thermal expansion coefficients are: $\alpha_a = 4.0\times10^{-6}$/K,$\alpha_b = 1.6\times10^{-6}$/K and α_c= 8.5 10^{-6}/K; another example of Yb^{3+}:YAB crystal has lattice constants: $a = b$=9.295Å and c = 7.243Å, then its thermal expansion coefficients are: $\alpha_a = 1.4\times10^{-6}$/K,$\alpha_c = 8.1\times10^{-6}$/K.

At very low temperature, the anharmonic interaction of lattice vibrations can be neglected, so the thermal conductivity of the materials is very high. The phonon mean free path l_p is also related to the order degree of the material. The order degree of a material can be seen from two aspects: first is whether the mass of lattice ions is uniform; if the masses of lattice ions are equal or approximately equal, then the phonon mean free path reaches its maximum. The larger the difference of the mass of lattice ion, the lower the lattice order, the shorter the mean free path l_p and the lower the thermal conductivity *K* of the crystal. An obvious example is the comparison of the thermal conductivity of yttrium aluminum garnet with that of the ruby. The density of yttrium aluminum garnet ($Y_3Al_5O_{12}$) is 4.55 g/cm^3 but the density of ruby (Al_2O_3) is only 3.98 g/cm^3. However, the thermal conductivity of the yttrium aluminum garnet is 11 W/m·K which is lower than half of the thermal conductivity 28 W/m·K of the ruby. One of the important reasons is the masses of aluminum ion and oxygen ion are very different from that of the yttrium ions while the ion mass difference is much smaller in ruby. The second factor that affects the order of a material is the defect in the material. A material with fewer defects will have higher order degree and so longer phonon mean free path and higher thermal conductivity, while a material with more defects only has shorter phonon mean free path and lower thermal conductivity. Especially when the temperature is not high, the main mechanism of phonon scattering is not the anharmonic lattice vibration but the phonon scattering by the impurities or defects, so that there is a great difference in the thermal conductivity of the material with and without impurities or defects. When active ions enter into the crystal lattice, due to the mass difference between the active ion and the host ion as well as the other defects introduced by the doping of the active ion, the mean free phonon path will be greatly reduced, so that the thermal conductivity is reduced appreciably. This can be seen by the comparison of thermal conductivity temperature dependence of pure YAG crystal and 0.5 at.% Nd^{3+}:YAG crystal, shown in Fig. 10.7.

The thermal properties of YAG, YAP, and other laser crystals are studied in detail by Aggarwal et al. [45]. Table 10.2 lists the data measured at temperature 300 K (the temperature coefficient of refractive index is measured at 1.06 μm), where Yb^{3+}:YAB data are quoted from Nikogosyan's book [46]. These data are not complete; some of these values are not consistent with those published in the literatures. Actually, due to the different doping concentration of active ion in the crystal, different measurement wavelength and temperature, different crystal integrity and optical homogeneity, as well as different measurement methods, the

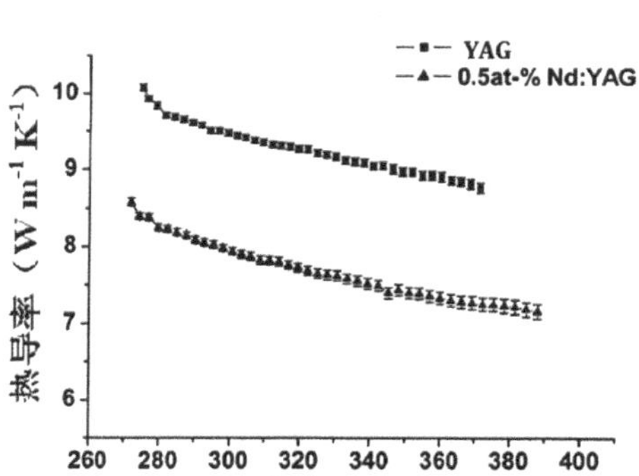

Fig. 10.7 Comparison of thermal conductivity temperature dependence of pure and 0.5 at.% Nd^{3+}:YAG crystals [44]

Table 10.2 Comparison of thermal performance parameters of several laser materials

Crystal	Expansion coefficient α ($10^{-6}K^{-1}$)	Temp. derivative of refractive index dn/dT ($10^{-6}K^{-1}$)	Conductivity K ($Wm^{-1}K^{-1}$)
YAG	6.14	7.8	11.2
Al_2O_3	4.4(⊥c), 6.8(//c)	12.8 (E⊥c), 11.8 (E//c)	28
$YAlO_3$	2.32(*a* axis), 8.0 8(*b* axis) 8.7(*c* axis)	7.7(E-*a* axis), 11.7(E-*b* axis) 8.3(E-*c* axis)	11.7(*a* axis), 10.0 (*b* axis) 13.3 (*c* axis)
YLF	14.31 (*a* axis), 10.05 (*c* axis)	−4.6 (E-*a* axis), −6.6 (E-*c* axis)	5.3 (*a* axis), 7.2 (*c* axis)
LuLF	13.6 (*a* axis), 10.8 (*c* axis)	−3.6 (E-*a* axis), −6.0 (E-*c* axis)	5.0 (*a* axis), 6.3 (*c* axis)
KGW_b axis	1.9		2.6
KYW_b axis	3.0		2.7
Yb:YAB	8.1(*c* axis), 4(⊥*c* axis) (Yb 1 at.%)		4.7 (Yb 5.6 at.%)

results obtained by different authors often to be different. However, some data published in the literature also do not conform to the principles of crystal physics. For example, the thermal expansion coefficient is a second-order symmetric tensor which is an isotropic physical quantity in the cubic YAG crystal, while there are reports of different thermal expansion coefficients in different directions (e.g. the data quoted in the literature [33]). Certainly, the non-uniform distribution of Nd^{3+} ions in the crystal or other inhomogeneity in the crystal growth process can lead to the deviation of the measured data.

Aggarwal et al. [45] obtained a series of the temperature function of thermal conductivity, thermal expansion coefficient, and the temperature coefficient of refractive index for laser crystals. For example, the relationship between expansion coefficient and temperature is

$$\alpha(T) = M_0 + M_1 T + M_2 T^2 + M_3 T^3$$

The parameters for YAG crystal are $M_0 = -1.8496$, $M_1 = 4.368 \times 10^{-2}$, $M_2 = -5.6844 \times 10^{-5}$, $M_3 = 0$.

The relationship between the temperature coefficient of refractive index and temperature is

$$\frac{dn}{dT}(T) = P_0 + P_1 T + P_2 T^2 + P_3 T^3$$

The parameters for YAG crystal are $P_0 = -3.946$, $P_1 = 5.294 \times 10^{-2}$, $P_2 = -4.5605 \times 10^{-5}$, $P_3 = 0$.

It is necessary to mention another important physical property of materials, that is, the hardness of materials. The end surface of laser materials is required to be processed into a surface with a high degree of parallelism and fineness; it requires the material to have a high hardness and therefore a high binding energy.

In short, starting from the requirements of laser device, a laser material must have a high thermal conductivity K, a high thermal diffusion coefficient κ, a high hardness H, a high tensile strength, and a high thermal shock resistance R_m. On the other hand, it is necessary to have a small coefficient of thermal expansion α and a small refractive index temperature coefficient. The thermal resistance and thermal expansion of an insulator are all resulted from the anharmonic effect of lattice vibration. The thermal conductivity decreases with the increase of the mass difference of the lattice ions while temperature coefficient of refractive index is proportional to the coefficient of thermal expansion α. The requirements of a small thermal expansion coefficient, a high thermal conductivity, and a small refractive index temperature coefficient are consistent. Because of these performance requirements, an ionic crystal with a higher binding energy, a higher melting point, and a smaller mass difference of the lattice ions should be adopted as a laser host material.

10.5 Laser Damage and Nonlinear Optical Properties

Many laser crystals have both laser emission and various nonlinear optical properties. This section mainly introduces the relationship between the nonlinear optical properties and laser damage of these crystals, but the nonlinear optical properties of laser crystals and their applications in laser operations such as self-frequency doubling, self-frequency mixing, and self-Raman lasers will be described in next chapter.

In high power or high energy laser operation, the nonlinear optical properties of a laser material can result in laser beam distortion, self-focusing, self-defocusing, [47, 48] and material damage. When laser intensity is very high, the nonlinear effect of dielectric polarization cannot be ignored and the dielectric coefficient of the material can be written as

$$\varepsilon = \varepsilon_0 + \varepsilon_2 E^2 \tag{10.68}$$

The corresponding refractive index has the following relation

$$n = n_0 + n_2 E^2 \tag{10.69}$$

where E is the electric field intensity of light field. When laser power is very high, the intensity of E is high enough, and then the second term of the above equation will produce a nonlinear effect. It should be noted that the intensity of laser beam often is the strongest at the central part (near the axis). If n_2 is positive, the effect of nonlinear term makes the refractive index of the material near the central of laser beam cross-section higher than that of its marginal part. The result of this nonlinear effect is that the laser beam just like is focused by a series of converging lens. If n_2 is negative, the effect of nonlinear term makes the refractive index of the material near the central of beam cross-section lower than that of the marginal part, then the result is that the laser beam just like is defocused by a series of divergent lens. The difference of the refractive index of different points on the cross section of laser beam caused by the nonlinear effect also produces a phase difference between the different points in the cross-section of laser beam; thus the wavefront distortion is produced and the linewidth of the laser beam is increased. The strong lasers can also generate two-photon absorption, stimulated Raman scattering, stimulated Brillouin scattering, and other nonlinear optical processes. All these effects will bring about the loss of laser power or energy. Therefore, the materials with weak nonlinear optical effect should be selected as the laser host materials of high power or high energy laser; for example, the crystal with central symmetry or low refractive index (by the Miller rule, if the linear polarization is small, then the nonlinear polarization is also small [49]). It is impossible to give a detailed description to a large number of nonlinear optical effects, but only choose self-focusing effect for further discussion, because it is directly related to the laser damage of the material.

The self-focusing of laser beam and its threshold can be described by a simple model. The main idea of this model is: when the refractive index $n_0 + n_2E^2$ inside the laser beam higher than refractive index n_0 outside the laser beam, total internal reflection will occur at the boundary of laser beam. This total internal reflection will cancel out the beam divergence due to diffraction and with the increase of the laser power the laser beam will be further converged to a smaller and smaller beam width. As shown in Fig. 10.8, the divergence angle of laser beam due to diffraction is δ_1, and the incidence angle of total internal reflection at the boundary of the beam is δ_2. Obviously, δ_1 and δ_2 are supplementary, that is

$$\sin^2\delta_1 + \sin^2\delta_2 = \sin^2\delta_1 + \cos^2\delta_2 = 1$$

If the beam width of a laser beam is D then the ideal divergence angle generated by the diffraction has the following relation

$$\sin\delta_1 \approx 1.22\lambda/n_0D \qquad (10.70)$$

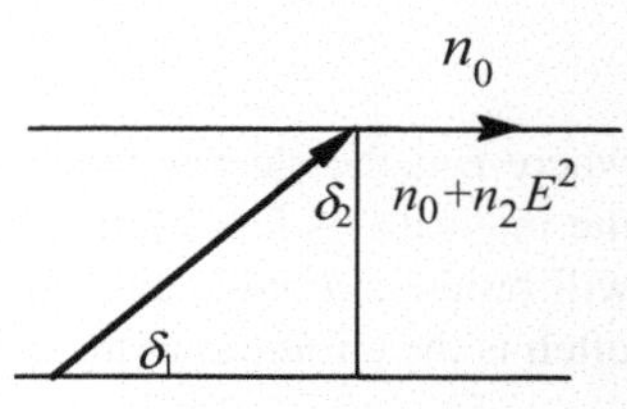

Fig. 10.8 Schematic diagram of laser beam self-focusing

According to refraction law and the refractive index of internal and external laser beam given above, it is concluded that the total internal reflection of a laser beam on the boundary must satisfy the following relation

$$\sin\delta_2 = n_0/(n_0 + n_2E^2) \tag{10.71}$$

The complementarily of δ_1 and δ_2 become

$$(1.22\lambda/n_0D)^2 + [n_0/\,(n_0 + n_2E^2)]^2 = 1 \tag{10.72}$$

The second term of the above equation can be approximately calculated as follows

$$n_0^2/(n_0 + n_2E^2)^2 = 1/(1 + E^2n_2/n_0)^2 \approx (1 - E^2n_2/n_0)^2 \approx 1 - 2E^2n_2/n_0 \tag{10.73}$$

then the following equation is established

$$(1.22\lambda/n_0D)^2 + 1 - 2E^2n_2/n_0 = 1 \tag{10.74}$$

Namely, when the electric field intensity of a laser beam is equal to $E^2 = (1.22\lambda)^2/(2n_0n_2D^2)$, the laser beam began to focus. Therefore the threshold electric field intensity of self-focusing is $E^2 \geq (1.22\lambda)^2/(2n_0n_2D^2)$. Because the sectional area of a laser beam is $\pi D^2/4$, the intensity of a laser beam is $n_0cE^2/8\pi$, so that the threshold power of self-focusing is

$$P_{TH} = (1.22\lambda)^2c/64n_2 \tag{10.75}$$

It can be seen that this threshold power is proportional to the square of laser wavelength and inversely proportional to nonlinear refractive index.

One can also approximately calculate the self-focusing threshold of a laser beam from Maxwell equation of classical electrodynamics [46]. This aspect of narrative can also refer to textbook published by Yariv [50] or monograph published by Akhmanov [51].

The appearance of self-focusing makes the diameter of a laser beam decrease and optical power density increase rapidly; therefore many physical processes introduced by strong laser including stimulated Raman scattering, stimulated Brillouin scattering, high-order harmonic generation, and a series of element excitation processes will take place. After absorbing strong laser energy, high temperature thermal effects such as the emission of electron, ion, or, molecule and the formation of plasma of a medium can be produced. Therefore, the laser beam self-focusing is closely related to the laser damage of a medium. The relationship between the self-focusing of laser beam and the damage of a material will be simply discussed in the following.

It is generally believed that laser damage is caused by the following three kinds of mechanism: first is the local fracture of a material caused by high-intensity ultrasonic phonon generated by stimulated Brillouin scattering; second is the defects or impurities in a material absorbing the laser energy then the local temperature rise up to tens of thousands degrees and thus the damage is caused by the explosion of heat shock wave or micro plasma; the third is that under extremely high laser power, the electric field intensity of laser radiation is close to or even higher than that of the atom itself, then the nonlinear effect has become very strong that the multi-photon absorption excites electrons in some atoms into the conduction band or even ionization. The resulting electrons accelerated by high-intensity laser field will generate the phenomenon of electron avalanche, similar to the dielectric breakdown under strong static electric field. No matter what kinds of mechanism, laser damage can occur only at sufficiently high laser power density. When laser beam is incident into an optical material, the power density of the laser may not reach the threshold of laser damage, but after going through a distance of the medium, the self-focusing effect will increase the power density of laser beam which entirely possible to reach or exceed the damage threshold. The material with positive temperature coefficient of refractive index ($dn/dT > 0$) may produce thermal lens effect due to the absorption of laser light by impurities and defects appeared. This is the reason of laser damage threshold on the incident surface is higher than that of the exit surface for some crystals and glasses.

One direct example of the relation of the damage threshold of laser self-focusing with laser wavelength is that after frequency doubling by using KDP or other nonlinear optical crystal the damage threshold is lower than that by using laser beam without frequency doubling. When the second harmonic laser of power 10^5 W focuses into ruby or sapphire crystal, some 1 cm long, 10 μm diameter filaments and a long string of cracks along the path of the light can be observed. However, the same power of laser without frequency doubling focus into the same materials cannot observe this phenomenon. As mentioned above, the damage threshold of laser self-focusing is proportional to the square of laser wavelength; obviously after frequency doubling, the damage threshold of laser self-focusing will be decreased four times, and this can explain the observed phenomenon. Of course, not all the laser damage of different materials results from beam self-focusing. By using laser pulse of single longitudinal mode and TEM_{00} transverse mode at wavelength 1064 nm with width of 9.9 ns, Do and Smith [52] observed the laser damage threshold of undoped and Nd^{3+}, Cr^{4+}, Yb^{4+}-doped YAG crystals and ceramics. The relevant parameters of YAG crystal were used to calculate damage threshold by laser self-focusing and damage threshold by stimulated Brillouin scattering. The results are higher than the actual observed damage threshold; while the multi-photon ionization and avalanche ionization induced by the light electric field heating electrons of conduction band can successfully illustrate the experimental data of single-pulse excitation. As for the damage threshold of continuous multiple pulse excitation lower than that of the single excitation, they considered that the cumulative damage may be due to the distribution of electrons to low

Table 10.3 The property requirements of solid-state laser material

Host material
1. Suitable size can be prepared easily and economically
2. Stable and durable during laser operation
3. Have chemical stability under light radiation, heating and mechanical processing
4. Better thermal properties:
(1) high thermal conductivity,
(2) high thermal diffusion coefficient
(3) small thermal expansion coefficient
5. Better mechanical properties:
(1) High stress rupture limit and high tensile strength
(2) Small stress optical coefficients
(3) High laser damage threshold (for the material used in high energy and high power laser)
(4) high hardness
6. Better optical quality:
(1) Minimum packages, defects, and other scattering centers
(2) Minimum absorption at laser and pump wavelength
(3) Small refractive index temperature coefficient
Active ion
1. Valence and radius of active ion can match that of the substituted host ion
2. Can be well-distributed in the host material
3. High ground state (but no excited state) absorption at pumping wavelength and minimum ground state or excited state absorption at laser wavelength
4. Small quantum defect to make high efficiency population of metastable level and reduce thermal loading
5. Having an appropriate metastable state lifetime to make full use of all the pump energy
6. Can generate radiation at laser wavelength with high quantum efficiency
7. Emission line with linewidth and intensity both meet the requirements of laser device.
8. Have suitable ion–ion interaction to achieve the maximum pumping absorption and minimal quenching

energy and long life self-trapping exciton states; their ionization is easier than that of the valence electrons.

Briefly, in order to satisfy the requirements of laser application, laser materials must have the appropriate mechanical, thermal, and optical properties. The property requirements of solid-state laser material are listed in Table 10.3 for reference.

References

1. V. Lupei, N. Pavel, T. Taira, Opt Comm. **201**, 431 (2002)
2. V. Lupei, N. Pavel, T. Taira, Appl. Phys. Lett. **2002**(81), 2677 (2002)
3. F. Auzel, in *2000 Proceeding of the 2nd International Symposium On Laser, Scintillator and Nonlinear Optical Materials* (Lyon, 2000), pp. 28–31
4. N.A. Tolstik, V.E. Kisel, N.V. Kuleshov et al., Appl. Phys. B **97**, 357 (2009)
5. Y.J. Chen, Y.F. Lin, J.H. Huang et al., Opt. Express **18**, 1370 (2010)
6. Y.J. Chen, Y.F. Lin, J.H. Huang et al., Opt. Express **21**, 18919 (2013)
7. K.N. Gorbachenyo, V.E. kisel, A.S. Yasukevich, et al., Opt. Lett. **38**, 2446 (2013)
8. Y.J. Chen, Y.F. Lin, J.H. Huang et al., Opt. Lett. **40**, 4927 (2015)
9. Y.J. Chen, Y.F. Lin, J.H. Huang et al., Opt. Express **23**, 72401 (2015)
10. Y.J. Chen, Y.F. Lin, J.H. Huang et al., Opt. Express **2017**(25), 17128 (2017)
11. J.H. Huang, Y.J. Chen, H. Wang et al., Opt. Express **2017**(25), 24001 (2017)
12. J. Huang, Y. Chen, Y. Lin et al., Opt. Lett. **43**, 1643 (2018)
13. Y.J. Chen, Y.F. Lin, J.H. Huang et al., Opt. Express **26**, 419 (2018)
14. Y.J. Chen, Y.F. Lin, Z.M. Yang et al., OSA Contin. **2**, 142 (2019)
15. N.A. Tolstik, S.V. Kurilchik, V.E. Kisel et al., Opt. Lett. **32**, 3233 (2007)
16. Y. Li, J. Feng, P. Li et al., Opt. Express **21**, 6082 (2013)
17. Y. Chen, Y. Lin, Y. Zou et al., Opt. Express **22**, 13969 (2014)
18. A.A. Lagalsky, V.E. Kisel, A.E. Jroshin et al., Opt. Lett. **33**, 83 (2008)
19. G.B. Loutts, C. Bonner, C. Meegoda et al., Appl. Phys. Lett. **71**, 303 (1997)
20. F. Auzel, in ed. by B. Di Bartolo, *Spectroscopy of Solid-State Laser-Type Materials.* (Plenum Press, New York, 1987)
21. V. Kushawaha, A. Banerjee, L. Major, Appl. Phys. B **56**, 239 (1993)
22. Y. Chen, L. Major, V. Kushawaha, Appl. Opt. **35**, 3203 (1996)
23. O. Musset, J.P. Boquillon, Appl. Phys. B **64**, 503 (1997)
24. O. Musset, J.P. Boquillon, Appl. Phys. B **65**, 13 (1997)
25. V. Kushawaha, Y. Yan, Y. Chen, Appl. Phys. B **62**, 533 (1996)
26. Z.D. Luo, X.Y. Chen, C.Y. Tu, et al., Chin. Phys. Lett. **17**, 888 (2000)
27. X. Mateos, P. Loiko, J.M. Serres et al., IEEE. J. Quant. Elect. **53**, 1 (2017)
28. Z.D. Luo, Y.D. Huang, Opt. Comm. **206**, 159 (2002)
29. Z.D. Luo, Y.D. Huang, M. Montes, D. Jaque, Appl. Phys. Lett. **85**, 715 (2004)
30. C. Jacinto, T. Catunda, D. Jaque et al., Phys. Rev. B **72**, 235111 (2005)
31. A. Penzkofer, Prog. Quantum Electron. **12**, 291 (1988)
32. Y.F. Chen, T.M. Huang et al., IEEE. IEEE J. Quantum Electron. **33**, 1424 (1997)
33. W. Koechner, *Solid-State Laser Engineering*, 6th edn. (Springer Verlag, Berlin, 1996)
34. M. Born, E. Wolf, *principle of Optics*, 5th edn. (Pergamon Press, Oxford, 1975)
35. X.B. Wang, X.J. Xu, X. Li, et al., Appl. Opt. **46**, 5237 (2007)
36. M.E. Innocenzi, H.T. Yura, C.L. Fincher, Appl. Phys. lett. **56**, 1831 (1990)
37. J.D. Foster, L.M. Osterink, J. Appl. Phys. **61**, 3656 (1970)
38. T. Chin, R.C. Morris, O. Kafri, et al., in *Technical Digest of the Conference on Lasers and Electro- Optics CLEO'86*. (California, San Francisco, 1986), p. 212
39. I. Moshe, S. Jackel, R. Lallouz et al., Proc. SPIE **1997**(3110), 238 (1997)
40. J.F. Nye, *Physical Properties of Crystals, their Representation by Tensors and Matrices* (Clarendon Press, Oxford, 1957)
41. G. Leibfried, E. Schloemann, Nachr Ak Wiss Gottingen. Math. Phys. **11a**, 71 (1954)
42. P.G. Klemens, in *Solid State Physics*, vol. 7, eds. Seitz, Turnbull, (Academic Press, New York, 1958)
43. C. Kittel, *Introduction to Physics*, 5th edn. (Wiley, New York, 1976)
44. B.S. Wang, H.H. Jiang et al., Eur. Phys. **39**, 23 (2007)
45. R.L. Aggarwal, D.J. Ripin, et al., J. Appl. phys. **98**, 103514 (2005)

46. D.N. Nikogosyan, *Nonlinear Optical Crystals-A Complete Survey* (Springer Science + Business Media, New York, 2005)
47. R. Menzel, *Photonics—Linear and Nonlinear Interactions of Laser Light and Matter* (Springer, Berlin, 2001)
48. Z.D. Luo, Optoelectronic Technology (Internal publication of the Fujian Institute of the Research on the Structure of Mater), First Issue 27, (1973) (in Chinese)
49. Y.R. Shen, *The Principles of Nonlinear Optics*. (Wiley, 1984)
50. A. Yariv, *Quantum Electronics*, 3rd edn. (Wiley, 1989)
51. S.A. Akhmanov, R.V. Khokhlov, A.P. Sukhorukov, *Laser Handbook*, vol. 2 (North–Holland Publishing Company, Amsterdam, 1972), p. P1151
52. B. Do, A. Smith, Appl. Opt. **48**, 3512 (2009)

Chapter 11
Nonlinear Optical Properties of Laser Crystals and Their Applications

As mentioned above, many laser crystals have nonlinear optical properties and so can double, mixing, or transform the frequency of laser beam by second harmonic (SH) and stimulated Raman scattering (SRS) effects. With the development of laser technology, the study on these effects and their applications has become increasingly extensive and in-depth because of its important significance in technology.

More laser applications require the devices to be more portable, more compact, and easier to be adjusted and used. To meet these requirements is the key to expand laser application, because small and portable devices are easy to be used in more occasions. To concentrate the generation and modulation function of laser beam in the same crystal, no doubt that it will bring the benefits of small, lightweight, easy to be adjusted and used. This requires that a crystal has both the performances of laser generation and the modulation of spatial and temporal characteristics. In this chapter, we first discuss the combination of laser emission with frequency doubling and mixing, then the combination of laser emission with SRS effect. The emphasis will be placed on the physical principle rather than technical aspect, a relatively simple model is adopted, and so only the most influential factors are considered while some problems that must be considered in practical application are neglected.

As it is well known, blue and green lasers are an urgent need in information technology and many other laser applications. At the start of the study, it is found that to directly generate the laser emission at such wavelengths is more difficult than to generate near-infrared laser emission and then to double or mix the frequency of fundamental laser emission. Technically, blue and green laser beams are generally produced by an infra laser beam generated by one laser crystal, then through frequency doubling, mixing, or optical parametric oscillation by another nonlinear crystal. If the second-order nonlinear optical coefficient of the laser crystal is large enough, then can the laser beam with double frequency be obtained by the frequency doubling effect of the laser crystal itself, that is, generating the so-called self-frequency doubling (SFD) laser beam? The answer is affirmative. Johnson [1] in 1969 and Dmitriev [2] in 1979 published, respectively, their laser SFD results by using Tm^{3+} and Nd^{3+} ion activated $LiNbO_3$ crystal. Of course, the SFD efficiency

Z. Luo and Y. Huang, *Physics of Solid-State Laser Materials*, Springer Series in Materials Science 289, https://doi.org/10.1007/978-981-32-9668-8_11

of this crystal was still very low. Dorozhkin [3] in 1981 first did the SFD experiment of Nd^{3+}:$YAl_3(BO_3)_4$ (NYAB) crystals. Because the concentration of Nd^{3+} ion was too high, the self-absorption at the wavelength 0.53 μm was serious, so he only obtained the SFD laser at 0.66 μm generated by the frequency doubling of fundamental laser at 1.32 μm. Lu et al. [4] in 1986 grew this crystal with a Nd^{3+} ion concentration lower than 20% atomic ratios and realized the 0.53 μm SFD laser output pumped by pulsed dye laser, but the doubling efficiency is still low. The author [5, 6] pointed out that the Nd^{3+} ion concentration in laser crystal influences not only fundamental laser output level and luminescence concentration quenching effect but also nonlinear optical coefficients, and all aspects of the effect must be taken into account to determine the optimum Nd^{3+} concentration in NYAB crystal. Both suitable Nd^{3+} concentration and suitable crystal growth method were adopted and the 1.06 → 0.53 μm frequency conversion of NYAB crystal pumped by an incoherent light source—a common pulse xenon lamp—was realized in 1988 [5, 6]. Since then, the SFD laser crystal attracted the attention of many scientists, and a lot of research work has been carried out on crystal spectral properties, energy level structure, laser experiment, and the application of diode pumping of NYAB and NGAB (Nd^{3+}:$GdAl_3(BO_3)_4$) crystals [7–24]. Later is the work of Bartschke [24] who obtained the green laser output of 225 mW by using an NYAB crystal pumped by 1.6 W LD laser and at the same time obtained the green laser output of 450 mW by using the same crystal pumped by 2.2 W Ti^{3+}:sapphire laser. It is shown that the SFD laser has practical application in laser technology. The SFD and self sum-frequency mixing (SSFM) properties of the Nd^{3+} and Yb^{3+} ions-activated MgO-$LiNbO_3$, GdCOB ($Ca_4GdO(BO_3)_3$), YdCOB (Ca_4YO $(BO_3)_3$), and YAB crystals were also studied by many authors [25–42]. 1.1 W green laser output was obtained from Yb^{3+}:YAB crystal, and its light–light efficiency reached 10% [40]. Moreover, a 510–545 nm green/yellow wide tuning laser was obtained in Yb^{3+}: YAB crystal [41]. GdCOB and YCOB crystals have smaller walk-off angle and large crystal can be grown by Czochralski method in a relatively short period of time. Among them, the performance of Yb^{3+}:YCOB is better than Nd^{3+}:YCOB, while Nd^{3+}GCOB is better than Yb3+:GCOB. A 1.35 W green laser output was obtained by the SFD of Nd^{3+}:GdCOB crystal with maximum optical′conversion efficiency of 17% [42] and it has application prospect of laser display, and so on. On the other hand, Capmany et al. [43, 44] used quasi-phase matching method to achieve SFD and SSFM laser output in the non-periodic polarization Nd^{3+}:$LiNbO_3$ crystal. This quasi-phase matching method can not only make use of the larger nonlinear optical coefficient d_{33} in $LiNbO_3$ crystal from the consideration of nonlinear optical conversion efficiency, but also take advantage of stronger fundamental polarization luminescence in a crystal from the angle of the fundamental laser generation.

In this chapter, NYAB crystal will be taken as an example to introduce the relationship between fundamental laser emission and frequency doubling (mixing) laser generation in a nonlinear laser crystal, the effect of active ion doping on the nonlinear coefficients of laser crystal, the selection of the optimum concentration of active ion, as well as how to efficiently generate self-doubling (mixing) laser and its relationship with crystal structure and general physical properties of crystal. In the end, the basic knowledge about the composite effect of laser and stimulated Raman scattering (self Raman frequency shift) is introduced.

11.1 Second-Order Nonlinear Optical Effect of Crystal

In general, nonlinear optical effects include frequency doubling, frequency mixing, optical parametric process, stimulated Raman scattering, stimulated Brillouin scattering, and so on. The following is a discussion starting from second-order nonlinear optical effect, especially frequency doubling effect and by using Maxwell equation of electromagnetic wave.

By the electrodynamics, in a dielectric medium (specific conductance $\sigma = 0$, electric current $\boldsymbol{i} = 0$), the magnetic and electric field vectors of electromagnetic wave satisfy the following equations

$$\nabla \times \boldsymbol{H} = \frac{1}{c}\frac{\partial \boldsymbol{D}}{\partial t} \tag{11.1}$$

$$\nabla \times \boldsymbol{E} = -\frac{1}{c}\frac{\partial \boldsymbol{H}}{\partial t} \tag{11.2}$$

$$\boldsymbol{D} = \boldsymbol{E} + 4\pi \boldsymbol{P} \tag{11.3}$$

If we only consider linear and second-order nonlinear polarization, polarization intensity $\boldsymbol{P}$ can be expressed as

$$\boldsymbol{P} = \chi^{(1)}\boldsymbol{E} + \chi^{(2)}\boldsymbol{E}\boldsymbol{E} \tag{11.4}$$

The second-order polarization susceptibility $\chi^{(2)}$ in (11.4) is a second-order tensor, and one component of $\boldsymbol{P}$ can be expressed as

$$P_i = \chi^{(1)}E_i + \chi_{ijk}^{(2)}E_jE_k \tag{11.5}$$

Because there is no free charge in the medium, $\nabla\cdot\boldsymbol{E} = \boldsymbol{0}$. By calculating the curl of both sides of (11.2) and using formula $\nabla \times \nabla \times \boldsymbol{E} = \nabla\nabla\cdot\boldsymbol{E} - \nabla^2\boldsymbol{E}$ and (11.1), the following can be obtained

$$\nabla^2\boldsymbol{E} = \frac{1}{c^2}\frac{\partial^2\boldsymbol{D}}{\partial t^2} \tag{11.6}$$

Discussing only the propagation of light waves along the z direction and assuming a plane wave approximation can be used, the electric field is constant in the x and y direction, and so the electric field intensity is taken as the following form

$$E_l^{(\omega_m)}(z,t) = \frac{1}{2} E_{ml}(z) \exp[i(\omega_m t - k_m z)] + c.c \tag{11.7}$$

where m in the above formula is the label of the frequency of different interaction laser beams (m can be 1, 2, 3) and l denotes the different electric field components of a laser beam (l can be i, j, k). In the calculation of (11.6), a slow-varying amplitude approximation should be used; that is, assuming that the variation of the light wave amplitude with z is small enough, the following inequality is satisfied

$$d^2 E_m(z)/dz^2 << k_m dE_m(z)/dz$$

The second-order derivative of $E_{ml}(z)$ with respect to z can be ignored, then $\nabla^2 E_{1i}^{(\omega_1)}(z,t)$ can be expressed as

$$\nabla^2 E_{1i}^{(\omega_1)}(z,t) \approx -\left[ik_1 \frac{dE_{1i}(z)}{dz} + \frac{1}{2} k_1^2 E_{1i}(z) \right] \exp[i(\omega_1 t - k_1 z)] + c.c \tag{11.8}$$

On the other hand, by using $\varepsilon = \left(1 + 4\pi\chi^{(1)}\right) = n^2$ and $k = n\omega/c$, the right-hand side of (11.6) becomes

$$-\frac{1}{2} k_1^2 E_{1i}(z) \exp[i(\omega_1 t - k_1 z)] - \frac{2\pi}{c^2} \omega_1^2 \chi_{ijk}^{(2)} E_{3j} E_{2k}^* \exp\{i[\omega_1 t - (k_3 - k_2) z]\} + c.c \tag{11.9}$$

It has been assumed that $\omega_1 = \omega_3 - \omega_2$ in the above equation. Let (11.8) is equal to (11.9) and eliminating the same terms on both sides of the equation obtained, the following equation can be obtained

$$\frac{dE_{1i}(z)}{dz} = -i \frac{2\pi\omega_1}{n_1 c} \chi_{ijk}^{(2)} E_{3j} E_{2k}^* \exp[-i(k_3 - k_2 - k_1) z] \tag{11.10}$$

Similarly, the following two equations can also be obtained

$$\frac{dE_{2k}(z)}{dz} = -i \frac{2\pi\omega_2}{n_2 c} \chi_{kij}^{(2)} E_{1i}^* E_{3j} \exp[-i(k_3 - k_2 - k_1) z] \tag{11.11}$$

$$\frac{dE_{3j}^*(z)}{dz} = +i \frac{2\pi\omega_3}{n_3 c} \chi_{jik}^{(2)} E_{1i}^* E_{2k}^* \exp[-i(k_3 - k_2 - k_1) z] \tag{11.12}$$

Remembering that light intensity is equal to the speed of light multiplied by its energy density

$$I = \frac{c}{n} \rho = \frac{c\varepsilon}{8\pi n} E^2 = \frac{cn}{8\pi} E^2$$

By using three equations of (11.10) to (11.12), it is clear that each laser beam is suitable for the following form of equation

$$\frac{dI(\omega,z)}{dz}=\frac{cn}{8\pi}\left[E(\omega,z)^*\frac{dE(\omega,z)}{dz}+E(\omega,z)\frac{dE(\omega,z)^*}{dz}\right]$$

Consider the situation that the laser beams 1, 2, 3 only have one electric field component and are not equal to zero. After determining the crystal orientation, the actual effect of tensor $\chi^{(2)}_{ijk}$ can be represented by tensor $\chi^{(2)}_{\text{eff}}$. The function relation between $\chi^{(2)}_{\text{eff}}$ and $\chi^{(2)}_{ijk}$ which depends on the symmetry of the crystal and the phase matching condition will not be discussed here. By using the relation between the complex amplitude phases quoted from YR Shen's book [45] in the case without incident light of frequency ω_3, the following equation can be obtained by (11.10)

$$\frac{dI(\omega_1)}{dz}=-\frac{4\pi\omega_1}{c}\sqrt{\frac{2\pi}{cn_1n_2n_3}}\chi^{(2)}_{\text{eff}}\sqrt{I(\omega_1)I(\omega_2)I(\omega_3)}\cos[(k_3-k_2-k_1)z] \quad (11.13)$$

Similarly

$$\frac{dI(\omega_2)}{dz}=-\frac{4\pi\omega_2}{c}\sqrt{\frac{2\pi}{cn_1n_2n_3}}\chi^{(2)}_{\text{eff}}\sqrt{I(\omega_1)I(\omega_2)I(\omega_3)}\cos[(k_3-k_2-k_1)z] \quad (11.14)$$

$$\frac{dI(\omega_3)}{dz}=+\frac{4\pi\omega_3}{c}\sqrt{\frac{2\pi}{cn_1n_2n_3}}\chi^{(2)}_{\text{eff}}\sqrt{I(\omega_1)I(\omega_2)I(\omega_3)}\cos[(k_3-k_2-k_1)z] \quad (11.15)$$

When $\omega_3 = \omega_1 + \omega_2$ it is obvious that

$$\frac{dI(\omega_3)}{dz}+\frac{dI(\omega_1)}{dz}+\frac{dI(\omega_2)}{dz}=0 \quad (11.16)$$

It means that in the process of a laser beam of frequency ω_1 interacts with a laser beam of frequency ω_2 to produce a laser beam of frequency ω_3, the total energy is conserved.

From the previous analysis, one can see clearly that the energy of laser beams with different frequencies can be converted to each other due to the existence of nonlinear polarization.

Now take a look at some of the characteristics of the second harmonic generation and the condition to achieve phase matching. For the frequency doubling process $\omega_1 = \omega_2 = \omega$, $\omega_3 = 2\omega$; in this case the second-order polarization susceptibilities in (11.10) to (11.12) have the relation $\chi^{(2)}_{jik} = \chi^{(2)}_{ijk} = \chi^{(2)}_{kij} = \chi^{(2)}_{\text{eff}}$; therefore (11.12) including phase factor can be written as

$$\frac{dE_{3j}(z)}{dz}=-i\frac{2\pi\omega}{n_3c}\chi_{eff}E_{1i}E_{2k}\exp(i\Delta kz) \quad (11.17)$$

where $\Delta k = k_3 - k_2 - k_1$. Suppose there is no second harmonic input, that is, $E_{3j}(0) = 0$ and the length of crystal is L. Under the small signal approximation, the efficiency of the double frequency conversion is very low, and the electric field intensity of the fundamental laser beam can be considered to be independent of z. Integrating (11.17) in this condition, the laser field at the output end of the crystal is obtained

$$E_{3j}(L) = -i\frac{2\pi\omega}{n_3 c}\chi_{eff}E_{1i}E_{2k}\frac{e^{i\Delta kL}-1}{i\Delta k}$$

Therefore

$$\left|E_{3j}(L)\right|^2 = \frac{4\pi^2\omega^2}{n_3^2c^2}\chi_{eff}^2E_{1j}^2E_{2k}^2L^2\frac{\sin^2(\Delta kL/2)}{(\Delta kL/2)^2}$$

Attention, in the case of frequency doubling, $I_2 \equiv I(\omega_3) = I(2\omega)$, $I_1 \equiv I(\omega_1) = I(\omega_2) = I(\omega)$, while in (11.17) and the above equation, the refractive index n_3 is denoted by n_2 and the refractive index at the fundamental wavelength is expressed as n_1, thus the second harmonic conversion efficiency can be obtained

$$\frac{I_2}{I_1} = 32\,\pi^3\omega^2\chi_{eff}^2L^2\frac{1}{c^3n_2n_1^2}\frac{\sin^2(\Delta kL/2)}{(\Delta kL/2)^2}\times I_1 \tag{11.18}$$

It can be seen that the second harmonic conversion efficiency is proportional to the intensity and the frequency square of the fundamental laser, the square of second-order nonlinear optical coefficient, and the square of crystal length. The factor $\sin^2(\Delta kL/2)/(\Delta kL/2)^2$ is determined by a phase matching condition. In the case of phase matching $\Delta k = 0$, this factor reaches its maximum value. When calculating the second harmonic intensity, the total loss of the fundamental wave is the sum of (11.13) and (11.14), which is two times that of the (11.13). A is used to express the product of the following series of parameters

$$A = (8\,\pi\omega_1/cn_1)\chi_{eff}(2\,\pi/cn_2)^{1/2} = (16\,\pi^2\nu_1/cn_1)\chi_{eff}^{(2)}(2\,\pi/cn_2)^{1/2}$$

Therefore the coupling equation can be written as

$$\begin{cases} \frac{dI_1}{dz} = -AI_1\sqrt{I_2} - (\kappa_1+\gamma_1)I_1 \\ \frac{dI_2}{dz} = AI_1\sqrt{I_2} - (\kappa_2+\gamma_2)I_2 \end{cases} \tag{11.19}$$

The prerequisite $\Delta \boldsymbol{k} = \boldsymbol{k}_3 - \boldsymbol{k}_2 - \boldsymbol{k}_1 = 0$ for the generation of second harmonic wave has been taken into account and at the same time, the absorption coefficient κ_1,κ_2 and the loss coefficient γ_1, γ_2 are added to take into account the absorption of fundamental and double frequency laser beams and the other losses.

Different ways to achieve $\Delta \boldsymbol{k} = 0$ is by corresponding to different types of phase matching. One can take the negative uniaxial crystal as an example to explain how to realize the type I and the type II phase matching. Wave vector $\boldsymbol{k}$ is the momentum of photon and $\Delta \boldsymbol{k} = 0$ is $\boldsymbol{k}_3 = \boldsymbol{k}_2 + \boldsymbol{k}_1$. The equality on both sides of this vector equation means the consistency of the momentum in number and direction, so phase matching is the conservation of momentum in nonlinear optical process. If this condition is satisfied, the second harmonic generation in fundamental wave propagation process will strengthen each other and form a stronger second harmonic laser beam. The wave vector is $n\omega/c$ in quantity and for the frequency doubling process, the condition of phase matching $k_2 = 2k_1$ is

$$n_2(\omega_2)\omega_2 = 2n_1(\omega_1)\omega_1$$

Substituting $\omega_2 = 2\omega_1$ then

$$n(2\omega_1) = n(\omega_1) \tag{11.20}$$

In the case of light propagating in a normal dispersion medium, the higher the frequency, the higher the refractive index, the above conditions cannot be satisfied. However, when there is birefraction phenomenon in the crystal, the refractive index of ordinary light (o light) is not equal to that of the extraordinary light (e light). If the fundamental light is o (or e) light while the second harmonic light is e (or o) light, then (11.20) can be satisfied. The way to realize the phase matching is to select a particular direction of the crystal so that (11.20) can be satisfied; it is called angle phase matching. In some of the nonlinear optical crystals, the temperature coefficient of the refractive index for ordinary light is different from that of the extraordinary light and so phase matching can be achieved by changing the temperature; it is called temperature phase matching.

Some specific problems of angle phase matching will be explained as follows. For a negative uniaxial crystal ($n_o > n_e$), if one chooses fundamental light as o light and second harmonic light as e light, the refractive index of high-frequency second harmonic light can be equal to that of the low-frequency fundamental light. This matching method is referred to as ooe type matching method. Obviously, for a positive uniaxial crystal, because $n_e > n_o$, one can only choose eeo matching method to satisfy the phase matching condition. It can be seen that, whether it is ooe or eeo matching method, the polarization directions of two fundamental light are parallel, so it is called a parallel phase matching or a type-I phase matching. Of course, one can still choose two fundamental waves as o light and e light, respectively, and the frequency doubling light in a negative uniaxial crystal as e light (oee type phase matching) and in an uniaxial positive crystal as o light (oeo type phase matching). In these two cases the polarization directions of two fundamental lights are all perpendicular to each other, so these phase matching are called orthogonal phase matching or type-II phase matching.

According to the phase matching conditions, the optimal angle of different types of phase matching can be deduced. The calculation formulas will be simply listed in the following. The phase matching angle for type-I phase matching of a negative uniaxial crystals is

$$(\theta_m)_I^- = \sin^{-1}\left[\left(\frac{n_2^e}{n_1^o}\right)^2 \times \frac{(n_2^o)^2 - (n_1^o)^2}{(n_2^o)^2 - (n_2^e)^2}\right]^{1/2} \tag{11.21}$$

The phase matching angle for type-II phase matching of a negative uniaxial crystals is

$$(\theta_m)_{II}^- = \sin^{-1}\left\{\frac{(2n_2^o)^2 \Big/ \left[n_1^e(\theta_m)_{II}^- + n_1^o\right]^2 - 1}{(n_2^o/n_2^e)^2 - 1}\right\}^{1/2} \tag{11.22}$$

The right side of (11.22) includes the unknown quantity, so one can use iterative method to work out the phase matching angle. For type-I and type-II phase matching of a positive uniaxial crystal, the corresponding phase matching angle formulas can be obtained as follows

$$(\theta_m)_I^+ = \sin^{-1}\left[\frac{1 - (n_1^o/n_2^o)^2}{1 - (n_1^o/n_1^e)^2}\right]^{1/2} \tag{11.23}$$

$$(\theta_m)_{II}^+ = \sin^{-1}\left[\frac{\left[n_1^o/(2n_2^o - n_1^o)\right]^2 - 1}{(n_1^o/n_1^e)^2 - 1}\right]^{1/2} \tag{11.24}$$

The angle θ_m in (11.21) to (11.24) is the angle between the wave vector of optical wave and the optical axis of crystal. It must be noted that when the direction of wave vector (direction of light power flow) and the direction of the optical axis of a uniaxial crystal is not equal to 90°, the traveling direction of the fundamental laser beam and that of the second harmonic laser beam will not completely coincide but is taken off a small angle δ called walk-off angle. For type-I phase matching of a negative uniaxial crystal, δ satisfies the following equation [46]

$$\tan\delta = \frac{(n_1^0)^2}{2}\left[(n_2^e)^{-2} - (n_2^0)^{-2}\right]\sin(2\theta_m) \tag{11.25}$$

As the light continues to travel in a crystal, the overlapping between fundamental laser beam and second harmonic laser beam becomes smaller and the frequency doubling efficiency decreases, which greatly limits the crystal size that can be used. It can be seen from (11.25) that $\delta = 0$ when $\theta_m = 90°$ and the unfavorable factor caused by walk-off angle disappears. Usually the phase matching of $\theta_m = 90°$ is called non-critical phase matching while the phase matching of $\theta_m \neq 90°$ is called critical phase matching. It can be seen from (11.21) that for the type-I phase matching of a negative uniaxial crystal, when $n_1^o = n_2^e$, $\theta_m = 90°$. Technically, the crystals with n_1^o close to n_2^e and $d(n_1^o - n_2^e)/dT \neq 0$ is generally chosen to realize non-critical phase matching by using the method of temperature regulation.

In addition to the above-mentioned laser power, frequency doubling coefficient, and crystal length, the other factors that affect the efficiency of frequency doubling should be taken into account. First is the allow angle breadth $\Delta\theta_m$; beyond it the frequency doubling efficiency is less than 50% of its maximum value, and can be obtained from the formula [46]

$$\Delta\theta_m = \frac{0.66\lambda_1 n_1^o}{L(n_2^o - n_2^e)\sin 2\theta_m} \tag{11.26}$$

where λ_1 is the wavelength of fundamental laser beam and L is the length of crystal. The frequency doubling efficiency will decrease when the wavelength of fundamental laser deviates from the wavelength of phase matching. If the wavelength deviation is $\delta\lambda$ and the frequency doubling efficiency reduced to half of its maximum value, then the laser spectral linewidth $\Delta\lambda = 2\delta\lambda$ and $\Delta\lambda$ can be expressed as [46]

$$\Delta\lambda = \frac{0.44\lambda_1}{L\left[\frac{\partial n_1^o}{\partial\lambda_1} - \frac{1}{2}\frac{\partial n_2^e(0)}{\partial\lambda_2}\right]} \tag{11.27}$$

Obviously, the smaller the divergence angle, the better the monochromaticity of a laser beam and the higher the efficiency of second harmonic generation; therefore one can also use spectral brightness to simultaneously describe the monochromaticity, directionality, and the power of a laser beam, that is, the power of laser beam per unit solid angle and unit wavelength interval. The spectral brightness can be improved by using a mode selection method of transverse mode and longitudinal mode. In addition, the change of crystal working temperature will cause the change of the refractive index of a crystal, which deviates from the phase matching point. Therefore, depending on the refractive index temperature coefficient of various crystals, each nonlinear optical crystal has a permissible frequency doubling temperature range ΔT [46]

$$\Delta T = \frac{0.44\lambda_1}{Ld(n_2^e - n_1^o)/dT} \tag{11.28}$$

The temperature change which affects the phase matching mainly comes from the change of environment temperature, but the light absorption of a crystal may also produce the temperature change. Particularly for a SFD laser crystal, because the doping of active ion, the total thermal effect or non-uniform thermal effect caused by the non-uniform absorption of pump light will reduce the efficiency of frequency doubling. This will be further discussed in the following. Similarly, the optical inhomogeneity in a crystal will also make parts of the region passing through by the laser beam does not satisfy the phase matching condition, thereby reducing the total conversion efficiency of frequency doubling.

A series of second harmonic laser parameters mentioned above are related to the refractive indexes of o-light and e-light for fundamental and second harmonic wave, while these refractive indexes also related to the active ion concentration of a crystal. The dispersion relations of the refractive index of two typical SFD laser crystals are listed below.

NYAB: The Nd^{3+} concentration is 5.6 at%, in the temperature range of 293 K < T < 473 K and wavelength range of 0.4 μm < λ < 0.7 μm. The dispersion relation is (the temperature T is in unit of Kelvin and the wavelength λ is in unit of μm) [19]

$$n_o^2 = 1 + \frac{172.4727}{\left(0.10985 + 7.7 \times 10^{-7}T - 2.38 \times 10^{-9}T^2\right)^{-2} - \lambda^{-2}}$$

$$n_e^2 = 1 + \frac{161.08069}{\left(0.10669 + 1.3 \times 10^{-6}T - 3.2 \times 10^{-9}T^2\right)^{-2} - \lambda^{-2}}$$

Yb^{3+}:YAB: The Yb^{3+} concentration is 8 at%. The dispersion relation at room temperature is (the wavelength λ is in unit of μm) [47]

$$n_o^2 = 3.1762 + 0.0013/\left(\lambda^2 - 0.1480\right) - 0.0971\lambda^2$$

$$n_e^2 = 2.8632 + 0.0090/\left(\lambda^2 - 0.0937\right) - 0.0083\lambda^2$$

Phase matching angle, walk-off angle, and other parameters can be calculated by (11.21) to (11.28), while the refractive indexes and phase matching parameters of a series of nonlinear optical crystals, including the frequency doubling crystal, can also be found in related manuals and books [48–50].

11.2 Relationship Between Fundamental and Second Harmonic Waves in SFD Laser Crystal

In principle, the SFD laser can be summed up in one sentence: the stimulated radiation emitted by the active ions forms a fundamental laser beam in an optical resonator, then a second harmonic laser beam is generated by the second-order nonlinear optical effect of the crystal, finally the second harmonic laser beam output by the output mirror of a laser resonator. However, it cannot be understood as a simple addition of laser effect and second harmonic effect. On the one hand, the efficiency of second harmonic conversion is proportional to the intensity of fundamental laser, and the increase of the intensity of fundamental laser is undoubtedly beneficial to the increase of the intensity of second harmonic laser. On the other hand, the conversion between fundamental laser and second harmonic lasers is not only mutually promoted but also restricted to each other. This is owing to the fact: first, the doping of ions into the host crystal has certain effect on the nonlinear optical coefficient of the crystal; secondly, a fundamental laser beam transformed into a second harmonic laser beam will form an important loss of the fundamental laser oscillation, that is, the generation of second harmonic laser has a restrictive function on the generation of fundamental laser; thirdly, the second harmonic laser absorbed by the active ion will weaken the intensity of second harmonic laser; fourthly, in the Nd^{3+} ion-doped crystal, second harmonic laser has a pumping effect on the active ions. The specific details of the above four aspects must be analyzed separately and then use a model to sum up the four kinds of effect and study the interaction between them.

First, the effect of the doping of active ion on the second-order nonlinear optical susceptibility of host crystal should be studied. Taking NYAB crystal as an example, the Y^{3+} ion in the crystal is substituted by doping Nd^{3+} ion. The radius of Nd^{3+} ion is slightly larger than that of the Y^{3+} ion (the Goldschmidt radius of Nd^{3+} ion is 0.115 nm and that of the Y^{3+} ion is 0.106 nm). The competition between increase effect and decrease effect on nonlinear optical susceptibility by the doping of Nd^{3+} ion determines the optimal concentration of Nd^{3+} ion. The nonlinear optical susceptibility of NYAB crystal can be seen as the summation of the nonlinear optical susceptibility of all the BO_3 groups in the crystal, while the nonlinear optical susceptibility of BO_3 group mainly comes from the contribution of π electron in the group. In NYAB crystal, there is one type of BO_3 group (type I group) in which the O^{2+} ion (O^{2+} (1)) only is directly adjacent to Al^{3+} ion. The other type of BO_3 group (type II group) has two O^{2+} (O^{2+} (2)) ions occupying, respectively, the vertices of two triangular cones in which the Y^{3+} ion is situated at the center. The π electron distribution of the BO_3 group of type II will be affected by the substitution of Nd^{3+} ion. When Nd^{3+} ion concentration is low, one Nd^{3+} ion replaces the Y^{3+} ion in one of the triangular cone, then the surrounding environment symmetry of BO_3 group is reduced, and the nonlinear optical susceptibility of this type of group is increased. The local space symmetry distortion of the crystal in the case of low Nd^{3+} ion concentration is less important and has not obvious influence on the summation of nonlinear optical susceptibility of different BO_3 groups in the crystal; therefore, the overall nonlinear optical susceptibility of the crystal will increase with the increase of Nd^{3+} ion concentration. However, when Nd^{3+} ion concentration increased to a certain level, the total variation of local space symmetry caused by the Nd^{3+} doping is obvious and is detrimental to the summation of nonlinear optical susceptibility of different BO_3 groups in the crystal. At the same time, when Nd^{3+} ion concentration increases to reach a certain level, there will be more BO_3 groups that have two Y^{3+} ions substituted by Nd^{3+} ions and so the number of low environment symmetry BO_3 group will be lower than that of the low Nd^{3+} ion concentration case. Hence, the number of BO_3 group with higher nonlinear optical susceptibility will be decreased with the increase of Nd^{3+} ion concentration, at the same time the local space group symmetry distortion will decrease the summing result of nonlinear optical susceptibility of different BO_3 groups. Consequently, when Nd^{3+} ion concentration exceeds a certain level (optimum concentration), the nonlinear optical susceptibility of NYAB crystal will be decreased with the increase of Nd^{3+} ion concentration. The test of frequency doubling effect of NYAB powder sample with different Nd^{3+} ion concentration basically reflects the competition of these two opposite effects.

The nonlinear coupling equations of the SFD process are discussed in detail below. First, study the term of second harmonic conversion in (11.19). In accordance with the condition of equal gain and loss under steady state, one has

$$\mathrm{A}I_1\sqrt{I_2} = (\gamma_2 + \kappa_2)I_2$$

where κ_2 is the absorption coefficient of second harmonic laser in crystal, γ_2 is the total loss including the output loss, the diffraction loss, and the walk out angle loss of second harmonic laser, in which the output loss is the main loss. One has the following expression for the intensity of second harmonic laser.

$$I_2 = \left(\frac{A}{\gamma_2 + \kappa_2}\right)^2 I_1^2 \tag{11.29}$$

Let $h = A^2/(\gamma_2 + \kappa_2)$ then $AI_1\sqrt{I_2} = hI_1^2$. By (11.19), if one ignores the small fundamental wave absorption and other small loss, the following equation resulted

$$\begin{cases} \frac{dI_1}{dz} = -hI_1^2 \\ \frac{dI_2}{dz} = hI_1^2 - (\kappa_2 + \gamma_2)I_2 \end{cases} \tag{11.30}$$

The effect of Nd^{3+} concentration on the intensity of second harmonic laser beam is included in the second equation. If the contribution of pumping intensity to fundamental laser intensity is taken into account in (11.30), a nonlinear-coupled equation set can be obtained, which includes the generation of fundamental laser and the transformation of fundamental laser into second harmonic laser. When the relationship between parameter A and Nd^{3+} concentration is obtained, then this equation can sum up the three aspects of the effects mentioned, which can be used to study the relation among several laser and nonlinear optics parameters. The pumping effect of the second harmonic wave is reflected in the inversion of particle number, which will be taken into account in the rate equation of population inversion in the next section.

It can be seen from (10.6) of Chap. 10 that the fundamental laser gain coefficient $\alpha_g = \sigma_{em}\,\Delta N$, because ΔN is proportional to the product of the pump absorption rate a_p and the fluorescence lifetimeτ_f of laser level, and so $\alpha_g \propto a_p \sigma_{em} \tau_f$. Look at the relation of pump absorption rate with Nd^{3+} ion concentration. By means of absorption spectra measurement, the absorption coefficient at wavelength 0.804 μm when Nd^{3+} ion concentration is 1%, $\kappa_{0.804}$ (1%Nd) = 1.44 cm^{-1} for NYAB crystal can be obtained, then the pump absorption rate of this crystal is proportional to the following factor

$$a_p \propto 1 - e^{-\kappa_{0.804}(1\%_{cc})N(\%)t_p}$$

The concentration of Nd^{3+} ion N (%) is in the unit of atomic percentage (it will be simply expressed as % in the following), and t_p is the crystal thickness of pump light penetration. The relation of the utilization rate of pumping light with Nd^{3+} ion concentration can be seen in Fig. 1 in reference [6]: when pumped with LD at wavelength 0.804 μm, if the crystal length is 0.5 cm, the concentration of Nd^{3+} ion is 8%, then the pump light utilization rate can reach 99.7%. It means that on further increase, the Nd^{3+} doping concentration has no practical benefit to the improvement of pump efficiency.

On the other hand, the fluorescence quenching effect of Nd^{3+} ion in a YAB crystal is very weak, which belongs to one kind of self-active laser crystal. The concentration quenching rate K_Q of this kind of crystal is proportional to the concentration of active ion

$$K_Q = C_g N = \frac{1}{\tau_f} - \frac{1}{\tau_r} \tag{11.31}$$

According to the experimental data of NYAB, the following relation can be obtained [6]

$$\tau_f = \left[3.3 \times 10^{-4} N(\%) + 1.66 \times 10^{-2}\right]^{-1} \mu s \tag{11.32}$$

Because one has the relation $\alpha_g \propto a_p \sigma_{em} \tau_f$ and emission cross-section σ_{em} is essentially independent of Nd^{3+} concentration, so that gain coefficient G can be expressed as

$$G = C_g \times \left[3.3 \times 10^{-4} N(\%) + 1.66 \times 10^{-2}\right]^{-1} \times \left[1 - e^{-\kappa_{0.804}(1\%)N(\%)t_p}\right] \tag{11.33}$$

where C_g is a constant independent of Nd^{3+} concentration and pump light penetration thickness t_p is in the unit of cm. It can be seen from Fig. 11.1 that regardless of the pump light penetration thickness, t_p is 0.3 or 0.5 cm, under the conditions of Nd^{3+} concentration <8%, the fundamental laser gain coefficient has been saturated.

Look at the absorption of second harmonic laser. If laser oscillation length is L (for LD end pumping $L = t_p$), the attenuation factor of second harmonic laser is obviously $\exp[-\kappa_d N_p L]$, where κ_d is the absorption coefficient for Nd^{3+} ion of 1% concentration at second harmonic wavelength (for NYAB crystal $\kappa_d = 0.27$ cm^{-1}). The attenuation factor of second harmonic laser by the absorption of Nd^{3+} ion in NYAB crystal is shown in Fig. 11.2. It can be seen that the attenuation factor of 15% Nd^{3+} concentration is four times larger than that of the 4% Nd^{3+} concentration for laser crystal with length of 0.5 cm. Although the second harmonic laser light intensity of powder sample measured for the 15% Nd^{3+} ion concentration reaches its maximum, it does not reach two times the corresponding value when the

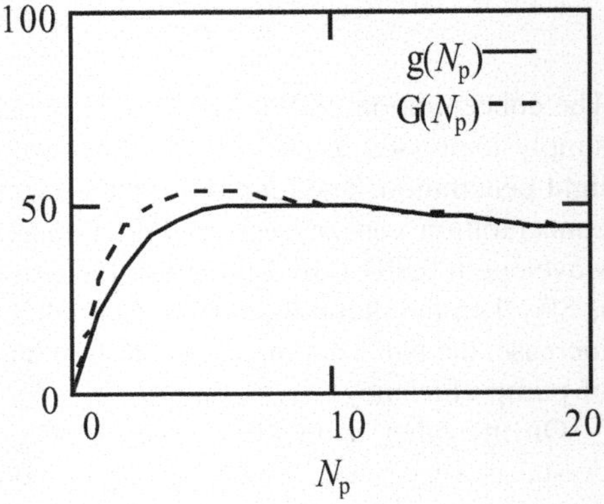

Fig. 11.1 Relationship between the atomic percentage concentration of Nd^{3+} ion and the fundamental laser gain coefficient of NYAB crystal: $g(N_p)$ corresponds to the pump light penetration thickness $t_p = 0.3$ cm; $G(N_p)$ corresponds to the pump light penetration thickness $t_p = 0.5$ cm

concentration is 4%. By (10.20) of Chap. 10, it is known that the power flux of the fundamental laser can be expressed as

$$\phi_f = \frac{\hbar\omega_p}{c\tau_c}\frac{n}{\sigma_p N_1 \tau_f \sigma_{em}}\left(\frac{\phi_{in}}{\phi_{th}} - 1\right) \tag{11.34}$$

The increase of fundamental laser flux by the increase of Nd^{3+} ion concentration is related to the ratio of input power flux to threshold power flux. It can be seen by (11.34) that when the pump power is constant, the increase of Nd^{3+} concentration does not lead to the laser power increase with the same multiple. Even though the fact that the second harmonic conversion constitutes an important loss of resonant cavity has not been considered, by taking into account the factors of fundamental laser power flux and the attenuation caused by the Nd^{3+} absorption of second harmonic laser, lower Nd^{3+} concentrations also should be selected. If the loss of the fundamental laser by the second harmonic conversion, the inhomogeneous thermal

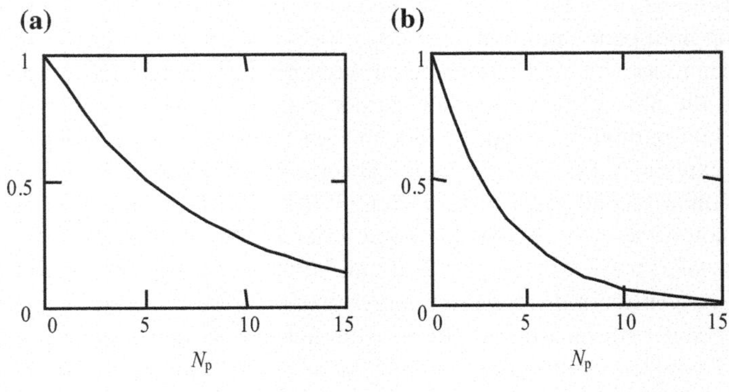

Fig. 11.2 Attenuation factor of second harmonic laser absorption by crystal NYAB; the horizontal coordinate is Nd^{3+} atomic percentage concentration N_p: **a** $L = 0.5$ cm; **b** $L = 1.0$ cm

effect, and the phase mismatch produced by the inhomogeneous of refractive index in higher Nd^{3+} ion doping level are considered, one has more reason to choose lower Nd^{3+} doping concentration.

Of course, to accurately calculate the optimal Nd^{3+} concentration, it is necessary to use the nonlinear optical coupling equation including fundamental laser generated, the conversion of fundamental laser into second harmonic laser and the attenuation of second harmonic laser, and so on, which will be introduced in the next section. Many useful information can be obtained by solving these equations. Obviously, the optimal doping concentration is a parameter related to pumping level and the parameters of laser device. Different pumping levels and different device parameters require different optimal doping concentrations. By the study of these nonlinear optical coupling equations, one can understand it clearly.

11.3 Nonlinear Optical Coupling Equation of SFD Laser

The simplest nonlinear laser-coupled equation of SFD is that the laser spatial distribution and the influence of walk-off angle are not taken into account [10]. The spatial distribution of laser pump light can be further included in [51], while the effect of walk-off angle can be calculated in different ways [52–54]. The theoretical model described below considers neither the spatial distribution of pump light, fundamental laser, and output laser nor the walk-off angle effect. The purpose is to let readers have a relatively simple and clear understanding of the physical process of SFD. In order to obtain a result, which can be compared with the experimental data, the above mentioned factors should be taken into account; the reader can refer to relevant literature, especially the theoretical model of Boyd and Kleinman [55]. However, it must be pointed out that the nonlinear optical coefficient of nonlinear laser crystal depends on the doping concentration of active ion, and discussion on the optimum doping concentration without considering this problem [53, 54] cannot obtain a correct result.

The rate equation set of the fundamental laser operation of a four-level system independent of spatial distribution is composed of a differential equation of Nd^{3+} ion population inversion density and a differential equation of fundamental laser light intensity I_1, and it can be written as

$$\begin{cases} \dfrac{d\Delta N}{dt} = WN_1 - (WN_1)_{\text{th}} - \dfrac{\sigma_{\text{em}} I_1}{\hbar\omega_1}\Delta N \\ \dfrac{dI_1}{dt} = \dfrac{c\sigma_{\text{em}} I_1}{n_1}\Delta N - \dfrac{I_1}{\tau_{\text{c}}} \end{cases} \tag{11.35}$$

In the above first equation, the total change rate of particle inversion number is equal to the increase rate of particle inversion number generated by pump light and minus the change rate of inversion number necessary to reach the threshold of laser oscillation and the change rate of inversion particle number produced by laser generation. Where $\Delta N = N_2 - (g_2/g_1)N_1$, $W = \eta B_{21} d(\omega)_{\text{p}}$ is the pumping rate, WN_1

indicates the contribution of the pumping to population inversion density, η is the efficiency of pumping, and B_{21} expresses Einstein coefficient in (5.53a) of Chap. 5. How to express energy density $d(\omega)_p$ as a function of light intensity will be discussed below. The second term $(WN_1)_{th}$ indicates the pumping rate at the laser threshold, which includes the pumping rate necessary to overcome the losses of spontaneous emission and fundamental laser light. The first term on the right hand side of the second equation of (11.35) is the contribution of stimulated radiation to laser intensity. In general laser rate equation this term should be divided by a factor of $1 + I_1/I_s$, due to the fact that the saturation intensity of the fundamental laser in the experiments discussed is $I_s = hc/\lambda\sigma_{em}\tau_f \approx 10^4$ W/cm^2, but the intensity of fundamental laser light discussed satisfy $I_1 << I_s$, so this factor can be omitted. The second term is the loss of the fundamental laser intensity in resonant cavity. Where σ_{em},τ_c, I_1, ω_1, and n_1 are the emission cross-section, the resonant cavity lifetime, the intensity, the frequency, and the refractive index of fundamental laser, respectively. In order to write down the rate equation set of fundamental laser generation and its second harmonic conversion, one should obviously combine equations of (11.30) and (11.35) in some way. First, one should rewrite (11.30) into time differential form, namely to use formula $dI/dt = dI/dz \times dz/dt$ to transform (11.30), where dz/dt is the speed of light in the medium, then the following equation set of the fundamental and second harmonic laser intensities in the case of steady output can be obtained

$$\begin{cases} \dfrac{dI_1}{dt} = -\dfrac{c}{n_1} h I_1^2 \\ \dfrac{dI_2}{dt} = \dfrac{c}{n_2} h I_1^2 - (\kappa_2 + \gamma_2)\dfrac{c}{n_2} I_2 \end{cases} \tag{11.36}$$

In the calculation of population inversion rate, for the total change rate of particle inversion number, in addition to those expressed in (11.35), the increase rate of particle inversion number generated by second harmonic wave should be considered. Combining (11.35) and (11.36) while taking into account that the second harmonic laser has pumping effect on active ion with pumping rate W_2, that is, W_2N_1 and the pumping rate of pump source is W_0, the following rate equation set for the fundamental laser of SFD laser crystal can be obtained

$$\begin{cases} \dfrac{d\Delta N}{dt} = (W_0 + W_2)N_1 - (W_0 N_1)_{th} - \dfrac{\sigma_{em} I_1}{\hbar\omega_1}\Delta N \\ \dfrac{dI_1}{dt} = \dfrac{c}{n_1}\sigma_{em}\Delta N I_1 - \dfrac{I_1}{\tau_c} - h\dfrac{c}{n_1} I_1^2 \end{cases} \tag{11.37}$$

Note, here ΔN is the population inversion density between the upper and the lower laser levels. Because the lower laser level is not ground state, the particle number density is very low, and it can be ignored, the particle number density of upper laser level is approximately equal to the population inversion density between the upper and the lower laser levels. N_1 is the particle density of ground state, that is, the active ion density of the crystal. Under the steady-state one has the following equation set

$$\begin{cases} (W_0 + W_2)N_1 - (W_0N_1)_{\text{th}} - \dfrac{\sigma_{\text{em}} I_1}{\hbar\omega_1}\Delta N = 0 \\ \sigma_{\text{em}}\Delta N I_1 - \dfrac{n_1}{c\tau_{\text{c}}} I_1 - h I_1^2 = 0 \end{cases} \quad (11.38)$$

W_2N_1, which involves the variable I_2, can be neglected in solving the first-order approximate solution. The following equation can be obtained by the first equation of (11.38)

$$\sigma_{\text{em}}\Delta N I_1 = \left[W_0N_1 - (W_0N_1)_{\text{th}}\right]\hbar\omega_1$$

A quadratic equation of I_1 with the following solution can be obtained

$$I_1 = \frac{-\frac{n_1}{c\tau_{\text{c}}} + \sqrt{\left(\frac{n_1}{c\tau_{\text{c}}}\right)^2 + 4h\left[W_0N_1 - (W_0N_1)_{\text{th}}\right]\hbar\omega_1}}{2h} \quad (11.39)$$

Because the light intensity must be positive, the negative solution has been given up.

The problem of how to use the second equation set of (11.36) to calculate SFD laser intensity will be discussed in the following. Pump energy density $d(\omega_{\text{p}})$ can be expressed as

$$d(\omega_{\text{p}}) = \frac{I(\omega_{\text{p}})}{(c/n_{\text{p}})} \quad (11.40)$$

In the problem discussed, the particle number of the upper laser level is much smaller than that of the ground state. The approximate assumption in Chap. 5 for the derivation of (5.11a) is valid. By using equations of (5.11a) and (5.53b), it is obvious that

$$B_{12}g(\omega) = \frac{c\kappa(\omega)}{\hbar\omega n N_1} \quad (11.41)$$

For the pumping laser with narrow linewidth, one has $I(\omega_{\text{p}}) = I_0\delta(\omega - \omega_{\text{p}})$, then the contribution of pumping laser to the variation rate of the particle number of the upper laser level can be obtained

$$W_0N_1 = \eta N_1 \int B_{12}g(\omega)d(\omega_{\text{p}})\text{d}\omega = \eta \int \frac{c}{\hbar\omega n_{\text{p}}}\frac{I_0\delta(\omega - \omega_{\text{p}})}{(c/n_p)}\kappa(\omega_{\text{p}})\text{d}\omega = \frac{\eta\kappa_{\text{p}}}{\hbar\omega_{\text{p}}}I_0 \quad (11.42)$$

where $\kappa_{\text{p}} \equiv \kappa(\omega_{\text{p}})$ and η is the pumping efficiency. For a certain volume of laser crystal, the intensity of pump light in different position is not the same, for example for diode laser end pumping, the pumping intensity is strongest on the incident surface of pumping light, deep into the interior of the crystal the pumping intensity is exponential decay. The average pumping rate per unit volume should be an integral divided by the volume of the pumping region in the crystal. When the length of a crystal is L, assuming that the intensity distribution of pump laser on the cross section of the crystal is uniform then the average pumping rate per unit volume is

$$\frac{\int W_0 N_1 \mathrm{d}v}{V} = \frac{\eta \int_0^L \kappa_\mathrm{p} I_0 e^{-\kappa_\mathrm{p} l} \mathrm{d}l}{\hbar\omega_\mathrm{p} L} = \frac{(1 - e^{-\kappa_\mathrm{p} L})}{\hbar\omega_\mathrm{p} L} \eta I_0 \tag{11.43}$$

The pump band is very close to the upper laser level, then the probability of non-radiative transition from pump band to the upper laser level is much higher than those of the radiative or non-radiative transitions from pump band to other lower energy levels and so $\eta \approx 1$ can be assumed. Similarly, the pumping effect of the double frequency laser on the crystal can be written as

$$W_2 N_1 = \frac{\kappa_2}{\hbar\omega_2} I_2 \tag{11.44}$$

The pump rate of pump source at the threshold can be expressed as follows

$$(W_0 N_1)_\mathrm{th} = \frac{\kappa_\mathrm{p} I_\mathrm{th}}{\hbar\omega_\mathrm{p}} \tag{11.45}$$

where I_th is the threshold pump intensity, that is, the threshold pump power per unit area. It is also the threshold power flux discussed in Chap. 10 and can be expressed as

$$I_\mathrm{th} \equiv \phi_\mathrm{th} = \frac{\hbar\omega_\mathrm{p}}{c\tau_\mathrm{c}} \frac{n_\mathrm{p}}{\sigma_\mathrm{p} N_1 \tau_\mathrm{f} \sigma_\mathrm{em}} \tag{11.46}$$

where σ_p is the absorption cross-section of the crystal at pump light wavelength and σ_em is the emission cross-section of the crystal at fundamental laser wavelength, therefore

$$(W_0 N_1)_\mathrm{th} = \frac{n_\mathrm{p}}{c\tau_\mathrm{c}\tau_\mathrm{f}\sigma_\mathrm{em}} \tag{11.47}$$

If laser output mirror has reflectivity R at pump light wavelength, then the corresponding pumping contribution of the reflected light should be added. By using the second equation of (11.39) and (11.29), as well as (11.47), the first-order approximate solution of SFD laser intensity can be obtained

$$I_2^{(1)} = \frac{\left[\sqrt{\left(\frac{n_1}{2c\tau_c\sqrt{h}}\right)^2 + \left(\frac{(1-e^{-\kappa_p L})(1+Re^{-\kappa_p L})}{L} \times \frac{I_0\omega_1}{\omega_p} - \frac{n_p\hbar\omega_1}{c\tau_c\tau_f\sigma_{em}}\right)} - \frac{n_1}{2c\tau_c\sqrt{h}}\right]^2}{(N_1\sigma_2 + \gamma_2)} \quad (11.48)$$

The second-order approximate solution of SFD laser intensity is obtained by adding the pumping effect of the SFD laser obtained in the first-order approximation.

$$I_2^{(2)} = \frac{\left[\sqrt{\left(\frac{n_1}{2c\tau_c\sqrt{h}}\right)^2 + \left(\frac{(1-e^{-\kappa_p L})(1+Re^{-\kappa_p L})}{L} \times \frac{I_0\omega_1}{\omega_0} - \frac{n_p\hbar\omega_1}{c\tau_c\tau_f\sigma_{em}}\right) + 0.5N_1\sigma_2 I_2^{(1)}} - \frac{n_1}{2c\tau_c\sqrt{h}}\right]^2}{(N_1\sigma_2 + \gamma_2)} \quad (11.49)$$

After $n-1$ iterations, the following is obtained

$$I_2^{(n)} = \frac{\left[\sqrt{\left(\frac{n_1}{2c\tau_c\sqrt{h}}\right)^2 + \left(\frac{(1-e^{-\kappa_p L})(1+Re^{-\kappa_p L})}{L} \times \frac{I_0\omega_1}{\omega_0} - \frac{n_p\hbar\omega_1}{c\tau_c\tau_f\sigma_{em}}\right) + 0.5N\sigma_2 I_2^{(n-1)}} - \frac{n_1}{2c\tau_c\sqrt{h}}\right]^2}{(N_1\sigma_2 + \gamma_2)} \quad (11.50)$$

According to (11.39) and (11.44), $W_2N_1 \times \hbar\omega_1 = N_1\sigma_2 I_2 \times \hbar\omega_1/\hbar\omega_2 = 0.5N_1\sigma_2 I_2$ representing the pumping effect of second harmonic laser has been added to (11.48) to (11.50). After several iterations, the convergence of double frequency laser intensity can be obtained. The intensity of output SFD laser is equal to the above laser intensity multiplied by the transmission of output mirror.

As mentioned in the previous section, the doping of a small amount of Nd^{3+} ion increases the nonlinear coefficient of NYAB crystal. According to the experimental data in terms of nonlinear optical coefficient measurement of a series of crystal powder samples, it can be assumed in the simulation calculation that $\chi^{(2)}_{eff}(N_p) = 6.1207 \times 10^{-9} \times (1 + N_p/30 - N_p^2/900)$esu, as is shown in Fig. 11.3. For NYAB crystal, the refractive index $n_1 = 1.755$, $n_2 = 1.708$ and in the case $\omega_1 = 1.77 \times 10^{15}\ s^{-1}$, $\omega_2 = 3.54 \times 10^{15}\ s^{-1}$, for type I (ooe) phase matching, an expression of $A = 5.73 \times 10^{-8} \times (1 + N_p/30 - N_p^2/900)$ can be obtained by the formula shown before the introduction of (11.19), where the Nd^{3+} concentration N_p is in the unit of atomic percentage while the product of active ion number N_1 per unit volume and absorption cross-section σ_p can be replaced by the product of N_p and the absorption coefficient $\kappa_p(1\%)$ of Nd^{3+} ion at unit atomic percentage concentration, as shown in Sect. 11.2. $\kappa_p = N_1\sigma_p \equiv N_p\kappa_p(1\%)$ and $\kappa_2 = N_1\sigma_2 \equiv N_p\kappa_2(1\%)$. According to the data given, for the semiconductor laser pumping at wavelength 0.804 μm, $\kappa_{0.804}(1\%) = 1.44\ cm^{-1}$ and $\kappa_2(1\%) = 0.27\ cm^{-1}$, so that $\kappa_p = 1.44N_p\ (cm^{-1})$ and $\kappa_2 = 0.27N_p\ (cm^{-1})$.

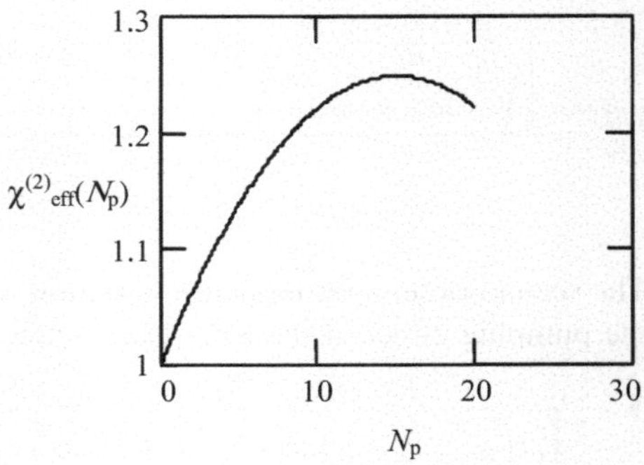

Fig. 11.3 The relationship between nonlinear optical coefficient $\chi^{(2)}_{eff}(N_p)$ and Nd^{3+} atomic percentage concentration N_p

Suppose the surfaces of a NYAB crystal with length $L = 0.5$ cm has 100% transmittance for the pump light, while the rear mirror and output mirror of the resonator with length L_c have reflectivity 85 and 15%, respectively, at the wavelength of second harmonic laser, the cavity loss coefficient per centimeter is $\gamma_2 = -\ln(0.15 \times 0.85)/2L_c \times (L_c/L) = 1.03/2L$. By including other losses, $\gamma_2 = 1.25/L$ can be assumed, therefore

$$h = \frac{A^2}{\kappa_2 + \gamma_2} = \frac{\left[5.73 \times 10^{-8} \times \left(1 + N_p/30 - N_p^2/900\right)\right]^2}{0.27N_p + 1.25/L}$$

If the crystal is placed in a resonant cavity consisting of a rear mirror M_1 and an output mirror M_2, the cavity length is $L_c = 50$ mm. The reflectivities of the rear mirror and the output mirror are all 99.75% at the wavelength of fundamental laser and the reflectivity of output mirror at the wavelength of pumping light is 70%. The single pass loss of the fundamental wave in the cavity mirrors is very low. By taking into account the other loss, it can be assumed that a round-trip loss in the cavity is 2%, that is, $\gamma_c = 0.02$, then the photon lifetime in the cavity can be calculated by$\tau_c = 2[L_c + (n_1 - 1)L]/(c\gamma_c)$.

In the calculation of SHG intensity of NYAB crystal, according to (11.48) to (11.50), the parameters are: fluorescence lifetime $\tau_f = (3.3 \times 10^{-4} \times N_p + 1.66 \times 10^{-2})^{-1}$ μs [6], emission cross-section at the wavelength of fundamental laser $\sigma_{em} = 4.5 \times 10^{-19}$ cm^2 [12], and the refractive index at pumping wavelength $n_0 = 1.77$. If the length of the NYAB crystal is 0.5 cm, the power of the pump laser is 400 mW and pumping light after focusing is a cylindrical beam with diameter 100 nm. In this case the curve of SFD efficiency for different Nd^{3+} concentration obtained is shown in Fig. 11.4.

The maximum efficiency of the SFD conversion of NYAB crystal published by Japanese authors [12] was 17.5%, which is much lower than 29% shown in Fig. 11.4. This is understandable, because the above calculation does not consider the walk-off angle effect and the fact that the diameter of pump light cannot maintain 100 nm in the whole crystal.

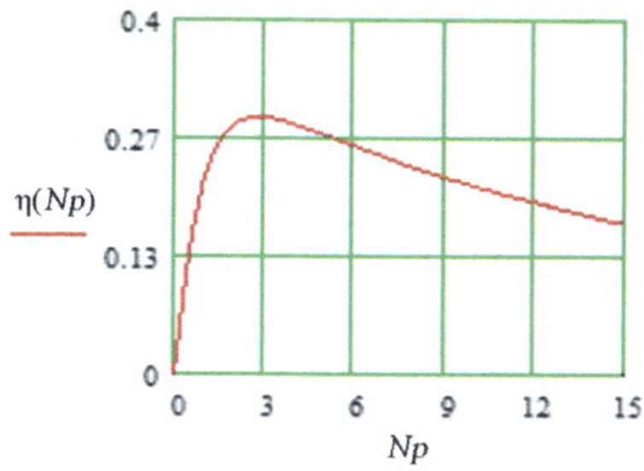

Fig. 11.4 Relation of SFD efficiency with atomic percentage concentration N_p of Nd^{3+} ion

Thus, even if one takes into account of the factor that the nonlinear optical coefficient is increased with the increase of Nd^{3+} concentration, the optimum concentration of Nd^{3+} ion is still in the range of 2–4%. Further calculations show that the optimal Nd^{3+} concentration does not change with the pump power, but is related to the length of the crystal, higher Nd^{3+} concentration is more suitable for shorter crystals. The intensity of thermal effect by the increase of Nd^{3+} concentration and the effect of phase mismatch produced by the inhomogeneous of refractive index on the conversion efficiency have not been taken into account in this calculation . Considering the effect of phase mismatch, the lower Nd^{3+} concentration is appropriate. It also can be seen by the calculation that with the increase of pump power, the efficiency of frequency doubling will be increased. For the 1 W power pumping (other conditions unchanged), if the above various influence effects do not calculate, the maximum theoretical efficiency up to 33.4% can be obtained.

It can be seen by this calculation that the loss of fundamental wave (mainly the loss of mirror) in the laser cavity has a great effect on the output of SFD laser. The above calculation result is obtained in the case of $\gamma_c = 0.02$. If $\gamma_c = 0.04$ is adopted then pumped by the same power the maximum theoretical conversion efficiency is only 26.4%. With the increase of crystal length, in order to obtain a best SFD efficiency, the lower Nd^{3+} concentration should be adopted. On the other hand, taking into account the effect of walk-off angle on the efficiency of second harmonic generation and with the increase of crystal length, to assume the beam width of pump laser remains 100 nm in the whole crystal the error will be bigger, and it can be seen that selecting the Nd^{3+} concentration of 4% and the crystal length of 0.3 cm are more appropriate. After calculating the effects of the spatial distribution of the pumping, fundamental and second harmonic laser beams, as well as the walk-off angle, the results in good agreement with the experimental observation values can be obtained. This information can be obtained by referring to the literatures [54–57].

11.4 Self Sum-Frequency Mixing Effect in Nonlinear Laser Crystal

Blue laser is an important light source in the field of optical information storage, lithography, laser printing, underwater laser communication, laser display, and laser medical treatment. By using the self sum-frequency mixing (SSFM) effect of nonlinear laser crystal to generate blue laser has the advantages of integration, easy adjustment, and high beam quality. The NYAB and Nd^{3+}:GdCOB crystals first achieve the blue laser output by this kind of SSFM effect.

The principle of SSFM laser generation will be described below. A simple rate equation model is introduced first, which does not consider the spatial distribution of laser beam and the effect of walk-off angle. The aim is to let the reader have a simple and clear understanding of the physical process of SSFM.

In the problem to be studied the pump beam besides acting as an excitation source of the fundamental laser (frequencyω_1) will also mix with the fundamental laser by SSFM effect to produce a laser beam of frequency $\omega_2 = \omega_1 + \omega_p$, here, ω_p is the frequency of pump beam. The corresponding wave vectors satisfy the phase matching relation $\boldsymbol{k}_2 = \boldsymbol{k}_1 + \boldsymbol{k}_p$. Referring to Sect. 11.1, nonlinear optical coupling equations can be written as

$$\frac{dI(\omega_p)}{dz} = -\frac{4\pi\omega_p}{c}\sqrt{\frac{2\pi}{cn_pn_1n_2}}\chi_{\text{eff}}^{(2)}\sqrt{I(\omega_p)I(\omega_1)I(\omega_2)}\exp\left[-i(k_2 - k_1 - k_p)z\right] \tag{11.51}$$

$$\frac{dI(\omega_1)}{dz} = -\frac{4\pi\omega_1}{c}\sqrt{\frac{2\pi}{cn_pn_1n_2}}\chi_{\text{eff}}^{(2)}\sqrt{I(\omega_p)I(\omega_1)I(\omega_2)}\exp\left[-i(k_2 - k_1 - k_p)z\right] \tag{11.52}$$

$$\frac{dI(\omega_2)}{dz} = +\frac{4\pi\omega_2}{c}\sqrt{\frac{2\pi}{cn_pn_1n_2}}\chi_{\text{eff}}^{(2)}\sqrt{I(\omega_p)I(\omega_1)I(\omega_2)}\exp\left[i(k_2 - k_1 - k_p)z\right] \tag{11.53}$$

By using a symbol B to represent $4\pi(2\pi/cn_pn_1n_2)^{1/2}\chi_{\text{eff}}^{(2)}$ and let $I_1 \equiv I(\omega_1)$, $I_2 \equiv I(\omega_2)$. Because I_0 is used to denote the pump beam intensity on the incident face, $I_p \equiv I(\omega_p)$ will be used to represent the pump beam intensity in the crystal (which has a relation with the incident laser intensity I_0 illustrated in the following), then the rate equations satisfying the phase matching condition can be written as

$$\frac{dI_p}{dz} = -\frac{\omega_p}{c}B\sqrt{I_pI_1I_2} \tag{11.54}$$

$$\frac{dI_1}{dz} = -\frac{\omega_1}{c} B\sqrt{I_p I_1 I_2} \tag{11.55}$$

$$\frac{dI_2}{dz} = +\frac{\omega_2}{c} B\sqrt{I_p I_1 I_2} \tag{11.56}$$

In steady state, the loss per unit crystal length caused by the generation of second harmonic laser is equal to the gain expressed by (11.56), therefore

$$\gamma_2 I_2 = (\omega_2/c) \times B \times \left[I_p I_1 I_2\right]^{1/2} \tag{11.57}$$

$$I_2 = \left(\frac{\omega_2 B}{c\gamma_2}\right)^2 I_p I_1 \tag{11.58}$$

The above rate equations become

$$\frac{dI_p}{dz} = -\frac{\omega_p \omega_2 B^2}{c^2 \gamma_2} I_p I_1 \tag{11.59}$$

$$\frac{dI_1}{dz} = -\frac{\omega_1 \omega_2 B^2}{c^2 \gamma_2} I_p I_1 \tag{11.60}$$

$$\frac{dI_2}{dz} = +\frac{\omega_2^2 B^2}{c^2 \gamma_2} I_p I_1 \tag{11.61}$$

By (11.59) it can be seen that in addition to the original absorption loss κ_p of the crystal, pump beam also subjected to a loss κ resulted from frequency mixing conversion. Taking the incident point of pump beam as zero, the pump beam intensity at the position z can be expressed as

$$I_p = I_0 e^{-(\kappa_p + \kappa)z} \tag{11.62}$$

where $\kappa = \frac{\omega_p \omega_2}{c^2 \gamma_2} B^2 I_1$. Combining the laser rate equations for a four-level system in steady state, the nonlinear optical coupling equations can be written. The corresponding equation of (11.38) for the fundamental laser is

$$\begin{cases} W_0 N_1 - (W_0 N_1)_{\text{th}} - \dfrac{\sigma_{em} I_1}{\hbar\omega_1} \Delta N = 0 \\ \dfrac{c}{n_1} \sigma \Delta N I_1 - \dfrac{1}{\tau_c} I_1 - \dfrac{\omega_1 \omega_2}{c\gamma_2 n_1} B^2 I_p I_1 = 0 \end{cases} \tag{11.63}$$

Intensity I_1 can be obtained by the above equation after averaging the pump beam intensity of (11.62) by the same way as (11.43) and taking into account the effect of output mirror reflectivity R on the pump beam; at the same time, the expressions of $W_0 N_1$ and $(W_0 N_1)_{\text{th}}$ take the similar forms of (11.42) and (11.47), respectively.

The fundamental laser intensity I_1 by multiple trips in the cavity will arrive at the following steady value

$$I_1 = \frac{c\left[W_0N_1 - (W_0N_1)_{\text{th}}\right]\hbar\omega_1}{\dfrac{n_1}{\tau_c} + \dfrac{\omega_1\omega_2B^2I_0\left[1 - e^{-(\kappa_p+\kappa)L}\right]}{c\gamma_2(\kappa_p+\kappa)L}} = \frac{\left[c\left[1 - \mathrm{e}^{-(\kappa_p+\kappa)L}\right]\left[1 + Re^{-(\kappa_p+\kappa)L}\right] \times I_0\omega_1\kappa_p/\omega_p - \dfrac{n_p\hbar\omega_1}{\tau_c\tau_f\sigma_{em}}\right]}{\dfrac{Ln_1}{\tau_c} + \dfrac{\omega_1\omega_2B^2I_0\left[1 - e^{-(\kappa_p+\kappa)L}\right]}{c\gamma_2(\kappa_p+\kappa)}} \tag{11.64}$$

The fundamental laser intensity I_1 should be solved by iterative method. By assuming $\kappa = 0$ in (11.64) to calculate the first-order approximation of I_1, and then using this first-order approximation of I_1 to calculate the second-order approximation of I_1, after several iterations, the results of I_1 and κ can be obtained. By the integral of equation (11.61), the intensity of the frequency mixing light inside the cavity can be obtained

$$I_2 = \left(\frac{\omega_2B}{c\gamma_2^{1/2}}\right)^2 \frac{I_0\left[1 - e^{-(\kappa_p+\kappa)L}\right]}{(\kappa_p+\kappa)} \frac{\left[c\left[1 - \mathrm{e}^{-(\kappa_p+\kappa)L}\right]\left[1 + Re^{-(\kappa_p+\kappa)L}\right] \times I_0\omega_1\kappa_p/\omega_p - \dfrac{n_p\hbar\omega_1}{\tau_c\tau_f\sigma_{em}}\right]}{\dfrac{Ln_1}{\tau_c} + \dfrac{\omega_1\omega_2B^2I_0\left[1 - e^{-(\kappa_p+\kappa)L}\right]}{c\gamma_2(\kappa_p+\kappa)}} \tag{11.65}$$

According to the above formula, the relationship between the efficiency of SSFM and Nd^{3+} concentration can be estimated.

Assuming that the intensity of SSFM laser output is $TI_2 = 0.85I_2$ and $\eta = 0.85I_2/I_0$, the photon cavity lifetime τ_c of the fundamental laser can be calculated by the value in Sect. 11.3 and the loss coefficient is estimated to be $\gamma_2 = 1.5/L$. When pump power and crystal length is 500 mW and 0.5 cm, respectively, and the reflectivity of output mirror at pump wavelength is $R = 0.7$, according to the above formula the calculated efficiency of blue laser is shown in Fig. 11.5. Obviously, this efficiency is much lower than that of the SFD green laser

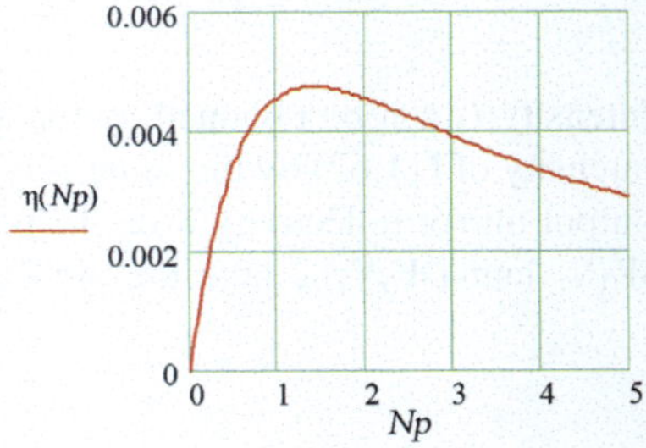

Fig. 11.5 Relation between efficiency of SSFM and Nd^{3+} atomic percentage concentration N_p

in the same condition. One of the important reasons is: as a component of SSFM process, the loss of the pumping light is very high.

The above calculation also ignores the two influence factors, that is, the effect of walk-off angle and pump beam is impossible to be focused into a very fine beam in the whole length of the crystal. Taking into account these factors, the efficiency of SSFM will be much lower. Mougel et al. [22] used GCOB crystals to obtain SSFM laser beam. The efficiency was only 0.2%. Although the above calculation shows that the SSFM efficiency will increase with the increase of crystal length, it is not good to use a longer crystal after considering the effect of walk-off angle and other loss factors. In addition, higher SSFM efficiency can be achieved only when the Nd^{3+} ion concentration is about 1%.

Considering the real space distribution of all the light beams and the effect of walk-off angle, a result near to the experimental observation value can be obtained. The following is a brief introduction of how to use the expression published by Boyd and Kleinman [55] and correct their formula of efficiency-decreasing factor to calculate the laser output power of SSFM, and discuss the relationship of SSFM efficiency with crystal length and Nd^{3+} ion concentration [56] (symbols of physical quantity used are the same as reference [56]).

In the case discussed, a nonlinear laser crystal with length L is placed as close as possible to the input mirror in a plano-concave resonator with reflectances R_1 and R_2 at fundamental wavelength for the input and output mirrors, respectively. The light is propagated along the crystal length direction z. The pump beam is focused at the entrance face of the crystal. With this design a maximum mode overlap among the pump, the fundamental and the SSFM lasers inside the cavity can be obtained within crystal length L. The pump beam is a focused Gaussian beam of TEM_{00} mode and the fundamental and the SSFM lasers are also Gaussian beams. The beam waists of pump beam and fundamental laser are w_{p0} and w_{c0}, respectively. The spot sizes are assumed to propagate according to the following expression

$$w_p^2(z) = w_{p0}^2\left(1 + \frac{\lambda_1^2 z^2}{\pi^2 w_{p0}^4 n_1^2}\right) \tag{11.66}$$

$$w_c^2(z) = w_{c0}^2\left(1 + \frac{\lambda_2^2 z^2}{\pi^2 w_{c0}^4 n_2^2}\right) \tag{11.67}$$

where λ_1 and λ_2 are the wavelengths of the pump beam and fundamental laser, respectively, and n_1 and n_2 are the corresponding refractive indexes. Let the intensity of the pump, the fundamental, and the SSFM lasers vary with coordinate z and radial coordinate r according to the following expression

$$I(r,z) = P(z)f(r,z) \tag{11.68}$$

$$f(r,z) = \frac{2}{\pi w^2(z)}\exp\left(-2r^2/w^2(z)\right) \tag{11.69}$$

where $P(z)$ is the power of the wave as a function of z and $f(r, z)$ is the normalized beam profile for the TEM_{00} mode. The pump power $P_p(z)$ as a function can be expressed as

$$P(z) = P_{p0} \exp(-\alpha_p z) \tag{11.70}$$

In the above expression P_{p0} is the pump power incident upon the crystal, α_p is the absorption coefficient at the pump wavelength. The rate equation of the fundamental laser is

$$\frac{dI_c}{dt} = \left(\frac{g_0(r,z)}{1 + 2I_c/I_s} - \alpha_c \right) I_c \tag{11.71}$$

where α_c is the distributed loss per unit length in the crystal and I_s denotes the saturation intensity of fundamental laser, $I_s = hc/\lambda_2 \sigma_e \tau_f$, the small signal gain coefficient with spatial distribution $g_0(r,z)$ is expressed as

$$g_0(r,z) = \frac{2\sigma_e \tau_f \alpha_p \lambda_1 P_{p0}}{\pi hc w_p^2(z)} \exp\left(-\alpha_p z - 2r^2/w_p^2(z)\right) \tag{11.72}$$

where σ_e, τ_f, are the emission cross-section of fundamental wave and the lifetime of upper laser level, respectively. After calculation, the circulating fundamental power P_c in the cavity can be expressed as

$$P_c = \frac{\pi \bar{w}_c^2 I_s}{4q} \left(\frac{2G}{2\delta L - \ln R + 2KP_{p0}} P_{p0} - 1 \right) \tag{11.73}$$

where δ is the internal loss of the crystal, which originates from the crystal defect, scattering, diffraction, and so on. K is defined by the following expression

$$K = \frac{2\omega_0^2 \chi_{eff}^2}{\pi \varepsilon_0 c^3 n_0^2 n_3} L k_0 \exp(-\alpha_p L/2) \times \left[\frac{(1-\zeta^2)(1-\gamma^2)}{1+\gamma\zeta} \right] h'(\sigma, \beta, \kappa, \xi, \mu)$$

where parameters γ, ζ and efficiency-decreasing factor $h'(\sigma, \beta, \kappa, \xi, \mu)$ can be found in (11.80) and (11.81). Taking into account the reflection factors that are due to the absence of antireflection coatings on the surfaces of the crystal, the effective reflectance of the two cavity mirrors with reflectances R_1 and R_2 is $R = [r^{1/2} + (R_1R_2)^{1/2}]^2[1 + (rR_1R_2)^{1/2}]^{-2}(r = (n^2 - 1)^2/(n^2 + 1)^2)$, and the factors G and q were expressed as

$$G = \frac{2\sigma_e \tau_f \lambda_1 \left[1 - \exp(-\alpha_p L)\right]}{\pi h c\left(\bar{w}_p^2 + \bar{w}_c^2\right)} \tag{11.74}$$

$$q = 1 - 0.92 \frac{\bar{w}_p^2}{\bar{w}_p^2 + \bar{w}_c^2} \tag{11.75}$$

where $\bar{w}_p^2$ and $\bar{w}_c^2$ are the mean beam spot size for the pump and fundamental lasers

$$\bar{w}^2 = \frac{\int_0^L w^2(z)dz}{L} \tag{11.76}$$

The next is to introduce the expression of P_3, the power of the frequency mixing laser in resonant cavity, by a heuristic approach of Boyd and Kleinman [55]

$$P_3 = \frac{2\omega_0^2 \chi_{eff}^2 P_{p0} P_c}{\pi \varepsilon_0 c^3 n_0^2 n_3} L k_0 \exp(-\alpha' L) \times \left[\frac{(1-\zeta^2)(1-\gamma^2)}{1+\gamma\zeta}\right] h(\sigma, \beta, \kappa, \xi, \mu) \tag{11.77}$$

where the parameters σ, β, κ, ξ, and μ represent phase mismatch, double refraction, absorption, strength of focusing, and focal position, respectively. The subscripts 1, 2, 3 of the physical quantities correspond to the pumping wave, the fundamental wave, and the frequency mixing wave. Here the parameters γ, ζ, ω_0, and n_0 is introduced to let $\omega_1 = \omega_0(1 - \gamma)$, $\omega_2 = \omega_0(1 + \gamma)$, $n_1 = n_0(1 - \zeta)$, and $n_2 = n_0(1 + \zeta)$, where $\gamma = (\omega_2 - \omega_1)/(\omega_2 + \omega_1)$ and $\zeta = (n_2 - n_1)/(n_2 + n_1)$. The phase matching condition $n_3\omega_3 = n_1\omega_1 + n_2\omega_2$ must be satisfied, so that $n_3 = n_0(1 + \gamma\zeta)$ and $k_0 = (k_1 + k_2)/2$. The efficiency-decreasing factor $h(\sigma,\beta,\kappa,\xi,\mu)$ introduced by Boyd and Kleinman is

$$h(\sigma, \beta, \kappa, \xi, \mu) = \frac{\exp(\mu\alpha' L)}{4\xi} \int \int_{-\xi(1-\mu)}^{\xi(1+\mu)} d\tau d\tau' \times \frac{\exp\left[-\kappa(\tau + \tau') + i\sigma(\tau - \tau') - \beta^2(\tau - \tau')^2\right]}{(1 + i\tau)(1 - i\tau')} \tag{11.78}$$

The above formula is applicable to the case of the source beams with the same confocal parameters (such as the case of frequency doubling). In this case the confocal parameters are $b = k(w_0)^2$, $\sigma = b\Delta k/2$ (Δk is the phase mismatch), $\xi = L/b$, $\mu = (L - 2f)/L$ (f is the focal position of the pump beam in the crystal, when pump beam is focused on the input surface $f = 0$, $\mu = 1$), $\kappa = \alpha b/2$, $\alpha = (\alpha_1 + \alpha_2 - \alpha_3)/2$, $\alpha' = (\alpha_1 + \alpha_2 + \alpha_3)/2$, $\beta = \rho k_1 w_{01}/2$ (ρ is the walk-off angle of the frequency mixing beam generated by the birefringence of crystal).

In the case of SFM with arbitrary confocal parameters (or beam waists) of pump and fundamental laser beams, the expression of efficiency-decreasing factor must be extended to the case of source beam with arbitrary waist (w_{10}, w_{20}) or confocal parameters $b_1 = k_1(w_{01})^2$ and $b_2 = k_2(w_{02})^2$. Introducing a so-called fictitious beam and let its wave vector k_0, confocal parameter b_0, and beam waist w_0 satisfy the following expression

$$k_0 = (k_1 + k_2)/2, b_0 = (k_1 b_1 + k_2 b_2)/2, w_0^2 = 2\left(k_1^2 w_{10}^2 + k_2^2 w_{20}^2\right)/(k_1 + k_2)^2 \tag{11.79}$$

The frequency and refractive index of the fictitious beam are defined as

$$\omega_0 = \omega_1/(1-\gamma) = \omega_2/(1+\gamma), \quad n_0 = n_1/(1-\zeta) = n_2/(1+\zeta) \tag{11.80}$$

Introducing $\beta_1 = b_1/b_0$ and $\beta_2 = b_2/b_0$ then the efficiency-decreasing factor $h'(\sigma, \beta, \kappa, \xi, \mu)$ can be expressed as

$$h'(\sigma,\beta,\kappa,\xi,\mu) = \frac{\exp(\mu\alpha L)}{4\xi\beta_1\beta_2}\int\int_{-\xi(1-\mu)}^{\xi(1+\mu)} \mathrm{d}\tau\mathrm{d}\tau' \times \frac{(1+i\tau)(1-i\tau')\exp\left[-\kappa(\tau+\tau')+i\sigma(\tau-\tau')-\beta^2(\tau-\tau')^2\right]}{(1+i\tau/\beta_1)(1-i\tau'/\beta_1)(1+i\tau/\beta_2)(1-i\tau'/\beta_2)} \tag{11.81}$$

When the two beams with the same confocal parameter are involved in the mixing, then $b_1 = b_2$, so that $\beta_1 = \beta_2 = 1$. Obviously (11.81) is the same as (11.78), that is, (11.78) is the expression of (11.81) in the special case.

If the efficiency-decreasing factor $h(\sigma,\beta,\kappa,\xi,\mu)$ is replaced by $h'(\sigma,\beta,\kappa,\xi,\mu)$ and consider that the output mirror has a transmittance T at the frequency mixing wavelength, the absorption coefficients of the crystal on the fundamental and frequency mixing beams can be neglected, then $\alpha = \alpha' = \alpha_p/2$; therefore the output power of frequency mixing wave will be

$$\begin{aligned} P_s &= P_3 T \\ &= P_{p0} P_c T \times \frac{2 w_0^2 \chi_{eff}^2}{\pi\varepsilon_0 c^3 n_0^2 n_3} L k_0 \exp(-\alpha_p L/2) \times \left[\frac{(1-\zeta^2)(1-\gamma^2)}{1+\gamma\zeta}\right] h'(\sigma,\beta,\kappa,\xi,\mu) \end{aligned} \tag{11.82}$$

It can be seen from (11.73) to (11.82) that the laser power output of the SSFM is a complicate function of crystal length, doping ion concentration, pump beam waist, fundamental cavity mode size, walk-off angle, phase mismatch, effective nonlinear coefficient, pump beam absorption coefficient, pump power, and internal loss. The relationship of the output power (efficiency) of NYAB crystal with several parameters is shown in Figs. 11.6, 11.7, 11.8.

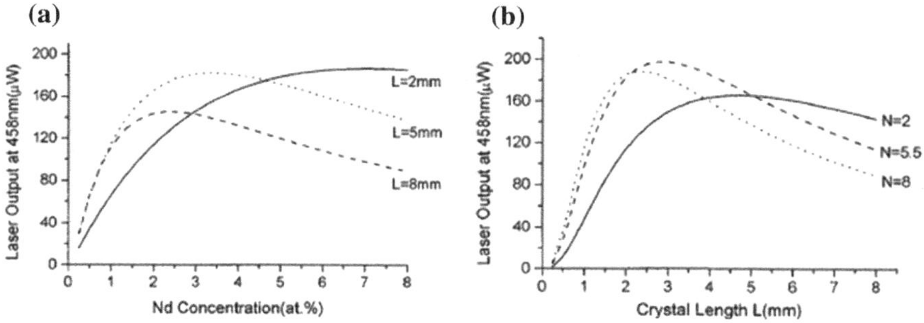

Fig. 11.6 Typical curves of the self-frequency mixing laser output of NYAB crystal as functions of **a** Nd^{3+} concentration for different crystal lengths; **b** crystal length for different Nd^{3+} concentrations [56]

Fig. 11.7 Dependences of the self-frequency mixing laser output on the walk-off angle. Dotted curves, output at the walk-off angle of $\rho = 2.4°$ for NYAB crystal [56]

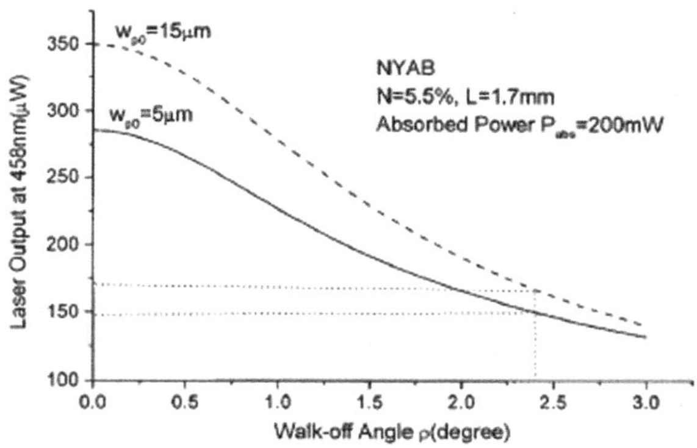

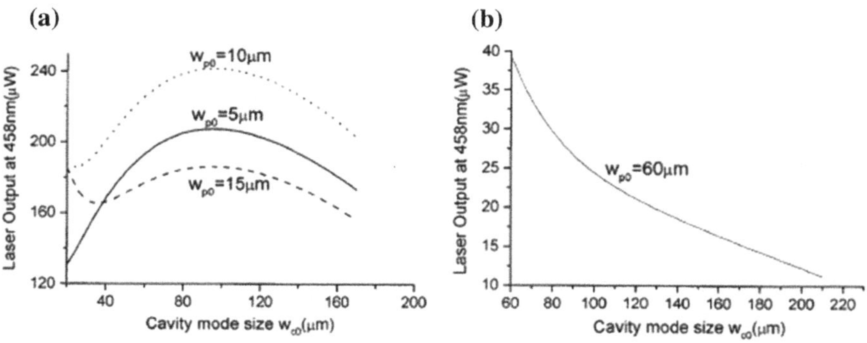

Fig. 11.8 Self-frequency mixing laser output of NYAB crystal as functions of **a** different pump beam waists in simulating a Ti sapphire pump and **b** at a fixed pump waist $w_{p0} = 60$ μm in simulating a LD pump, $N = 5.5$ at% and $L = 1.7$ mm for the NYAB crystal. The absorbed pump power is 200 m [56]

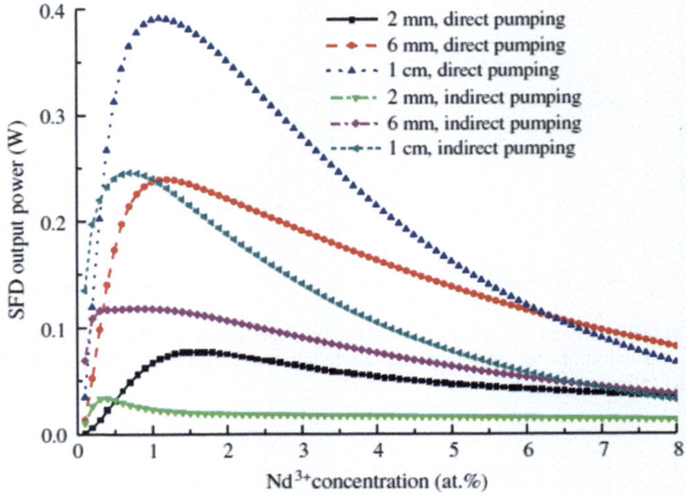

Fig. 11.9 Output power of SFD of NYAB crystal versus Nd^{3+} concentration for the different crystal length under direct and indirect pumping [54]

As mentioned in Chap. 10, the thermal load of the fundamental laser operation of current Nd^{3+} ion-doped laser crystals can be reduced by pumping directly into the upper laser level $^4F_{3/2}$. This method can also be used in the operation of SFD and SSFM; the problem of small pumping absorption coefficient can be solved by appropriate increase in the active ion concentration to maintain sufficiently high absorption efficiency. Especially for the SSFM, according to the above calculation, it can be seen that direct pumping to the upper laser level, being another component in the mixing process, the pump laser beam can have higher intensity, although owing to the weaker $^4I_{9/2} \rightarrow {}^4F_{3/2}$ absorption, the fundamental laser intensity may be lower for certain concentration of activate ion. By appropriate choice of the active ion concentration, the new pumping method can still achieve high enough conversion efficiency. On the other hand, by using direct pumping method the thermal load is smaller and the temperature rise of the crystal will be lower and so bring about the improvement of the actual output efficiency. Figure 11.9 is a comparison of the theoretical results of SFD output power of NYAB crystal for the direct pumping and the indirect pumping.

11.5 Stimulated Raman Scattering Effect of Laser Crystal

The laser crystal with strong enough third-order nonlinear optical effect is possible to transform a fundamental laser beam into a series of Raman laser emission lines at different wavelengths by stimulated Raman scattering (SRS) effect and have been realized in Nd^{3+}:KGW [57] and Nd^{3+}:$Gd_2(MoO_4)_3$ [58] crystals. SRS process has no phase matching requirement as those in the second-order nonlinear optical processes of frequency doubling, mixing, or optical parametric oscillation; thereby has no instability caused by temperature, angle, and wavelength variations, which shows that it has advantages in producing a series of special laser wavelength. The main properties of this effect and its relationship with crystal structure characteristics will be briefly introduced in this section.

It is well known that Raman scattering is a non-elastic interaction between light and optical lattice vibration (or photon and optical phonon). Photon can emit one or several phonons (Stokes scattering) or absorb one or several phonons (anti-Stokes scattering) in Raman scattering. Raman scattering cross-section is usually calculated by perturbation theory of quantum mechanics while the semi-classical theory is used to discuss some features of this effect.

The dipole moment of a medium can be expressed as $\mu = \mu_I + \mu_P$, where μ_I is the dipole moment induced by external electric field and μ_P is its intrinsic dipole moment. These two kinds of polarization can be changed by lattice vibration. μ_I is generally expressed as $\mu_I = \alpha E$ (for simplicity the scalar form is written). The lattice vibration will cause the variation of induced polarization of a medium and it can be expanded with respect to vibration coordinate u in first-order approximation as

$$\alpha = \alpha_0 + (\partial\alpha/\partial u)_0 u + \ldots$$

On the other hand, the intrinsic dipole moment of the medium can also be changed by the lattice vibration, which can be expanded respect to coordinates u in the same approximation as

$$\mu_p = \mu_p^0 + \left(\partial\mu_p/\partial u\right)_0 u$$

so that

$$\mu = \mu_p^0 + \alpha E = \mu_p^0 + \alpha_0 E + \left(\partial\mu_p/\partial u\right)_0 u + (\partial\alpha/\partial u)_0 uE$$

It is well known that the energy operator of dipole moment interacting with the electric field $\boldsymbol{E}$ of external radiation field is

$$H_{\mathrm{i}} = -\boldsymbol{\mu}\cdot\boldsymbol{E}$$

Using the scalar form, it can be directly written as

$$H_{\mathrm{i}} = -\mu_{\mathrm{p}}^0 E - \alpha_0 E^2 - (\partial\mu_{\mathrm{P}}/\partial u)_0 uE - (\partial\alpha/\partial u)_0 uE^2 \tag{11.83}$$

The third and fourth terms of the above equation are the interaction energy of lattice vibration with external electric field. Written in operator form, the electric field E of the photon can be expressed as (4.5b) in Chap. 4, that is a linear combination of photon creation and annihilation operators and can be simply expressed as

$$E = \sum_{k\alpha}\left[c_1(\boldsymbol{k}\alpha)a_{k\alpha} + c_2(\boldsymbol{k}\alpha)a_{k\alpha}^{+}\right] \tag{11.84}$$

On the other hand, the displacement u of lattice vibration can also be expressed as a formula in Chap. 6 in the form of linear combination of phonon creation and annihilation operators written as

$$u = \sum_{qj} \left[c_1(\boldsymbol{q}j) b_{\boldsymbol{q}j} + c_2(\boldsymbol{q}j) b_{\boldsymbol{q}j}^{+} \right] \tag{11.85}$$

The letters $\boldsymbol{k}$ and $\boldsymbol{q}$ are used to represent the wave vector of a photon and a phonon, respectively. The transition matrix element of the interaction of photon with phonon is $\langle i|H_{\mathrm{i}}|f\rangle$. Only the third and the fourth terms of (11.83) are involved in the change of photon and phonon states. The third term involves the matrix element $\left\langle i \middle| c_1(\boldsymbol{k}\alpha) c_2(\boldsymbol{q}j) a_{\boldsymbol{k}_\alpha} b_{qj}^{+} \middle| f \right\rangle$ corresponding to absorb a photon and emit a phonon. The phonon frequency is in the infrared range, so it is called infrared absorption. The fourth term involves matrix element $\left\langle i \middle| c_1(\boldsymbol{k}\alpha) c_2(\boldsymbol{k}'\alpha') c_2(\boldsymbol{q}j) a_{\boldsymbol{k}_\alpha} a_{k'\alpha}^{+} b_{\boldsymbol{q}j}^{+} \middle| f \right\rangle$ corresponding to absorb a photon of mode $\boldsymbol{k}\alpha$ (frequencyω_{k}) and emit a photon of mode $\boldsymbol{k}'\alpha'$ (frequency $\omega_{\mathrm{k}'}$) at the same time emit a phonon of mode $\boldsymbol{qj}$ (frequencyω_{q}), where $\omega_{\mathrm{k}} > \omega_{\mathrm{k}'}$. It is Stokes Raman scattering. Equation (11.83) also includes another form of matrix elements, that is $\left\langle i \middle| c_1(\boldsymbol{k}\alpha) c_2(\boldsymbol{k}'\alpha') c_1(\boldsymbol{q}j) a_{\boldsymbol{k}_\alpha}^{+} a_{k'\alpha} b_{\boldsymbol{q}j} \middle| f \right\rangle$ corresponding to absorb a photon of mode $\boldsymbol{k}'\alpha'$ (frequency$\omega_{\mathrm{k}'}$), emit a photon of mode $\boldsymbol{k}\alpha$ (frequency ω_k), and at the same time absorb a phonon of mode $\boldsymbol{qj}$ (frequencyω_{q}); it is the inverse process of Stokes scattering and is called anti-Stokes scattering. Referring to (4.2), the calculation of the matrix element of photon creation and annihilation operator shows that the probability of positive process is proportional to $[p(\omega_{\mathrm{k}'}) + 1]p(\omega_{\mathrm{k}})(\partial\alpha/\partial u)_0^2$ while the probability of inverse process is proportional to $[p(\omega_k) + 1]p(\omega_{\mathrm{k}'})(\partial\alpha/\partial u)_0^2$, where $p(\omega_k)$ and $p(\omega_{\mathrm{k}'})$ are the photon numbers of incident light and scatter light, respectively. In the general Raman scattering process, because of $p(\omega_{\mathrm{k}'}) << p(\omega_k)$, the probability of the inverse process is much lower than that of the positive process. In this spontaneous Raman scattering process the medium has no amplification effect on the scattering light of frequency $\omega_{\mathrm{k}'}$. It can be shown [59] that the third-order nonlinear optical coefficient of a medium is the negative imaginary part of the refraction index of Raman scattered light. When the medium has a larger χ^3, it will have a sufficiently large amplification effect on the Raman scattering light; therefore under a strong pumping, the gain of the Raman scattering light is enough to overcome its loss and can generate a stimulated Raman emission.

The important difference of the SRS with the spontaneous Raman scattering is the scattering light of SRS is a coherent light and the intensity of scattered light increases exponentially with its distance through a medium. In addition, because the gain factor of stimulated scattering is proportional to the incident light intensity in the unit solid angle, if the gain medium is placed inside a cavity it has the so-called "cleanup effect" which makes the light beam quality of the scattering light better than that of the incident light. Therefore, the poor pump light quality can be converted to Gaussian laser beam output of the Raman laser. A detail discussion on these important nonlinear optical effects can be referred to the literatures [60] and [61].

Raman lasers utilize the SRS effect of the materials to shift the laser frequency to a series of Stokes and anti-Stokes laser lines. In order to make the Raman laser have a higher efficiency, it is necessary to have a higher total Raman gain G. It can be proved that the light intensity $I_s(l)$ of the first Stokes Raman scattering has the following relation with the length l of the Raman medium the light passed

$$I_{\mathrm{s}}(l) = I_{\mathrm{s}}(0)\exp\left(g_{\mathrm{R}}I_{\mathrm{p}}l\right) \tag{11.86}$$

where I_p is the pump intensity and g_R is the Raman gain coefficient. The total Raman gain G of the medium obviously is

$$G = g_{\mathrm{R}}Il \tag{11.87}$$

Raman gain coefficient g_R can also be expressed as

$$g_{\mathrm{R}} \propto \rho_{\mathrm{n}}(\partial\sigma/\partial\Omega)/v_{\mathrm{s}}^3\Delta v_{\mathrm{s}} \tag{11.88}$$

In the above expression ρ_n is the phonon density, $(\partial\sigma/\partial\Omega)$ is called the differential scattering cross-section of Raman scattering (from previous derivation, it is shown that this cross-section is proportional to $(\partial\alpha/\partial u)_0^2$) and v_s is the frequency of Raman scattering light with linewidth Δv_s. Therefore, the requirement of Raman laser on its working material is: the Raman linewidth should be narrow, the differential cross-section should be high. Of course, the material must be transparent in the spectral region of Raman emission. From this basic requirement, we can study the relationship of SRS performance with the structure and the chemical composition of materials.

Raman line broadening is induced by the anharmonic interaction of a crystal, in which one phonon decays into two other phonons. One can easily imagine that Raman spectral linewidth will be narrow for the materials with weak anharmonic interaction. It is well known that the thermal conductivity and the thermal expansion properties of a crystal are involved with the anharmonic interaction. Therefore, the requirement for narrow Raman linewidth to the materials is consistent with that of the high thermal conductivity and small thermal expansion. In other words, materials with high thermal conductivity and small thermal expansion have narrow Raman line. It is easy to imagine that the width of Raman line is proportional to the coefficient of linear expansion α_{L} and inversely proportional to specific heat C_{v} and compressibility K_{c}

$$\Delta v_s \propto \alpha_L/K_cV_c \tag{11.89}$$

On the other hand, the high gain coefficient requires the materials have high values of polarizability α and high values of $\partial\alpha/\partial u$. Two cases can satisfy these requirements. First case is α has a relatively high value and also has a great variation with the vibration; second case is α has not as high as the first case but has a greater variation with the vibration.

The crystals consist of molecule groups having the so-called "breathing vibrational mode" which belong to irreducible representation A_{1g} having a maximum polarizability variation $\partial\alpha/\partial u$, also have a lowest-lying energy level of charge transfer to act as an intermediate state to generate a high value of polarizability tensor, because the value of polarizability tensor is inversely proportional to the square of energy interval between initial state and intermediate state. Therefore, the tungstate, molybdate, vanadate, and nitrate have both high values of α and $\partial\alpha/\partial u$, thus have a high Raman scattering gain.

The second case is like the crystals such as diamond. They have the advantage of the covalent crystal and have a higher value of $\partial\alpha/\partial u$ than that of the ionic crystals although without low-lying charge transfer level of molecular group to act as intermediate state to result in a high value of polarizability. In addition, as will be mentioned below, the very narrow Raman spectral linewidth of diamond makes it have a very high Raman gain coefficient. The data of Raman gain coefficient, Raman linewidth, and Raman frequency shift of some typical materials are given in Table 11.1.

Stankov and Marowsky [57] used Nd^{3+}:KGW crystal pumped by flash lamp and electro-optic Q switched to obtain 1067 nm fundamental wave laser, then by the self SRS effect directly produce a first Stokes Raman line at 1180.8 nm, a second Stokes line at 1321.5 nm, and a first anti-Stokes Raman line at 973.5 nm. By the frequency doubling and mixing effects of nonlinear optical crystal $LiIO_3$, laser emissions of 355.6, 367, 378, 393, 408, 423, 440, 533.5, 560, 590, 620, and 660 nm, altogether 12 different wavelengths of ultraviolet and visible light were obtained. Chang [65] demonstrated a continuous-wave self-Raman laser with high

Table 11.1 Relevant data of several materials with high gain coefficient of Raman scattering

Molecular formula	Gain coefficient (cm/GW)	Linewidth (cm^{-1})	Raman shift (cm^{-1})
$PbWO_4$ [62]	8.4 (532 nm)		≈900
$SrWO_4$ [63]	5.0 (1064 nm)	2.7	921
$BaWO_4$ [63]	8.5 (1064 nm)	1.6	926
$La_2(WO_4)_3$ [66]	30 (532 nm, estimated value)		940
$KGd(WO_4)_2$ [61]	6 (1064 nm)	5.9	901.5
$SrMoO_4$ [64]	5.7 (1064 nm)		886
$Li_3Ba_2Gd_3(MoO_4)_8$ [61]	4.7 (1064 nm)	3.1	768
YVO_4 [72]	≈4.5 (1064 nm)	<3.0	890,814
$GdVO_4$ [73]	≈4.5 (1064 nm)	<3.0	882,807
$Ba(NO_3)_2$ [61]	11 (1046 nm)		1047
$CaCO_3$ [61]	13 (1085 nm)	2.3	1085
$NaNO_3$ [61]	47 (1066 nm)	2.0	1066
Diamond [86]	12.5 (1064 nm)	1.5	1332

The pump wavelengths of Raman gain coefficients are indicated in the followed brackets. For the diamond, if the pump wavelength is 532 nm the gain coefficient can reach 64 cm/GW [86]

conversion efficiency by using Yb^{3+}:KGW as Raman crystal, under the diode pump power of 7.8 W. The CW Raman output power of 1.7 W was attained which corresponds to the conversion efficiency of diode laser emission to Raman laser is 21.8%.

The laser, nonlinear optics, and Raman scattering properties of β'-$Gd_2(MoO_4)_3$ crystal were carefully studied by Kaminskii et al. [58]. The Curie temperature of this crystal is about 159 °C; its structure belongs to space group Pba2 called β' phase at the temperature lower than Curie point. It is a positive biaxial crystal with 2 V angle about 10°and has both high second-order and third-order nonlinear optical effects. Owing to the fact that only small size of crystal can be obtained, Kaminskii et al. used a ultrashort pulsed Nd^{3+}:YAG laser at wavelength 1.06415μm as the excitation source. After two-stage amplification, its energy reached about 10 mJ. The crystal was placed at the focus of a lens with $f = 500$ mm. The beam waist was 75μm after focusing. With excitation in such a high laser power density, the SRS spectra can be very clearly observed. The SRS spectra excited by the laser at wavelength 1.06415μm are shown in Fig. 11.10. The SRS spectral lines excited by laser beam at 0.53207 μm generated by frequency doubling of KD*P crystal are shown in Fig. 11.11. If the fundamental laser at wavelength 1.06415 μm is along the phase matching direction of $Gd_2(MoO_4)_3$ crystal, then the laser line at wavelength 0.53207 μm and its two Stokes lines and one anti-Stokes line can be observed (as shown in Fig. 11.12). Obviously, this is because β'-$Gd_2(MoO_4)_3$ crystal has a high enough second-order nonlinear optical coefficient, the laser beam generated by Nd^{3+}:YAG crystal is directly frequency doubled by β'-$Gd_2(MoO_4)_3$ crystal then to produce the Raman frequency shifted line around 0.53207 μm.

Many tungstate crystals, such as $PbWO_4$, $CaWO_4$, $CdWO_4$, $NaY(WO_4)_2$, and $La_2(WO_4)_3$, can be used to generate SRS lines. The total Stokes conversion

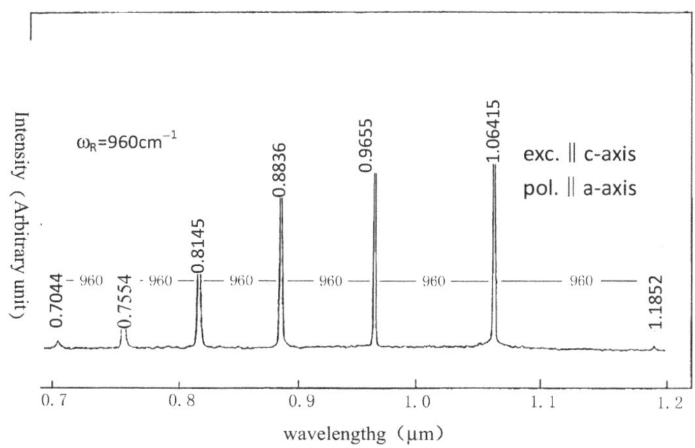

Fig. 11.10 Observed SRS spectra in β′-$Gd_2(MoO_4)_3$ crystal by using 1.06415 μm laser as excitation source [58]

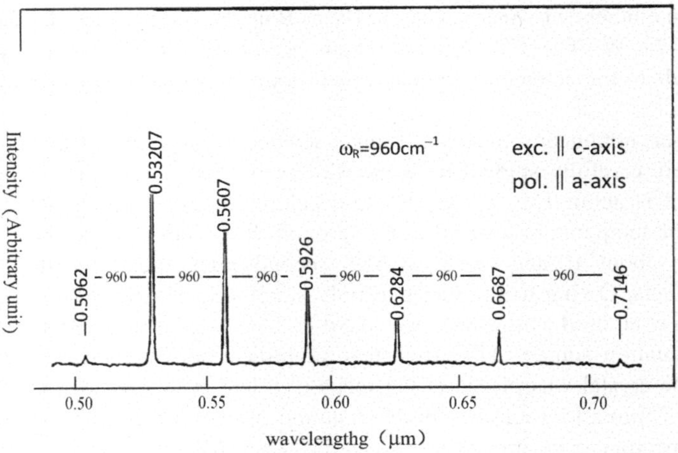

Fig. 11.11 Observed SRS spectra in β'-Gd_2 $(MoO_4)_3$ crystal by using 0.53207 μm laser as excitation source [58]

Fig. 11.12 Observed 0.53207μm laser line and their two Stokes Raman lines and one anti Stokes Raman line obtained by 1.06415 μm excitation laser beam along the phase matching direction of β'-Gd_2 $(MoO_4)_3$ crystal [58]

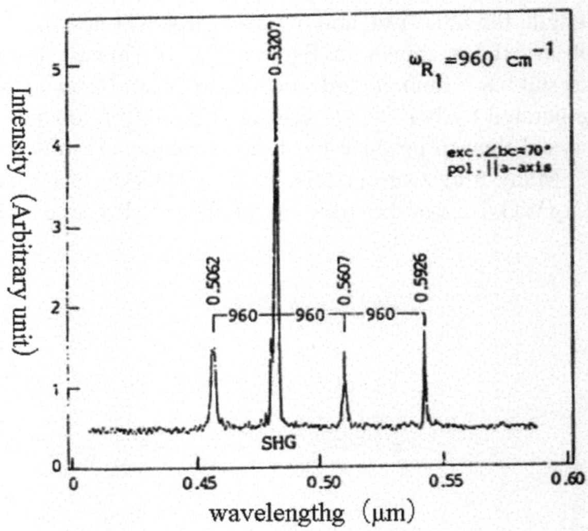

efficiency of tetragonal $NaY(WO_4)_2$ and $CaWO_4$ crystals at pumping power density close to their optical damage threshold at λ = 0.53207 mm reached 50%. For $PbWO_4$ it even reached 55% [66], while the fibrous $La_2(WO_4)_3$ crystal of 80 mm length pumped by power density about 0.165 GW/cm^2, the conversion efficiency of first Stokes stimulated Raman shift reached 6.8% [67]. Because $La_2(WO_4)_3$ is one of the tungstate crystals with high Raman gain coefficient and as being shown in (11.87) that the total Raman gain coefficient G is proportional to the crystal length l, it can be seen that the fibrous $La_2(WO_4)$ crystal doped with rare earth ions can become a good self Raman laser material.

The nanosecond Raman laser by using $PbWO_4$ crystal as laser material has a conversion efficiency up to 20%, which is expected to be used in the laser radar for the detection of ozone [68]. The conversion efficiency is very closely related to the crystal orientation, when the polarization direction of the excitation light is parallel to the *c*-axis of the crystal, and conversion efficiency can reach its maximum but if the polarization direction of the excitation light is vertical to the *c*-axis of the crystal, the SRS conversion cannot be observed. The Raman gain coefficients of Nd^{3+}:$PbMoO_4$ crystals are two to three times larger than those of Nd^{3+}:KGW and Nd^{3+}:$GdVO_4$ crystals. Under LD pumping of power 1.5 W, the output power of the free-running mode at 1056 nm reached 0.65 W with slope efficiency up to 53% was achieved. By using different LiF:F_2^- saturable absorbers, a maximum pulse energy of 11 J at 1.4 ns pulse duration and maximum output power of 0.35 W (10 μJ) with 7 ns pulse width in the passively *Q*-switched mode was obtained. The self-Raman-laser output pulse energy of the first Stokes line at 1163 nm was measured to be as high as 6 μJ and the pulse duration was estimated to be shorter than 500 ps [69].

Owing to the fact that third-order nonlinear optical effect is not limited by the structure types of the material, the SRS effect has been observed in many of the second-order nonlinear optical crystals such as β barium metaborate BBO [70] and COB [71] crystals. Typical laser crystals YVO_4 and $GdVO_4$ can become SRS crystals with good performance, conversion efficiency of fundamental laser to Raman laser of these two vanadate crystals can be higher than 60% [72, 73] and the continuous, Q-switched as well as mode-locked picosecond SRS laser have been realized [74–78]. A fundamental laser beam and a self SRS laser beam generated by Nd^{3+}:$GdVO_4$ crystal through the frequency sum mixing of LBO crystal to output a laser beam of power 5.3 W at 559 nm wavelength [79] and by controlling Raman cascade within an intracavity Raman laser crystal Nd^{3+}:GdVO4, laser output at 532 nm (green), 559 nm (lime), 586 nm (yellow), and 620 nm (red) in watt level were generated [80].

In recent years, the research of Raman lasers has developed rapidly. The wavelength range has been widened, the operation pattern includes continuous, nanosecond, picosecond, and femtosecond pulses, and the output energy, power, and efficiency are increased greatly at the same time. A large number of articles published in this field are too numerous to be enumerated. However, limited by thermal lens effect, it is difficult to obtain higher energy and power output of infrared laser by the self Raman effect of laser crystal such as tungsten and vanadate or visible laser by frequency doubling and mixing.

Two Raman lasers which have been developed rapidly in recent years should be mentioned.

First is the fiber Raman laser. Because the fiber has a very small loss coefficient and considerable total Raman gain coefficient in a wide wavelength range, it benefits from the aforementioned total Raman gain and is proportional to the length of Raman medium, and so using optical fiber as a Raman laser material has a unique advantage. By using silicon fiber as a Raman gain medium pumped by Yb fiber laser at wavelength 1070 nm, Raman laser output of 153 W at wavelength 1120 nm has been obtained in 2009. The light to light efficiency reached 85% [81]. In 2013, the continuous laser output power of cascaded Raman fiber laser has reached 301 W with conversion efficiency 42% of pump light at 975 nm to Raman laser output at 1480 nm [82], while the Yb fiber laser output of power 1.54 KW at wavelength 1080 nm transform into Raman laser of power 1.28 KW at wavelength 1120 nm has been reported in 2014 [83]. The improvement of technology is expected to further increase the power level, the wavelength range, and the beam quality of Raman laser output.

Another is diamond Raman laser. The thermal conductivity of diamond is as high as 2000 W/mK [84], which is higher than that of the above Raman crystals for two or three orders of magnitude. Its room temperature coefficient of thermal expansion is about 1×10^{-6}/K [85]. This reflects the fact that the anharmonic effect of this crystal is very weak. As previously described, the anharmonic effect of a crystal is related to its linewidth of Raman line, the very weak anharmonic effect of diamond results in a very narrow Raman linewidth which is favorable to the output characteristics of a diamond Raman laser. The high thermal conductivity and low thermal expansion coefficient of this crystal are beneficial to avoid thermal lens, thermal birefringence, and stress destruction, thus eliminate the obstacles to improve the output power and energy. On the other hand, this crystal has excellent light transmittance in the wavelength range of 0.23–2.5μm, and its absorption coefficient at 1.064 μm is less than 0.006 cm^{-1} [86]. A diamond crystal pumped by CW Nd^{3+}:YVO_4 laser of power 31 W at wavelength 1064 nm has realized the continuous Raman laser output of power 10.1 W at wavelength 1240 nm and a slope efficiency as high as 49.7% [87]. Diamond Raman laser with output average power up to 24.5 W, slope efficiency reached 57% [88], average power reached hundred watt level, and conversion efficiency up to 34% have also been reported [89]. The first Stokes shift of diamond as large as 1332 cm^{-1} also enables a diamond Raman laser through direct or cascade mode to output laser emission in very wide wavelength range. Sabella et al. [90] used OPO oscillator to output laser emission at 2.5μm, then pump the diamond crystal. Through the tuning of the oscillator wavelength, they obtained the laser output in the wavelength range of 3.38–3.80μm.

The output of diamond Raman laser has extended all over the deep UV [91], visible [92, 93], near-infrared, [87–89] and mid infrared band [90]. Their powers have attained 750 W with efficiency and beam quality so far unperturbed by nonlinear or thermal-induced side effects [94]. It can be expected that with the deepening of research work, Raman laser materials and devices with more different wavelengths, higher power and energy as well as higher efficiency will appear in the field of laser application.

References

1. L.F. Johnson, A.A. Ballman, J. Appl. Phys. **40**, 297 (1969)
2. V.G. Dmitriev, E.V. Raevskii, N.M. Rubin et al., Sov. Tech. Phys. Lett. **5**, 1440 (1979)
3. L.M. Dorozhkin, I.I. Kuratev, N.I. Leonyuk et al., Sov. Tech. Phys. Lett. **7**, 1297 (1981)
4. B.S. Lu, J. Wang, H.F. Pan et al., Chin. Phys. Lett. **3**, 413 (1986)
5. Z.D. Luo, A.D. Jiang, Y.C. Huang et al., Chin. Phys. Lett. **6**, 440 (1989)
6. Z.D. Luo, Prog. Nat. Sci. **4**, 505 (1994)
7. Z.D. Luo, J.T. Lin, A.D. Jiang et al., SPIE **1989**(1104), 132 (1989)
8. B.S. Lu, J. Wang, H.F. Pan et al., J. Appl. Phys. **66**, 6052 (1989)
9. I. Schutz, I. Freitag, R. Wallenstein, Opt. Commun. **77**, 221 (1990)
10. Z.D. Luo, X.Y. Fang, Y.D. Huang, Opt. Commun. **81**, 59 (1991)
11. Amano S. LEOS'91, paper ELT5.3
12. S. Amano, T. Mochizuki, Nonlinear Opt. **1**, 297 (1991)
13. R. E. Stone, S.C. Wang, *Toptical Meeting in Compact Blue-Green Laser* (Santa Fe, paper THE 5, 1992)
14. H. Hemmati, IEEE Quant. Elecron. **28**, 1169 (1992)
15. M.W. Qiu, Y.X. Fan, R.C. Schlecht et al., Mol. Cryst. Liq. Cryst. Sci. Technol-Sec B: Nonlin. Opt. **4**, 31 (1993)
16. D. Jaque, J. Capmany, Z.D. Luo et al., J. Phys.: Condens. Matter **9**, 9715 (1997)
17. D. Jaque, J. Capmany, F. Molero et al., Appl. Phys. Lett. **73**, 3659 (1998)
18. D. Jaque, J. Capmany, Sole J. Garcia, Appl. Phys. Lett. **75**, 325 (1999)
19. D. Jaque, J. Capmany, J. Rams, J. Appl. Phys. **87**, 1042 (2000)
20. A. Brenier, G. Boulon, J. Lumin. **86**, 125 (2000)
21. M.L. Huang, Y.J. Chen, X.Y. Chen et al., Opt. Commun. **204**, 333 (2002)
22. M.L. Huang, Y.J. Chen, X.Y. Chen et al., Opt. Commun. **208**, 163 (2002)
23. Y.J. Chen, M.L. Huang, Y.D. Huang et al., Opt. Commun. **218**, 379 (2003)
24. J. Bartschke, R. Knappe, K.-J. Boller et al., IEEE J. Quant. Elect. **33**, 2295 (1997)
25. A. Cordova-Plaza, T.Y. Fan, M.J.F. Digonnet et al., Opt. Lett. **1988**(13), 209 (1988)
26. F. Mougel, E. Auge, G. Aka et al., Appl. Phys. B **67**, 533 (1998)
27. F. Mougel, G. Aka, A. Kahn-Harari et al., Opt. Mater. **13**, 293 (1999)
28. J. Wallace, Laser Focus World **35**, 24 (1999)
29. F. Mougel, K. Dardenne, G. Aka et al., J. Opt. Soc. Am. B **16**, 164 (1999)
30. E. Montoya, J. Capmany et al., Appl. Phys. Lett. **74**, 3113 (1999)
31. P. Wang, P. Dekker et al., Opt. Lett. **25**, 731 (2000)
32. F. Druon, F. Auge, F. Balembois et al., J. Opt. Soc. Am. B **17**, 18 (2000)
33. D.A. Hammons, M. Richardson, B.H.T. Chai et al., IEEE J. Quant. Elect. **36**, 991 (2000)
34. A. Brenier, J. Lumin. **91**, 121 (2000)
35. A. Brenier, C. Tu et al., Opt. Lett. **27**, 240 (2002)
36. D.J. Jaque, Opt. Soc. Am B **19**, 1326 (2002)
37. G. Aka, A. Brenier, Opt. Mater. **22**, 89 (2003)
38. P. Dekker, P.A. Burns et al., J. Opt. Soc. Am. B **20**, 706 (2003)
39. J. Li, J.Y. Wang, H.J. Zhang, J. Synth. Cryst. **34**, 778 (2005). (in Chinese)
40. P. Dekker, J.M. Dawes, J.A. Piper et al., Opt. Commun. **195**, 431 (2001)
41. P. Dekker, J.M. Dawes, P. Burns et al., *Advanced Solid-State Laser* (Conference 2002, Tech. Dig. 41)
42. J. Wang, H. Zhang, Z. Wamg et al., Opt. Exp. **18**, 11058 (2010)
43. J. Capmany, V. Bermudez, D. Callejo et al., Appl. Phys. Lett. **76**, 1225 (2000)
44. J. Capmany, Appl. Phys. Lett. **76**, 144 (2001)
45. Y.R. Shen, *The Principles of Nonlinear Optics* (Wiley, 1984)
46. W. Koechner, *Solid-State Laser Engineering*, 5th edn. (Springer, Berlin, 1999)
47. H. Jiang, J. Li, J. Wang et al., J. Cryst. Grow. **233**, 248 (2001)
48. P.F. Bordui, M.M. Fejer, Annu. Rev. Matter Sci. **23**, 321 (1993)

49. V.G. Dmitriev, G.G. Gurzadyan, D.N. Nikogosyan, Prop. Nonlinear Opt. Crys. **64**, 67–288 (1999)
50. D.N. Nikogosyan, *Nonlinear Optical Crystals: A Complete Survey* (Springer, 2005)
51. LuoZD ZhaoTJ, Acta Optica Sinica **15**, 1199 (1995). (in chinese)
52. Z. Huang, X. Gong, Y. Huang et al., IEEE J. Quant. Elect. **40**, 1441 (2004)
53. A. Brenier, Opt. Comm. **141**, 221 (1997)
54. Z. Huang, J. Opt. Soc. Am. B **30**, 33 (2013)
55. G.D. Boyd, D.A. Kleinman, J. Appl. Phys. **39**, 3597 (1968)
56. X.Y. Chen, Z.D. Luo, Y.D. Huang, J. Opt. Soc. Am. B **18**, 646 (2001)
57. K.A. Stankov, G. Marowsky, Appl. Phys. B **61**, 213 (1995)
58. A.A. Kaminskii, A.V. Butashin, H.-J. Eichler et al., Optical Mater. **7**, 59 (1997)
59. P.N. Butcher, *Nonlinear Optical Phenomena* (Ohio State University, 1965)
60. M. Schubert, B. Wilhelmi, *Nonlinear Opticals and Quantum Electronics* (Wiley, New York, 1986)
61. R.C. Powell, *Physics of Solid-State Laser Materials* (Springer, New York, 1998)
62. W. Chen, Y. Inagawa, T. Omatsu et al., Opt. Commun. **194**, 401 (2001)
63. L.I. Ivleva, T.T. Basiev, I.S. Voronina et al., Opt. Mater. **23**, 439 (2003)
64. H.H. Yu, Z. Li, A.J. Lee et al., Opt. Lett. **36**, 579 (2011)
65. M.T. Chang, W.Z. Zhuang, K.W. Su et al., Opt. Express **21**, 24590 (2013)
66. A.A. Kaminskii, H.J. Eichler, K. Ueda et al., Appl. Opt. **38**, 4533 (1999)
67. Y. Urada, S. Wada, H. Tashiro et al., Appl. Phys. Lett. **75**, 636 (1999)
68. A.A. Kaminskii, C.L. McCray, H.R. Lee et al., Opt. Commun. **183**, 277 (2000)
69. T.T. Bssiev, S.V. Vassiliev, M.E. Dorosgenko et al., Opt. Lett. **31**, 65 (2006)
70. A.A. Kaminskii, S.N. Bagaev, A.M. Yurkin et al., Dokl. Akad. Nauk SSSR **367**, 468 (1999)
71. A.A. Kaminskii, G. Aka, J. Findeisen et al., Laser Phys. **8**, 752 (1998)
72. A.A. Kaminskii, H.J. Eichler, K. Ueda et al., Phys. Stat. Sol. (a) **181**, R19 (2000)
73. A.A. Kaminskii, K. Ueda, H.J. Eichler et al., Opt. Commun. **194**, 201 (2001)
74. X. Li, A.J. Lee, H.M. Pask, Opt. Lett. **36**, 1428 (2011)
75. A.J. Lee, H.M. Pask, P. Dekker et al., Opt. Express **16**, 21988 (2008)
76. H.Y. Zhu, Y.M. Duan, G. Zhang, Appl. Phys. **103**, 559 (2011)
77. X.Y. Wang, J. Dang, X.J. Wang et al., Opt. Lett. **41**, 3559 (2016)
78. M. Weitz, C. Theobaid, R. Wallentein et al., Appl. Phys. Lett. **92**, 091122 (2008)
79. A.J. Lee, H.M. Pask, D.J. Spence et al., Opt. Lett. **35**, 682 (2010)
80. A.J. Lee, H.M. Pask, D.J. Spence, Opt. Lett. **37**, 3840 (2012)
81. Y. Feng, L.R. Taylor, D.B. Calia, Opt. Express **17**, 23678 (2009)
82. V.R. Supradeepa, J.W. Nicholson, Opt. Lett. **38**, 2538 (2013)
83. L. Zhang, C. Liu, H. Jiang et al., Opt. Express **22**, 18483 (2014)
84. J.R. Olson, R.O. Pohl, J.W. Vandersande et al., Phys. Rev. B **47**, 14850 (1993)
85. S. Stoupin, Y.V. Shvyd'ko, Phys. Rev. Lett. **104**, 085901 (2010)
86. W. Lubeigt, V.G. Savitski, G.M. Bonner et al., Opt. Express **19**, 6938 (2011)
87. O. Kitzler, A. Mckay, R.P. Mildren et al., Opt. Lett. **37**, 2790 (2012)
88. J.M. Feve, K.E. Shortoff, M.J. Bahn, Opt. Express **19**, 915 (2011)
89. R.J. Williams, O. Kitzler, A. Makay et al., Opt. Lett. **39**, 4152 (2014)
90. A. Sabella, J.A. Piper, R.P. Mildren, Opt. Lett. **39**, 4037 (2014)
91. E. Granados, D.J. Spence, R.P. Mildren, Opt. Express **19**, 10857 (2011)
92. R.P. Mildren, A. Sabella, Opt. Lett. **34**, 2811 (2009)
93. D.C. Parrotta, A.J. Kemp, IEEE J. Select. Topics Quant Elect. **19**, 1400108 (2013)
94. R.J. Williams, O. Kitzler, Z. Bai, IEEE J. Select. Topics Quant Elect. **24**, 1602214 (2018)

Chapter 12
Apparent Crystal Field Model of Laser Glass and Its Application

Similar to crystals, glasses also have important applications in the area of optics and photoelectronics. The laser glasses discussed in this chapter are one important part in the photoelectronic ones, and have more important applications in laser oscillation and amplification as well as transmission of optical information. The glass fiber, waveguide, and film have wide applications in the micromation and integration of photoelectronic devices. The preparation, chemical and physical properties, and research advancement have been reviewed and the relations between the compositions of glasses and the spectral properties of active ions, including the Judd–Ofelt intensity parameters, lifetime of the multiplets, and energy transfer probabilities, have also been analyzed [1–4].

The same as in crystals, a multiplet of rare earth ions in glasses generally splits into several crystal field energy levels owing to the symmetry broken around the ions. The disorder of the structure in glasses causes difficulty in both the theoretical and experimental analyses of the optical transition between crystal field levels. The crystal field levels can be determined from the peak positions of low temperature absorption and emission spectra for crystals. However, the structural disorder of glasses induces the inhomogeneous broadening of spectral lines, so the overlapping among the spectral lines is serious even at low temperature and the positions of crystal field levels are very difficult to be determined. Generally, this kind of work is almost focused on the rare earth ions with simple structure of crystal field levels, such as Eu^{3+} and Yb^{3+} [5, 6]. For the ions widely used in laser materials, such as Nd^{3+}, Er^{3+}, and Pr^{3+}, the research of the level positions and transitions between the levels in glasses is challenging. The positions and widths of crystal field levels, which depend on the composition and structure of host materials, will determine the peak positions and widths of spectral lines that resulted from the transitions between the crystal field levels, and so the laser properties of the glasses. The studies of the crystal field levels of rare earth ions in glasses, like a bridge connecting the composition and structure of glasses with their spectral and laser properties, are so important for the development of laser glasses.

Z. Luo and Y. Huang, *Physics of Solid-State Laser Materials*, Springer Series in Materials Science 289, https://doi.org/10.1007/978-981-32-9668-8_12

In this chapter, the structure and spectral characteristics of glasses are introduced first. From the variety of the symmetries around the rare earth ions, it can be found that the group chain method in the crystal field theory described in Chap. 3 is effective for the analysis of spectral properties of rare earth ions in glasses, especially, for the relation between the properties with the structure of glasses. Therefore, the group chain method will be used to analyze the spectra of rare earth ions in glasses and some initial results will be presented.

12.1 Structure and Spectral Characteristics of Glasses

The structure of glasses is disordered in long range and the array of atoms can be described as a continuous but non-periodical three-dimensional network. However, the existence of short-range order is also well known. The meaning of short-range order is that the array of atoms is more or less symmetrical in a local region (generally less than 1.0 nm), but the lattice for each atom is not exactly uniform. The short-range order has happened mostly around the elements of network former, such as the local order around the Si, B, and P in oxide glasses. The structural units having short-range order are generally polyhedrons constructed by strong covalent bonds and definite nearest ligands, such as the SiO_4 tetrahedron and BO_3 triangle. The structure of SiO_2 glass is shown in Fig. 12.1. The SiO_4 tetrahedron has definite geometrical structure, but the bond angles, such as O–Si–O, can vary in a certain range. The tetrahedrons connect each other by the corners (bridge oxygen), so the bond angles of Si–O–Si can vary in a large range. The variation of the bond angles α (O–Si–O) and β (Si–O–Si) causes the disordered three-dimensional network structure of SiO_2 glass and the SiO_4 tetrahedron is the basic structural unit in the network. Some compounds, for example, SiO_2, GeO_2, B_2O_3, and P_2O_5, can exist lonely in one amorphous phase and become a simple glass former. Some compounds, such as TeO, Al_2O_3, and WO_3, and other compounds added for forming glasses are called conditional glass formers. The other compounds change the network of glasses, and generally, the anion connecting two cations of the network former can become non-bridge, that is, it only connects with one cation but not two cations and plays the role as a bridge in the network forming process. The cations of network modifier, such as the alkali metal, alkaline earth, and other ions with higher valence, occupy random positions close to the non-bridge anions. This is the continuous random network model proposed by Zachariasen [7].

With the development of the measurement methods such as extended X-ray absorption fine structure (EXAFS), neutron diffraction, and X-ray diffraction, the ligand environment around particular ions in glass former and modifier can be analyzed. First, the experimental results have revealed that the nearest ligand environment around the cations in glass modifier is much more explicit than the anticipation of the model mentioned above, furthermore, similar to the local ligand environment in crystals. Secondly, it has been found that modifiers do not distribute homogeneously in glasses, but form rich areas inhomogeneously in glass or

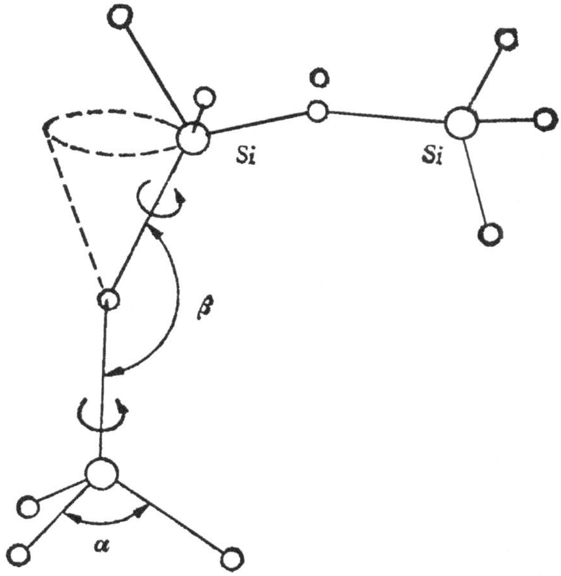

Fig. 12.1 The simple structure model of SiO_2 glass

separate the rich areas of glass former. This is very important for understanding the fluorescence quenching effect of rare earth ions in glasses. Lastly, it has also been found that the coordination number around cations and the distance between ions have only slight change with the concentration. Consequently, the structure of glass contains the disorder in long distance, order in short distance around the glass formers, and order in middle distance around cations of the glass modifiers.

On the basis of the experimental results, Greaves [8] proposed a modified random network model. The model shows that the cations of glass former and modifier link with the bridge oxygen and non-bride oxygen in disorder, and the ligand environment around the cations is orderly. The ligand number and the distance between ions are invariable. Huang and Cormack [9, 10] confirmed the model through the calculation of molecular dynamics. The glass structures described in the normal and modified continuous random network models are compared in Fig. 12.2.

Besides the two popular models mentioned above, there are several other models describing the structure of inorganic glass [4]. Whichever model is adopted, it is common that the ligand environments around rare earth ions in glass are different from each other more or less.

In spectral experiments, one obvious result caused by the different environments of rare earth is the broadening of full-width at half-maximum (FWHM) of spectral lines. In the spectra of Er^{3+} ions in $LiYF_4$ crystals recorded at 13 K, the FWHM of the spectral lines corresponding to the transitions between crystal field energy levels is between 1 and 4 cm^{-1} in general. However, recorded at the same temperature, the FWHM of spectral lines for Er^{3+}-doped ZBLAN (ZrF_4-BaF_2-LaF_3-AlF_3-NaF) is in

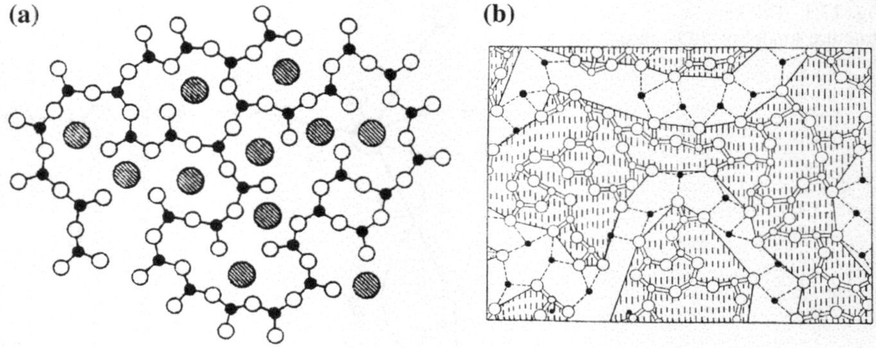

Fig. 12.2 Two models of glass structure: **a** the continuous random network model; **b** the modified continuous random network model

a range from 20 to larger than 100 cm^{-1}. Although the ligands of Er^{3+} ions in both the crystal and glass are the F^- ions, the coordination number of Er^{3+} ions and the bond length and angle between Er^{3+} and F^- ions are variable in the glass. As a result, the peak positions of spectral lines are different for Er^{3+} ions at different locations. Generally, the measured spectral lines of rare earth ions are the summation of the individual spectral lines of the ions at different locations. This is why the FWHMs of spectral lines for Er^{3+} ions in the ZBLAN glass are much larger than those in the $LiYF_4$ crystal.

In many oxide and fluoride glasses, the cations in the modifier display the characteristics of ionic bond. Therefore, the Pauling's rules in crystal chemistry for the ionic structural architecture can be used to estimate the local environments [11] and the following conclusion can be obtained: the coordination anion number is determined by the radius ratio of the rare earth to the ligand ion. A rare earth ion with radius ratio in a range from 0.414 to 0.732 should occupy a six-coordinated position, while in a range from 0.732 to 0.904 the rare earth cation may occupy up to a nine-coordinated position. Rare earth ions decrease in radius from about 0.120 to 0.085 nm with increasing atomic number due to the lanthanide contraction effect. In oxide glasses, the ligand ions are essentially oxygen anions O^{2-} with an ionic radius of about 0.140 nm. Consequently, the ratio decreases from about 0.85 to 0.61 as the rare earth atomic number increases. A notable feature in the ratio variation is that it covers the transitional value 0.732, which represents the transition from a six-coordinated state to higher coordination. Distinct change in local structure can be expected as the 4f subshell of rare earth is filled and the ionic radius crosses the transition value. For example, Er^{3+} should be six-coordinated by oxygen O^{2-} ($R_{Er3+}/R_{O2-} = 0.628 < 0.732$), while the Nd^{3+} in oxide glass is capable of 7–9-coordinated states ($R_{Nd3+}/R_{O2-} = 0.743 > 0.732$). This has been confirmed in the rare earth silicate crystal compounds, where smaller rare earth cations (Er^{3+}, Tm^{3+}, Yb^{3+}, etc.) have coordination number 6 (or 7 at very high rare earth concentration) and the larger cations (Nd^{3+}, Pr^{3+}, Eu^{3+}, etc.) have coordination numbers from 7 to 9.

According to the modified random network model, the same behavior should apply for rare earth-doped oxide glasses, although with greater distortion due to the long-range disorder. Many previous results of different types of experimental spectroscopy on the local environments of rare earth ions in various glasses support this ion size ratio-dependent characteristic.

Since the nearest neighbor ions (O^{2-} for oxide glasses) have the strongest influence on the central rare earth ions, their properties (e.g. bridging or non-bridging) should be analyzed to further assess the regularity of the local environments of rare earth ions in glass. A detailed analysis of EXAFS measurements confirmed that the regularity of the site occupied by the rare earth ion (i.e. the range of distances to ligand ion) increases as polymerization decreases (dissociation of the network by introducing the modifiers). This is caused, in part, by the increasing number of non-bridged oxygen ions in the rare earth's coordination sphere with decreasing polymerization. Further, the regularity of the rare earth site also increases as the size of the rare earth ions decreases. This is due to the fact that the rare earth ions with smaller radius and higher crystal field-strength are easier to coordinate with non-bridging oxygen.

In addition, the separation between rare earth ions plays a key role in their spectral behavior. The modified random network model predicted the existence of correlation and therefore clustering among the modifying cations in glass contrary to the continuous random network model, which predicts a homogeneous distribution of the rare earth. Upon application of the principles of crystal chemistry, it follows that more glass-modifying cation species in a glass would induce a larger spacing among the rare earth ions since larger structural units are preferably formed over the smaller ones, and the number of essentially different types of constituent in a system tends to be small (Pauling's rule of parsimony). This means that different species of modifying elements tend to be evenly distributed with respect to one another within the modifier-rich regions.

In summary, from crystal chemistry principles on the bases of modified random network model for glass structure, the local environments of the rare earth ions are mainly influenced by the following three factors: the number of coordinating anions around the rare earth ion is essentially governed by the radius ratio of the central cation to the ligand anion; the regularity of the site is dominated by both the degree of depolymerization of the network and the size of the rare earth ion; the extra-modifying species are helpful in enlarging spacings among rare earth ions. However, the condition of predominant ionic nature between a rare earth ion and its ligand anion must be satisfied in order to utilize these conclusions.

These ideas demonstrate that a considerable change in environment is expected for different rare earth ions in a certain glass host, as well as the same rare earth ion in different hosts. In particular, we could expect a marked difference between the elements in the two ends of the lanthanide series. This is contrary to the opinion that all the rare earth ions should have similar local structures in a glass; therefore contrary to the conclusion that the study of one rare earth ion's local structure would represent all the other ions' local structures in the same glass. These results also disagree with the pictures described by the conventional continuous random network model for glass, where the rare earth ion could have coordination numbers from 6 to 12 and distribute uniformly throughout the glass.

According to the crystal chemistry model of the local environments of various rare earth ions in glasses, the following points about the spectral features of the various rare earth ions in oxide glasses can be generally inferred. Since the smaller rare earth ions (Er^{3+}, Tm^{3+}, Yb^{3+}, etc.) possess only six-coordinated sites and have higher site regularities, the homogeneity of their spectra will be higher than those of the larger rare earth ions (Nd^{3+}, Pr^{3+}, Eu^{3+}, etc.), because the larger rare earth ions could have different types of sites with 7–9-coordinated and lower site regularities. The fluorescence decay characteristics for smaller rare earth ions in oxide glasses should be more or less simple exponential, because only one six-coordinated site is available rather than a variety of coordinated site, and this six-coordinated site exhibits higher site regularity. The fluorescence decay characteristics for the larger rare earth ions are expected to be non-exponential because of the much larger site-to-site differences seen by these ions. In fact, the fluorescence decay of the $^4F_{3/2} \rightarrow {}^4I_{11/2}$ transition in Nd^{3+} in oxide glasses has always been found in a non-simple exponential function with higher spectral inhomogeneity, whereas the fluorescence decay of the $^4F_{13/2} \rightarrow {}^4I_{15/2}$ transition in Er^{3+} in oxide glasses has exhibited a near-simple exponential characteristic with much lower spectral inhomogeneity.

Using the crystal chemistry model, it is also able to realize the phenomenon that the alumina-doped silica glass accommodates more rare earth element than the pure silica glass. First, the structural role of Al^{3+} ions in the pure silica-glass needs to be examined. In the alumina-doped silica, owing to the lack of extra free O^{2-} ions, Al^{3+} ions will act as a glass-modifying element (in six-coordinated state) rather than going to the silica-glass network as a glass-forming element (four-coordinated state). This view is supported by the evidence in alumino-silicate crystals and alkali-free alumino-silicate glasses, where all Al^{3+} ions have a six-coordinated structural configuration. The doped rare earth ions in silica glass also behave as glass modifiers. Thus, according to the modified random network model of glass structure, the modifier-rich region in the rare earth and alumina co-doped silica glass will consist of both Al^{3+} and rare earth ions. Rare earth ions will be preferably partitioned by Al^{3+} and forming Al–O–RE bonds, rather than only separated by a O^{2-} ion and forming RE–O–RE bonds in the modifier-rich regions. Subsequently, larger spacing among rare earth elements is present in the alumina-doped silica host than in the non-alumina-containing silica host. The spectral behavior of the rare earth ion in the alumina-doped silica would thus be similar to that of the rare earth-doped multi-component oxide glasses. In contrast with the germania-doped silica host, the germania is forming part of the network along with silica; thus, there is no effective modifying element in the modifier-rich region for partitioning rare earths. As a consequence, more pronounced correlation and clustering effects are present and earlier quenching is observed in the luminescent properties of the germania-modified silica than those of the alumina-modified silica.

In terms of rare earth-doped heavy metal fluoride (HMF) glasses, the model mentioned above is also applicable because of the strong ionic nature of the fluoride glasses. The structures of HMF glasses have been extensively studied and various structural models have been proposed. It has been generally recognized that classification of glass formers and modifiers in HMF glasses is much more ambiguous than that in the conventional multi-component oxide glasses. The fact that LaF_3 and AlF_3 can stabilize HMF glasses indicates that the LaF_3 and AlF_3 act as intermediate elements in the HMF glasses. Using the crystal chemistry model, the following qualitative conclusions about rare earth local structures in the HMF glasses can be introduced. All the rare earth ions in the HMF glasses are capable of having many types of 7–9-coordinated sites owing to the smaller radius of F^- ion (about 0.119 nm) inducing large rare earth ion to fluorine ion ratios (>0.732). Then, because of the much higher ionicity between the rare earth cation and the fluorine anion and much weaker polymerized network in the HMF glasses compared with oxide glasses, the regularity of the site occupied by the rare earth ion in the HMF glasses is much higher than that in the oxide glasses, and increases as the rare earth ion radius decreases. For example, Tm^{3+} ions in silica-based glasses possess one distinct type of site with a broad distribution of bond lengths or angles (low regularities), whereas in ZBLAN glass they are found to possess more than one distinct type of sites with a narrow distribution of bond lengths and angles (high regularities). Owing to the role of rare earth fluorides (ReF_3) as intermediate element in these glasses, they are able to be both in and out of network positions. This implies that rare earth ions in HMF glasses are more evenly distributed. This behavior is in distinct contrast with oxide glasses, in which rare earth ions play only the glass modifier role. Further, the multi-component nature of these HMF glasses allows the rare earth ion–ion separation characteristic to be more remarkable than that in the common silica-based oxide glasses. The fact that fluorescence quenching of Er^{3+} in the ZBLAN glass occurs at higher Er^{3+}-doping level than in the oxide glasses is in accord with the prediction by the crystal chemistry model. Similarly, it can be predicted that the rare earth concentration-dependent fluorescence quenching effect is also less in fluorophosphate glasses, because the rare earth is also an intermediate element in this glass.

Clearly, the crystal chemistry model for the local structure of rare earth ions in glasses can well describe and predict the spectral properties of the active ions, including the concentration-dependent fluorescence quenching effect.

12.2 Apparent Crystal Field Hamiltonian for Rare Earth Ions in Non-crystal Host

The rare earth-doped non-crystal and polycrystal play more important role in the photoelectronic area. Besides the popular oxide and fluoride glasses, the rare earth-doped sulfide glass can also be prepared as bulk, fiber, and film devices and become scholarly interests due to its low phonon energy, high refractivity, and transparent in infrared. The Er^{3+} and Pr^{3+}-doped fiber amplifiers are the key element in optical fiber communications. Different from that in the crystals, local environment of rare earth ions in the above hosts are variant.

Local structures around rare earth ions in non-crystal materials can be investigated using, for example EXAFS, neutron or X-ray scattering, and fluorescence line narrowing (FLN) laser spectroscopy. Unlike the case of crystal, except the method of FLN, only the average result, that is, the average number of coordination around rare earth ion, the average distance between the rare earth ion and the ligands, and the "average" local symmetry are obtained. The method of FLN still cannot determine the local structures exactly. Consequently, the direct relation between the point symmetry of the site occupied by a rare earth ion and its energy level splitting cannot be built through crystal field Hamiltonian like in crystals.

Strictly, a rare earth ion cannot occupy a position with point symmetry higher than C_1 in non-crystal host. It means that radiative transition polarized in any direction is permitted without retraction coming from point group symmetry selection rule. Furthermore, the lack of long-range correlation between sites prevents the use of polarization behavior as a criterion to identify a given transition. Nevertheless, certain features common to the entire set of spectra indicate that some degree of local order persists in the immediate environment of the emitting ion, which has been proved in the above analysis of glass structure.

Conventionally, the analysis of crystal field levels for rare earth ions in non-crystal host by means of the crystal field theory is generally focused on the Eu^{3+} ions with small J number on the basis of the model proposed by Brecher and Riseberg [5]. In this model, the C_{2v} point group is adopted to describe the "average" local symmetry, that is, the "average" crystal field Hamiltonian. The major reasons for adoption of the C_{2v} are as follows. First, it is the highest symmetry in which full splitting of the $^7F_1$ and $^7F_2$ levels is allowed. Secondly, it is a subgroup of almost all the point groups with higher symmetry in which rare earth ions are normally found, enabling application of the standard technique of descending symmetries. And finally, it is the lowest symmetry for which simple crystal field calculations can be routinely performed at that time. This model can explain the crystal field splitting of Eu^{3+} ions in glasses very well. Although the model was proposed for Eu^{3+} ions in silicate glasses, it has subsequently been applied to the analysis of rare earth ions in many kinds of non-crystal hosts [12–14].

To explain the model clearly, Brecher and Riseberg assumed the geometrical structure of the coordination around Eu^{3+} ions in silicate and fluoride glasses [5, 15]. However, experimental results have shown that the ligand environments, that is, the point symmetries, around the rare earth ions in non-crystal hosts are variant and do not support the assumption of single local environment. Consequently, the physical basis of choosing C_{2v} to describe the crystal field of rare earth ions in glasses is not very sound. Moreover, the model only allows four transitions between $^5D_0$ and $^7F_2$ multiplets for Eu^{3+} ions, but the whole five transitions have been observed in spectral experiments. Many spectral experiments have also revealed that all the transitions between crystal field levels are allowed for rare earth ions in non-crystal hosts as in crystals with low symmetry, although the structure of energy levels has the character of high point symmetry [16, 17]. This also cannot be explained by the above model.

To relate the spectral properties of rare earth ions and the effect of ligand ions in non-crystal materials more close to the reality, it is necessary to establish a simple and practical crystal field model with sound physical basis from the real ligand conditions around the rare earth ions in non-crystal hosts. The model is expected to describe the spectral properties and the average results of energy level splitting, and called apparent crystal field model. Owing to the fact that the crystal field is variable in non-crystal host, the group chain scheme introduced in Chap. 3 may be a simple and effective way for this purpose.

The group–subgroup relation of the 32 point groups (imbeddings of the crystallographic point groups [18]) is shown in Fig. 12.3. The distortion from high to low symmetries step by step is displayed clearly in the figure. With the high point symmetries descending to low ones, the representations of group O_3 or SO_3 reduce to their subgroups. The eigenfunctions of crystal field levels and the crystal field Hamiltonian H_{cf} are expressed by a series of the irreducible representations of the group chain from group to subgroup step by step. Suppose that the active ions occupy the positions having the symmetry of point group G, then the H_{cf} is invariant under the action of any elements of G, that is, it transforms as the scalar irreducible representation. As an example, for the condition of $G = D_4$, considering the group chain to be $SO_3 \supset O \supset D_4, H_{\mathrm{cf}}$, H_{cf} can be expressed as

$$H_{\mathrm{cf}} = \sum_{k,\mu} C_{\mu}^{k} b_{\mu}^{k} = C_{3}^{2} b_{3}^{2} + C_{1}^{4} b_{1}^{4} + C_{3}^{4} b_{3}^{4} + C_{1}^{6} b_{1}^{6} + C_{3}^{6} b_{3}^{6} \tag{12.1}$$

The parameters in this expression are defined in Chap. 3.

When the symmetry of environment descent to its subgroup G_1, the terms in $H_{\mathrm{cf}}^{G_1}$ should belong to identical representation of G_1. Furthermore, the group–subgroup branching rule tells us that the identical representation of G is still identical representation of G_1 when symmetry goes down from G to G_1. Therefore, all the terms

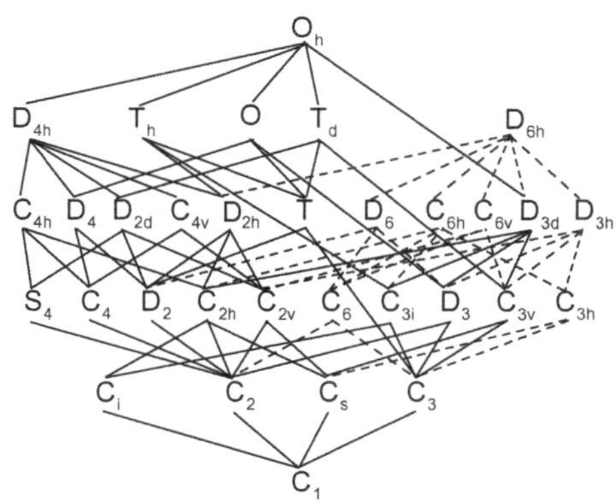

Fig. 12.3 Group–subgroup relation of the 32 point groups [18]

in H_{cf}^{G} will appear in $H_{cf}^{G_1}$ such as can be seen from Table 3.17. The terms of H_{cf} with point group of higher symmetry, for example, O, will appear in most of the H_{cf} with different point groups of lower symmetry. The lower symmetry position is actually some kind of distortion from high symmetry position and the distortion potential is always smaller than the high symmetry potential.

For rare earth ions in a glass, the terms of H_{cf} of rare earth ions in higher symmetry positions will appear "repeatedly" in most of the H_{cf} of rare earth ions in different lower symmetry positions and then take a more important part than other terms belonging only to the particularly lower symmetry position in the effective H_{cf}. Contrary to the choice of C_{2v} which is a subgroup of almost all the higher symmetry point groups [5], we consider a high symmetry point group with many subgroups of lower symmetry to describe the apparent crystal field of rare earth ions in glasses. The corresponding crystal field Hamiltonian is called the apparent crystal field Hamiltonian H_{cf}^{a}.

For a given rotation group G, its isomorphic groups and its corresponding rotation-inversion direct product group $G_i = G \times C_i$ have the same mathematical expression of H_{cf} [19]. We first choose O, including T_d and O_h, which have the same expression of H_{cf}, as the high-point symmetry group (see Fig. 12.3). The point symmetry D_6 (or C_{6v}, D_{3h}, D_{6h}) is not considered because positions with symmetry of 6-fold rotation axis are rarely observed for active ions in crystal [20] and so in glasses. In the group chain scheme, the H_{cf} of O point symmetry H_{cf}^{O} has only two terms, one for $k = 4$ and the other for $k = 6$, which will express the major terms of the corresponding rank of H_{cf} for rare earth ions in glass H_{cf}^{a}. The term of $k = 2$ only appears in crystal field with symmetry of group D_4 (including C_{4v}, D_{2d}, and D_{4h}), D_3 (including C_{3v} and D_{3d}) and lower symmetry groups, so there are two possible choices for this term. After these considerations, the H_{cf}^{a} may be written in a "simplified" group chain [21, 22] $SO_3 \supset O \supset D_4$ as

$$H_{cf(D_4)}^{a} = C_3^2 b_3^2 + C_1^4 b_1^4 + C_1^6 b_1^6 \tag{12.2}$$

or $SO_3 \supset O \supset D_3$ as

$$H_{cf(D_3)}^{a} = C_5^2 b_5^2 + C_1^4 b_1^4 + C_1^6 b_1^6 \tag{12.3}$$

In these equations, the terms $C_1^4 b_1^4$ and $C_1^6 b_1^6$ belong to the H_{cf}^{O} and still exist in the H_{cf} of its subgroup D_4 and D_3. The other terms of $k = 4$ and 6 in D_4 and D_3, which do not belong to identical representation of point group O, can be neglected because these components of crystal field are perturbation of crystal field of symmetry O. As a result, in the expressions of H_{cf}^{a}, only three crystal field parameters C_3^2 (or C_5^2), C_1^4, and C_1^6, one for each k, will be fitted by the values of crystal field splitting data.

From (9.3.7) in [18], it can be easily shown that the relationship between the crystal field parameters of the conventional JM scheme B_{kq} and the group chain scheme C_μ^k are:

For D_4 point symmetry:

$$\begin{aligned} C_3^2 &= -B_{20}, C_1^4 = \frac{1}{2}\sqrt{\frac{7}{3}}B_{40} + \sqrt{\frac{5}{6}}B_{44}, C_3^4 = \frac{1}{2}\sqrt{\frac{5}{3}}B_{40} - \sqrt{\frac{7}{6}}B_{44}, \\ C_1^6 &= -\frac{1}{2\sqrt{2}}B_{60} + \frac{\sqrt{7}}{2}B_{64}, C_3^6 = \frac{1}{2}\sqrt{\frac{7}{2}}B_{60} + \frac{1}{2}B_{64} \end{aligned} \tag{12.4}$$

and the relationship for D_3 point symmetry has been shown in (3.31).

As having introduced above, the crystal field theory has been widely used to analyze the energy level splitting of rare earth ions in crystals. However, a quantitative comparison of crystal field strengths for rare earth ions in different materials cannot be made directly from the crystal field parameters. Therefore, as having mentioned in Chap. 9, the concept of scalar crystal field strength was introduced by Auzel [23, 24]. The relation between the scalar strength N_v and the crystal field parameters is

$$N_v = \left[\sum_{k,q}\left(\frac{4\pi}{2k+1}\right)\left(B_{kq}\right)^2\right]^{1/2} = \left[\sum_{k,\mu}\left(\frac{4\pi}{2k+1}\right)\left(C_\mu^k\right)^2\right]^{1/2} \tag{12.5}$$

The strength of crystal field for rare earth ions in crystals and glasses can be analyzed by means of the parameter N_v. Whatever the symmetry, the strengths of crystal field for rare earth ions between different crystals, furthermore, between crystal and non-crystal can be analyzed and compared quantificationally [13, 14, 23–25]. Subsequently, Leavitt [26] proposed the invariants of quadratic rotation as

$$s_k = \left[\frac{1}{2k+1}\sum_q\left(B_{kq}\right)^2\right]^{1/2} = \left[\frac{1}{2k+1}\sum_\mu\left(C_\mu^k\right)^2\right]^{1/2} \tag{12.6}$$

Obviously, these invariants express the contributions of components $k = 2$, 4, and 6 to the whole strength of crystal field. The relationship between the invariants and the scalar strength is

$$N_v = \left[4\pi\sum_{k=2,4,6}\left(s_k\right)^2\right]^{1/2} \tag{12.7}$$

Auzel and Malta [24] related the invariant s_k with the maximum splitting of multiplet ΔE, that is, the gap between the highest and lowest crystal field levels in multiplet ${}^{2S+1}L_J$, as

$$(\Delta E)^2 = \frac{12 g_a^2}{g(g_a+2)(g_a+1)} \sum_{k=2,4,6} \left|\left\langle [SLJ] \| U^{(k)} \| [SLJ] \right\rangle\right|^2 \left|\left\langle f \| C^{(k)} \| f \right\rangle\right|^2 (s_k)^2 \tag{12.8}$$

where $g = (2J + 1)$, g_a is the effective degeneracy of the multiplet: $g_a = g/2$ if J is a half-integer. Furthermore, for the multiplets with $J \geq 3$ and the difference of $|\langle [SLJ] \| U^{(k)} \| [SLJ] \rangle|^2$ for different k less than a factor of 10, the maximum splitting of multiplet and the scalar strength of crystal field can be related by

$$\Delta E = \left[\frac{3 g_a^2}{\pi\, g(g_a+2)(g_a+1)}\right]^{1/2} \left[\prod_{k=2,4,6} \left|\left\langle [SLJ] \| U^{(k)} \| [SLJ] \right\rangle\right| \left|\left\langle f \| C^{(k)} \| f \right\rangle\right|\right]^{1/3} N_v \tag{12.9}$$

As an example, for the common active ion Nd^{3+} in solid-state laser materials, the scalar strength of crystal field can be estimated from the maximum splitting of $^4I_{9/2}$ multiplet in various hosts.

In Chap. 9, the analysis of self-activated Nd^{3+} laser crystal on the basis of the scalar strength of crystal field introduced by Auzel has been mentioned. The relation between the splitting of the $^4I_{9/2}$ multiplet and the scalar strength of crystal field for Nd^{3+} in a series of crystal and non-crystal materials is shown in Fig. 12.4. The relation obeys (12.9). Besides applications in the analysis of spectra and energy levels, the scalar strength of crystal field can be related with other properties. For example, though there isn't a general theory of melting points of solids (T_m), it may be related with N_v by [27]

$$T_m \approx 0.48\, N_v + 300 \tag{12.10}$$

where the intercept (300 °C) corresponds to the B_{00} term not contained in N_v. The parameter N_v is also related with the hardness and cohesive energy of crystals [27].

The authors and Auzel compared the strengths of crystal field for different rare earth ions in a host by means of N_v [28, 29]. From Fig. 12.5, it can be found that there is a linear relationship between the scalar strength of Nd^{3+} and Er^{3+} in different crystals. An empirical relation between the crystal field strengths of Nd^{3+} ions with electron configuration $4f^3$ and rare earth ions with $4f^q$ has been proposed by Auzel [29].

$$N_V(4f^q) = N_V(4f^3) - 0.034(q-3) N_V(4f^3)$$

Therefore, from the crystal field strength of one rare earth ion in a host, the crystal field strength of other rare earth ions in the host can be estimated. This is more beneficial to the Yb^{3+}-doped materials. It is obvious that the scalar crystal

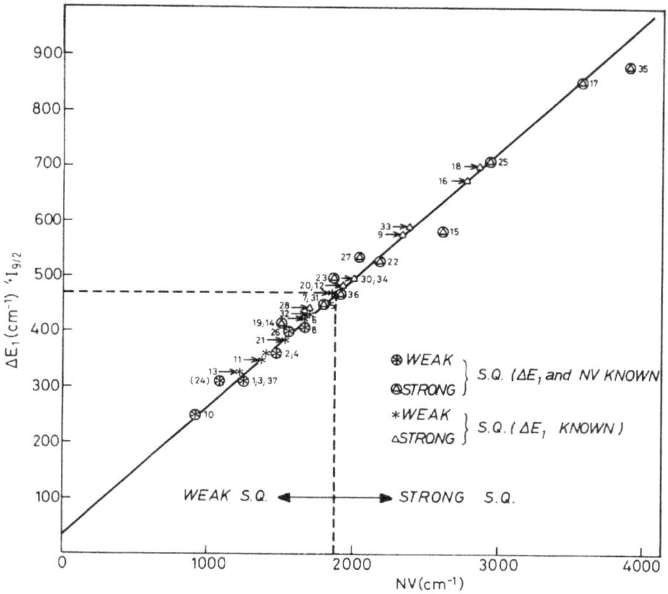

1-NdP_5O_{14}, 2-$LiNd(PO_3)_4$, 3-$Na_2Nd_2Pb_6(PO_4)Cl_2$, 4-$Na_5Nd(WO_4)_4$, 5-$Na(La,Nd)(WO_4)_2$, 6-$K_5Nd(WO_4)_4$, 7-$K_5Nd(MoO_4)_4$, 8-$NdAl_3(BO_3)_4$, 9-$Na_{18}Nd(BO_3)_7$, 10-$NdCl_3$, 11-$(Nd,La)Ta_7O_{19}$, 12-$LiNdNbO_4$, 13-tellurite glass, 14-Nd_2O_2S, 15-CaF_2, 16-$YAlO_3(La,Nd)$, 17-$Y_3Al_5O_{12}(La,Nd)$, 18-$Ba_2MgGe_2O_7$, 19-$NdNb_5O_{14}$, 20-silicate flint glass, 21-$Gd_{1.7}Nd_{0.3}(MoO_4)_3$, 22-$Li(Y,Nd)F_4$, 23-$(Nd,La)F_3$, 24-$Nd(C_2H_5ZO_4)_3$, 25-$Ca_5(PO_4)_3F$, 26-$KY_3F_{10}$, 27-$Nd_2S_3$, 28-$YVO_4$, 29-$PbMoO_4$, 30-$LiNbO_3$, 31-phosphate glass, 32-$KY(MoO_4)_2$, 33-$CaAl_4O_7$, 34-La_2O_3, 35-LaOF, 36-$CaWO_4$, 37-$NdCl_3$ in ice

Fig. 12.4 Maximum splitting of multiplet ${}^4I_{9/2}(Nd^{3+})$ versus N_v for different host materials [27]

field strength has important application in the analysis of rare earth ion-doped laser materials.

In the apparent crystal field model mentioned above, there is only one term for each k in the expressions of H_{cf}^{a} in (12.2) and (12.3). Therefore, for rare earth ions in non-crystal host, the invariant of quadratic rotation in (12.8) can be substituted directly by the crystal field parameter

$$(\Delta E)^2 = \frac{12 g_a^2}{g(g_a+2)(g_a+1)} \sum_{k=2,4,6} \frac{\left|\langle [SLJ] \| U^{(k)} \| [SLJ] \rangle\right|^2 \left|\langle f \| C^{(k)} \| f \rangle\right|^2}{2k+1} \left(C_\mu^k\right)^2 \tag{12.11}$$

In the conventional analysis of spectra and crystal field for rare earth ions in crystals, the crystal field parameters are related with the crystal field splittings of multiplets through the crystal field Hamiltonian. The parameters can be obtained by calculating the eigenvalues of perturbation matrix and fitting with the values of splittings recorded in low temperature spectral experiments. Now, on the basis of

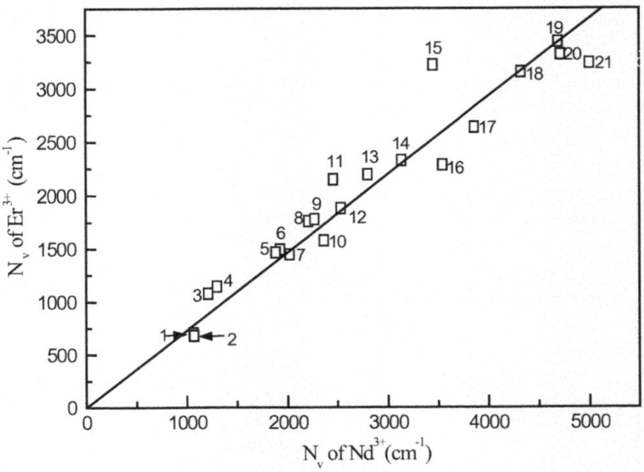

1-$LaCl_3$, 2-$LaBr_3$, 3-$ThGeO_4$, 4-$R(C_2H_5SO_4)_3{\cdot}9H_2O$, 5-$YPO_4$, 6-$LiRP_4O_{12}$, 7-$LuPO_4$, 8-$YVO_4$, 9-$La(MoO_4)_2$, 10-$LaF_3$, 11-$CsCdBr_3$, 12-$CaWO_4$, 13-$LiYF_4$, 14-$Cs_2NaRCl_{16}$, 15 -$Bi_4Ge_3O_{12}$, 16-$YAlO_3$, 17-$CaF_2$, 18-$Y_3Ga_5O_{12}$, 19-$Y_2O_3$, 20-$Y_3Sc_2Al_3O_{12}$, 21-$Y_3Al_5O_{12}$

Fig. 12.5 Crystal field strength of Nd^{3+} and Er^{3+} ions in different crystals

the apparent crystal field model, the crystal field parameters for rare earth ions in non-crystal hosts can be related with the maximum splittings of multiplets by (12.11) and the absolute values of the parameters can be obtained simply by linear fitting process. The calculation is simple and the result is explicit; furthermore, the requirement of spectral experiment is not excessive. As mentioned above, the large inhomogeneous spectral line widths of rare earth ions in non-crystal hosts cause a lot of trouble to identification of the spectral lines, which related to the transitions between crystal field levels; so it is very difficult to determine all the splittings in a multiplet. Especially, for the rare earth ions having important application in solid-state laser materials, such as Er^{3+} and Nd^{3+}, the determination is more difficult due to the large J value of the multiplets related to laser operation and so the large number of crystal field levels. By means of the apparent crystal field model, the crystal field parameters for rare earth ions in non-crystal hosts can be calculated from only the maximum splittings but not every splittings in the multiplets. It means that the parameters may be obtained after the width of spectral band corresponding to the transitions between multiplets is measured at low temperature. Then, the crystal field parameters can be used to analyze the properties of spectra and laser of non-crystal materials.

Conventionally, the disorder of non-crystal materials causes a lot of trouble to the analysis of crystal field and spectra of rare earth ions doped inside. Quite the contrary, on the basis of the apparent crystal field model, the disorder makes the calculation and analysis simpler than those in crystals. Auzel adopted a higher icosahedral symmetry further to describe the apparent crystal field of non-crystal and make the calculation more simple [30]. It can also be used to analyze the spectral characteristics of rare earth ion in glasses but cannot deal with the subtle crystal field splittings of a multiplet.

12.3 Crystal Field Level Analysis for Er^{3+} Ions in Three Typical Glasses

Silicate, germanate, and ZBLAN glasses are three kinds of host materials widely used in solid-state laser. Especially, the ZBLAN is the medium of the Er^{3+}-doped fiber amplifier, which is one of the key elements in the DWDM technique of optical communications. In the following, the crystal field splittings of Er^{3+} ions in the three glasses are systematically analyzed on the basis of the apparent crystal field model [21, 22].

The samples used in the measurement of spectra, that is, silicate, germanate, and ZBLAN glasses of formulas SiO_2–Na_2O, GeO_2–BaO–K_2O, and ZrF_4–BaF_2–LaF_3–AlF_3–NaF, respectively, were cooled to 13 K. Absorption spectra of Er^{3+} ions were recorded in a range from 400 to 1600 nm corresponding to the transitions from ${}^4I_{15/2} \rightarrow {}^2G_{9/2}$ to ${}^4I_{15/2} \rightarrow {}^4I_{13/2}$. A 488 nm Ar^+ ion laser was used to excite Er^{3+} ions to their $4F_{7/2}$ multiplet. Fluorescence spectra corresponding to the ${}^4S_{3/2} \rightarrow {}^4I_{15/2}$, ${}^4S_{3/2} \rightarrow {}^4I_{13/2}$, ${}^4F_{9/2} \rightarrow {}^4I_{15/2}$, ${}^4I_{11/2} \rightarrow {}^4I_{15/2}$, and ${}^4I_{13/2} \rightarrow {}^4I_{15/2}$ transitions were measured.

The absorption and emission spectra recorded at 13 K were resolved into different components with Voigt-type profile. This profile was chosen because the linewidth may contain homogeneous and inhomogeneous broadenings with Lorenzian and Gaussian profiles, respectively. The central position, FWHM, and area of each separated spectral line were determined. At 13 K, the higher levels in multiplets have much lower populations and the spectra are predominated by the transitions from the lowest levels in multiplets. As an example, Fig. 12.6 shows the fitted Voigt components in the emission spectral bands of the transition from ${}^4S_{3/2}$ to ${}^4I_{15/2}$ for Er^{3+} ions in the silicate, germanate, and ZBLAN. The spectral lines corresponding to transitions from the lower crystal field level (level 1 in the figure) in ${}^4S_{3/2}$ to all the crystal field levels (levels 1–8 in the figure) in ${}^4S_{15/2}$ are pointed out.

Previous studies on Eu^{3+} ions in glasses have shown that, due to the low-point symmetries of the sites occupied by rare earth ions, the multiplets of the ions split thoroughly [7, 15]. For Er^{3+} ions, the J multiplet splits into $(2J + 1)/2$ crystal field levels after taking into account the Kramers degeneracy. Owing to the large inhomogeneous width, some of the spectral lines mix in one Voigt profile and make the number of observable lines, corresponding to the transitions between some couples of multiplets, less than the number of expected spectral lines. In view of this, the FWHM, the area under spectral line corresponding to the intensity of transition, and the expected number of spectral lines are considered. For a line with

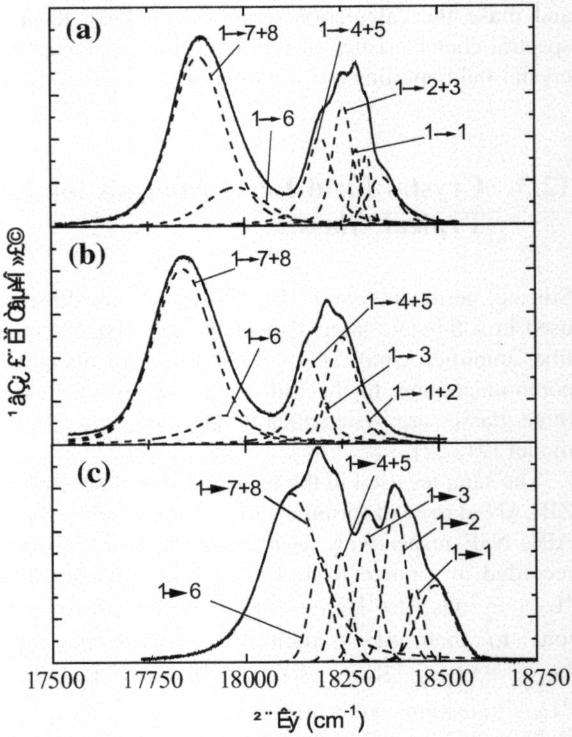

Fig. 12.6 Voigt profile fit to the emission spectra of $^4S_{3/2} \rightarrow {}^4I_{15/2}$ transitions of Er^{3+} at 13 K. **a** Silicate, **b** germanate, and **c** ZBLAN

extremely large values of FWHM and area, more than one transition can be reasonably involved in the line. Some of the mixing lines will be separated after all the related experimental spectra are resolved. Finally, the central positions and widths of the crystal field levels for Er^{3+} ions in silicate, germanate, and ZBLAN glasses obtained from the above resolving are listed in Table 12.1. The uncertainty on the crystal field level positions determined from the experimental spectra is 12 cm^{-1} for all the three glasses.

When two levels are overlapped together, another choice is also given in the table and signed by "*": we assume that the two levels have the same shape and that their central positions are separated by a distance as large as possible before they can be separated by the procedure mentioned above, then we have (Rayleigh criterion)

$$w = 0.7(w_1 + w_2), w_1 = w_2$$

so that

Table 12.1 Positions and widths of crystal field levels of Er^{3+} ions in silicate, germanate and ZBLAN glasses from experimental spectra at 13 K (cm^{-1}).

Manifold	Position		Width	
		*		*
$^4I_{15/2}$	0		0.5	
	23		11	
	50		15	
	70		38	
	139		45	
	295		147	
	414	386(√)	133	95(√)
	414	443(√)		95(√)
$^4I_{13/2}$	6498		23	
	6503		24	
	6540		24	
	6559		26	
	6582		47	
	6760		81	
	6820		68	
$^4I_{11/2}$	10,205		13	
	10,223		18	
	10,240	10,228(√)	54	39(√)
	10,240	10,252(√)		39(√)
	10,338(√)	10,321	79(√)	56
	10,338(√)	10,355		56
$^4I_{9/2}$	12,384		79	
	12,438		49	
	12,543		98	
	12,614	12,601(√)	60	43(√)
	12,614	12,627(√)		43(√)
$^4F_{9/2}$	15,203		28	
	15,232		39	
	15,348		58	
	15,392		55	
	15,494		60	
$^4S_{3/2}$	18,291		35	
	18,348		99	
$^2H_{11/2}$	19,129		24	
	19,161		44	
	19,210		36	
	19,268		52	
	19,301		20	
	19,323		24	

(continued)

Table 12.1 (continued)

Manifold	Position		Width	
		*		*
$^4F_{7/2}$	20,456		30	
	20,522		55	
	20,572		32	
	20,613		99	
$^4F_{5/2}$	22,145	22,136(√)	44	31(√)
	22,145	22,154(√)		31(√)
	22,265		52	
$^4F_{3/2}$	22,509		61	
	22,610		75	
$^2G_{9/2}$	24,440		64	
	24,516		61	
	24,555		49	
	24,606		40	
	24,642		38	
(b)				
$^4I_{15/2}$	0		18	
	35		29	
	63		21	
	95	82(√)	60	43(√)
	95	108(√)		43(√)
	274		161	
	377		135	
	457		94	
$^4I_{13/2}$	6488		5	
	6500		20	
	6530		10	
	6564		8	
	6595		40	
	6754(√)	6719	160(√)	114
	6754(√)	6788		114
$^4I_{11/2}$	10,202		7	
	10,224		65	
	10,224			
	10,224			
	10,330	10,316(√)	64	46(√)
	10,330	10,343(√)		46(√)
$^4I_{9/2}$	12,386		102	
	12,432		32	
	12,542		80	
	12,615(√)	12,605	47(√)	34
	12,615(√)	12,625		34

(continued)

Table 12.1 (continued)

Manifold	Position		Width	
		*		*
$^4F_{9/2}$	15,183		12	
	15,219		28	
	15,337		48	
	15,382		31	
	15,480		54	
$^4S_{3/2}$	18,262		35	
	18,329		43	
$^2H_{11/2}$	19,108		36	
	19,145		14	
	19,186		18	
	19,255		47	
	19,280		0	
	19,309		6	
$^4F_{7/2}$	20,425		31	
	20,498		48	
	20,549		14	
	20,586		100	
$^4F_{5/2}$	22,119	22,108(√)	48	34(√)
	22,119	22,129(√)		34(√)
	22,242		25	
$^4F_{3/2}$	22,496		72	
	22,594		54	
$^2G_{9/2}$	24,427		107	
	24,509		25	
	24,549		25	
	24,593		17	
	24,629		38	
(c)				
$^4I_{15/2}$	0		40	
	47		54	
	111		51	
	154		30	
	194		57	
	226		82	
	308(√)	275	154(√)	110
	308(√)	341		110
$^4I_{13/2}$	6542		29	
	6569		32	
	6603		18	
	6631		33	

(continued)

Table 12.1 (continued)

Manifold	Position		Width	
		*		*
	6696	6677(√)	87	62(√)
	6696	6715(√)		62(√)
	6780		81	
$^4I_{11/2}$	10,228		20	
	10,289	10,267(√)	59	30(√)
	10,289	10,285(√)		30(√)
	10,289	10,289(√)		30(√)
	10,289	10,293(√)		30(√)
	10,289	10,311(√)		30(√)
$^4I_{9/2}$	12,413		4	
	12,472		35	
	12,538		40	
	12,617(√)	12,604	61(√)	44
	12,617(√)	12,630		44
$^4F_{9/2}$	15,271		26	
	15,301		44	
	15,361		47	
	15,398		8	
	15,449		62	
$^4S_{3/2}$	18,429		11	
	18,525		46	
$^2H_{11/2}$	19,160		18	
	19,203		3	
	19,235		0	
	19,267		0	
	19,309(√)	19,300	42(√)	30
	19,309(√)	19,318		30
$^4F_{7/2}$	20,541		20	
	20,586		18	
	20,643	20,619(√)	114	81(√)
	20,643	20,667(√)		81(√)
$^4F_{5/2}$	22,244		24	
	22,273	22,264(√)	44	31(√)
	22,273	22,282(√)		31(√)
$^4F_{3/2}$	22,571		30	
	22,664		69	
$^2G_{9/2}$	24,545		0	
	24,582		18	
	24,631		17	
	24,689	24,679(√)	47	34(√)
	24,689	24,699(√)		34(√)

$$w = 1.4\,w_1$$

where w is the total width determined by spectral line while w_1 and w_2 are, respectively, the widths of the two single levels. As a result, the central positions of the two levels are

$$xc_1 = xc - 0.3w_1, xc_2 = xc + 0.3w_2$$

where xc is the central position of the mixing level, while xc_1 and xc_2 are the central positions of the two levels, respectively. After the apparent crystal field analysis presented in the following, one of these two data for overlapping levels will be chosen and signed by "($\surd$)" in Table 12.1.

To relate the crystal field splitting of Er^{3+} ions with the apparent crystal field, the parameters C^k_μ were first estimated through fitting the crystal field splittings listed in Table 12.1 directly by the apparent crystal field Hamiltonian expressed in (12.2) or (12.3) in the conventional way [31]. Simultaneously, (12.11) for the scalar crystal field strength could also be used to derive the absolute values of parameters C^k_μ by fitting the experimental values of ΔE obtained from Table 12.1 by a least-squares fitting procedure. Then, the signs of the values were determined by comparing the crystal field splittings calculated from the parameters C^k_μ by (12.2) or (12.3) and the experimental data. In the calculations, the intermediate coupling wavefunctions given by Weber [32] and the values of reduced matrix elements $\langle \alpha SL \| U^{(k)} \| \alpha SL \rangle$ provided in [33] were used to obtain the values of $\langle [SLJ] \| U^{(k)} \| [SLJ] \rangle$; and the J-mixing was not taken into account.

To verify the assumption on H^{a}_{cf}, a series of Hamiltonian H_{cf} from point symmetry O (including T_d and O_h), D_4 (including C_{4v}, D_{2d}, and D_{4h}), D_3 (including C_{3v} and D_{3d}), to D_2 (including C_{2v} and D_{2h}), which contain all the possible crystal field parameters, were also adopted to fit the crystal field splittings of Er^{3+} ions in the three glasses. The root-mean-square (RMS) deviations of all the calculations are shown in Fig. 12.7a, b, and c, respectively, for Er^{3+} ions in silicate, germanate, and ZBLAN. Because some of the crystal field levels cannot be resolved in the experiment, different possible choices of level structure should be considered and the ranges of the RMS deviations are represented by error bars in Fig. 12.7. On the x-axis, the various point symmetries used in the analysis are distributed: the positive direction represents the distortion along the chain of $O \supset D_4 \supset D_2$, and the negative direction represents the distortion along the chain of $O \supset D_3$. The apparent crystal

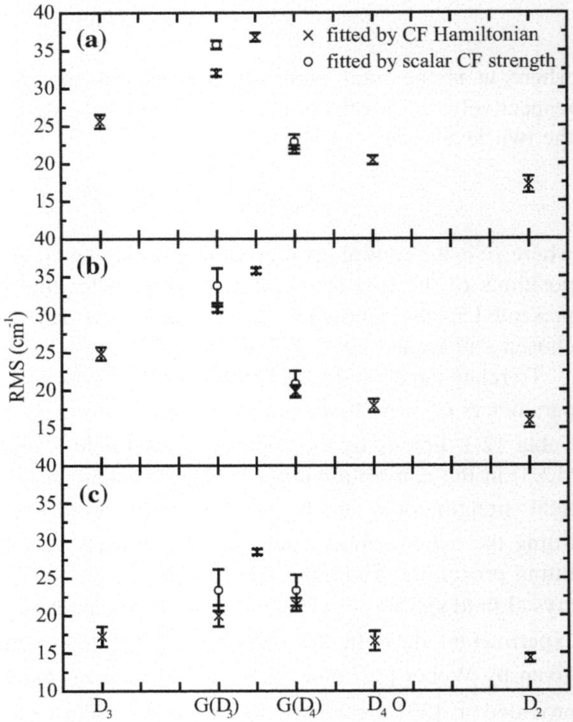

Fig. 12.7 RMS deviations between the experimental crystal field splittings and those calculated by crystal field Hamiltonian of O, D_4, D_3, D_2 point symmetries and the apparent crystal field model proposed for non-crystal, $H^{a}_{cf(D_4)}$ and $H^{a}_{cf(D_3)}$. **a** Silicate, **b** germanate, and **c** ZBLAN. The ranges of RMS, for all possible choices of level structure, are shown as error bars

field Hamiltonians are represented by $H^{a}_{cf(D_4)}$ and $H^{a}_{cf(D_3)}$, locating in the group chains $O \supset D_4 \supset D_2$ and $O \supset D_3$, respectively. For the convenience of presentation, the distance between the representations of the different symmetries has arbitrarily been chosen as the difference of the numbers of crystal field parameters included in the fitting process, for example, there are two crystal field parameters for O point symmetry and five for D_4, so the distance between them is 3 units.

For the silicate and germanate glasses, although there is only one more adjustable parameter, C^2_3, in $H^{a}_{cf(D_4)}$ than in H^{O}_{cf}, the RMS deviations shown in Fig. 12.7a, b have remarkably smaller values in the $H^{a}_{cf(D_4)}$. Moreover, $H^{a}_{cf(D_4)}$, $H^{D_4}_{cf}$, and $H^{D_2}_{cf}$ have similar values of RMS although the numbers of adjustable parameters in the Hamiltonian are 3, 5, and 9, respectively. On the other hand, the values of RMS show that $H^{a}_{cf(D_4)}$ is much better than $H^{a}_{cf(D_3)}$ for expressing the apparent crystal field Hamiltonian of Er^{3+} ions in the oxide glasses. As a whole, the RMS deviations have smaller values in the direction $O \supset D_4 \supset D_2$ than in $O \supset D_3$, especially when the number of fitting parameters is considered. From Fig. 12.7a, b, it can be concluded that the Er^{3+} ions in the oxide glasses occupy preferentially the positions with point symmetry of D_4, D_{2d}, C_{4v}, D_{4h} and their subgroups, that is, the positions with the 4- and twofold rotation axes rather the threefold rotation axis. Figure 12.7a, b also show that the parameters C_{μ}^{k} could be obtained through fitting the maximum splittings of multiplets by (12.11). Then all crystal field splittings could be calculated by (12.2) from the parameters. The RMS deviations between the calculated and experimental data of crystal field splitting are almost the same as those of fitting all the crystal field splittings in the conventional way by (12.2). Moreover, the C_{μ}^{k} obtained by these two different procedures are listed in Table 12.2 and have very similar values. Consequently, the crystal field splittings of Er^{3+} ions in the oxide glasses calculated by $H_{\mathrm{cf}(D_4)}^{\mathrm{a}}$ in which the parameters C_{μ}^{k} were obtained from the scalar crystal field strength analysis (12.11) are listed and compared with the data obtained from the low temperature spectra in Table 12.3. After the analysis, the positions of the levels appearing mixed in spectral experiments were determined (see the levels signed "(√)" in Table 12.1).

For the ZBLAN glass, Fig. 12.7c demonstrates that, different from the oxide glasses, the RMS deviations have similar values in both directions of the group chains $O \supset D_4 \supset D_2$ and $O \supset D_3$; especially, the deviations do not decrease remarkably from H_{cf}^{O} to apparent glass Hamiltonian $H_{\mathrm{cf}(D_4)}^{\mathrm{a}}$ or $H_{\mathrm{cf}(D_3)}^{\mathrm{a}}$. Therefore, it can be supposed that Er^{3+} ions in the ZBLAN glass can equally occupy positions with any kind of point symmetries, that is, symmetries distorted not only from D_4, $D_{4\mathrm{h}}$ and their isomorphic groups but also from D_3, $D_{3\mathrm{d}}$ and their isomorphic groups. Comparing the experimental data for crystal field splittings and the data calculated by $H_{\mathrm{cf}}^{\mathrm{a}}$, the parameters C_{μ}^{k} in $H_{\mathrm{cf}}^{\mathrm{a}}$ could be obtained through fitting the maximum splittings of multiplet by (12.11) or fitting all of the crystal field splittings in conventional way by (12.2) or (12.3). The RMS deviations are different for these two ways (see Fig. 12.7c), and the C_{μ}^{k} got by these two different procedures are listed in Table 12.4. In the oxide glasses, the values of C_{μ}^{k} got by these two different procedures are very similar (see Table 12.2); however, in the ZBLAN glass, the values are different. The above results show that the symmetry of the apparent crystal field for the ZBLAN glass is higher than that for the oxide glasses, so the crystal field cannot be expressed by $H_{\mathrm{cf}(D_4)}^{\mathrm{a}}$ or $H_{\mathrm{cf}(D_3)}^{\mathrm{a}}$ alone.

The crystal field splittings calculated by $H_{\mathrm{cf}(D_4)}^{\mathrm{a}}$ and $H_{\mathrm{cf}(D_3)}^{\mathrm{a}}$ with parameters C_{μ}^{k} obtained from the scalar crystal field strength analysis (12.11) are listed and compared with the experimental data in Table 12.5. In this table, the average values of the crystal field splittings calculated by $H_{\mathrm{cf}(D_4)}^{\mathrm{a}}$ and $H_{\mathrm{cf}(D_3)}^{\mathrm{a}}$ are also given out.

Table 12.2 Parameters C_{μ}^{k} and N_{v} of Er^{3+} ions in silicate and germanate glasses (cm^{-1})

	Silicate		Germanate	
	Fitting ΔE	Fitting crystal field level	Fitting ΔE	Fitting crystal field level
C_3^2	683.3	672.2	706.2	669.1
C_1^4	2477.7	2535.6	2274.5	2290.5
C_1^6	−95.2	−96.1	−471.5	−506.6
N_v	3123	3182	2948	2949

Table 12.3 Comparison of experimental and calculated crystal field splittings of Er^{3+} ions in silicate and germanate glasses (cm^{-1})

Multiplet	Silicate		Germanate	
	Exp.	Cal.	Exp.	Cal.
$^4I_{15/2}$	−175.7	−181.6	−174.7	−168.6
	−152.3	−151.7	−139.3	−149.6
	−125.8	−129.9	−111.6	−128.5
	−105.9	−65.3	−92.3	−57.2
	−36.6	−10.2	−66.7	−45.8
	119.2	83.3	99.7	84.1
	210.1	222.3	202.5	223.6
	266.9	233.1	282.5	241.9
$^4I_{13/2}$	−110.9	−133.6	−109.5	−121.3
	−105.9	−117.0	−97.5	−117.5
	−68.8	−76.5	−67.9	−66.6
	−49.8	−26.9	−34.1	−29.7
	−26.8	35.1	−2.7	24.8
	151.1	150.2	155.8	154.0
	211.1	168.6	155.8	156.3
$^4I_{11/2}$	−58.9	−76.0	−53.9	−74.2
	−41.0	−58.9	−31.3	−54.3
	−35.6	−25.4	−31.3	−21.6
	−12.4	−6.6	−31.3	−11.8
	74.0	73.9	60.2	75.6
	74.0	93.1	87.6	86.4
$^4I_{9/2}$	−134.6	−106.6	−132.0	−112.3
	−80.6	−98.9	−86.0	−103.6
	24.4	47.0	24.0	29.7
	82.5	62.1	97.0	88.5
	108.3	96.3	97.0	97.7
$^4F_{9/2}$	−131.1	−118.5	−137.2	−107.9
	−101.7	−82.6	−101.2	−73.9
	14.3	10.4	16.8	−4.2
	58.3	34.2	61.8	28.9
	160.3	156.5	159.8	157.1
$^4S_{3/2}$	−28.3	−37.9	−33.4	−39.1
	28.3	37.9	33.4	39.1
$^2H_{11/2}$	−103.0	−90.3	−105.8	−91.6
	−71.0	−80.4	−68.8	−82.0
	−22.0	−41.5	−27.8	−18.0
	36.0	57.5	41.2	50.1
	69.0	63.9	66.2	56.2
	91.0	90.9	95.2	85.3

(continued)

Table 12.3 (continued)

Multiplet	Silicate		Germanate	
	Exp.	Cal.	Exp.	Cal.
$^4F_{7/2}$	−84.8	−90.6	−89.7	−99.9
	−18.8	−32.5	−16.4	−21.7
	31.3	52.3	34.6	40.7
	72.3	70.8	71.6	80.8
$^4F_{5/2}$	−49.4	−42.9	−51.4	−42.0
	−30.6	−1.0	−30.8	0.6
	80.0	43.8	82.2	41.4
$^4F_{3/2}$	−50.5	−55.4	−49	−57.2
	50.5	55.4	49	57.2
$^2G_{9/2}$	−111.8	−94.6	−114.4	−99.4
	−35.8	−81.2	−32.4	−84.1
	3.2	36.6	7.6	23.0
	54.2	46.5	51.6	66.1
	90.2	92.7	87.6	94.6
RMS dev.		21.9		19.3

Table 12.4 Parameters C^k_μ and N_v of Er^{3+} ions in ZBLAN glass (cm^{-1})

	Fitting ΔE	Fitting crystal field level by $H^{glass}_{cf(D_4)}$	Fitting crystal field level by $H^{glass}_{cf(D_3)}$
C^2_μ	503.8	509.6	583.5
C^4_1	1597.6	1376.6	1295.8
C^6_1	−367.0	−482.9	−497.9
N_v	2081	1877	1855

The RMS deviations of 21.3 cm^{-1} for $H^a_{cf(D_4)}$ and 21.6 cm^{-1} for $H^a_{cf(D_3)}$ decreasing to 18.9 cm^{-1} for the average result shows that, in the analysis of apparent crystal field in ZBLAN glass, the combination of $H^a_{cf(D_4)}$ and $H^a_{cf(D_3)}$ may be a suitable way.

The values of the scalar crystal field strength parameter N_v calculated by (12.5) are listed in Tables 12.2 and 12.4. For Er^{3+} ions in the oxide glasses, silicate and germanate, the values are around 3000 cm^{-1}; while in the ZBLAN glass, the values are only around 2000 cm^{-1}. It shows that the first neighbor anions take a crucial part in the crystal field strength and the crystal field for the oxide is generally much stronger than that for the fluoride.

In the conventional way, a "pure" and low-point symmetry was tried to describe the average environment of rare earth ions in glasses. On the basis of the fact that the disordered structure of glass leads to site-to-site variations at the rare earth ion sites, an apparent crystal field Hamiltonian for the rare earth ions has been introduced by means of the group chain scheme. Although this is a simple model and only three parameters C^k_μ of the Hamiltonian are considered, it has been used to

Table 12.5 Comparison of experimental and calculated crystal field splittings of Er^{3+} ions in ZBLAN glass (cm^{-1})

Multiplet	Exp.	Cal.		
		$H_{cf(D_4)}^{glass}$	$H_{cf(D_3)}^{glass}$	Average
$^4I_{15/2}$	−168.3	−119.1	−150.1	−134.6
	−121.6	−109.6	−113.1	−111.4
	−57.2	−89.5	−54.1	−71.8
	−14.3	−40.9	−49.6	−45.3
	25.2	−34.0	−34.6	−34.3
	57.3	60.4	116.0	88.2
	139.5	158.6	138.6	148.6
	139.5	174.1	146.9	160.5
$^4I_{13/2}$	−103.0	−85.1	−120.0	−102.6
	−76.0	−83.7	−64.4	−74.1
	−42.4	−46.5	−35.4	−40.9
	−14.4	−21.7	−29.3	−25.5
	31.9	17.0	80.7	48.8
	69.3	110.0	82.2	96.1
	134.6	110.0	86.2	98.1
$^4I_{11/2}$	−51.1	−52.6	−72.4	−62.5
	−11.8	−38.1	−30.4	−34.3
	6.2	−15.0	−15.8	−15.4
	10.2	−8.9	36.4	13.7
	14.2	54.0	38.2	46.1
	32.2	60.8	44.1	52.4
$^4I_{9/2}$	−118.4	−80.3	−82.2	−81.3
	−59.4	−74.1	−72.8	−73.4
	6.6	16.1	17.8	17.0
	85.6	65.7	67.9	66.8
	85.6	72.6	69.3	70.9
$^4I_{9/2}$	−85.0	−75.7	−90.9	−83.3
	−55.0	−51.8	−57.0	−54.4
	5.0	−4.2	15.4	5.6
	42.0	20.1	46.5	33.3
	93.0	111.7	86.0	98.9
$^4S_{3/2}$	−47.8	−27.9	−27.9	−27.9
	47.8	27.9	27.9	27.9
$^2H_{11/2}$	−87.2	−65.2	−63.1	−64.1
	−44.2	−58.5	−60.3	−59.4
	−12.2	−10.8	−11.3	−11.0
	19.8	35.0	34.9	35.0
	61.8	39.3	39.1	39.2
	61.8	60.1	60.7	60.4

(continued)

Table 12.5 (continued)

Multiplet	Exp.	Cal.		
		$H^{glass}_{cf(D_4)}$	$H^{glass}_{cf(D_3)}$	Average
$^4F_{7/2}$	−62.3	−71.8	−58.7	−65.3
	−17.3	−14.6	−16.7	−15.6
	15.4	28.0	−1.0	13.5
	64.2	58.4	76.4	67.4
$^4F_{5/2}$	−19.3	−29.8	−17.1	−23.4
	0.3	0.5	−17.0	−8.2
	19.1	29.2	34.1	31.6
$^4F_{3/2}$	−46.5	−40.8	−40.8	−40.8
	46.5	40.8	40.8	40.8
$^2G_{9/2}$	−82.2	−71.1	−74.5	−72.8
	−45.2	−60.1	−58.9	−59.5
	3.8	12.5	17.4	14.9
	51.7	49.5	55.7	52.6
	71.9	69.2	60.3	64.7
RMS dev.		21.3	21.6	18.9

explain the spectral properties of rare earth ions in non-crystal materials in a rather realistic way.

The fitting results have shown that the Hamiltonian $H^{a}_{cf(D_4)}$ is more suitable than $H^{a}_{cf(D_3)}$ for describing the apparent crystal field of Er^{3+} ions in the silicate and germanate oxide glasses. This means that the Er^{3+} ions in the oxide glasses could not equally occupy positions with any kind of point symmetries. This conclusion is in agreement with those mentioned in Sect. 12.1, that is, rare earth ions can only play a role as network modifier in oxide glasses but cannot joint the network of host, and the Er^{3+} ions in oxide glasses can only have coordination number 6 [11]. Because of the strong ionic nature of the fluoride glasses, the rare earth fluorides play a role as an intermediate element in the glasses. They are able to be both in and out of network positions. On the other hand, all rare earth ions in the fluoride glasses are capable of having 7–9-coordinated sites owing to the smaller radius of F^- ion [11]. As a result, the rare earth ions in fluoride glasses possess more than one distinct type of sites. This implies that rare earth ions in fluoride glasses may be more evenly distributed. As an apparent result, the Hamiltonian for fluoride glass may have "higher" point symmetry than for oxide glass.

By means of the group chain scheme, the apparent crystal field Hamiltonian has the characteristic of high-point symmetry close to the O point group. The cubic crystal field terms (C^4_1 and C^6_1) give the major contribution to the crystal field strength when the term from the low symmetry (C^2_3 or C^2_5) allows the complete removal of degeneracy. Such a consideration explains why transitions between crystal field levels of rare earth ions are allowed in glasses as in crystal with low-point symmetry, but at the same time the structure of levels displays the feature of high-point symmetry [34].

In conclusion, on the basis of the apparent crystal field model the crystal field splittings of Er^{3+}-doped glasses, which have important applications in the area of photoelectronics, can be analyzed quite completely and systematically. Obviously, this is a simple method for analyzing energy level splitting of rare earth ions in disordered hosts. The values of the splittings are essential for predicting properties of spectra and laser of rare earth-doped glasses [35–38].

References

1. F. Gan, *Optical and Spectral Properties of Glass (in Chinese)* (Shanghai Scientific & Technical Publishers, Shanghai, 1992)
2. F. Gan, *Photonic Glasses and Their Applications (in Chinese)* (Shanghai Scientific & Technical Publishers, Shanghai, 2011)
3. M.J. Weber, Laser excited fluorescence spectroscopy in glass, in *Laser Spectroscopy of Solids*, ed. by W.M. Yen, P.M. Selzer (Springer, New York, 1981), pp. P189–P239
4. H. Scholze, *Glass: Nature, Structure, and Properties* (Springer, New York, 1991)
5. C. Brecher, L.A. Riseberg, Phys. Rev. B **13**, 81 (1976)
6. G. Lei, J.E. Anderson, M.I. Buchwald et al., Phys. Rev. B **57**, 7673 (1998)
7. W.H. Zachariasen, J. Am. Chem. Soc. **54**, 3841 (1932)
8. C.N. Greaves, J. Non-Cryst. Solids **71**, 203 (1985)
9. C. Huang, A.N. Cormack, J. Chem. Phys. **93**, 8180 (1990)
10. C. Huang, A.N. Cormack, J. Chem. Phys. **95**, 3634 (1990)
11. J. Wang, W.S. Brecklesby, J.R. Lincoln et al., J. Non-Cryst. Solids **163**, 261 (1993)
12. T.F. Belliveau, D.J. Simkin, J. Non-Cryst. Solids **110**, 127 (1989)
13. M.J. Lochhead, K.L. Bray, Phys. Rev. B **52**, 15763 (1995)
14. G. Pucker, K. Gatterer, H.P. Fritzer et al., Phys. Rev. B **53**, 6225 (1996)
15. C. Brecher, L.A. Riseberg, Phys. Rev. B **21**, 2607 (1980)
16. C.C. Robinson, J.T. Fournier, J. Phys. Chem. Solids **31**, 895 (1970)
17. J.T. Fournier, R.H. Bartram, J. Phys. Chem. Solids **31**, 2615 (1970)
18. P.H. Butler, *Point Group Symmetry Application: Method and Tables* (Plenum, New York, 1981)
19. C.A. Morrison, R.P. Leavitt, Spectroscopic properties of triply ionized lanthanides in transparent host crystals, in *Handbook on the Physics and Chemistry of Rare Earths*, ed. by K. A. Gschneidner, Jr., L. Eyring (North-Holland Publishing Company, Amsterdam, 1982), pp. 461–692
20. A.A. Kaminskii, *Laser Crystals-Their Physics and Properties* (Springer, London, 1990)
21. Y.D. Huang, M. Mortier, F. Auzel, Opt. Mater. **15**, 243 (2001)
22. Y.D. Huang, M. Mortier, F. Auzel, Opt. Mater. **17**, 501 (2001)
23. F. Auzel, Mater. Res. Bull. **14**, 223 (1979)
24. F. Auzel, O.L. Malta, J. Phys. (Paris) **44**, 201 (1983)
25. F. Auzel, J. Chavignon, D. Meichenin et al., J. Non-Cryst. Solids **161**, 109 (1993)
26. R.P. Leavitt, J. Chem. Phys. **77**, 1661 (1982)
27. F. Auzel, Materials for ionic solid state lasers, in *Spectroscopy of Solid-State Laser-Type Materials*, ed. by B. Di Bartolo (Plenum Press, London, 1987), pp. P293–P341
28. Y. Huang, Z. Luo, Y. Chen et al., J. Chin. Rare Earth Soc. (in Chinese) **19**, 614 (2001)
29. F. Auzel, Opt. Mater. **19**, 89 (2002)
30. F. Auzel, Mol. Phys. **102**, 1113 (2004)

31. Z. D. Luo, Y. D. Huang, J. Phys.: Condens. Matter **6**, 3737 (1994)
32. M.J. Weber, Phys. Rev. **157**, 262 (1967)
33. C.W. Neilson, G.F. Koster, *Spectroscopic Coefficients for p^n, d^n and f^n Configurations* (MIT, Massachusetts, 1974)
34. L. Bigot, A.M. Jurdyc, B. Jacquier et al., Phys. Rev. B **66**, 214204 (2002)
35. M. Gorjan, M. Marincek, M. Copic, J. Opt. Soc. Am. B **27**, 2784 (2010)
36. A. Malouf, O. Henderson-Sapir, M. Gorjan et al., IEEE J. Quantum Electron. **52**, 1600412 (2016)
37. F. Maes, V. Fortin, M. Bernier et al., IEEE J. Quantum Electron. **53**, 1600208 (2017)
38. J. Li, H. Luo, Y. Liu et al., IEEE J. Sel. Topics Quantum Electron. **20**, 0900414 (2014)

Appendix A
Character Tables for Point-Symmetry Group

In this appendix the tables of irreducible representation character are given by using three different representation labels adopted by Milliken, Bethe, and Butler for the convenient of comparison. The spatial basis functions are only given for the usage in this book.

$C_1=(1)$

$C_1 \times C_i = C_i = S_2 = (\bar{1})$

C_1	E	$\bar{E}$	Bases
A Γ_1^+	1	1	R_x, R_y, R_z
Γ_1^-	1	1	x, y, z
Γ_2^+	1	−1	
Γ_2^-	1	−1	

$C_2=(2)$和 $C_s=C_{1h}=(m)$

$C_2 \times C_i = C_{2h} = (2/m) = \left(\frac{2}{m}\right)$

		C_s	E	σ_z	$\bar{E}$	$\bar{\sigma}_z$	C_2	C_s
		C_2	E	C_{2z}	$\bar{E}$	$\bar{C}_{2z}$	Bases	Bases
A'; A	Γ_1	0	1	1	1	1	z, R_z	$x, y; R_z,$
	Γ_3	$\frac{1}{2}$	1	$-i$	−1	i		
	Γ_4	$-\frac{1}{2}$	1	i	−1	$-i$		
A''; B	Γ_2	1	1	−1	1	−1	$x, y; R_x, R_y$	$z; R_x, R_y$

Note The first and the second column on the left are the Milliken symbols correspond to the point groups of the first and the second row respectively, the following tables where the Milliken symbols of two groups being different are also so arranged

Z. Luo and Y. Huang, *Physics of Solid-State Laser Materials*, Springer Series in Materials Science 289, https://doi.org/10.1007/978-981-32-9668-8

C_3=(3)

$C_3 \times C_i = C_{3i} = S_6 = (\bar{3})$

		C_3	E	C_{3z}	C_{3z}^{-1}	$\bar{E}$	$\bar{C}_{3z}$	$\bar{C}_{3z}^{-1}$	Bases
A	Γ_1	0	1	1	1	1	1	1	$z;R_z$
	Γ_4	$\frac{1}{2}$	1	ω	ω^5	−1	ω^4	ω^2	
	Γ_5	$-\frac{1}{2}$	1	ω^5	ω	−1	ω^2	ω^4	
E	Γ_2	1	1	ω^2	ω^4	1	ω^2	ω^4	$x,y;R_x,R_y$
	Γ_3	−1	1	ω^4	ω^2	1	ω^4	ω^2	
	Γ_6	$\frac{3}{2}$	1	−1	−1	−1	1	1	

Note $\omega = e^{-i\pi/3}$, $\omega^3 = -1$, here Γ_2 and Γ_3 are defined as those of Butler, which are different from Koster et al defined in "Properties of the thirty-two point groups", there are similar differences in the following tables

C_4=(4)和 $S_4 = (\bar{4})$

$C_4 \times C_i = C_{4h} = (4/m) = \left(\frac{4}{m}\right)$

		S_4	E	S_4^{-1}	C_{2z}	S_4	$\bar{E}$	$\bar{S}_4^{-1}$	$\bar{C}_{2z}$	$\bar{S}_4$	C_4	S_4
		C_4	E	C_{4z}	C_{2z}	C_{4z}^{-1}	$\bar{E}$	$\bar{C}_{4z}$	$\bar{C}_{2z}$	$\bar{C}_{4z}^{-1}$	Bases	Bases
A	Γ_1	0	1	1	1	1	1	1	1	1	$z;\ R_z$	R_z
	Γ_5	$\frac{1}{2}$	1	ω	$-i$	ω^7	−1	ω^5	i	ω^3		
	Γ_6	$-\frac{1}{2}$	1	ω^7	i	ω	−1	ω^3	$-i$	ω^5		
E	Γ_3	1	1	$-i$	−1	i	1	$-i$	−1	i	$x,y;R_x,R_y$	$x,y;R_x,R_y$
	Γ_4	−1	1	i	−1	$-i$	1	i	−1	$-i$		
	Γ_8	$\frac{3}{2}$	1	ω^3	i	ω^5	−1	ω^7	$-i$	ω		
	Γ_7	$-\frac{3}{2}$	1	ω^5	$-i$	ω^3	−1	ω	i	ω^7		
B	Γ_2	2	1	−1	1	−1	1	−1	1	−1	xy	$z;\ xy$

$\omega = e^{-i\pi/4}$, $\omega^2 = -i$

C_6=(6)和 $C_{3h} = (\bar{6})$

$C_6 \times C_i = C_{6h} = (6/m) = \left(\frac{6}{m}\right)$

		C_{3h}	E	S_{3z}^{-1}	C_{3z}	σ_z	C_{3z}^{-1}	S_{3z}	$\bar{E}$	$\bar{S}_{3z}^{-1}$	$\bar{C}_{3z}$	$\bar{\sigma}_z$	$\bar{C}_{3z}^{-1}$	$\bar{S}_{3z}$	C_{3h}	C_6
		C_6	E	C_{6z}	C_{3z}	C_{2z}	C_{3z}^{-1}	C_{6z}^{-1}	$\bar{E}$	$\bar{C}_{6z}$	$\bar{C}_{3z}$	$\bar{C}_{2z}$	$\bar{C}_{3z}^{-1}$	$\bar{C}_{6z}^{-1}$	Bases	Bases
A'; A	Γ_1	0	1	1	1	1	1	1	1	1	1	1	1	1	r	$r;\ z;\ R_z$
	Γ_7	$\frac{1}{2}$	1	ω	ω^2	$-i$	ω^{10}	ω^{11}	−1	ω^7	ω^8	i	ω^4	ω^5		
	Γ_8	$-\frac{1}{2}$	1	ω^{11}	ω^{10}	i	ω^2	ω	−1	ω^5	ω^4	$-i$	ω^8	ω^7		
E'; E_1	Γ_5	1	1	ω^2	ω^4	−1	ω^8	ω^{10}	1	ω^2	ω^4	−1	ω^8	ω^{10}	$x,y,$	$x,y;$ R_x,R_y
	Γ_6	−1	1	ω^{10}	ω^8	−1	ω^4	ω^2	1	ω^{10}	ω^8	−1	ω^4	ω^2		
	Γ_{12}	$\frac{3}{2}$	1	$-i$	−1	i	−1	i	−1	i	1	$-i$	1	$-i$		
	Γ_{11}	$-\frac{3}{2}$	1	i	−1	$-i$	−1	$-i$	−1	$-i$	1	i	1	i		
E''; E_2	Γ_3	2	1	ω^4	ω^8	1	ω^4	ω^8	1	ω^4	ω^8	1	ω^4	ω^8	R_x,R_y	
	Γ_2	−2	1	ω^8	ω^4	1	ω^8	ω^4	1	ω^8	ω^4	1	ω^8	ω^4		
	Γ_{10}	$\frac{5}{2}$	1	ω^5	ω^{10}	$-i$	ω^2	ω^7	−1	ω^{11}	ω^4	i	ω^8	ω		
	Γ_9	$-\frac{5}{2}$	1	ω^7	ω^2	i	ω^{10}	ω^5	−1	ω	ω^8	$-i$	ω^4	ω^{11}		
A''; B	Γ_4	3	1	−1	1	−1	1	−1	1	−1	1	−1	1	−1	z	

$\omega = e^{-i\pi/6}$, $\omega^3 = -i$; $C_{3z} = C_{6z}^2$, $C_{2z} = C_{6z}^3$

$D_2=V=(222)$ and $C_{2v}=(mm2)$

$$D_2 \times C_i = D_{2h} = V_h(mmm) = \left(\frac{2}{m}\frac{2}{m}\frac{2}{m}\right)$$

		C_{2v}	E	$C_2, \overline{C}_2$	$\sigma, \overline{\sigma}$	$\sigma', \overline{\sigma}'$	$\overline{E}$	D_2	C_{2v}
		D_2	E	$C_2, \overline{C}_2$	$C_2', \overline{C}_2'$	$C_2'', \overline{C}_2''$	$\overline{E}$	Bases	Bases
A_1; A_1	Γ_1	0	1	1	1	1	1	r ; xyz	z
	Γ_5	$\frac{1}{2}$	2	0	0	0	−2		
A_2; B_1	Γ_3	$\tilde{0}$	1	1	−1	−1	1	z; R_z	xy; R_z
B_1; B_2	Γ_2	1	1	−1	1	−1	1	y; R_y	x; R_y
B_2; B_3	Γ_4	$\tilde{1}$	1	−1	−1	1	1	x; R_x	y; R_x

$D_3=(32)$和 $C_{3v}=(3m)$

$$D_3 \times C_i = D_{3i} = D_{3d} = (\overline{3}m) = \left(\overline{3}\frac{2}{m}\right)$$

		C_{3v}	E	$2C_3$	3σ	$\overline{E}$	$2\overline{C}_3$	$3\overline{\sigma}$	D_3	C_{3v}
		D_3	E	$2C_3$	$3C_2$	$\overline{E}$	$2\overline{C}_3$	$3\overline{C}_2$	Bases	Bases
A_1	Γ_1	0	1	1	1	1	1	1	r	r ; z
E'	Γ_4	$\frac{1}{2}$	2	1	0	−2	−1	0		
A_2	Γ_2	$\tilde{0}$	1	1	−1	1	1	−1	z; R_z	R_z
E	Γ_3	1	2	−1	0	2	-1	0	$x,y;R_x,R_y$	$x,y;R_x,R_y$
E'' {	Γ_5	$\frac{3}{2}$	1	−1	$-i$	−1	1	i		
	Γ_6	$-\frac{3}{2}$	1	−1	i	−1	1	$-i$		

$D_4=(422)$, $C_{4v}=(4mm)$ and $D_{2d} = V_d = (\overline{3}2m)$

$$D_4 \times C_i = D_{4i} = D_{4h} = (4/mmm) = \left(\frac{4}{m}\frac{2}{m}\frac{2}{m}\right)$$

		D_{2d}	E	$2S_4$	$C_2, \overline{C}_2$	$2C_2', 2\overline{C}_2'$	$2\sigma_d, 2\overline{\sigma}_d$	$\overline{E}$	$2\overline{S}_4$			
		C_{4v}	E	$2C_4$	$C_2, \overline{C}_2$	$2\sigma_v, 2\overline{\sigma}_v$	$2\sigma_d, 2\overline{\sigma}_d$	$\overline{E}$	$2\overline{C}_4$	D_4	C_{4v}	D_{2d}
		D_4	E	$2C_4$	$C_2, \overline{C}_2$	$2C_2', 2\overline{C}_2'$	$2C_2'', 2\overline{C}_2''$	$\overline{E}$	$2\overline{C}_4$	Bases	Bases	Bases
A_1	Γ_1	0	1	1	1	1	1	1	1	r	r; z	r
E'	Γ_6	$\frac{1}{2}$	2	$\sqrt{2}$	0	0	0	−2	$-\sqrt{2}$			
A_2	Γ_2	$\tilde{0}$	1	1	1	−1	-1	1	1	z; R_z	R_z	R_z
E	Γ_5	1	2	0	−2	0	0	2	0	$R_x,R_y;x,y$	$R_x,R_y;x,y$	$R_x,R_y;x,y$
E''	Γ_7	$\frac{3}{2}$	2	$-\sqrt{2}$	0	0	0	−2	$\sqrt{2}$			
B_1	Γ_3	2	1	−1	1	1	−1	1	−1	(x^2-y^2)	(x^2-y^2)	(x^2-y^2)
B_2	Γ_4	$\tilde{2}$	1	−1	1	−1	1	1	−1	xy	xy	xy ; z

D_6=(622), C_{6v}=(6*mm*) and $D_{3h} = (\bar{6}m2)$

$$D_6 \times C_i = D_{6i} = D_{6h} = (6/mmm) = \left(\frac{6}{m}\frac{2}{m}\frac{2}{m}\right)$$

		D_{3h}	E	$2S_3$	$2C_3$	$\sigma_h, \bar{\sigma}_h$	$3C_2', 3\bar{C}_2'$	$3\sigma_v, 3\bar{\sigma}_v$	$\bar{E}$	$2\bar{S}_3$	$2\bar{C}_3$			
		C_{6v}	E	$2C_6$	$2C_3$	$C_2, \bar{C}_2$	$3\sigma_d, 3\bar{\sigma}_d$	$3\sigma_v, 3\bar{\sigma}_v$	$\bar{E}$	$2\bar{C}_6$	$2\bar{C}_3$	D_6	C_{6v}	D_{3h}
		D_6	E	$2C_6$	$2C_3$	$C_2, \bar{C}_2$	$3C_2', 3\bar{C}_2'$	$3C_2'', 3\bar{C}_2''$	$\bar{E}$	$2\bar{C}_6$	$2\bar{C}_3$	Bases	Bases	Bases
A_1	Γ_1	0	1	1	1	1	1	1	1	1	1	r	r or z	r
E'	Γ_7	$\frac{1}{2}$	2	$\sqrt{3}$	1	0	0	0	−2	$-\sqrt{3}$	−1			
A_2	Γ_2	$\tilde{0}$	1	1	1	1	−1	−1	1	1	1	z or R_z	R_z	R_z
E_1	Γ_5	1	2	1	−1	−2	0	0	2	1	−1	R_x, R_y x,y	R_x, R_y x,y	R_x, R_y
E'''	Γ_9	$\frac{3}{2}$	2	0	−2	0	0	0	−2	0	2			
E_2	Γ_6	2	2	−1	−1	2	0	0	2	−1	−1	x^2-y^2, xy	x^2-y^2, xy	x,y
E''	Γ_8	$\frac{5}{2}$	2	$-\sqrt{3}$	1	0	0	0	−2	$\sqrt{3}$	−1			
B_1	Γ_3	3	1	−1	1	−1	1	−1	1	−1	1	y^3-3yx^2	x^3-3xy^2	zR_z
B_2	Γ_4	$\tilde{3}$	1	−1	1	−1	−1	1	1	−1	1	x^3-3xy^2	y^3-3yx^2	z

Note The Milliken symbols of point groups C_{6v} and D_6 are as shown above; the Milliken symbol of point group D_{3h} are: $\Gamma_1 \rightarrow A_1'$, $\Gamma_2 \rightarrow A_2'$, $\Gamma_3 \rightarrow A_1''$, $\Gamma_4 \rightarrow A_2''$, $\Gamma_5 \rightarrow E''$, $\Gamma_6 \rightarrow E'$

T=(23)

$$T \times C_i = T_i = T_h = (m3) = \left(\frac{2}{m}\bar{3}\right)$$

	T	E	$4C_3$	$4C_3^{-1}$	$3C_2, 3\bar{C}_2$	$\bar{E}$	$4\bar{C}_3$	$4\bar{C}_3^{-1}$	Bases
A_1 Γ_1	0	1	1	1	1	1	1	1	r or xyz
E' Γ_5	$\frac{1}{2}$	2	1	1	0	−2	−1	−1	
T Γ_4	1	3	0	0	−1	3	0	0	R_x, R_y, R_z, x,y,z
E'' Γ_6	$\frac{3}{2}$	2	ω^2	ω^4	0	−2	ω^5	ω	
E''' Γ_7	$-\frac{3}{2}$	2	ω^4	ω^2	0	−2	ω	ω^5	
E { Γ_2	2	1	ω^2	ω^4	1	1	ω^2	ω^4	$1/\sqrt{2}(u+v)$
E { Γ_3	−2	1	ω^4	ω^2	1	1	ω^4	ω^2	$1/\sqrt{2}(u-v)$

$\omega = e^{-i\pi/3}$, $\omega^3 = -1$; $u = 3z^2 - r^2$, $v = \sqrt{3}(x^2 - y^2)$

O=(432) and $T_d = (\bar{4}3m)$

$O \times C_i = O_i = O_h = (m3m)$

		T_d	E	$8C_3$	$6S_4$	$3C_2, 3\bar{C}_2$	$6\sigma_d, 6\bar{\sigma}_d$	$\bar{E}$	$8\bar{C}_3$	$6\bar{S}_4$	O	T_d
		O	E	$8C_3$	$6C_4$	$3C_2, 3\bar{C}_2$	$6C_2', 6\bar{C}_2'$	$\bar{E}$	$8\bar{C}_3$	$6\bar{C}_4$	Bases	Bases
A_1	Γ_1	0	1	1	1	1	1	1	1	1	r	r or xyz
E'	Γ_6	$\frac{1}{2}$	2	1	$\sqrt{2}$	0	0	−2	−1	$-\sqrt{2}$		
T_1	Γ_4	1	3	0	1	−1	−1	3	0	1	R_x, R_y, R_z, x, y, z	R_x, R_y, R_z
U'	Γ_8	$\frac{3}{2}$	4	−1	0	0	0	−4	1	0		
E	Γ_3	2	2	−1	0	2	0	2	−1	0	$(2z^2-x^2-y^2)$, $\sqrt{3}(x^2-y^2)$	$(2z^2-x^2-y^2)$, $\sqrt{3}(x^2-y^2)$
T_2	Γ_5	$\tilde{1}$	3	0	−1	−1	1	3	0	−1	yz, xz, xy	x, y, z
E''	Γ_7	$\frac{\tilde{1}}{2}$	2	1	$-\sqrt{2}$	0	0	−2	−1	$\sqrt{2}$		
A_2	Γ_2	$\tilde{0}$	1	1	−1	1	−1	1	1	−1	xyz	$R_xR_yR_z$

Appendix B
Correlation Table of Group–Subgroup

In the following tables, the orbital angular momentum quantum number will be used to label the irreducible representation of SO_3 group, but the subscripts of Bethe symbols are used to label those of point groups. When more than one irreducible representation of subgroup corresponds to one irreducible representation, right subscripts are used to distinguish. Sign of + and − in front of the symbols of irreducible representation of subgroup means that corresponding $2jm$ factor equal to +1 and −1, respectively. For the point group having central symmetry the symbol with superscript + represents even irreducible representation and the symbol with superscript − represents odd irreducible representation.

Z. Luo and Y. Huang, *Physics of Solid-State Laser Materials*, Springer Series in Materials Science 289, https://doi.org/10.1007/978-981-32-9668-8

$SO_3 \to O$
$(O_3 \to O_h)$

$0 \to +1$
$1/2 \to +6$
$1 \to +4$
$3/2 \to +8$
$2 \to +3+5$
$5/2 \to +7+8$
$3 \to +2+4+5$
$7/2 \to +6+7+8$
$4 \to +1+3+4+5$
$9/2 \to +6+8_0+8_1$
$5 \to +3+4_0+4_1+5$
$11/2 \to +6+7+8_0+8_1$
$6 \to +1+2+4+3+5_0+5_1$
$13/2 \to +6+7_0+7_1+8_0+8_1$
$7 \to +2+3+4_0+4_1+5_0+5_1$
$15/2 \to +6+7+8_0+8_1+8_2$
$8 \to +1+3+3+4+4+5_0+5_1$

$O \to T$
$T_d \to T$
$(O_h \to T_h)$

$1 \to +1$
$2 \to -1$
$3 \to +2+3$
$4 \to +4$
$5 \to -4$
$6 \to +5$
$7 \to -5$
$8 \to +6+7$

$O_h \to T_d$

$1^+ \to +1$
$1^- \to +2$
$2^+ \to +2$
$2^- \to +1$
$3^+ \to +3$
$3^- \to +3$
$4^+ \to +4$
$4^- \to +5$
$5^+ \to +5$
$5^- \to +4$
$6^+ \to +6$
$6^- \to +7$
$7^+ \to +7$
$7^- \to +6$
$8^+ \to +8$
$8^- \to +8$

$O \to D_4$
$T_d \to D_{2d}$
$(O_h \to D_{4h})$

$1 \to +1$
$2 \to +3$
$3 \to +1+3$
$4 \to +2+5$
$5 \to +5+4$
$6 \to +6$
$7 \to +7$
$8 \to +6+7$

$O \to D_3$
$T_d \to C_{3v}$
$(O_h \to D_{3d})$

$1 \to +1$
$2 \to +2$
$3 \to +3$
$4 \to +2+3$
$5 \to +1+3$
$6 \to +4$
$7 \to +4$
$8 \to +4+5+6$

$T \to D_2$
$(T_h \to D_{2h})$

$1 \to +1$
$2 \to +1$
$3 \to +1$
$4 \to +2+3+4$
$5 \to +5$
$6 \to +5$
$7 \to +5$

$T \to C_3$
$(T_h \to C_{3i})$

$1 \to +1$
$2 \to +2$
$3 \to +3$
$4 \to -1+2+3$
$5 \to +4-5$
$6 \to +4+6$
$7 \to -5-6$

$D_{4h} \to D_{2d}$

$1^+ \to +1$
$1^- \to +3$
$2^+ \to +2$
$2^- \to +4$
$3^+ \to +3$
$3^- \to +1$
$4^+ \to +4$
$4^- \to +2$
$6^+ \to +6$
$6^- \to +7$
$5^+ \to +5$
$5^- \to +5$
$7^+ \to +7$
$7^- \to +6$

$D_{4h} \to C_{4v}$

$1^+ \to +1$
$1^- \to +2$
$2^+ \to +2$
$2^- \to +1$
$3^+ \to +3$
$3^- \to +4$
$4^+ \to +4$
$4^- \to +3$
$5^+ \to +5$
$5^- \to +5$
$6^+ \to +6$
$6^- \to +6$
$7^+ \to +7$
$7^- \to +7$

$D_{2h} \to C_{2v}$

$1^+ \to +1$
$1^- \to +3$
$2^+ \to +2$
$2^- \to +4$
$3^+ \to +3$
$3^- \to +1$
$4^+ \to +4$
$4^- \to +1$
$5^+ \to +5$
$5^- \to +5$

$D_{3d} \to C_{3v}$

$1^+ \to +1$
$1^- \to +2$
$2^+ \to +2$
$2^- \to +1$
$3^+ \to +3$
$3^- \to +3$
$4^+ \to +4$
$4^- \to +4$
$5^+ \to +5$
$5^- \to +6$
$6^+ \to +6$
$6^- \to +5$

$D_4 \to D_2$
$C_{4v} \to C_{2v}$
$D_{2d} \to D_2$
$D_{2d} \to C_{2v}^*$
$(D_{4h} \to D_{2h})$

$1 \to +1$
$2 \to +3$
$3 \to +1$
$4 \to +3$
$5 \to +2+4$
$6 \to +5$
$7 \to +5$

* for **$D_{2d} \to C_{2v}$**, the representation 3 and 4 of **D_{2d}** should be exchanged

$D_4 \to C_4$
$C_{4v} \to C_4$
$D_{2d} \to S_4$
$(D_{4h} \to D_{2h})$

$1 \to +1$
$2 \to -1$
$3 \to +2$
$4 \to -2$

5→+3–4
6→+5–6
7→+8–7

$D_3 \to C_3$
$C_{3v} \to C_3$
($D_{3d} \to C_{3i}$)

1→+1
2→–1
3→+2+3
4→+4–5
5→+6
6→–6

$D_3 \to C_2$
$C_{3v} \to C_s$
($D_{3d} \to C_{2h}$)

1→+1
2→+2
3→–1–2
4→+3–4
5→+3
6→–4

$D_2 \to C_2$
$C_{2v} \to C_2$
($D_{2h} \to C_{2h}$)

1→+1
2→+2
3→–1
4→–2
5→+3–4

$C_4 \to C_2$
$S_4 \to C_2$
($C_{4h} \to C_{2h}$)

1→+1
2→+1
3→+2
4→+2
5→+3
6→+4
7→+3
8→+4

$C_{4h} \to S_4$

1^+→+1
1^-→+2
2^+→+2
2^-→+1
3^+→+3
3^-→+4
4^+→+4
4^-→+3
5^+→+5
5^-→+7
6^+→+6
6^-→+8
7^+→+7
7^-→+5
8^+→+8
8^-→+6

$C_3 \to C_1$
($C_{3i} \to C_i$)

1→+1
2→+1
3→+1
4→+2
5→+2
6→+2

$C_{2h} \to C_s$

1^+→+1
1^-→+2
2^+→+2
2^-→+1
3^+→+3
3^-→+4
4^+→+4
4^-→+3

$C_2 \to C_1$
$C_s \to C_1$
($C_{2h} \to C_1$)

1→+1
2→+1
3→+2
4→+2

Appendix C
Multiplication Tables for Some Point Groups

In the following tables, the subscripts of Bethe's symbol are used to represent irreducible representation, and the direct summations of irreducible representation are denoted by the symbol ⊕.

Point group C_2 and C_s

1	2	3	4	Subscript i of Γ_i
1	2	3	4	1
	1	4	3	2
		2	1	3
			2	4

Point group C_3

1	2	3	4	5	6	Subscript i of Γ_i
1	2	3	4	5	6	1
	3	1	6	4	5	2
		2	5	6	4	3
			2	1	3	4
				3	2	5
					1	6

Point group C_4, S_4

1	2	3	4	5	6	7	8	Subscript i of Γ_i
1	2	3	4	5	6	7	8	1
	1	4	3	7	8	5	6	2
		2	1	8	5	6	7	3
			2	6	7	8	5	4
				3	1	4	2	5
					4	2	3	6
						3	1	7
							4	8

Point group D_2, C_{2v}

1	2	3	4	5	Subscript i of Γ_i
1	2	3	4	5	1
	1	4	3	5	2
		1	2	5	3
			1	5	4
				1⊕2⊕3⊕4	5

Z. Luo and Y. Huang, *Physics of Solid-State Laser Materials*, Springer Series in Materials Science 289, https://doi.org/10.1007/978-981-32-9668-8

Point group D_3, C_{3v}

1	2	3	4	5	6	Subscript i of Γ_i
1	2	3	4	5	6	1
	1	3	4	6	5	2
		$1\oplus2\oplus3$	$4\oplus5\oplus6$	4	4	3
			$1\oplus2\oplus3$	3	3	4
				2	1	5
					2	6

Point group D_4, C_{4v}, D_{2d}

1	2	3	4	5	6	7	Subscript i of Γ_i
1	2	3	4	5	6	7	1
	1	4	3	5	6	7	2
		1	2	5	7	6	3
			1	5	7	6	4
				$1\oplus2\oplus3\oplus4$	$6\oplus7$	$6\oplus7$	5
					$1\oplus2\oplus5$	$3\oplus4\oplus5$	6
						$1\oplus2\oplus5$	7

Point group T

1	2	3	4	5	6	7	Subscript i of Γ_i
1	2	3	4	5	6	7	1
	3	1	4	6	7	5	2
		2	4	7	5	6	3
			$1\oplus2\oplus3\oplus3\oplus4$	$5\oplus6\oplus7$	$5\oplus6\oplus7$	$5\oplus6\oplus7$	4
				$1\oplus4$	$2\oplus4$	$3\oplus4$	5
					$3\oplus4$	$1\oplus4$	6
						$2\oplus4$	7

Point group O T_d

1	2	3	4	5	6	7	8	Subscript i of Γ_i
1	2	3	4	5	6	7	8	1
	1	3	5	4	7	6	8	2
		$1\oplus2\oplus3$	$4\oplus5$	$4\oplus5$	8	8	$6\oplus7\oplus8$	3
			$1\oplus3\oplus4\oplus5$	$2\oplus3\oplus4\oplus5$	$6\oplus8$	$7\oplus8$	$6\oplus7\oplus8\oplus8$	4
				$1\oplus3\oplus4\oplus5$	$7\oplus8$	$6\oplus8$	$6\oplus7\oplus8\oplus8$	5
					$1\oplus4$	$2\oplus5$	$3\oplus4\oplus5$	6
						$1\oplus4$	$3\oplus4\oplus5$	7
							$1\oplus2\oplus3\oplus4\oplus 4\oplus5\oplus5$	8

Appendix D
Squared Reduced-Matrix Elements of Unit Operator for $J \rightarrow J'$ Transition in Rare Earth Ions

Squared reduced-matrix elements of unit operator for $J \rightarrow J'$ transitions in rare earth ions expressed as $|\langle 4f^{n}[\alpha SL]J\|U^{(k)}\|4f^{n}[\alpha' S' L']J'\rangle|^{2}$ commonly used in laser materials are given in this appendix, which the intermediate coupling has been taken into account and the following relation is assumed $|\langle 4f^{n}[\alpha SL]J\|U^{(k)}\|4f^{n}[\alpha' S' L']J'\rangle|^{2} = |\langle 4f^{n}\ [\alpha' S' L']J'\|U^{(k)}\|4f^{n}[\alpha SL]J\rangle|^{2}$. These data are cited from the book published by A.A. Kaminskii: "*Laser Crystals, their Physics and Properties*. Heidelberg: Springer-Verlag, 1981" and "*Crystalline Lasers: Physical Processes and Operating Schemes*. Boca Raton: CRC Press, 1996".

Pr^{3+}:

$^{2S+1}L_J$	$^{2S'+1}L'_{J'}$	$E_{JJ'}$ (cm^{-1})	$k = 2$	$k = 4$	$k = 6$
$^3H_5$	$^3H_4$	2050	0.1096	0.2035	0.6106
	$^3H_5$	2200	0.1080	0.2328	0.6421
$^3H_6$	$^3H_4$	4250	0.0002	0.0322	0.1407
$^3F_2$	$^3H_6$	700	0	0.0167	0.3040
	$^3H_5$	2900	0	0.2978	0.6597
	$^3H_4$	4950	0.5079	0.4048	0.1196
$^3F_3$	$^3F_2$	1380	0.0212	0.0508	0
	$^3H_6$	2100	0	0.3182	0.9460
	$^3H_5$	4300	0.6285	0.3468	0
	$^3H_4$	6350	0.0658	0.3487	0.7002
$^3F_4$	$^3F_3$	450	0.0262	0.0735	0.013
	$^3F_2$	1850	0.0015	0.0027	0.0948
	$^3H_6$	2550	0.6330	0.6805	0.5165
	$^3H_5$	4750	0.0368	0.3371	0.5278
	$^3H_4$	6800	0.0162	0.0528	0.4901

(continued)

Z. Luo and Y. Huang, *Physics of Solid-State Laser Materials*, Springer Series in Materials Science 289, https://doi.org/10.1007/978-981-32-9668-8

(continued)

$^{2S+1}L_J$	$^{2S'+1}L'_{J'}$	$E_{JJ'}$ (cm^{-1})	$k = 2$	$k = 4$	$k = 6$
$^1G_4$	$^3F_4$	3050	0.0683	0.1203	0.2844
	$^3F_3$	3500	0.0026	0.0031	0.0452
	$^3F_2$	4900	0.0002	0.0138	0.0032
	$^3H_6$	5600	0.1927	0.1905	0.1865
	$^3H_5$	7800	0.0307	0.0715	0.3344
	$^3H_4$	9850	0.0019	0.0044	0.0119
$^1D_2$	$^1G_4$	6950	0.3865	0.0493	0.0844
	$^3F_4$	10,000	0.5144	0.0004	0.0147
	$^3F_3$	14,950	0	0.0009	0.0025
	$^3F_2$	16,300	0	0.0050	0.0301
	$^3H_6$	17,000	0.0036	0.0132	0.0046
	$^3H_5$	21,250	0.0081	0.0447	0.0203
	$^3H_4$	16,800	0.0200	0.0165	0.0493
$^3P_0$	$^1D_2$	3900	0.0134	0	0
	$^1G_4$	10,800	0	0.0425	0
	$^3F_4$	13,850	0	0.1213	0
	$^3F_3$	14,350	0	0	0
	$^3F_2$	15,700	0.2943	0	0
	$^3H_6$	16,400	0	0	0.0726
	$^3H_5$	18,600	0	0	0
	$^3H_4$	20,650	0	0.1713	0
$^3P_1$	$^3P_0$	600	0	0	0
	$^1D_2$	4500	0.0749	0	0
	$^1G_4$	11,400	0	0.0605	0
	$^3F_4$	14,450	0	0.2852	0
	$^3F_3$	14,950	0.5714	0.1964	0
	$^3F_2$	16,300	0.2698	0	0
	$^3H_6$	17,000	0	0	0.1246
	$^3H_5$	19,200	0	0.2857	0.0893
	$^3H_4$	21,250	0	0.1721	0
$^1I_6$	$^3H_4$	21,250	0.0081	0.0447	0.0203
$^3P_2$	$^1I_6$	1250	0	0.0257	0.1405
	$^3P_1$	1250	0.4232	0	0
	$^3P_0$	1850	0.1929	0	0
	$^1D_2$	5700	0.0011	0.0718	0
	$^1G_4$	12,650	0.5640	0.0341	0.0184
	$^3F_4$	15,700	0.5233	0.1170	0.0072
	$^3F_3$	16,150	0.2584	0.3082	0
	$^3F_2$	17,550	0.0323	0.3001	0
	$^3H_6$	18,250	0	0.5010	0.0544
	$^3H_5$	20,450	0	0.1888	0.1316
	$^3H_4$	22,500	6×10^{-5}	0.0362	0.1373
$^1S_0$	$^3H_4$	46,750	0	0.0064	0

Nd^{3+}:

$^{2S+1}L_J$	$^{2S'+1}L'_{J'}$	$E_{JJ'}$ (cm^{-1})	$k = 2$	$k = 4$	$k = 6$
$^4I_{11/2}$	$^4I_{9/2}$	1850	0.0195	0.1073	1.1653
$^4I_{13/2}$	$^4I_{11/2}$	2000	0.0256	0.1353	1.2379
	$^4I_{9/2}$	3850	0.0001	0.0136	0.4558
$^4I_{15/2}$	$^4I_{13/2}$	2100	0.0196	0.1189	1.4511
	$^4I_{11/2}$	4100	0	0.0109	0.4184
	$^4I_{9/2}$	5950	0	0.0001	0.0453
$^4F_{3/2}$	$^4I_{15/2}$	5400	0	0	0.0288
	$^4I_{13/2}$	7500	0	0	0.2085
	$^4I_{11/2}$	9500	0	0.1136	0.4104
	$^4I_{9/2}$	11,350	0	0.2293	0.0548
$^4F_{5/2}$	$^4F_{3/2}$	1000	0.0795	0.0523	0
	$^4I_{15/2}$	6400	0	0	0.2309
	$^4I_{13/2}$	8500	0	0.1805	0.4025
	$^4I_{11/2}$	10,500	0	0.1681	0.0364
	$^4I_{9/2}$	12,350	0.0010	0.2371	0.3972
$^2H(2)_{9/2}$	$^4F_{5/2}$	100	0.0065	0.0315	0.0050
	$^4F_{3/2}$	1100	0	0.0144	0.0217
	$^4I_{15/2}$	6500	0	0.2120	0.0765
	$^4I_{13/2}$	8600	0.0387	0.0061	0.1155
	$^4I_{11/2}$	10,600	0.0028	0.0004	0.0246
	$^4I_{9/2}$	12,450	0.0092	0.0080	0.1155
$^4F_{7/2}$	$^2H(2)_{9/2}$	1050	0.0059	0.0342	0.0036
	$^4F_{5/2}$	1150	0.0674	0.0547	0.0902
	$^4F_{3/2}$	2150	0.0066	0.0792	0
	$^4I_{15/2}$	7550	0	0.1545	0.6209
	$^4I_{13/2}$	9650	0	0.3282	0.0001
	$^4I_{11/2}$	11,650	0.0006	0.2373	0.3094
	$^4I_{9/2}$	13,500	0.0010	0.0423	0.4246
$^4S_{3/2}$	$^4F_{7/2}$	–	0	0.0004	0
	$^2H(2)_{9/2}$	1050	0	0.0044	0.0002
	$^4F_{5/2}$	1150	0	0.0006	0
	$^4F_{3/2}$	2150	0.0004	0	0
	$^4I_{15/2}$	7550	0	0	0.3302
	$^4I_{13/2}$	9650	0	0	0.3307
	$^4I_{11/2}$	11,650	0	0	0.2033
	$^4I_{9/2}$	13,500	0	0.0027	0.2352

(continued)

(continued)

$^{2S+1}L_J$	$^{2S'+1}L'_{J'}$	$E_{JJ'}$ (cm^{-1})	$k = 2$	$k = 4$	$k = 6$
$^4F_{9/2}$	$^4S_{3/2}$	1150	0	0.0026	00013
	$^4F_{7/2}$	1150	0.1473	0.0076	0.1040
	$^2H(2)_{9/2}$	2200	0.0486	0.0026	0.0056
	$^4F_{5/2}$	2300	0.0109	0.0504	0.1082
	$^4F_{3/2}$	3300	0	0.0048	0.1137
	$^4I_{15/2}$	8700	0	0.0050	0.4668
	$^4I_{13/2}$	10,800	0.0035	0.2188	0.5165
	$^4I_{11/2}$	12,800	0.0001	0.0345	0.3678
	$^4I_{9/2}$	14,650	0.0009	0.0092	0.0417
$^2H(2)_{11/2}$	$^4F_{9/2}$	1150	0.0892	0.0248	0.0910
	$^4S_{3/2}$	2300	0	0.0568	0.0017
	$^4F_{7/2}$	2300	0.0950	0.0929	0.0760
	$^2H(2)_{9/2}$	3350	0.0676	0.0039	0.2229
	$^4F_{5/2}$	3450	0	0.0029	0.0236
	$^4F_{3/2}$	4450	0	0.0002	0.0081
	$^4I_{15/2}$	9850	0.1311	0.0695	0
	$^4I_{13/2}$	11,950	0.0045	0.0172	0.0028
	$^4I_{11/2}$	13,950	0.0043	0.0093	0.0063
	$^4I_{9/2}$	15,800	0.0001	0.0027	0.0104
$^4G_{5/2}$	$^2H(2)_{11/2}$	1400	0	0.0003	0.0146
	$^4F_{9/2}$	2550	0.0025	0.0064	0.1338
	$^4S_{3/2}$	3700	0.0009	0.1771	0
	$^4F_{7/2}$	3700	0.0339	0.0612	01267
	$^2H(2)_{9/2}$	4750	0.0010	0.0138	0.0017
	$^4F_{5/2}$	4850	0.2681	0.1292	0
	$^4F_{3/2}$	5850	0.4830	0.0444	0
	$^4I_{15/2}$	11,250	0	0	0.0968
	$^4I_{13/2}$	13,350	0	0.0349	0.0472
	$^4I_{11/2}$	15,350	0	0.2892	0.0104
	$^4I_{9/2}$	17,200	0.8979	0.4093	0.0359
$^2G(1)_{7/2}$	$^4G_{5/2}$	–	0.0002	0.1493	0.0874
	$^2H(2)_{11/2}$	1400	0.0066	0	0.3467
	$^4F_{9/2}$	2550	0.0264	0.0520	0.2705
	$^4S_{3/2}$	3700	0.0008	0.0834	0
	$^4F_{7/2}$	3700	0.0389	0.1018	0.1636
	$^2H(2)_{9/2}$	4750	0.0076	0.0302	0.1822
	$^4F_{5/2}$	4850	0.2683	0.0069	0.0825
	$^4F_{3/2}$	5850	0.0817	0.0449	0
	$^4I_{15/2}$	11,250	0	0.0017	0.1048
	$^4I_{13/2}$	13,350	0	0.1025	0.0383
	$^4I_{11/2}$	13,530	0.4416	0.1844	0.0484
	$^4I_{9/2}$	17,200	0.0757	0.1848	0.0314

(continued)

(continued)

$^{2S+1}L_J$	$^{2S'+1}L'_{J'}$	$E_{JJ'}$ (cm^{-1})	$k = 2$	$k = 4$	$k = 6$
$^2K_{13/2}$	$^2G(1)_{7/2}$	1550	0	0.6615	0.0013
	$^4G_{5/2}$	1550	0	0.0019	0.0081
	$^2H(2)_{11/2}$	2950	0.0014	0.0043	0.4823
	$^4F_{9/2}$	4100	0.0437	0.0042	0.2821
	$^4S_{3/2}$	5250	0	0	0.0007
	$^4F_{7/2}$	5250	0.1407	0.0653	0.0091
	$^2H(2)_{9/2}$	6300	0.1201	0.0061	1.1455
	$^4F_{5/2}$	6400	0	0.0028	0.0049
	$^4F_{3/2}$	7400	0	0	0.0073
	$^4I_{15/2}$	12,800	0	0.0010	0.0068
	$^4I_{13/2}$	14,900	0.0034	0.0002	0.0025
	$^4I_{11/2}$	16,900	0.0002	0	0.0171
	$^4I_{9/2}$	18,750	0.0069	0.0002	0.0312
$^4G_{7/2}$	$^2K_{13/2}$	250	0	0.5688	0.0014
	$^2G(1)_{7/2}$	1800	0.0575	0.0005	0.0377
	$^4G_{5/2}$	1800	0	0.2246	0.0053
	$^2H(2)_{11/2}$	3200	0.0016	0.0208	0.1166
	$^4F_{9/2}$	4350	0.0006	0.0089	0.0074
	$^4S_{3/2}$	5500	0.0021	0.1929	0
	$^4F_{7/2}$	5500	0.1641	0.0671	0.0031
	$^2H(2)_{9/2}$	6550	0.0518	0.0503	0.3898
	$^4F_{5/2}$	6650	0.2359	0.0015	0.1124
	$^4F_{3/2}$	7650	0.0983	0.0584	0
	$^4I_{15/2}$	13,050	0	0.0270	0.0027
	$^4I_{13/2}$	15,150	0	0.2296	0.0576
	$^4I_{11/2}$	17,150	0.5314	0.0959	0.0120
	$^4I_{9/2}$	19,000	0.0550	0.1571	0.0553
$^4G_{9/2}$	$^4G_{7/2}$	450	0.0401	0.1400	0.1343
	$^2K_{13/2}$	700	0.0195	0.0402	0.1080
	$^2G(1)_{7/2}$	2250	0.0195	0.2246	0.0053
	$^4G_{5/2}$	2250	0	0.10676	0.2531
	$^2H(2)_{11/2}$	3650	0.0025	0.0327	0.2089
	$^4F_{9/2}$	4800	0.1298	0.1604	0.0272
	$^4S_{3/2}$	5950	0	0.2010	0.0009
	$^4F_{7/2}$	5950	0.5747	0	0.0894
	$^2H(2)_{9/2}$	7000	0.0381	0.0485	0.0909
	$^4F_{5/2}$	7100	0.1987	0.1060	0.0016
	$^4F_{3/2}$	8100	0	0.0610	0.1492
	$^4I_{15/2}$	13,500	0	0.1580	0.2346
	$^4I_{13/2}$	15,600	0.9950	0.3890	0.0123
	$^4I_{11/2}$	17,600	0.1447	0.3577	0.0552
	$^4I_{9/2}$	19,450	0.0046	0.0609	0.0406

(continued)

(continued)

$^{2S+1}L_J$	$^{2S'+1}L'_{J'}$	$E_{JJ'}$ (cm^{-1})	$k = 2$	$k = 4$	$k = 6$
$^2K_{15/2}$	$^4G_{9/2}$	1450	0	0.1135	0.0228
	$^4G_{7/2}$	1900	0	0.0200	0.0025
	$^2K_{13/2}$	2150	0.0068	0.4290	0.0221
	$^2G(1)_{7/2}$	3700	0	0.0354	0.0128
	$^4G_{5/2}$	3700	0	0	0.0053
	$^2H(2)_{11/2}$	5100	0.1511	0.0014	1.5109
	$^4F_{9/2}$	6250	0	0.0003	0.2020
	$^4S_{3/2}$	7400	0	0	0.0028
	$^4F_{7/2}$	7400	0	0	0.0084
	$^2H(2)_{9/2}$	8450	0	0.4724	0.4238
	$^4F_{5/2}$	8550	0	0	0.0050
	$^4F_{3/2}$	9550	0	0	0.0097
	$^4I_{15/2}$	14,950	0.0112	0.0002	0.0243
	$^4I_{13/2}$	17,050	0.0004	0.0003	0.0186
	$^4I_{11/2}$	19,050	0.0020	0.0003	0.0034
	$^4I_{9/2}$	20,900	0	0.0052	0.0143
$^2G(1)_{9/2}$	$^2K_{15/2}$	150	0	0.9742	0.0182
	$^4G_{9/2}$	1600	0.0018	0.0351	0.1866
	$^4G_{7/2}$	2050	0.0147	0.0578	0.0231
	$^2K_{13/2}$	2300	0.0147	0.0888	0.0543
	$^2G(1)_{7/2}$	3850	0.0050	0.0631	0.0298
	$^4G_{5/2}$	3850	0.0001	0.0282	0.0308
	$^2H(2)_{11/2}$	5250	0.0562	0.0344	0.5917
	$^4F_{9/2}$	6400	0.1026	0.0088	0.3021
	$^4S_{3/2}$	7550	0	0.0800	0
	$^4F_{7/2}$	7550	0.1218	0.0127	0.0453
	$^2H(2)_{9/2}$	8600	0.0015	0.0009	0.1452
	$^4F_{5/2}$	8700	0.0416	0.0353	0.0057
	$^4F_{3/2}$	9700	0	0.0190	0.0265
	$^4I_{15/2}$	15,100	0	0.1424	0.0044
	$^4I_{13/2}$	17,200	0.2854	0.0275	0.0727
	$^4I_{11/2}$	19,200	0.0238	0.0434	0.0452
	$^4I_{9/2}$	21,050	0.0010	0.0148	0.0139

(continued)

(continued)

$^{2S+1}L_J$	$^{2S'+1}L'_{J'}$	$E_{JJ'}$ (cm^{-1})	$k = 2$	$k = 4$	$k = 6$
$^2D(1)_{3/2}$	$^2G(1)_{9/2}$	100	0	0.0089	0.0311
	$^2K_{15/2}$	250	0	0	0.3531
	$^4G_{9/2}$	1700	0	0.0029	0.0262
	$^4G_{7/2}$	2150	0.0216	0.0477	0
	$^2K_{13/2}$	2400	0	0	0.0788
	$^2G(1)_{7/2}$	3950	0.0087	0.0839	0
	$^4G_{5/2}$	3950	0.0003	0.0038	0
	$^2H(2)_{11/2}$	5350	0	0.1148	0.1804
	$^4F_{9/2}$	6500	0	0.0003	0.0156
	$^4S_{3/2}$	7650	0.0068	0	0
	$^4F_{7/2}$	7550	0.0018	0.0003	0
	$^2H(2)_{9/2}$	8750	0	0.0255	0.0960
	$^4F_{5/2}$	8800	0.0041	0.	0
	$^4F_{3/2}$	9800	0.0046	0.	0
	$^4I_{15/2}$	15,200	0	0	0.0081
	$^4I_{13/2}$	17,300	0	0.0016	0.0325
	$^4I_{11/2}$	19,300	0.0238	0.0434	0.0452
	$^4I_{9/2}$	21,150	0	0.0189	0.0002
$^4G_{11/2}$	$^2D(1)_{3/2}$	300	0	0.0312	0.0003
	$^2G(1)_{9/2}$	400	0.0997	0.0104	0.3721
	$^2K_{15/2}$	550	0.0094	0.0168	0.0919
	$^4G_{9/2}$	2000	0.0444	0.5128	0.1365
	$^4G_{7/2}$	2450	0.0028	0.0283	0.3827
	$^2K_{13/2}$	2700	0.0002	0.0084	0.0232
	$^2G(1)_{7/2}$	4250	0.0107	0.1266	0.0040
	$^4G_{5/2}$	4250	0	0.0091	0.0930
	$^2H(2)_{11/2}$	5650	0.0010	0.0843	0.0111
	$^4F_{9/2}$	6800	0.9477	0.2103	0.0398
	$^4S_{3/2}$	7950	0	0.3242	0.0005
	$^4F_{7/2}$	7950	0.1736	0.1640	0.1418
	$^2H(2)_{9/2}$	9000	0.1747	0	0.0033
	$^4F_{5/2}$	9100	0	0.0618	0.1959
	$^4F_{3/2}$	10,100	0	0.0015	0.0995
	$^4I_{15/2}$	15,500	1.5282	0.8891	0.1600
	$^4I_{13/2}$	17,600	0.1282	0.3502	0.1608
	$^4I_{11/2}$	19,600	00044	0.0645	0.0563
	$^4I_{9/2}$	21,450	0	0.0053	0.0080

(continued)

(continued)

$^{2S+1}L_J$	$^{2S'+1}L'_{J'}$	$E_{JJ'}$ (cm^{-1})	$k = 2$	$k = 4$	$k = 6$
$^2P_{1/2}$	$^4G_{11/2}$	1500	0	0	0.0001
	$^2D(1)_{3/2}$	1850	0.0349	0	0
	$^2G(1)_{9/2}$	1950	0	0.0002	0
	$^2K_{15/2}$	2100	0	0	0
	$^4G_{9/2}$	3550	0	0.0010	0
	$^4G_{7/2}$	4000	0	0.0029	0
	$^2K_{13/2}$	4250	0	0	0.1761
	$^2G(1)_{7/2}$	5800	0	0.0096	0
	$^4G_{5/2}$	5800	0.0327	0	0
	$^2H(2)_{11/2}$	7200	0	0	0.1628
	$^4F_{9/2}$	8350	0	0.0034	0
	$^4S_{3/2}$	9500	0.0167	0	0
	$^4F_{7/2}$	9500	0	0.0183	0.1418
	$^2H(2)_{9/2}$	10,550	0	0.0874	0
	$^4F_{5/2}$	10,650	0.0093	0	0
	$^4F_{3/2}$	11,650	0.0128	0	0
	$^4I_{15/2}$	17,050	0	0	0
	$^4I_{13/2}$	19,150	0	0	0
	$^4I_{11/2}$	21,150	0	0	0.0003
	$^4I_{9/2}$	23,000	0	0.0367	0
$^2D(1)_{5/2}$	$^2P_{1/2}$	750	0.0300	0	0
	$^4G_{11/2}$	2300	0	0.0345	0.0008
	$^2D(1)_{3/2}$	2600	0.2011	0.0030	0
	$^2G(1)_{9/2}$	2700	0.0383	0.1987	0.0186
	$^2K_{15/2}$	2850	0	0	0.4977
	$^4G_{9/2}$	4300	0.0002	0.0055	0.0052
	$^4G_{7/2}$	4750	0.0004	0.0552	0.0671
	$^2K_{13/2}$	5000	0	0.0040	0.1766
	$^2G(1)_{7/2}$	6550	0.0089	0.0681	0.1674
	$^4G_{5/2}$	6550	0.0016	0.0007	0
	$^2H(2)_{11/2}$	7950	0	0.2650	0.0152
	$^4F_{9/2}$	9100	0.0006	0.0094	0.0114
	$^4S_{3/2}$	10,250	0.0096	0	0
	$^4F_{7/2}$	10,250	0.0003	0.0188	0.0061
	$^2H(2)_{9/2}$	11,300	0.0125	0.2248	0.0736
	$^4F_{5/2}$	11,400	0.0011	0.0005	0
	$^4F_{3/2}$	12,400	0.0015	0.0015	0
	$^4I_{15/2}$	17,800	0	0	0.0003
	$^4I_{13/2}$	19,900	0	0.0052	0.0170
	$^4I_{11/2}$	21,900	0	0	0.0025
	$^4I_{9/2}$	23,750	0	0.0002	0.0021

(continued)

(continued)

$^{2S+1}L_J$	$^{2S'+1}L'_{J'}$	$E_{JJ'}$ (cm^{-1})	$k = 2$	$k = 4$	$k = 6$
$^2P_{3/2}$	$^2D(1)_{5/2}$	2350	0.0050	0.0001	0
	$^2P_{1/2}$	3100	0.0026	0	0
	$^4G_{11/2}$	4650	0	0.0031	0.0001
	$^2D(1)_{3/2}$	4950	0.0058	0	0
	$^2G(1)_{9/2}$	5050	0	0.0910	0.1466
	$^2K_{15/2}$	5200	0	0	0.0043
	$^4G_{9/2}$	6650	0	0.0602	0.0590
	$^4G_{7/2}$	7100	0.0045	0.0086	0
	$^2K_{13/2}$	7350	0	0	0.2481
	$^2G(1)_{7/2}$	7850	0.0001	0	0
	$^4G_{5/2}$	7850	0.0066	0	0
	$^2H(2)_{11/2}$	10,300	0	0.0031	0.0125
	$^4F_{9/2}$	11,450	0	0.0596	0.0539
	$^4S_{3/2}$	12,600	0.0073	0	0
	$^4F_{7/2}$	12,600	0.0009	0.0003	0
	$^2H(2)_{9/2}$	13,650	0	0.0130	0.0732
	$^4F_{5/2}$	13,750	0.0023	0.0044	0
	$^4F_{3/2}$	14,750	0.0031	0	0
	$^4I_{15/2}$	20,150	0	0	0.0029
	$^4I_{13/2}$	22,250	0	0	0.0090
	$^4I_{11/2}$	24,250	0	0.0172	0.0008
	$^4I_{9/2}$	26,100	0	0.0014	0.0008
$^4D_{3/2}$	$^2P_{3/2}$	2100	0.0118	0.	0
	$^2D(1)_{5/2}$	4450	0.0268	0.0072	0
	$^2P_{1/2}$	5200	0.1086	0	0
	$^4G_{11/2}$	6750	0	0.0039	0.2047
	$^2D(1)_{3/2}$	7050	0	0	0
	$^2G(1)_{9/2}$	7150	0	0	0.0346
	$^2K_{15/2}$	7300	0	0	0.0014
	$^4G_{9/2}$	8750	0	0.0002	0.0904
	$^4G_{7/2}$	9200	0.1363	0.0125	0
	$^2K_{13/2}$	9450	0	0	0.0024
	$^2G(1)_{7/2}$	11,000	0.2824	0.0173	0
	$^4G_{5/2}$	11,000	0.1786	0.0007	0
	$^2H(2)_{11/2}$	12,400	0	0.0427	0.0135
	$^4F_{9/2}$	13,550	0	0.2023	0.0007
	$^4S_{3/2}$	14,700	0.2297	0	0
	$^4F_{7/2}$	14,700	0.2326	0.0657	0
	$^2H(2)_{9/2}$	15,750	0	0.0209	0.0033
	$^4F_{5/2}$	15,850	0.0595	0.2095	0
	$^4F_{3/2}$	16,850	0.1430	0	0

(continued)

(continued)

$^{2S+1}L_J$	$^{2S'+1}L'_{J'}$	$E_{JJ'}$ (cm^{-1})	$k = 2$	$k = 4$	$k = 6$
	$^4I_{15/2}$	22,250	0	0	0.0091
	$^4I_{13/2}$	24,350	0	0	0.0245
	$^4I_{11/2}$	26,350	0	0.2942	0.0084
	$^4I_{9/2}$	28,200	0	0.1959	0.0169
$^4D_{5/2}$	$^4I_{9/2}$	28,350	0.0001	0.0567	0.0275
$^2I_{11/2}$	$^4I_{9/2}$	28,500	0.0049	0.0146	0.0034
$^4D_{1/2}$	$^4I_{9/2}$	28,750	0	0.2584	0
$^2L_{15/2}$	$^4I_{9/2}$	29,150	0	0.0248	0.0097
$^2I_{13/2}$	$^4I_{9/2}$	29,850	0.0001	0.0013	0.0017
$^4D_{7/2}$	$^4I_{9/2}$	30,400	0	0.0036	0.0080
$^2L_{17/2}$	$^4I_{9/2}$	30,600	0	0.0010	0.0013
$^2H(1)_{9/2}$	$^4I_{9/2}$	32,450	0.0001	0.0085	0
$^2D(2)_{3/2}$	$^4I_{9/2}$	33,350	0	0.0112	0.0011
$^2H(1)_{11/2}$	$^4I_{9/2}$	33,800	0.0001	0.0001	0.0002
$^2D(2)_{5/2}$	$^4I_{9/2}$	34,350	0.0007	0.0006	0.0034
$^2F(2)_{5/2}$	$^4I_{9/2}$	38,350	0.0021	0.0033	0
$^2F(2)_{7/2}$	$^4I_{9/2}$	39,800	0	0.0004	0.0007
$^2G(2)_{9/2}$	$^4I_{9/2}$	47,550	0	0.0015	0.0001

Sm^{3+}

$^{2S+1}L_J$	$^{2S'+1}L'_{J'}$	$E_{JJ'}$ (cm^{-1})	$k = 2$	$k = 4$	$k = 6$
$^6H_{7/2}$	$^6H_{5/2}$	1100	0.2085	0.1986	0.0822
$^6H_{9/2}$	$^6H_{7/2}$	1200	0.2983	0.1652	0.0104
	$^6H_{5/2}$	2300	0.0221	0.1340	0.3117
$^6H_{11/2}$	$^6H_{9/2}$	1300	0.3451	0.2226	5×10^{-5}
	$^6H_{7/2}$	2500	0.0336	0.1760	0.2588
	$^6H_{5/2}$	3600	0	0.0262	0.2678
$^6H_{13/2}$	$^6H_{11/2}$	1500	0.3412	0.3450	0.1129
	$^6H_{9/2}$	2800	0.0280	0.1696	0.3968
	$^6H_{7/2}$	4000	0	0.0276	0.2768
	$^6H_{5/2}$	5100	0	0.0011	0.0605
$^6F_{1/2}$	$^6H_{13/2}$	1100	0	0	0.1026
	$^6H_{11/2}$	2600	0	0	0.3343
	$^6H_{9/2}$	3900	0	0.1485	0
	$^6H_{7/2}$	5100	0	0.1395	0
	$^6H_{5/2}$	6200	0.1947	0	0

(continued)

(continued)

$^{2S+1}L_J$	$^{2S'+1}L'_{J'}$	$E_{JJ'}$ (cm^{-1})	$k = 2$	$k = 4$	$k = 6$
$^6F_{3/2}$	$^6F_{1/2}$	300	0.0182	0	0
	$^6H_{13/2}$	1400	0	0	0.4017
	$^6H_{11/2}$	2900	0	0.2100	0.0279
	$^6H_{9/2}$	4200	0	0.1183	0.3742
	$^6H_{7/2}$	5400	0.2443	0.1164	0
	$^6H_{5/2}$	6500	0.1427	0.1356	0
$^6H_{15/2}$	$^6F_{3/2}$	300	0	0	0.0600
	$^6F_{1/2}$	600	0	0	0
	$^6H_{13/2}$	1700	0.2371	0.4010	0.6819
	$^6H_{11/2}$	3200	0.0148	0.1133	0.4900
	$^6H_{9/2}$	4500	0	0.0104	0.1907
	$^6H_{7/2}$	5700	0	0.0006	0.0394
	$^6H_{5/2}$	6800	0	0	0.0031
$^6F_{5/2}$	$^6H_{15/2}$	400	0	0	0.3316
	$^6F_{3/2}$	700	0.0239	0.0250	0
	$^6F_{1/2}$	1000	0.0125	0	0
	$^6H_{13/2}$	2100	0	0.2014	0.2866
	$^6H_{11/2}$	3600	0	0.2319	0.2035
	$^6H_{9/2}$	4900	0.3358	0.1046	0.0398
	$^6H_{7/2}$	6100	0.1980	0.0538	0.4172
	$^6H_{5/2}$	7200	0.0365	0.2704	0
$^6F_{7/2}$	$^6F_{5/2}$	1000	0.0393	0.0049	0.0054
	$^6H_{15/2}$	1400	0	0.1104	0.6524
	$^6F_{3/2}$	1700	0.0133	0.0076	0
	$^6F_{1/2}$	2000	0	0.0227	0
	$^6H_{13/2}$	3100	0	0.4062	5×10^{-5}
	$^6H_{11/2}$	4600	0.4617	0.0370	0.3017
	$^6H_{9/2}$	5900	0.2140	0.1243	0.2447
	$^6H_{7/2}$	7100	0.0391	0.2824	8×10^{-5}
	$^6H_{5/2}$	8200	0.0044	0.1151	0.3984
$^6F_{9/2}$	$^6F_{7/2}$	1000	0.0520	0.0041	0.0419
	$^6F_{5/2}$	2000	0.0130	0.0335	0.0347
	$^6H_{15/2}$	2400	0	0.5734	0.7566
	$^6F_{3/2}$	2700	0	0.0286	0.0128
	$^6F_{1/2}$	3000	0	0.0060	0
	$^6H_{13/2}$	4100	0.7048	0.0293	0.4178
	$^6H_{11/2}$	5600	0.2264	0.3614	0.0092
	$^6H_{9/2}$	6900	0.0377	0.3323	0.1687
	$^6H_{7/2}$	8100	0.0032	0.1269	0.4513
	$^6H_{5/2}$	9200	9×10^{-5}	0.0190	0.3557

(continued)

(continued)

$^{2S+1}L_J$	$^{2S'+1}L'_{J'}$	$E_{JJ'}$ (cm^{-1})	$k = 2$	$k = 4$	$k = 6$
$^6F_{11/2}$	$^6F_{9/2}$	1400	0.0532	0.0676	0.0189
	$^6F_{7/2}$	2400	0.0065	0.0456	0.0325
	$^6F_{5/2}$	3400	0	0.0188	0.0500
	$^6H_{15/2}$	3800	1.0105	0.7661	0.3489
	$^6F_{3/2}$	4100	0	0.0030	0.0499
	$^6F_{1/2}$	4400	0	0	0.0313
	$^6H_{13/2}$	5500	0.1581	0.5983	0.7011
	$^6H_{11/2}$	7000	0.0190	0.2577	0.7461
	$^6H_{9/2}$	8300	0.0014	0.0674	0.5189
	$^6H_{7/2}$	9500	4×10^{-5}	0.0102	0.2322
	$^6H_{5/2}$	10,600	0	0.0006	0.0527

Eu^{3+}:

$^{2S+1}L_J$	$^{2S'+1}L'_{J'}$	$E_{JJ'}$ (cm^{-1})	$k = 2$	$k = 4$	$k = 6$
$^7F_1$	$^7F_0$	400	0	0	0
$^7F_2$	$^7F_1$	600	0.0518	0	0
	$^7F_0$	1000	0.1372	0	0
$^7F_3$	$^7F_2$	900	0.1852	0.2121	0
	$^7F_1$	1500	0.2087	0.1266	0
	$^7F_0$	1900	0	0	0
$^7F_4$	$^7F_3$	1000	0.3833	0.1337	0.1554
	$^7F_2$	1900	0.2207	0.0066	0.0317
	$^7F_1$	2500	0	0.1731	0
	$^7F_0$	2900	0	0.1382	0
$^7F_5$	$^7F_4$	1000	0.5611	0.0131	0.4430
	$^7F_3$	2000	0.1737	0.2522	0.3817
	$^7F_2$	2900	0	0.3150	0.2088
	$^7F_1$	3500	0	0.1190	0.0544
	$^7F_0$	3900	0	0	0
$^7F_6$	$^7F_5$	1100	0.5212	0.6455	0.1217
	$^7F_4$	2100	0.0846	0.5150	0.2719
	$^7F_3$	3100	0	0.2306	0.4120
	$^7F_2$	4000	0	0.0476	0.4696
	$^7F_1$	4600	0	0	0.3765
	$^7F_0$	5000	0	0	0.1442

(continued)

(continued)

$^{2S+1}L_J$	$^{2S'+1}L'_{J'}$	$E_{JJ'}$ (cm^{-1})	$k = 2$	$k = 4$	$k = 6$
$^5D_0$	$^7F_6$	12,300	0	0	0.0005
	$^7F_5$	13,400	0	0	0
	$^7F_4$	14,400	0	0.0030	0
	$^7F_3$	15,400	0	0	0
	$^7F_2$	16,300	0.0035	0	0
	$^7F_1$	16,900	0	0	0
	$^7F_0$	17,300	0	0	0
$^5D_1$	$^5D_0$	1700	0	0	0
	$^7F_6$	14,000	0	0	0.0006
	$^7F_5$	15,100	0	0.0009	5×10^{-5}
	$^7F_4$	16,100	0	0.0031	0
	$^7F_3$	17,100	0.0042	0.0024	0
	$^7F_2$	18,000	0.0009	0	0
	$^7F_1$	18,600	0.0028	0	0
	$^7F_0$	19,000	0	0	0
$^5D_2$	$^5D_1$	2500	0.0105	0	0
	$^5D_0$	4200	0.0112	0	0
	$^7F_6$	16,500	0	2×10^{-5}	0.0002
	$^7F_5$	17,600	0	0.0018	2×10^{-5}
	$^7F_4$	18,600	0.0024	0.0002	0.0001
	$^7F_3$	19,600	0.0023	0.0026	0
	$^7F_2$	20,500	0.0018	0.0015	0
	$^7F_1$	21,100	0.0002	0	0
	$^7F_0$	21,500	0.0009	0	0
$^5D_3$	$^5D_2$	2400	0.0300	0.0172	0
	$^5D_1$	4900	0.0136	0.0074	0
	$^5D_0$	6600	0	0	0
	$^7F_6$	18,900	0	7×10^{-5}	2×10^{-5}
	$^7F_5$	20,000	0.0002	0.0014	2×10^{-6}
	$^7F_4$	21,000	0.0036	0.0004	0.0002
	$^7F_3$	22,000	0.0009	0.0004	0.0002
	$^7F_2$	22,900	0.0003	0.0021	0
	$^7F_1$	23,500	0.0005	0.0012	0
	$^7F_0$	23,900	0	0	0

Tb^{3+}:

$^{2S+1}L_J$	$^{2S'+1}L'_{J'}$	$E_{JJ'}$ (cm^{-1})	$k = 2$	$k = 4$	$k = 6$
$^7F_5$	$^7F_6$	2100	0.5122	0.6426	0.1181
$^7F_4$	$^7F_5$	1400	0.5415	0.0122	0.4360
	$^7F_6$	3500	0.0868	0.5140	0.2667
$^7F_3$	$^7F_4$	900	0.3474	0.1288	0.1450
	$^7F_5$	2300	0.1621	0.2358	0.3512
	$^7F_6$	4400	0	0.2142	0.3804
$^7F_2$	$^7F_3$	700	0.1685	0.1936	0
	$^7F_4$	1600	0.2211	0.0060	0.0326
	$^7F_5$	3000	0	0.3129	0.2077
	$^7F_6$	5100	0	0.0479	0.4676
$^7F_1$	$^7F_2$	500	0.0512	0	0
	$^7F_3$	1200	0.1928	0.1204	0
	$^7F_4$	2100	0	0.1707	0
	$^7F_5$	3500	0	0.1175	0.0544
	$^7F_6$	5600	0	0	0.3738
$^7F_0$	$^7F_1$	200	0	0	0
	$^7F_2$	700	0.1388	0	0
	$^7F_3$	1400	0	0	0
	$^7F_4$	2300	0	0.1386	0
	$^7F_5$	3700	0	0	0
	$^7F_6$	5800	0	0	0.1440
$^5D_4$	$^7F_0$	14,700	0	0.0022	0
	$^7F_1$	14,900	0	0.0025	0
	$^7F_2$	15,400	0.0016	0.0004	8×10^{-5}
	$^7F_3$	16,100	0.0137	0.0010	0.0007
	$^7F_4$	17,000	0.0003	0.0019	0.0015
	$^7F_5$	18,400	0.0139	0.0010	0.0026
	$^7F_6$	20,500	0.0007	0.0013	0.0011
$^5D_3$	$^5D_4$	6300	0.0501	0.0274	0.0063
	$^7F_0$	21,000	0	0	0
	$^7F_1$	21,200	0.0017	0.0030	0
	$^7F_2$	21,700	0.0016	0.0030	0
	$^7F_3$	22,400	0.0005	0.0111	0.0002
	$^7F_4$	23,300	0.0073	0.0002	0.0003
	$^7F_5$	24,700	0.0008	0.0028	0.0016
	$^7F_6$	26,800	0	0.0006	0.0014

Dy^{3+}:

$^{2S+1}L_J$	$^{2S'+1}L'_{J'}$	$E_{JJ'}$ (cm^{-1})	$k = 2$	$k = 4$	$k = 6$
$^6H_{13/2}$	$^6H_{15/2}$	3400	0.2286	0.3879	0.6525
$^6H_{11/2}$	$^6H_{13/2}$	2250	0.2464	0.4412	0.0359
	$^6H_{15/2}$	5650	0.0673	0.0409	0.5891
$^6H_{9/2}$	$^6H_{11/2}$	1900	0.3280	0.1702	0.0077
	$^6H_{13/2}$	4150	0.0276	0.1647	0.3766
	$^6H_{15/2}$	7550	0	0.0157	0.2003
$^6F_{11/2}$	$^6H_{9/2}$	–	0.0039	0.1009	0.4608
	$^6H_{11/2}$	1900	0.0022	0.1907	0.7845
	$^6H_{13/2}$	4150	0.2274	0.4342	0.7254
	$^6H_{15/2}$	7550	0.8749	0.8360	0.1909
$^6H_{7/2}$	$^6F_{11/2}$	1350	0.0006	0.0256	0.1374
	$^6H_{9/2}$	1350	0.2974	0.1574	0.0095
	$^6H_{11/2}$	3250	0.0309	0.1466	0.3047
	$^6H_{13/2}$	5500	0	0.0265	0.2709
	$^6H_{15/2}$	8900	0	0.0007	0.0415
$^6F_{9/2}$	$^6H_{7/2}$	–	0.0025	0.1193	0.4088
	$^6F_{11/2}$	1350	0.0138	0.0162	0.0151
	$^6H_{9/2}$	1350	0.0356	0.3155	0.1816
	$^6H_{11/2}$	3250	0.2382	0.3663	0.0121
	$^6H_{13/2}$	5500	0.6199	0.0246	0.3811
	$^6H_{15/2}$	8900	0	0.5628	0.6747
$^6H_{5/2}$	$^6F_{9/2}$	1100	6×10^{-5}	0.0180	0.3172
	$^6H_{7/2}$	1100	0.2047	0.1871	0.0784
	$^6F_{11/2}$	2450	0	0.0023	0.0183
	$^6H_{9/2}$	2450	0.0228	0.1319	0.2970
	$^6H_{11/2}$	4350	0	0.0225	0.2812
	$^6H_{13/2}$	6600	0	0.0010	0.0606
	$^6H_{15/2}$	10,000	0	0	0.0032
$^6F_{7/2}$	$^6H_{5/2}$	800	0.0031	0.1256	0.3907
	$^6F_{9/2}$	1900	0.0450	0.0049	0.0481
	$^6H_{7/2}$	1900	0.0430	0.2747	0
	$^6F_{11/2}$	3250	0.0333	0.0272	0.0792
	$^6H_{9/2}$	3250	0.2182	0.1272	0.2475
	$^6H_{11/2}$	5150	0.4029	0.0527	0.2756
	$^6H_{13/2}$	7400	0	0.4138	0
	$^6H_{15/2}$	10,800	0	0.1167	0.6619

(continued)

(continued)

$^{2S+1}L_J$	$^{2S'+1}L'_{J'}$	$E_{JJ'}$ (cm^{-1})	$k = 2$	$k = 4$	$k = 6$
$^6F_{5/2}$	$^6F_{7/2}$	1400	0.0395	0.0061	0.0077
	$^6H_{5/2}$	2200	0.0353	0.2690	0
	$^6F_{9/2}$	3300	0.0133	0.0307	0.0288
	$^6H_{7/2}$	3300	0.1940	0.0542	0.3982
	$^6F_{11/2}$	4650	0	0.0455	0.0834
	$^6H_{9/2}$	4650	0.3313	0.1045	0.0425
	$^6H_{11/2}$	6550	0	0.1953	0.1690
	$^6H_{13/2}$	8800	0	0.1998	0.2747
	$^6H_{15/2}$	12,200	0	0	0.3385
$^6F_{3/2}$	$^6F_{5/2}$	800	0.0222	0.0273	0
	$^6F_{7/2}$	2200	0.0148	0.0071	0
	$^6H_{5/2}$	3000	0.1318	0.1279	0
	$^6F_{9/2}$	4100	0	0.0312	0.0092
	$^6H_{7/2}$	4100	0.0430	0.2747	0
	$^6F_{11/2}$	5450	0	0.0164	0.0302
	$^6H_{9/2}$	5450	0	0.1228	0.3315
	$^6H_{11/2}$	7350	0	0.1852	0.0340
	$^6H_{13/2}$	9600	0	0	0.3776
	$^6H_{15/2}$	13,000	0	0	0.0606
$^6F_{1/2}$	$^6F_{3/2}$	400	0.0152	0	0
	$^6F_{5/2}$	1200	0.0113	0	0
	$^6F_{7/2}$	2600	0	0.0260	0
	$^6H_{5/2}$	3400	0.1902	0	0
	$^6F_{9/2}$	4500	0	0.0072	0
	$^6H_{7/2}$	4500	0	0.1330	0
	$^6F_{11/2}$	5850	0	0	0.0048
	$^6H_{9/2}$	5850	0	0.1460	0
	$^6H_{11/2}$	7750	0	0	0.3336
	$^6H_{13/2}$	10,000	0	0	0.0991
	$^6H_{15/2}$	13,400	0	0	0

Ho^{3+}:

$^{2S+1}L_J$	$^{2S'+1}L'_{J'}$	$E_{JJ'}$ (cm^{-1})	$k = 2$	$k = 4$	$k = 6$
$^5I_7$	$^5I_8$	5050	0.0249	0.1344	1.5210
$^5I_6$	$^5I_7$	3400	0.0319	0.1336	0.9308
	$^5I_8$	8550	0.0083	0.0383	0.6917
$^5I_5$	$^5I_6$	2250	0.0438	0.1705	0.5729
	$^5I_7$	6050	0.0027	0.0226	0.8887
	$^5I_8$	11,150	0	0.0098	0.0936

(continued)

(continued)

$^{2S+1}L_J$	$^{2S'+1}L'_{J'}$	$E_{JJ'}$ (cm^{-1})	$k = 2$	$k = 4$	$k = 6$
$^5I_4$	$^5I_5$	2050	0.0313	0.1238	0.9100
	$^5I_6$	4600	0.0022	0.0281	0.6640
	$^5I_7$	8100	0	0.0033	0.1568
	$^5I_8$	13,150	0	0	0.0077
$^5F_5$	$^5I_4$	2250	0.0001	0.0059	0.0039
	$^5I_5$	4300	0.0068	0.0271	0.1649
	$^5I_6$	6850	0.0102	0.1213	0.4995
	$^5I_7$	10,350	0.0177	0.3298	0.4340
	$^5I_8$	15,400	0	0.4277	0.5685
$^5S_2$	$^5F_5$	3000	0	0.0110	0.0036
	$^5I_4$	5250	0.0014	0.0262	0.2795
	$^5I_5$	7300	0	0.0043	0.1062
	$^5I_6$	9850	0	0.0206	0.1540
	$^5I_7$	13,350	0	0	0.4096
	$^5I_8$	18,450	0	0	0.2270
$^5F_4$	$^5S_2$	–	0.0002	0.0147	0.0052
	$^5F_5$	3000	0.1944	0.0923	0.0080
	$^5I_4$	5250	0.0001	0.0234	0.2587
	$^5I_5$	7300	0.0018	0.1314	0.4655
	$^5I_6$	9850	0.0012	0.2580	0.1697
	$^5I_7$	13,350	0	0.1988	0.0324
	$^5I_8$	18,450	0	0.2402	0.7079
$^5F_3$	$^5F_4$	2100	0.0971	0.0289	0.0957
	$^5S_2$	2100	0.0061	0.0009	0
	$^5F_5$	5100	0.0388	0.0820	0.0873
	$^5I_4$	7350	0.0003	0.0980	0.3964
	$^5I_5$	9400	0	0.2154	0.0173
	$^5I_6$	11,950	0	0.0904	0.2170
	$^5I_7$	15,450	0	0.2483	0.2275
	$^5I_8$	20,550	0	0	0.3464
$^5F_2$	$^5F_3$	450	0.0459	0.0002	0
	$^5F_4$	2550	0.0067	0.0268	0.0270
	$^5S_2$	2550	0.0016	0.0028	0
	$^5F_5$	5600	0	0.0082	0.1172
	$^5I_4$	7950	0.0004	0.1810	0.0273
	$^5I_5$	9850	0	0.0427	0.3125
	$^5I_6$	12,450	0	0.1287	0.1532
	$^5I_7$	15,590	0	0	0.1760
	$^5I_8$	21,000	0	0	0.1918

(continued)

(continued)

$^{2S+1}L_J$	$^{2S'+1}L'_{J'}$	$E_{JJ'}$ (cm^{-1})	$k = 2$	$k = 4$	$k = 6$
$^3K_8$	$^5F_2$	250	0	0	0.0005
	$^5F_3$	750	0	0	0.0062
	$^5F_4$	2850	0	0.0079	0.0004
	$^5S_2$	2850	0	0	0.0002
	$^5F_5$	5850	0	0.0263	0.0195
	$^5I_4$	8100	0	0.0045	0.0012
	$^5I_5$	10,150	0	0.0004	0.0041
	$^5I_6$	12,700	0.0059	0.0041	0.0138
	$^5I_7$	16,200	0.0019	0.0046	0.0406
	$^5I_8$	21,250	0.0209	0.03,294	0.1579
$^5G_6$	$^3K_8$	850	0.0005	0.0027	0.2616
	$^5F_2$	1150	0	0.0257	0.1144
	$^5F_3$	1600	0	0.0666	0.1595
	$^5F_4$	3700	0.2551	0.2392	0.1347
	$^5S_2$	3700	0	0.3250	0.0031
	$^5F_5$	6700	1.1528	0.3576	0.0329
	$^5I_4$	8950	0.0005	0.0013	2×10^{-5}
	$^5I_5$	11,000	0.0165	0.0307	0.0133
	$^5I_6$	13,500	0.0093	0.0837	0.1094
	$^5I_7$	17,050	0.1523	0.04,374	0.2588
	$^5I_8$	22,150	1.5180	0.8548	0.1416
$^5F_1$	$^5G_6$	–	0	0	0.0530
	$^3K_8$	850	0	0	0
	$^5F_2$	1150	0.0496	0	0
	$^5F_3$	1600	0.0080	0.0604	0
	$^5F_4$	3700	0	0.0470	0
	$^5S_2$	3700	0.0096	0	0
	$^5F_5$	6700	0	3×10^{-5}	0.1139
	$^5I_4$	8950	0	0.1442	0
	$^5I_5$	11,000	0	0.14042	0.1710
	$^5I_6$	13,550	0	0	0.2398
	$^5I_7$	17,050	0	0	0.0570
	$^5I_8$	22,150	0	0	0
$^5G_5$	$^5F_1$	1750	0	0.0078	0.0574
	$^5G_6$	1750	0.0617	0.2595	00.617
	$^3K_8$	2650	0	0.0093	0.0007
	$^5F_2$	2900	0	0.1292	0.0022
	$^5F_3$	3500	0.1927	0.1029	00.319
	$^5F_4$	5450	0.2984	0.0216	0.1446
	$^5S_2$	5450	0	0.1292	0.0018
	$^5F_5$	8500	0.3540	0.0426	0.1198
	$^5I_4$	10,750	10^{-5}	0.0100	0.0427
	$^5I_5$	12,750	0.0036	0.0689	0.0598
	$^5I_6$	15,350	0.1374	0.1857	0.0806
	$^5I_7$	18,850	0.6185	0.0308	0.1090
	$^5I_8$	23,900	0	0.0548	0.1610

(continued)

(continued)

$^{2S+1}L_J$	$^{2S'+1}L'_{J'}$	$E_{JJ'}$ (cm^{-1})	$k = 2$	$k = 4$	$k = 6$
$^5G_4$	$^5G_5$	1900	0.0082	01650	0.0224
	$^5F_1$	3700	0	00564	0
	$^5G_6$	3700	0.0042	02045	0.2885
	$^3K_8$	4550	0	00800	0.0017
	$^5F_2$	4800	0.2149	00002	0.0732
	$^5F_3$	5300	0.2233	00215	0.0489
	$^5F_4$	7400	0.4086	01102	0.0214
	$^5S_2$	7400	0.0246	02293	0.0097
	$^5F_5$	10,400	0.0298	00141	0.2246
	$^5I_4$	12,600	0.0154	01093	0.0545
	$^5I_5$	14,700	0.2268	02775	0.0322
	$^5I_6$	17,250	0.7156	00246	0.0009
	$^5I_7$	20,750	0	02990	0.0653
	$^5I_8$	25,800	0	0.0346	0.0348
$^3K_7$	$^5G_4$	250	0	0.0139	0.0042
	$^5G_5$	2150	0.0163	00059	0.2302
	$^5F_1$	3900	0	0	0.0018
	$^5G_6$	3900	7×10^{-5}	0.0002	0.1060
	$^3K_8$	4800	0.0877	7×10^{-5}	0.1252
	$^5F_2$	5050	0	0	0.0104
	$^5F_3$	5500	0	0.0074	0.0041
	$^5F_4$	7600	0	0.0153	0.0089
	$^5S_2$	7600	0	0	0.0260
	$^5F_5$	10,650	0.0015	0.0113	0.0141
	$^5I_4$	12,900	0	0.0007	0.0069
	$^5I_5$	14,900	0.0068	0.0042	0.0312
	$^5I_6$	17,500	0.0011	0.0058	0.0716
	$^5I_7$	20,950	0.0056	0.0062	0.0050
	$^5I_8$	26,050	0.0058	0.0046	0.0338

Er^{3+}:

$^{2S+1}L_J$	$^{2S'+1}L'_{J'}$	$E_{JJ'}$ (cm^{-1})	$k = 2$	$k = 4$	$k = 6$
$^4I_{13/2}$	$^4I_{15/2}$	6500	0.0195	0.1173	1.4316
$^4I_{11/2}$	$^4I_{13/2}$	3600	0.0331	0.1708	1.0864
	$^4I_{15/2}$	10,100	0.0282	0.0003	0.3953
$^4I_{9/2}$	$^4I_{11/2}$	2150	0.0030	0.0674	0.1271
	$^4I_{13/2}$	5750	0.0004	0.0106	0.7162
	$^4I_{15/2}$	12,250	0	0.1732	0.0099
$^4F_{9/2}$	$^4I_{9/2}$	2900	0.1279	0.0059	0.0281

(continued)

(continued)

$^{2S+1}L_J$	$^{2S'+1}L'_{J'}$	$E_{JJ'}$ (cm^{-1})	$k = 2$	$k = 4$	$k = 6$
$^4F_{9/2}$	$^4I_{11/2}$	5050	0.0704	0.0112	1.2839
	$^4I_{13/2}$	8650	0.0101	0.1533	0.0714
	$^4I_{15/2}$	15,150	0	0.5354	0.4619
$^4S_{3/2}$	$^4F_{9/2}$	3200	0	0.0003	0.0264
	$^4I_{9/2}$	6100	0	0.0788	0.2542
	$^4I_{11/2}$	8250	0	0.0042	0.0739
	$^4I_{13/2}$	11,850	0	0	0.3462
	$^4I_{15/2}$	18,350	0	0	0.2211
$^2H(2)_{11/2}$	$^4S_{3/2}$	800	0	0.1988	0.0101
	$^4F_{9/2}$	4000	0.3629	0.0224	0.0022
	$^4I_{9/2}$	6900	0.2077	0.0662	0.2858
	$^4I_{11/2}$	9050	0.0357	0.1382	0.0371
	$^4I_{13/2}$	12,650	0.0230	0.0611	0.0527
	$^4I_{15/2}$	19,150	0.7125	0.4123	0.0925
$^4F_{7/2}$	$^2H(2)_{11/2}$	1150	0.1229	0.0153	0.4017
	$^4S_{3/2}$	1950	0.0001	0.0058	0
	$^4F_{9/2}$	5150	0.0121	0.0342	0.
	$^4I_{9/2}$	8050	0.0163	0.0954	0.4277
	$^4I_{11/2}$	10,200	0.0035	0.2648	0.1515
	$^4I_{13/2}$	13,800	0	0.3371	0.0001
	$^4I_{15/2}$	20,300	0	0.1468	0.6266
$^4F_{5/2}$	$^4F_{7/2}$	1650	0.0765	0.0503	0.1015
	$^2H(2)_{11/2}$	2800	0	0.0586	0.1825
	$^4S_{3/2}$	3600	0.0082	0.0040	0
	$^4F_{9/2}$	6800	0.0004	0.2415	0.3575
	$^4I_{9/2}$	9700	0.0107	0.0576	0.0102
	$^4I_{11/2}$	11,850	0	0.0979	0.0028
	$^4I_{13/2}$	15,450	0	0.1783	0.3429
	$^4I_{15/2}$	21,950	0	0	0.2233
$^4F_{3/2}$	$^4F_{5/2}$	350	0.0618	0.0350	0
	$^4F_{7/2}$	2000	0.0026	0.0584	0
	$^2H(2)_{11/2}$	3150	0	0.0005	0.0030
	$^4S_{3/2}$	3950	0.0260	0	0
	$^4F_{9/2}$	7150	0	0.0040	0.0595
	$^4I_{9/2}$	10,050	0	0.2299	0.0558
	$^4I_{11/2}$	12,200	0	0.0927	0.4861
	$^4I_{13/2}$	15,800	0	0	0.0345
	$^4I_{15/2}$	22,300	0	0	0.1272

(continued)

(continued)

$^{2S+1}L_J$	$^{2S'+1}L'_{J'}$	$E_{JJ'}$ (cm^{-1})	$k = 2$	$k = 4$	$k = 6$
$^2G(1)_{9/2}$	$^4F_{3/2}$	2100	0	0.0208	0.0087
	$^4F_{5/2}$	2450	0.0124	0.0259	0.0063
	$^4F_{7/2}$	4100	0.1058	0.0488	0.0240
	$^2H(2)_{11/2}$	5250	0.0308	0.1828	0.0671
	$^4S_{3/2}$	6050	0	0.0019	0.0025
	$^4F_{9/2}$	9250	0.0055	0.0314	0.0369
	$^4I_{9/2}$	12,150	0.0147	0.0062	0.0043
	$^4I_{11/2}$	14,300	0.0428	0.0824	0.1128
	$^4I_{13/2}$	17,900	0.0780	0.1194	0.3535
	$^4I_{15/2}$	24,400	0	0.0190	0.2255
$^4G_{11/2}$	$^2G(1)_{9/2}$	2000	0.2906	0.1170	0.1328
	$^4F_{3/2}$	4100	0	0.0234	0.0923
	$^4F_{5/2}$	4450	0	0.0378	0.0815
	$^4F_{7/2}$	6100	0.0877	0.1287	0.0159
	$^2H(2)_{11/2}$	7250	0.0004	0.1539	0.0494
	$^4S_{3/2}$	8050	0	0.1302	0.0044
	$^4F_{9/2}$	11,250	0.4252	0.0368	0.0122
	$^4I_{9/2}$	14,150	0.0716	0.0131	0.0235
	$^4I_{11/2}$	16,300	0.0003	0.0496	0.0134
	$^4I_{13/2}$	19,900	0.1013	0.2651	0.2594
	$^4I_{15/2}$	26,400	0.9181	0.5261	0.1171
$^4G_{9/2}$	$^4G_{11/2}$	950	0.0005	0.2021	0.1639
	$^2G(1)_{9/2}$	2950	0.0269	0	0.0452
	$^4F_{3/2}$	5050	0	0.1750	0.1089
	$^4F_{5/2}$	5400	0.1630	00824	0.0028
	$^4F_{7/2}$	7050	0.6062	0.0088	0.1243
	$^2H(2)_{11/2}$	8200	0.0218	0.3274	0.1496
	$^4S_{3/2}$	9000	0	0.1551	0.0100
	$^4F_{9/2}$	12,200	0.2201	0.3121	0.3765
	$^4I_{9/2}$	15,100	0.0051	0.0042	0.0027
	$^4I_{11/2}$	17,250	0.0894	0.1524	0.0144
	$^4I_{13/2}$	20,850	1.0908	0.3520	0.0160
	$^4I_{15/2}$	27,350	0	0.2415	0.1234

(continued)

(continued)

$^{2S+1}L_J$	$^{2S'+1}L'_{J'}$	$E_{JJ'}$ (cm^{-1})	$k = 2$	$k = 4$	$k = 6$
$^2K_{15/2}$	$^4G_{9/2}$	350	0	0.0114	0.0598
	$^4G_{11/2}$	1300	0.0965	0.0595	0.0706
	$^2G(1)_{9/2}$	3300	0	0.7106	0.0708
	$^4F_{3/2}$	5400	0	0	0.0001
	$^4F_{5/2}$	5750	0	0	0.0461
	$^4F_{7/2}$	7400	0	0.0001	0.0002
	$^2H(2)_{11/2}$	8550	0.0977	0.0001	1.1458
	$^4S_{3/2}$	9350	0	0	0.0032
	$^4F_{9/2}$	12,550	0	0.0776	0.0125
	$^4I_{9/2}$	15,450	0	0.2221	0.1003
	$^4I_{11/2}$	17,600	0.0468	0.0018	0.2488
	$^4I_{13/2}$	21,200	0.0001	0.0016	0.0361
	$^4I_{15/2}$	27,700	0.0219	0.0041	0.0758
$^2G(1)_{7/2}$	$^2K_{15/2}$	150	0	0.1154	0.0026
	$^4G_{9/2}$	500	0.0041	0.1891	0.1582
	$^4G_{11/2}$	1450	0.0150	0.0604	0.0193
	$^2G(1)_{9/2}$	3450	0.0145	0.0056	0.0205
	$^4F_{3/2}$	5550	0.0941	0.0314	0
	$^4F_{5/2}$	5900	0.3716	0.0023	0.0378
	$^4F_{7/2}$	7550	0.1239	0.0424	0.0071
	$^2H(2)_{11/2}$	8700	0.0019	0.0344	0.2672
	$^4S_{3/2}$	9500	0.0445	0.1594	0
	$^4F_{9/2}$	12,700	0.0003	0.0078	0.0128
	$^4I_{9/2}$	15,600	0.1586	0.3607	0.2204
	$^4I_{11/2}$	17,750	0.4934	0.2708	0.1674
	$^4I_{13/2}$	21,350	0	0.1009	0.0512
	$^4I_{15/2}$	27,850	0	0.0174	0.1163
$^2P_{3/2}$	$^2G(1)_{7/2}$	3700	0.0125	0.0004	0
	$^2K_{15/2}$	3850	0	0	0.0268
	$^4G_{9/2}$	4200	0	0.0125	0.0053
	$^4G_{11/2}$	5150	0	0.0266	0.0107
	$^2G(1)_{9/2}$	7150	0	0.2083	0.2591
	$^4F_{3/2}$	9250	0.0123	0	0
	$^4F_{5/2}$	9600	0.0173	0.0433	0
	$^4F_{7/2}$	11,250	0.0211	0.0076	0
	$^2H(2)_{11/2}$	12,400	0	0.0168	0.0263
	$^4S_{3/2}$	13,200	0.0813	0	0
	$^4F_{9/2}$	16,400	0	0.0464	0.0060
	$^4I_{9/2}$	19,300	0	0.0461	0.0041
	$^4I_{11/2}$	21,450	0	0.0995	0.0400
	$^4I_{13/2}$	25,050	0	0	0.1478
	$^4I_{15/2}$	31,550	0	0	0.0172

(continued)

(continued)

$^{2S+1}L_J$	$^{2S'+1}L'_{J'}$	$E_{JJ'}$ (cm^{-1})	$k = 2$	$k = 4$	$k = 6$
$^2K_{13/2}$	$^4I_{15/2}$	33,000	0.0032	0.0029	0.0152
$^4G_{5/2}$	$^4I_{15/2}$	33,300	0	0	0.0026
$^2P_{1/2}$	$^4I_{15/2}$	33,350	0	0	0
$^4G_{7/2}$	$^4I_{15/2}$	33,900	0	0.0334	0.0028
$^2D(1)_{5/2}$	$^4I_{15/2}$	34,700	0	0	0.0227
$^2H(2)_{9/2}$	$^4I_{15/2}$	36,450	0	0.0500	0.0001
$^4D_{5/2}$	$^4I_{15/2}$	38,450	0	0	0.0267
$^4D_{7/2}$	$^4I_{15/2}$	39,050	0	0.8919	0.0291
$^2I_{11/2}$	$^4I_{15/2}$	40,900	0.0002	0.0285	0.0034
$^2L_{17/2}$	$^4I_{15/2}$	41,600	0.0047	0.0663	0.0328
$^4D_{3/2}$	$^4I_{15/2}$	42,150	0	0	0.0126
$^2D(1)_{3/2}$	$^4I_{15/2}$	42,850	0	0	0.0002
$^2I_{13/2}$	$^4I_{15/2}$	43,600	0.0055	0.0171	0.0050
$^4D_{1/2}$	$^4I_{15/2}$	46,950	0	0	0
$^2H(1)_{9/2}$	$^4I_{15/2}$	47,750	0	0.0038	0.0001
$^2L_{15/2}$	$^4I_{15/2}$	47,750	0.0002	0.0027	0.0021
$^2D(2)_{5/2}$	$^4I_{15/2}$	48,900	0	0	0.0096

Tm^{3+}:

$^{2S+1}L_J$	$^{2S'+1}L'_{J'}$	$E_{JJ'}$ (cm^{-1})	$k = 2$	$k = 4$	$k = 6$
$^3H_4$ ($^3F_4$)	$^3H_6$	5600	0.249	0.118	0.608
$^3H_5$	$^3H_4$ ($^3F_4$)	2500	0.011	0.48	0.004
	$^3H_6$	8100	0.107	0.231	0.638
$^3F_4$ ($^3H_4$)	$^3H_5$	4450	0.089	0.125	0.905
	$^3H_4$ ($^3F_4$)	6950	0.129	0.133	0.213
	$^3H_6$	12,550	0.527	0.718	0.228
$^3F_3$	$^3F_4$ ($^3H_4$)	1800	0.002	0.0005	0.167
	$^3H_5$	6250	0.629	0.347	0
	$^3H_4$ ($^3F_4$)	8750	0.081	0.344	0.264
	$^3H_6$	14,350	0	0.316	0.841
$^3F_2$	$^3F_3$	650	0.004	0.075	0
	$^3F_4$ ($^3H_4$)	2450	0.311	0.056	0.044
	$^3H_5$	6900	0	0.29	0.583
	$^3H_4$ ($^3F_4$)	9400	0.287	0.163	0.074
	$^3H_6$	15,000	0	0.0005	0.258

(continued)

(continued)

$^{2S+1}L_J$	$^{2S'+1}L'_{J'}$	$E_{JJ'}$ (cm^{-1})	$k = 2$	$k = 4$	$k = 6$
$^1G_4$	$^3F_2$	6100	0.0060	0.0717	0.0417
	$^3F_3$	6700	0.0105	0.0733	0.3056
	$^3F_4$ ($^3H_4$)	8500	0.0034	0.0194	0.0718
	$^3H_5$	12,950	0.0738	0.0059	0.5423
	$^3H_4$ ($^3F_4$)	15,500	0.1583	0.0042	0.3783
	$^3H_6$	21,050	0.0481	0.0752	0.0119
$^1D_2$	$^1G_4$	6750	0.1905	0.1722	0.0009
	$^3F_2$	12,850	0.0647	0.3055	0
	$^3F_3$	13,450	0.1607	0.0680	0
	$^3F_4$ ($^3H_4$)	15,250	0.5631	0.0935	0.0225
	$^3H_5$	19,700	0	0.0011	0.0185
	$^3H_4$ ($^3F_4$)	22,250	0.1240	0.0116	0.2266
	$^3H_6$	27,800	0	0.3079	0.0926
$^3P_0$	$^1D_2$	6900	0.0253	0	0
	$^1G_4$	13,650	0	0.0525	0
	$^3F_2$	19,700	0.3543	0	0
	$^3F_3$	20,350	0	0	0
	$^3F_4$ ($^3H_4$)	22,150	0	0.2715	0
	$^3H_5$	26,600	0	0	0
	$^3H_4$ ($^3F_4$)	29,100	0	0.0204	0
	$^3H_6$	34,700	0	0	0.0757
$^1I_6$	$^3P_0$	1400	0	0	0.0190
	$^1D_2$	8300	0	0.0520	0.8448
	$^1G_4$	15,050	0.2128	1.2488	0.6302
	$^3F_2$	21,100	0	0.0410	0.3545
	$^3F_3$	21,750	0	0.0028	0.0074
	$^3F_4$ ($^3H_4$)	23,550	0.0622	0.5013	0.3921
	$^3H_5$	28,000	0.0010	0.0021	0.0056
	$^3H_4$ ($^3F_4$)	30,500	0.0669	0.3112	0.0992
	$^3H_6$	36,100	0.0107	0.0392	0.0136
$^3P_1$	$^1I_6$	–	0	0	0.0011
	$^3P_0$	1400	0	0	0
	$^1D_2$	8300	0.4444	0	0
	$^1G_4$	15,050	0	0.0050	0
	$^3F_2$	21,100	0.1370	0	0
	$^3F_3$	21,750	0.5714	0.1964	0
	$^3F_4$ ($^3H_4$)	23,550	0	0.1074	0
	$^3H_5$	28,000	0	0.2857	0.0893
	$^3H_4$ ($^3F_4$)	30,500	0	0.4055	0
	$^3H_6$	36,100	0	0	0.1239

(continued)

(continued)

$^{2S+1}L_J$	$^{2S'+1}L'_{J'}$	$E_{JJ'}$ (cm^{-1})	$k = 2$	$k = 4$	$k = 6$
$^3P_2$	$^3P_1$	1800	0.1822	0	0
	$^1I_6$	1800	0	0.0981	0.6808
	$^3P_0$	3200	0.1247	0	0
	$^1D_2$	10,100	0.0004	0.1560	0
	$^1G_4$	16,850	0.5964	0.0009	0.1055
	$^3F_2$	22,900	0.0045	0.0433	0
	$^3F_3$	23,550	0.1445	0.2343	0
	$^3F_4$ ($^3H_4$)	25,350	0.1359	0.0858	2×10^{-5}
	$^3H_5$	29,800	0	0.1979	0.1895
	$^3H_4$ ($^3F_4$)	32,300	0.2811	0.0096	0.0064
	$^3H_6$	37,900	0	0.2713	0.0228

Appendix E
3jm Factors for Some Group Chains

For the convenience of the table list, the symbol of point group irreducible representation adopted the subscript of Bathe symbol (i.e. $\Gamma_1 \Rightarrow 1, \Gamma_2 \Rightarrow 2, \Gamma_3 \Rightarrow 3, \ldots$). The irreducible representations of group G and its subgroup H denoted by λ_i and μ_i, respectively, then the 3jm factor of group chain $G \supset H$ expressed as follows

$$\begin{matrix} \lambda_1 & \lambda_2 & \lambda_3 & & & & \\ \mu_1 & \mu_2 & \mu_3 & s & a_1 & a_2 & a_3 \end{matrix}$$

where a_i denotes the multiple decomposition of irreducible representation, s denotes the multiple coupling of subgroup H. As an example, some of the 3jm factors of $SO_3 \supset O$ group chain will be used to illustrate the meaning of the following table. In the table of group chain $SO_3 \supset O$ it can be found the black numbers **3 5/2 3/2** which characterize the irreducible representations of SO_3 group and below this line there is a line of

$$4 \quad 7 \quad 8 \quad {\scriptstyle 0\,000+} \quad +\sqrt{3}/\sqrt{2.7}$$

It means that

$$\begin{pmatrix} 3 & 5/2 & 3/2 \\ 4 & 7 & 8 \end{pmatrix}_0 = +\sqrt{3}/\sqrt{2 \times 7}$$

In this 3jm factor the irreducible representations of O group are $1, \tilde{\frac{1}{2}}, \frac{3}{2}$ of Butler symbol and $\Gamma_4, \Gamma_7, \Gamma_8$ of Bethe while the symbol in this book is simply 4, 7, 8. The multiple coupling of subgroup is $s = 0$ and the multiple decomposition of irreducible representation $a_1 = 0, a_2 = 0, a_3 = 0$ and the sign change brought by the column permutation are given after the multiple decomposition symbol, the column permutation + and − express that the column exchange the sign of the 3jm factor does not change and change, respectively. Particular attention should be paid to the

Z. Luo and Y. Huang, *Physics of Solid-State Laser Materials*, Springer Series in Materials Science 289, https://doi.org/10.1007/978-981-32-9668-8

points between the numbers of the numerical value of the *3jm* factor. It expresses the meaning of multiplication, rather than the decimal point.

The next two lines are

$$\begin{matrix} 3 & 5/2 & 3/2 & & & & \\ 5 & 8 & 8 & 0 & 0 & 0 & 0+ \end{matrix} \equiv \begin{pmatrix} 3 & 5/2 & 3/2 \\ \Gamma_5 & \Gamma_8 & \Gamma_8 \end{pmatrix}^{SO_3}_{00} = \frac{5}{2\sqrt{21}}$$

with

$$\begin{matrix} 3 & 5/2 & 3/2 & & & & \\ 5 & 8 & 8 & 0 & 0 & 0 & 0- \end{matrix} \equiv \begin{pmatrix} 3 & 5/2 & 3/2 \\ \Gamma_5 & \Gamma_8 & \Gamma_8 \end{pmatrix}^{SO_3}_{10} = \frac{-1}{2\sqrt{21}}$$

For the situation of absence of multiple coupling of subgroup or multiple decomposition of irreducible representation, the corresponding marks will not appear in the table.

The $3jm$ factor has two kinds of symmetry, that is, the column permutation symmetry and the complex conjugate symmetry. After two column permutation operation, the value of $3jm$ factor remains unchanged; for example, the $3jm$ factor of $D_2 \supset C_2$ group chain

$$\begin{pmatrix} 5 & 5 & 1 \\ 4 & 3 & 1 \end{pmatrix}^{D_2}_{C_2} = -\frac{1}{\sqrt{2}}$$

After column permutation one have

$$\begin{pmatrix} 5 & 1 & 5 \\ 4 & 1 & 3 \end{pmatrix} = \begin{pmatrix} 1 & 5 & 5 \\ 1 & 3 & 4 \end{pmatrix} = \begin{pmatrix} 5 & 5 & 1 \\ 3 & 4 & 1 \end{pmatrix} = \frac{1}{\sqrt{2}}$$

If the complex conjugate will bring changes of the sign of $3jm$ factor then an asterisk will be added to the upper right corner of it, as an example

$$\begin{pmatrix} 5 & 5 & 1 \\ 4 & 3 & 1 \end{pmatrix}^{*} = \left(-\frac{1}{\sqrt{2}}\right)^{*} = -\begin{pmatrix} 5 & 5 & 1 \\ 3 & 4 & 1 \end{pmatrix}$$

This result is the same as the results of the permutation. For the $3jm$ factor of $D_3 \supset C_3$ group chain

$$\begin{pmatrix} 6 & 3 & 4 \\ 6 & 3 & 5 \end{pmatrix} = +\begin{pmatrix} 5 & 3 & 4 \\ 6 & 2 & 4 \end{pmatrix}^{*} = \left(\frac{1}{\sqrt{2}}\right)^{*} = \frac{1}{\sqrt{2}}$$

$$\begin{pmatrix} 6 & 3 & 4 \\ 6 & 2 & 4 \end{pmatrix} = -\begin{pmatrix} 5 & 3 & 4 \\ 6 & 3 & 5 \end{pmatrix}^{*} = -\left(\frac{i}{\sqrt{2}}\right)^{*} = \frac{i}{\sqrt{2}}$$

Some of the 3jm Factor of SO_3—O Group Chain

2	**3/2**	**3/2**		
3	8	8	0 000+	$-\sqrt{2}/\sqrt{5}$
5	8	8	0 000+	$-\sqrt{3}/\sqrt{5}$
5	8	8	1 000-	0
9/2	**9/2**	**2**		
8	6	3	0 000-	$-1/5\sqrt{11}$
8	6	3	0 100-	$-4\sqrt{7}/5\sqrt{3.11}$
8	6	5	0 000-	$-\sqrt{2}/\sqrt{3.11}$
8	6	5	0 100-	0
8	8	3	0 000+	$+4\sqrt{2.3}/5\sqrt{5.11}$
8	8	3	0 100+	$+\sqrt{2.7}/5\sqrt{5.11}$
8	8	3	0 110+	$+8\sqrt{2}/5\sqrt{3.5.11}$
8	8	5	0 000+	$-9/5\sqrt{5.11}$
8	8	5	0 100+	$+4\sqrt{7}/5\sqrt{3.5.11}$
8	8	5	0 110+	$-2.3/5\sqrt{5.11}$
8	8	5	1 000-	0
8	8	5	1 100-	$-2\sqrt{7}/\sqrt{3.5.11}$
8	8	5	1 110-	0
9/2	**9/2**	**4**		
6	6	1	0 000+	$+7\sqrt{2.7}/3\sqrt{3.5.11.13}$
6	6	4	0 000-	0
8	6	4	0 000+	$+3\sqrt{7}/\sqrt{2.5.11.13}$
8	6	4	0 100+	$+2\sqrt{2}/\sqrt{3.5.11.13}$
8	6	3	0 000-	$+4/\sqrt{5.11.13}$
8	6	3	0 100-	$-2\sqrt{7}/3\sqrt{3.5.11.13}$
8	6	5	0 000-	$+19/\sqrt{2.3.5.11.13}$
8	6	5	0 100-	$-2\sqrt{2.7}/3\sqrt{5.11.13}$
8	8	1	0 000+	$-\sqrt{3.7}/\sqrt{5.11.13}$
8	8	1	0 100+	$+4/\sqrt{5.11.13}$
8	8	1	0 110+	$+2\sqrt{7}/3\sqrt{3.5.11.13}$
8	8	4	0 000-	0
8	8	4	0 100-	$+4/\sqrt{11.13}$
8	8	4	0 110-	0
8	8	4	1 000-	0
8	8	4	1 100-	$+1/\sqrt{11.13}$
8	8	4	1 110-	0
8	8	3	0 000+	$+\sqrt{2.3}/\sqrt{11.13}$
8	8	3	0 100+	0
8	8	3	0 110+	$-8.2\sqrt{2}/3\sqrt{3.11.13}$
8	8	5	0 000+	$-2.3/5\sqrt{11.13}$
8	8	5	0 100+	$-4\sqrt{7}/5\sqrt{3.11.13}$
8	8	5	0 110+	$-4\sqrt{13}/3.5\sqrt{11}$
8	8	5	1 000-	0
8	8	5	1 100-	$-\sqrt{7}/\sqrt{3.11.13}$
8	8	5	1 110-	0
6	**9/2**	**9/2**		
1	6	6	0 000+	$+8/\sqrt{3.5.11.13}$
1	8	8	0 000+	$+4\sqrt{2.3}/5\sqrt{5.11.13}$
1	8	8	0 010+	$-3\sqrt{7}/5\sqrt{2.5.11.13}$
1	8	8	0 011+	$-8.4\sqrt{2}/5\sqrt{3.5.11.13}$
4	6	6	0 000-	0
4	8	6	0 000+	$+\sqrt{2.3.7}/5\sqrt{11.13}$
4	8	6	0 010+	$+3.7/5\sqrt{2.11.13}$
4	8	8	0 000-	0
4	8	8	0 010-	$-\sqrt{3}/\sqrt{5.11.13}$
4	8	8	0 011-	0
4	8	8	1 000-	0
4	8	8	1 010-	$-3\sqrt{3}/\sqrt{5.11.13}$
4	8	8	1 011-	0
3	8	6	0 000-	$+\sqrt{2.7}/5\sqrt{11.13}$
3	8	6	0 010-	$+8\sqrt{2}/5\sqrt{3.11.13}$
3	8	8	0 000+	$+8\sqrt{3.7}/5\sqrt{5.11.13}$
3	8	8	0 010+	$-\sqrt{11}/5\sqrt{5.13}$
3	8	8	0 011+	$+8.2\sqrt{7}/5\sqrt{3.5.11.13}$
5	8	6	0 000-	$-\sqrt{2}/\sqrt{3.5.13}$
5	8	6	0 010-	$+3/\sqrt{2.5.7.13}$
5	8	6	0 100-	$-2\sqrt{2.3}/5\sqrt{11.13}$
5	8	6	0 110-	$-2\sqrt{2.13}/5\sqrt{7.11}$
5	8	8	0 000+	0
5	8	8	0 010+	$-5/\sqrt{3.7.13}$

5	8	8	0 011+	0
5	8	8	0 100+	$+2.3.7/5\sqrt{5.11.13}$
5	8	8	0 110+	$-8\sqrt{3}/5\sqrt{5.7.11.13}$
5	8	8	0 111+	$-7/5\sqrt{5.11.13}$
5	8	8	1 000-	0
5	8	8	1 010-	$-1/\sqrt{3.7.13}$
5	8	8	1 011-	0
5	8	8	1 100-	0
5	8	8	1 110-	$-2\sqrt{3}/\sqrt{5.7.11.13}$
5	8	8	1 111-	0
2	8	8	0 000-	0
2	8	8	0 010-	$-1/\sqrt{2.13}$
2	8	8	0 011-	0
4	**3**	**3**		
1	4	4	0 000+	$+1/\sqrt{2.11}$
1	5	5	0 000+	$-1/3\sqrt{2.11}$
1	2	2	0 000+	$-\sqrt{2}/\sqrt{3.11}$
4	4	4	0 000-	0
4	5	4	0 000+	$+1/\sqrt{11}$
4	5	5	0 000-	0
4	2	5	0 000+	$-\sqrt{5}/\sqrt{2.3.11}$
3	4	4	0 000+	$+\sqrt{5}/\sqrt{2.7.11}$
3	5	4	0 000-	$+1/\sqrt{2.7.11}$
3	5	5	0 000+	$-\sqrt{5.7}/3\sqrt{2.11}$
5	4	4	0 000+	$-\sqrt{3.5}/\sqrt{7.11}$
5	5	4	0 000-	$+2/\sqrt{7.11}$
5	5	5	0 000+	$+\sqrt{5}/\sqrt{3.7.11}$
5	2	4	0 000+	$+1/\sqrt{2.7.11}$
4	**4**	**4**		
1	1	1	0 000+	$+7\sqrt{2}/\sqrt{3.11.13}$
4	4	1	0 000+	$+7\sqrt{3}/\sqrt{2.11.13}$
4	4	4	0 000-	0
3	4	4	0 000+	$+\sqrt{5.7}/3\sqrt{2.11.13}$
3	3	1	0 000+	$+2/3\sqrt{3.11.13}$
3	3	3	0 000+	$-8.2\sqrt{2.5}/3\sqrt{3.7.11.13}$
5	4	4	0 000-	$+\sqrt{5.7}/\sqrt{3.11.13}$
5	5	1	0 000+	$-\sqrt{13}/3\sqrt{2.11}$
5	5	4	0 000-	0
5	5	3	0 000+	$+5\sqrt{5}/3\sqrt{2.7.11.13}$
5	5	5	0 000+	$-\sqrt{5}/\sqrt{3.7.11.13}$

$O-D_3$ and T_d-D_{3v}

6	**6**	**1**	0	
4	4	1	+	+1

4	**6**	**6**	0	
2	4	4	+	$+1/\sqrt{3}$
3	4	4	+	$+\sqrt{2}/\sqrt{3}$

8	**4**	**6**	0	
4	2	4	-	$+1/\sqrt{3}$
4	3	4	-	$-1/\sqrt{2.3}$
5	3	4	+	$+1/2$

8	**8**	**1**	0	
4	4	1	+	$+1/\sqrt{2}$
6	5	1	+	$+1/2$

8	**8**	**4**	0	
4	4	2	+	$+1/\sqrt{2.3}$
4	4	3	+	0
5	4	3	-	$+1/3\sqrt{2}-i/3$
5	5	2	+	$+1/2.3\sqrt{3}+i\sqrt{2}/3\sqrt{}$

8	**8**	**4**	1	
4	4	2	+	0
4	4	3	+	$-1/\sqrt{3}$
5	4	3	-	$+1/3\sqrt{2}+i/2.3$
5	5	2	+	$+2/3\sqrt{3}-i/3\sqrt{2.3}$

3	**8**	**6**	0	
3	4	4	+	$+1/\sqrt{2}$
3	5	4	-	$-1/2\sqrt{3}+i/\sqrt{2.3}$

3	**8**	**8**	0	
3	4	4	-	0
3	5	4	+	$+1/\sqrt{2.3}+i/2\sqrt{3}$

5	**8**	**6**	0	
1	4	4	-	$+1/\sqrt{3}$
3	4	4	+	$+1/\sqrt{2.3}$
3	5	4	-	$-1/2.3-i\sqrt{2}/3$

5	**8**	**8**	0	
1	4	4	+	$-1/\sqrt{2.3}$
1	6	5	+	$+1/2\sqrt{3}$
3	4	4	-	0
3	5	4	+	$+1/3\sqrt{2}-i/3$

5	**8**	**8**	1	
1	4	4	-	0
1	6	5	-	$+i/\sqrt{2.3}$
3	4	4	+	$+1/\sqrt{3}$
3	5	4	-	$+1/3\sqrt{2}+i/2.3$

7	**8**	**4**	0	
4	4	2	-	$-1/\sqrt{3}$
4	4	3	-	$-1/\sqrt{2.3}$

7	**8**	**4**	0	
4	5	3	+	$-1/2.3+i\sqrt{2}/3$

7	**3**	**8**	0	
4	3	4	+	$-1/\sqrt{2}$
4	3	5	-	$-1/2\sqrt{3}+i/\sqrt{2.3}$

7	**5**	**6**	0	
4	1	4	+	$+1/\sqrt{3}$
4	3	4	-	$+\sqrt{2}/\sqrt{3}$

7	**5**	**8**	0	
4	1	4	-	$-1/\sqrt{3}$
4	3	4	+	$+1/\sqrt{2.3}$
4	3	5	-	$-1/2$

7	**7**	**1**	0	
4	4	1	+	+1

7	**7**	**4**	0	
4	4	2	+	$-1/\sqrt{3}$
4	4	3	+	$+\sqrt{2}/\sqrt{3}$

2	**8**	**8**	0	
2	4	4	+	$-1/\sqrt{2}$
2	5	5	+	$+1/2.3+i\sqrt{2}/3$

2	**7**	**6**	0	
2	4	4	+	+1

Appendix F
Clebsch-Gordan Coefficients of the Cubic Point Group with Trigonal Bases

$A_2 \times E$

M_1	M_2 \ Γ M	E u_+	E u_-
e_2	u_+	$-i$	0
	u_-	0	i

$A_2 \times T_1$

M_1	M_2 \ Γ M	T_2 x_+	T_2 x_-	T_2 x_0
	a_+	1	0	0
e_2	a_-	0	1	0
	a_0	0	0	1

$A_2 \times T_2$

M_1	M_2 \ Γ M	T_1 a_+	T_1 a_-	T_1 a_0
	x_+	-1	0	0
e_2	x_-	0	-1	0
	x_0	0	0	-1

$E \times E$

M_1	M_2 \ Γ M	A_1 e_1	A_2 e_2	E u_+	E u_-
u_+	u_+	0	0	0	-1
	u_-	$-1/\sqrt{2}$	$-i/\sqrt{2}$	0	0
u_-	u_+	$-1/\sqrt{2}$	$i/\sqrt{2}$	0	0
	u_-	0	0	1	0

Z. Luo and Y. Huang, *Physics of Solid-State Laser Materials*, Springer Series in Materials Science 289, https://doi.org/10.1007/978-981-32-9668-8

$E\times T_1$

M_1	M_2	Γ / M	T_1: a_+	T_1: a_-	T_1: a_0	T_2: x_+	T_2: x_-	T_2: x_0
	a_+		0	$1/\sqrt{2}$	0	0	$i/\sqrt{2}$	0
u_+	a_-		0	0	$-1/\sqrt{2}$	0	0	$-i/\sqrt{2}$
	a_0		$1/\sqrt{2}$	0	0	$i/\sqrt{2}$	0	0
	a_+		0	0	$-1/\sqrt{2}$	0	0	$i/\sqrt{2}$
u_-	a_-		$-1/\sqrt{2}$	0	0	$i/\sqrt{2}$	0	0
	a_0		0	$1/\sqrt{2}$	0	0	$-i/\sqrt{2}$	0

$E\times T_2$

M_1	M_2	Γ / M	T_1: a_+	T_1: a_-	T_1: a_0	T_2: x_+	T_2: x_-	T_2: x_0
	x_+		0	$-i/\sqrt{2}$	0	0	$1/\sqrt{2}$	0
u_+	x_-		0	0	$i/\sqrt{2}$	0	0	$-1/\sqrt{2}$
	x_0		$-i/\sqrt{2}$	0	0	$1/\sqrt{2}$	0	0
	x_+		0	0	$-i/\sqrt{2}$	0	0	$-1/\sqrt{2}$
u_-	x_-		$-i/\sqrt{2}$	0	0	$-1/\sqrt{2}$	0	0
	x_0		0	$i/\sqrt{2}$	0	0	$1/\sqrt{2}$	0

$T_1\times T_1$

M_1	M_2	Γ / M	A_1: e_1	E: u_+	E: u_-	T_1: a_+	T_1: a_-	T_1: a_0	T_2: x_+	T_2: x_-	T_2: x_0
	a_+		0	0	$-1/\sqrt{3}$	0	0	0	0	$-\sqrt{2}/\sqrt{3}$	0
a_+	a_-		$1/\sqrt{3}$	0	0	0	0	$i/\sqrt{2}$	0	0	$-1/\sqrt{6}$
	a_0		0	$-1/\sqrt{3}$	0	$i/\sqrt{2}$	0	0	$1/\sqrt{6}$	0	0
	a_+		$1/\sqrt{3}$	0	0	0	0	$-i/\sqrt{2}$	0	0	$-1/\sqrt{6}$
a_-	a_-		0	$1/\sqrt{3}$	0	0	0	0	$\sqrt{2}/\sqrt{3}$	0	0
	a_0		0	0	$-1/\sqrt{3}$	0	$-i/\sqrt{2}$	0	0	$1/\sqrt{6}$	0
	a_+ a_+		0	$-1/\sqrt{3}$	0	$-i/\sqrt{2}$	0	0	$1/\sqrt{6}$	0	0
a_0	a_-		0	0	$-1/\sqrt{3}$	0	$i/\sqrt{2}$	0	0	$1/\sqrt{6}$	0
	a_0		$-1/\sqrt{3}$	0	0	0	0	0	0	0	$-\sqrt{2}/\sqrt{3}$

$T_1 \times T_2$

	Γ	A_2	E		T_1			T_2		
M_1	M_2 / M	e_2	u_+	u_-	a_+	a_-	a_0	x_+	x_-	x_0
a_+	x_+	0	0	$-i/\sqrt{3}$	0	$\sqrt{2}/\sqrt{3}$	0	0	0	0
	x_-	$1/\sqrt{3}$	0	0	0	0	$1/\sqrt{6}$	0	0	$i/\sqrt{2}$
	x_0	0	$i/\sqrt{3}$	0	$-1/\sqrt{6}$	0	0	$i/\sqrt{2}$	0	0
a_-	x_+	$1/\sqrt{3}$	0	0	0	0	$1/\sqrt{6}$	0	0	$-i/\sqrt{2}$
	x_-	0	$-i/\sqrt{3}$	0	$-\sqrt{2}/\sqrt{3}$	0	0	0	0	0
	x_0	0	0	$-i/\sqrt{3}$	0	$-1/\sqrt{6}$	0	0	$-i/\sqrt{2}$	0
a_0	x_+	0	$i/\sqrt{3}$	0	$-1/\sqrt{6}$	0	0	$-i/\sqrt{2}$	0	0
	x_-	0	0	$-i/\sqrt{3}$	0	$-1/\sqrt{6}$	0	0	$i/\sqrt{2}$	0
	x_0	$-1/\sqrt{3}$	0	0	0	0	$\sqrt{2}/\sqrt{3}$	0	0	0

$T_2 \times T_2$

	Γ	A_1	E		T_1			T_2		
M_1	M_2 / M	e_1	u_+	u_-	a_+	a_-	a_0	x_+	x_-	x_0
x_+	x_+	0	0	$1/\sqrt{3}$	0	0	0	0	$\sqrt{2}/\sqrt{3}$	0
	x_-	$-1/\sqrt{3}$	0	0	0	0	$-i/\sqrt{2}$	0	0	$1/\sqrt{6}$
	x_0	0	$1/\sqrt{3}$	0	$-i/\sqrt{2}$	0	0	$-1/\sqrt{6}$	0	0
x_-	x_+	$-1/\sqrt{3}$	0	0	0	0	$i/\sqrt{2}$	0	0	$1/\sqrt{6}$
	x_-	0	$-1/\sqrt{3}$	0	0	0	0	$-\sqrt{2}/\sqrt{3}$	0	0
	x_0	0	0	$1/\sqrt{3}$	0	$i/\sqrt{2}$	0	0	$-1/\sqrt{6}$	0
x_0	x_+	0	$1/\sqrt{3}$	0	$i/\sqrt{2}$	0	0	$-1/\sqrt{6}$	0	0
	x_-	0	0	$1/\sqrt{3}$	0	$-i/\sqrt{2}$	0	0	$-1/\sqrt{6}$	0
	x_0	$1/\sqrt{3}$	0	0	0	0	0	0	0	$\sqrt{2}/\sqrt{3}$

Appendix G
Integral Numerical Value Associated with the Thermal Effect of the Spectra

$$\left(\frac{T}{T_{\mathrm{D}}}\right)^7 \int_0^{T_{\mathrm{D}}/T} \frac{x^6 e^x}{(e^x - 1)^2} \mathrm{d}x$$

T_D/T	0.0	0.1	0.2	0.3	0.4	0.5	0.6	0.7	0.8	0.9
0		19988.1×10^{-3}	4988.1×10^{-3}	2210.3×10^{-3}	1238.1×10^{-3}	788.2×10^{-3}	543.8×10^{-3}	396.4×10^{-3}	300.8×10^{-3}	235.3×10^{-3}
1	188.5×10^{-3}	153.9×10^{-3}	127.6×10^{-3}	107.1×10^{-3}	90.98×10^{-3}	77.95×10^{-3}	67.31×10^{-3}	58.52×10^{-3}	51.18×10^{-3}	44.99×10^{-3}
2	39.73×10^{-3}	35.23×10^{-3}	31.35×10^{-3}	27.99×10^{-3}	25.06×10^{-3}	22.49×10^{-3}	20.24×10^{-3}	18.25×10^{-3}	16.48×10^{-3}	14.92×10^{-3}
3	13.52×10^{-3}	12.28×10^{-3}	11.16×10^{-3}	10.16×10^{-3}	9.26×10^{-3}	8.44×10^{-3}	7.71×10^{-3}	7.05×10^{-3}	6.44×10^{-3}	5.90×10^{-3}
4	5.40×10^{-3}	4.95×10^{-3}	4.54×10^{-3}	4.17×10^{-3}	3.83×10^{-3}	3.52×10^{-3}	3.24×10^{-3}	2.98×10^{-3}	2.74×10^{-3}	2.53×10^{-3}
5	2.33×10^{-3}	2.15×10^{-3}	1.98×10^{-3}	1.83×10^{-3}	1.69×10^{-3}	1.56×10^{-3}	1.44×10^{-3}	1.33×10^{-3}	1.23×10^{-3}	1.14×10^{-3}
6	1.05×10^{-3}	9.77×10^{-4}	9.05×10^{-4}	8.38×10^{-4}	7.77×10^{-4}	7.20×10^{-4}	6.68×10^{-4}	6.19×10^{-4}	5.75×10^{-4}	5.34×10^{-4}
7	4.96×10^{-4}	4.60×10^{-4}	4.28×10^{-4}	3.98×10^{-4}	3.70×10^{-4}	3.45×10^{-4}	3.21×10^{-4}	2.98×10^{-4}	2.78×10^{-4}	2.59×10^{-4}
8	2.41×10^{-4}	2.25×10^{-4}	2.10×10^{-4}	1.96×10^{-4}	1.83×10^{-4}	1.71×10^{-4}	1.59×10^{-4}	1.49×10^{-4}	1.39×10^{-4}	1.30×10^{-4}

(continued)

Z. Luo and Y. Huang, *Physics of Solid-State Laser Materials*, Springer Series in Materials Science 289, https://doi.org/10.1007/978-981-32-9668-8

(continued)

T_D/T	0.0	0.1	0.2	0.3	0.4	0.5	0.6	0.7	0.8	0.9
9	1.22×10^{-4}	1.14×10^{-4}	1.07×10^{-4}	1.00×10^{-4}	9.37×10^{-5}	8.79×10^{-5}	8.24×10^{-5}	7.73×10^{-5}	7.25×10^{-5}	6.80×10^{-5}
10	6.39×10^{-5}	6.00×10^{-5}	5.63×10^{-5}	5.30×10^{-5}	4.98×10^{-5}	4.68×10^{-5}	4.41×10^{-5}	4.15×10^{-5}	3.90×10^{-5}	3.68×10^{-5}
11	3.47×10^{-5}	3.27×10^{-5}	3.08×10^{-5}	2.91×10^{-5}	2.74×10^{-5}	2.59×10^{-5}	2.44×10^{-5}	2.31×10^{-5}	2.18×10^{-5}	2.06×10^{-5}
12	1.95×10^{-5}	1.84×10^{-5}	1.75×10^{-5}	1.65×10^{-5}	1.56×10^{-5}	1.48×10^{-5}	1.40×10^{-5}	1.33×10^{-5}	1.26×10^{-5}	1.19×10^{-5}
13	1.13×10^{-5}	1.08×10^{-5}	1.03×10^{-5}	9.73×10^{-6}	9.25×10^{-6}	8.79×10^{-6}	8.36×10^{-6}	7.95×10^{-6}	7.56×10^{-6}	7.20×10^{-6}
14	6.85×10^{-6}	6.52×10^{-6}	6.21×10^{-6}	5.92×10^{-6}	5.64×10^{-6}	5.37×10^{-6}	5.12×10^{-6}	4.89×10^{-6}	4.67×10^{-6}	4.46×10^{-6}
15	4.25×10^{-6}	4.06×10^{-6}	3.88×10^{-6}	3.71×10^{-6}	3.55×10^{-6}	3.39×10^{-6}	3.24×10^{-6}	3.10×10^{-6}	2.97×10^{-6}	2.84×10^{-6}
16	2.72×10^{-6}	2.60×10^{-6}	2.49×10^{-6}	2.39×10^{-6}	2.29×10^{-6}	2.19×10^{-6}	2.10×10^{-6}	2.02×10^{-6}	1.93×10^{-6}	1.86×10^{-6}
17	1.78×10^{-6}	1.71×10^{-6}	1.64×10^{-6}	1.58×10^{-6}	1.51×10^{-6}	1.46×10^{-6}	1.40×10^{-6}	1.34×10^{-6}	1.29×10^{-6}	1.24×10^{-6}
18	1.20×10^{-6}	1.15×10^{-6}	1.11×10^{-6}	1.06×10^{-6}	1.03×10^{-6}	9.87×10^{-7}	9.50×10^{-7}	9.15×10^{-7}	8.82×10^{-7}	8.50×10^{-7}
19	8.19×10^{-7}	7.90×10^{-7}	7.61×10^{-7}	7.34×10^{-7}	7.08×10^{-7}	6.83×10^{-7}	6.59×10^{-7}	6.36×10^{-7}	6.14×10^{-7}	5.93×10^{-7}
20	5.72×10^{-7}	5.52×10^{-7}	5.34×10^{-7}	5.16×10^{-7}	4.98×10^{-7}	4.81×10^{-7}	4.65×10^{-7}	4.50×10^{-7}	4.35×10^{-7}	4.20×10^{-7}

$$10^5 \times \left(\frac{T}{T_D}\right)^4 \int_0^{T_D/T} \frac{x^3}{e^x - 1} dx$$

T_D/T	0.0	0.1	0.2	0.3	0.4	0.5	0.6	0.7	0.8	0.9
0		321000	154499	99110.5	71498.7	54997.5	44051.3	36279.0	30489.9	26022.7
1	22480.5	19610.5	17244.3	15265.4	13590.3	12157.5	10922.3	9848.7	8909.3	8083.1
2	7352.1	6703.8	6124.1	56068.4	5140.7	4722.0	4343.5	4005.4	3690.4	3407.9
3	3150.8	2915.9	2701.9	2505.8	2326.0	2160.9	2009.5	1870.0	1741.8	1623.3
4	1514.4	1413.8	1320.6	1234.4	1154.7	1080.8	1012.3	948.6	889.7	835.6
5	783.9	736.6	692.4	651.1	612.8	577.2	543.8	512.6	483.5	456.0
6	431.0	407.2	385.0	364.2	344.6	326.1	308.9	292.8	277.6	263.4
7	250.0	237.4	225.6	211.4	204.0	194.0	184.6	175.9	167.6	159.7
8	152.3	145.4	139.0	132.7	126.5	120.9	115.6	110.6	105.8	101.3
9	97.0	93.0	89.1	85.3	81.9	78.6	75.4	72.5	69.6	66.9
10	64.3	61.8	59.4	57.1	55.0	53.2	51.1	49.2	47.4	45.7
11	44.2	42.6	41.1	39.7	38.3	37.0	35.7	34.5	33.4	32.2
12	31.3	30.3	29.4	28.4	27.5	26.6	25.7	24.8	24.1	23.4
13	22.6	21.9	21.2	20.6	20.0	19.5	18.9	18.4	17.8	17.3
14	16.7	16.3	15.8	15.4	15.0	14.6	14.2	13.8	13.5	13.1
15	12.8	12.4	12.0	11.7	11.3	11.1	10.8	10.6	10.3	10.1

Index

Z. Luo and Y. Huang, *Physics of Solid-State Laser Materials*, Springer Series in Materials Science 289, https://doi.org/10.1007/978-981-32-9668-8

The manufacturer's authorised representative in the EU is Springer Nature Customer Service Centre GmbH, Europaplatz 3, 69115 Heidelberg, Germany. If you have any concerns regarding our products, please contact ProductSafety@springernature.com

Printed and bound by CPI Group (UK) Ltd, Croydon, CR0 4YY

30/09/2026

02225507-0018